Student's Companion to Stryer's BIOCHEMISTRY

Student's Companion to Stryer's
BIOCHEMISTRY

Richard I. Gumport, Ana Jonas, and Richard Mintel

Department of Biochemistry
College of Medicine at Urbana-Champaign
University of Illinois

Carl Rhodes

Southwestern Graduate School of Biomedical Sciences
The University of Texas
Southwestern Medical Center at Dallas

Expanded Solutions to Text Problems contributed by

Roger E. Koeppe

Department of Biochemistry
Oklahoma State University

W. H. FREEMAN AND COMPANY / NEW YORK

ISBN 0-7167-2075-2

Printed in the United States of America

1 2 3 4 5 6 7 8 9 0 KP 8 9

Contents

Acknowledgments vii
To the Student ix

PART I MOLECULAR DESIGN OF LIFE

CHAPTER 1. Prelude 1
2. Protein Structure and Function 9
3. Exploring Proteins 23
4. DNA and RNA: Molecules of Heredity 37
5. Flow of Genetic Information 52
6. Exploring Genes: Analyzing, Constructing, and Cloning DNA 67

PART II PROTEIN CONFORMATION, DYNAMICS, AND FUNCTION

CHAPTER 7. Oxygen-transporting Proteins: Myoglobin and Hemoglobin 81
8. Introduction to Enzymes 95
9. Mechanisms of Enzyme Action 115
10. Control of Enzymatic Activity 129
11. Connective-Tissue Proteins 143
12. Introduction to Biological Membranes 155

PART III GENERATION AND STORAGE OF METABOLIC ENERGY

CHAPTER 13. Metabolism: Basic Concepts and Design 171
14. Carbohydrates 185
15. Glycolysis 197
16. Citric Acid Cycle 215
17. Oxidative Phosphorylation 231
18. Pentose Phosphate Pathway and Gluconeogenesis 251
19. Glycogen Metabolism 267
20. Fatty Acid Metabolism 283
21. Amino Acid Degradation and the Urea Cycle 301
22. Photosynthesis 317

PART IV BIOSYNTHESIS OF MACROMOLECULAR PRECURSORS

CHAPTER 23. Biosynthesis of Membrane Lipids and Steroid Hormones 333
24. Biosynthesis of Amino Acids and Heme 349
25. Biosynthesis of Nucleotides 369
26. Integration of Metabolism 387

PART V GENETIC INFORMATION: Storage, transmission, and expression

CHAPTER 27. DNA: Structure, Replication, and Repair 405
28. Gene Rearrangements: Recombination and Transposition 421
29. RNA Synthesis and Splicing 433
30. Protein Synthesis 449
31. Protein Targeting 469
32. Control of Gene Expression in Procaryotes 485
33. Eucaryotic Chromosomes and Gene Expression 503
34. Viruses and Oncogenes 521

PART VI MOLECULAR PHYSIOLOGY: Interaction of information, conformation, and metabolism in physiological processes

CHAPTER 35. Molecular Immunology 541
36. Muscle Contraction and Cell Motility 555
37. Membrane Transport 573
38. Hormone Action 589
39. Excitable Membranes and Sensory Systems 609

Expanded Solutions to Text Problems 625

Acknowledgments

This book has its origins in a curriculum guide that was initiated over fifteen years ago for first-year medical students studying biochemistry at the College of Medicine of the University of Illinois at Urbana-Champaign. A number of colleagues have contributed to the guide over the years. We wish to thank John Clark, Lowell Hager, Walter Mangel, William McClure, and Robert Switzer for their efforts in helping to develop the curriculum guide. Special thanks go to George Ordal and James Kaput, our fellow teachers of biochemistry at the College of Medicine, for their continuing contributions. We also thank William Sorlie for educating us to the value of learning objectives. Thanks are also due to the students who took the time to criticize the guide. Finally, we appreciate the support provided our teaching efforts by the Department of Biochemistry and the College of Medicine. In particular, we thank the Word Processing Center for their careful and cheerful preparation of our written materials.

Richard I. Gumport
Ana Jonas
Richard Mintel

Much of my motivation for writing problems comes from organizing and teaching courses in biochemistry, first at Stanford University School of Medicine and then at Washington University in St. Louis. Dale Kaiser and Robert Lehman, in the Department of Biochemistry at Stanford, deserve my special thanks for their continued encouragement and support. More recently, I have taught introductory biochemistry to medical students in the Summer Matriculation Program at Stanford. I am grateful to Gilbert Martinez, who for the past three summers has provided inspirational assistance as a co-lecturer and as a source of new approaches to problems in biochemistry.

Carl Rhodes

To the Student

The woods are lovely, dark, and deep,
But I have promises to keep,
And miles to go before I sleep,
And miles to go before I sleep.

ROBERT FROST

Opening a comprehensive biochemistry text for the first time can be a daunting experience for a neophyte. There is so much detailed material that it is natural to wonder if you can possibly master it all in one or two semesters of study. Of course, you can't learn everything, but experience indicates that you can, indeed, learn the fundamental concepts in an introductory biochemistry course. We have written this *Student's Companion to Stryer's Biochemistry* to ease your entry into the exciting world of biochemistry.

Your goal is to "know" and "understand" biochemistry. Unfortunately, awareness of these grand goals offers no practical help in reaching them, because they represent such high-level and complex intellectual processes. In addition, it is difficult for you to know to what extent you have attained them. We have found that, by subdividing these goals into simpler ones and expressing them in terms of demonstrable behaviors, you can begin to approach them and, in addition, can readily assay your progress toward reaching them. Thus, a part of each chapter consists of Learning Objectives that ask you to do things that will help you begin to understand biochemistry. When you can master the objectives, you are well on your way to learning the material in the chapter. It is important to add a cautionary note here. Being able to respond to all the objectives adequately does *not* mean that you know biochemistry, for they are a limited sampling of all the possible objectives, and more to the point, they do not explicitly require such higher level activities as creation, analysis, integration, synthesis, problem solving, evaluation, application, and appreciation. These more advanced skills will develop to varying levels as you continue your studies of biochemistry beyond the introductory stage.

Each chapter in the *Companion* consists of an introduction, the *Learning Objectives* with the *key words* italicized, a *Self-Test, Problems, Answers to Self-Test,* and *Answers to Problems.* The introduction sets the scene, places the material in the chapter in the context of what you have learned before, and reminds you of material you may need to review in order to understand what follows. The Learning Objectives are presented in the order that the information they encompass occurs in Stryer's *Biochemistry.* Key words, which are important concepts or vocabulary, are italicized in the objectives. A selection of Self-Test questions requiring primarily information recall is followed by a Problems section, in which more complex skills are tested. Next, the answers to the Self-Test questions and the Problems are given. Finally, at the end of the book, explicit solutions to the problems at the end of the corresponding chapters of *Biochemistry* are presented.

There are many ways to use the *Companion,* and as you begin your studies you will develop the "system" that is best for you. Over fifteen years of experience teaching introductory biochemistry to first-year medical students has suggested one pathway that you should consider. Start by reviewing the prerequisite chapters mentioned in the introduction and skim the Learning Objectives to obtain an overview of what you are to learn. Some students like also to skim the Self-Test questions at this time to form an impression of the levels of difficulty and the kinds of questions that will be asked. Next, read the chapter in *Biochemistry,* using the Learning Objectives to help direct you to the essential concepts. Note the key words, and look up any you don't know. Then attempt to meet the objectives. When you cannot satisfy an objective, reread the relevant section of the text. You should now take the Self-Test to check your ability to recall and apply what you have learned. Finally, solve the Problems, which have been designed to further test your ability to apply the knowledge you have gained. Solving the problems is essential, because it reinforces, expands, and solidifies your knowledge. It is not sufficient simply to read the problems and look at the Answers to see if you would have done them in the way given. You must struggle through the solutions yourself to benefit from the Problems. As you are using the *Companion* you will, or course, be integrating what you have learned from your studies and your lectures or laboratory excercises.

Besides helping you to learn biochemistry, you will find the *Companion* useful in studying for examinations. Go over each of the Learning Objectives in the chapters covered by an examination and ensure yourself that you can respond to them knowledgeably. Similarly, review the key words. Decide which topics you feel uncertain of and read about them again. This protocol, coupled with a review of your lecture and reading notes, will prepare you well for examinations.

It is important to talk about biochemistry with others in order to learn how to pronounce the scientific terms and names and to help crystallize your thinking. Also, realize that although biochemistry has a sound foundation and we understand much about the chemistry of life, many of our concepts are hypotheses that will require modification or refinement as more experimental evidence accrues. Alternative and sometimes contradictory explanations exist for many biochemical observations. You should not regard the material in *Biochemistry* or the *Companion* as dogma, and you should, whenever possible, attempt to read about any given topic in at least two sources. Try to follow up topics that particularly interest you by reading about them in the scientific literature. References are given in *Biochemistry,* and your instructor can also help you locate research and review articles. In this way, you can begin to appreciate the diversity of opinion and emphasis that exists in biochemistry.

The authors welcome comments from readers, especially any drawing our attention to errors in the text. Comments should be sent to:

Professor Richard I. Gumport
Department of Biochemistry
University of Illinois
1209 West California Street
Urbana, Illinois 61801

Student's Companion to Stryer's BIOCHEMISTRY

CHAPTER 1

Prelude

The introductory chapter of *Biochemistry* begins by highlighting the impact of biochemistry on modern biology and medicine, which sets the stage for the study of this exciting branch of science. Stryer then presents the three types of atomic models that will be used to depict molecular structure. Since biochemistry is the study of the molecular basis of life, molecular dimensions, energies, and time scales are important to know; they are presented next. Then follows a discussion of the forces involved in the interactions of biomolecules. These interactions are responsible for the enormous variety of biological structures and biological processes found in nature. Because the great majority of biochemical processes occur in water, the properties of water and their effects on biomolecules are also described. Finally, Stryer outlines the six-part organization of the text.

When you have mastered this chapter, you should be able to complete the following objectives.

1. Discuss the most important achievements of *biochemistry* in the elucidation of the *molecular basis of life* and in the advancement of modern biology and medicine.

2. Explain the uses of the different *molecular models*.

3. Use the metric scale in relation to the *dimensions* of *biomolecules*, *assemblies* of biomolecules, and *cells*.

4. Give the ranges of *time* and the *energy scales* for biochemical processes.

5. List the three kinds of *noncovalent bonds* that mediate interactions of biomolecules and describe their characteristics.

6. Describe how the properties of *water* affect the interactions among biomolecules.

7. Explain the origin of *hydrophobic attractions* between *nonpolar molecules* and give examples of their importance in biochemical interactions.

SELF-TEST

1. Which of the following molecular patterns or processes are common to both bacteria and humans?

   (a) Development of tissues
   (b) Information flow from proteins to DNA
   ✓(c) Same energy currency
   (d) Same genetic information
   ✓(e) Similar biomolecules

2. Match the types of molecular models in the left column with the appropriate applications in the right column.

   (a) Space-filling models __2__

   (b) Ball-and-stick models __3__

   (c) Skeletal models __1__

   (1) show the bond framework in macromolecules.
   (2) indicate the volume occupied by a biomolecule.
   (3) show the bonding arrangement in small biomolecules.

3. Hydrogen atoms are frequently omitted from ball-and-stick models and skeletal models of biomolecules. Explain why.

---

4. Assume that amino acids, the building blocks of proteins, can be represented by spheres having a radius of 3 Å. Approximately how many amino acids could make up a spherical protein having a radius of 2 nm?

$VOL \propto d^3$ $1\,nm = 10\,Å$ $\dfrac{(20)^3}{3^3} = \dfrac{8000}{27} \approx 300$

    (a) 3 amino acids
    (b) 30 amino acids
✓(c) 300 amino acids
    (d) 3000 amino acids

5. If the protein in Question 4 were extended so that the amino acids were in linear sequence, how long would it be in nanometers?

$300 \times 3Å = 900\,Å = 90\,nm$

✓(a) 90 nm
    (b) 180 nm
    (c) 900 nm
    (d) 1800 nm

6. Could one see the extended protein in Question 5 with a light microscope? Explain.

NO. AS LIGHT MICROSCOPE RESOLVES only to 2000 Å

7. The shape of a red blood cell can be approximated by a disc 7 μm in diameter and 3 μm thick. If the cell density is close to 1.30 g/ml, how much does a single red blood cell weigh?

$V_0 = \pi r^2 L = \pi \left(\dfrac{7}{2}\right)^2 3 = \pi \dfrac{49}{2}3$

$1\mu m = 10^{-4}\,cm$
$1\mu m^3 = 10^{-12}\,cm^3$

$\approx \pi \cdot 3 \times 12 \times (10^{-4}\,\mu m)^3 = 108 \times 10^{-12}\,cm^3 \times 1.3$

$1\,ml = 1\,cc$

$= 150 \times 10^{-12}$
$= 1.5 \times 10^{-10}$

    (a) 1.5×10^{-10} g
    (b) 1.5×10^{-8} g
    (c) 1.5×10^{-6} g
    (d) 1.5×10^{-4} g

8. Most enzyme-catalyzed reactions occur in

    (a) picoseconds.
    (b) nanoseconds.
    (c) microseconds.
✓(d) milliseconds.
    (e) seconds.

9. What would you expect the generation time of a virus to be? Explain.

10. Ultraviolet light, which has an energy of around 100 kcal/mol, can be damaging to tissues. Explain why.

11. Arrange the following in order of increasing energy content.

    (a) Protein molecule
    (b) Glucose molecule
    (c) Covalent C–C bond
    (d) Hydrogen bond

12. Convert the usable energy present in ATP (~12 kcal/mol) into joules.

---

13. For the bonds or interactions in the left column, indicate all the characteristics in the right column that are appropriate.

(a) Electrostatic interaction _____

(b) The hydrogen bond _____

(c) The van der Waals bond _____

(d) Hydrophobic interaction _____

(1) requires nonpolar species.
(2) involves charged species only.
(3) requires polar or charged species.
(4) involves either O and H or N and H atoms.
(5) involves nonspecific atoms.
(6) is also called a salt bridge.
(7) only exists in water.
(8) is optimal at the van der Waals contact distance.
(9) has an energy between 3 and 7 kcal/mol.
(10) has an energy of around 1 kcal/mol.
(11) is weakened in water.

14. The properties of water include

(a) the ability to form hydrophobic bonds with itself.
(b) a disordered structure in the liquid state.
(c) a low dielectric constant.
(d) being a strong dipole, with the negative end at the O atom.
(e) a diameter of 5 Å.

15. Biological membranes are made up of phospholipids, detergentlike molecules with long nonpolar chains attached to a polar head group. When isolated phospholipids are placed in water, they associate spontaneously to form membranelike structures. Explain this phenomenon.

PROBLEMS

1. As found in the answer to Question 7 of the Self-Test, the volume of a red blood cell is 1.15×10^{-10} cm³ or 1.15×10^{-16} m³. A hemoglobin molecule has a diameter of 65 Å.

(a) If a red blood cell were completely filled with hemoglobin molecules, how many hemoglobin molecules would it hold?

(b) Referring again to Question 7 of the Self-Test, the weight of the red cell is 1.5×10^{-10} g. The molecular weight of a hemoglobin molecule is 65,000 g. Using your answer for part (a), estimate the weight of a red blood cell completely filled with hemoglobin.

(c) To calculate the number of hemoglobin molecules in a red blood cell in a more realistic way, use the molecular weight of hemoglobin and the weight of the red blood cell from Question 7 to calculate the number of hemoglobin molecules in a cell, assuming that 80% of the weight of the cell is due to hemoglobin.

2. As will be seen in succeeding chapters, enzymes provide a specific binding site for substrates where one or more chemical steps can be carried out. Often these sites are designed so as to exclude water. Suppose that at a binding site, a negatively charged substrate interacts with a positively charged atom of an enzyme.

(a) Using Coulomb's equation, show how the presence of water might affect the interaction. What sort of environment might be preferable for an ionic interaction?

(b) How would an ionic interaction be affected by the distance between the oppositely charged atoms?

3. The melting points of straight-chain hydrocarbons increase in rough proportion to their length. Relate these observations to the three types of noncovalent interactions discussed in Chapter 1.

4. In some proteins the contact distance between an amide hydrogen and a carbonyl oxygen that are participating in hydrogen bonding is somewhat less than that expected from adding their respective van der Waals contact distances. What feature of hydrogen bonding allows the two atoms to be closer to each other?

5. Consider a spherical eucaryotic cell that is 20 μm in diameter and a virus that is a cube with sides of 200 Å.

(a) Estimate the maximum number of virus particles that could adhere to the surface of the cell.

(b) Most viruses interact with host cells by binding to specific receptor molecules on the cell surface. If there are 1000 receptors that are uniformly distributed on the cell, what is the maximum separation distance between each of the receptors?

ANSWERS TO SELF-TEST

1. c, e

2. (a) 2 (b) 3 (c) 1

3. The ball-and-stick model and skeletal model are intended to show the bonding arrangements and the backbone configurations of biomolecules; the inclusion of the very abundant hydrogen atoms would obscure the very features desired in these models.

4. c. The volume of the protein is calculated from $\frac{4}{3}\pi r^3$. This volume is then divided by the volume of an amino acid.

5. b. The diameter of an amino acid is multiplied by the number of amino acids in the protein.

6. No. Although the resulting length of 1800 Å is close to the resolution limit of the light microscope, the width of only 6 Å would not permit resolution.

7. a. The volume of a cylinder is $\pi r^2 \times h$, so the volume of the red blood cell is 115 μm^3. Converting to cubic centimeters (which are the same as milliliters) gives 1.15×10^{-10} cm^3. To obtain the weight, multiply the volume by the cell density.

8. d

9. The generation time of a virus should be longer than the synthesis time of a protein, since several proteins and other macromolecules are required to form a virus, but shorter than the generation time of a bacterium, which is a much more complex entity. Therefore, the generation time of a virus should be between 10 seconds and 20 minutes.

10. An energy of 100 kcal/mol is in the range of the energy of many covalent bonds; therefore, ultraviolet light can break down covalent bonds and, consequently, can damage the structures of some essential biomolecules.

11. d, c, b, a

12. 50,200 joules/mol

13. (a) 2, 6, 9, 11 (b) 3, 4, 9, 11 (c) 5, 8, 10 (d) 1, 7

14. d

15. When the nonpolar chains of the individual phospholipids are exposed to water, they form a cavity in the water network and order the water molecules around themselves. The ordering of the water molecules requires energy. By associating with each other, through hydrophobic interactions, the nonpolar chains of phospholipids release the ordered water and hence the energy required to order the water. This stabilizes the entire system, and membranelike structures form.

ANSWERS TO PROBLEMS

1. (a) The volume of a sphere is $\frac{4}{3}\pi r^3$, so the volume of a hemoglobin molecule is

$$1.33 \times 3.142 \times (3.25 \times 10^{-9}\ \text{m})^3 =$$
$$143.5 \times 10^{-27}\ \text{m}^3 = 1.43 \times 10^{-25}\ \text{m}^3$$

To find the number of hemoglobin molecules in the red blood cell, we divide the volume of the red blood cell by the volume of a hemoglobin molecule:

$$\frac{1.15 \times 10^{-16}\ \text{m}^3}{1.43 \times 10^{-25}\ \text{m}^3} = 8.0 \times 10^8\ \text{molecules}$$

(b) To find the weight of a molecule of hemoglobin, we divide the molecular weight of hemoglobin by Avogadro's number:

$$\frac{65,000\ \text{g}}{6.023 \times 10^{-23}} = 1.07 \times 10^{-19}\ \text{g}$$

Now, 8×10^8 molecules would weigh about 8×10^{-11} grams!

(c) Another estimate of the number of hemoglobin molecules in a red blood cell is found by first calculating the weight of the cell due to hemoglobin:

$$(1.5 \times 10^{-10} \text{ g}) \times 0.8 = 1.2 \times 10^{-10} \text{ g}$$

The number of molecules in the cell is

$$\frac{1.2 \times 10^{-10} \text{ g}}{1.07 \times 10^{-19} \text{ g/molecule}} = 1.3 \times 10^9 \text{ molecules}$$

2. (a) The magnitude of the electrostatic attraction would be diminished by the presence of water because D, the dielectric constant, is relatively high for water. Inspection of Coulomb's equation shows that higher values of D will reduce the force of the attraction. Lower values, such as those for hydrophobic molecules like hexane, allow a higher value for F. We shall see that many enzyme active sites are lined with hydrophobic residues, creating an environment that enhances ionic interaction.

(b) Inspection of Coulomb's equation also reveals that the force between two oppositely charged atoms will vary inversely with the square of the distance between them.

3. Straight-chain hydrocarbons, like hexane and decane, do not form hydrogen bonds nor are they normally ionized, but they do interact through van der Waals forces. Because a homogeneous group of straight-chain molecules provides many opportunities for steric complementarity, a large number of bonds are formed. The longer the chains are, the larger the number of bonds that are formed per molecule. Thus, for longer molecules, greater amounts of thermal energy are required to disrupt the bonds and cause the molecular assembly to melt.

4. Both atoms have partial charges that attract each other. The single electron of the hydrogen atom is partially shifted to the nitrogen atom to which the hydrogen is covalently bound. As a result, the distance between the electronic shells of the hydrogen and the carbonyl oxygen is reduced, allowing them to approach each other more closely.

5. (a) The surface area of a sphere is πD^2, so the surface area of the eucaryotic cell is

$$3.14 \times (20 \times 10^{-6} \text{ m})^2 = 1.26 \times 10^{-9} \text{ m}^2$$

The surface area of a face of the virus cube is

$$(200 \times 10^{-10} \text{ m})^2 = 4 \times 10^{-16} \text{ m}^2$$

To find the number of virus particles that could adhere to the surface of the cell, we divide the surface area of the cell by the surface area of one face of the virus cube:

$$\frac{1.26 \times 10^{-9} \text{ m}^2}{4 \times 10^{-16} \text{ m}^2} = 3.15 \times 10^6 \text{ virus particles per cell}$$

(b) The maximum area available for each receptor is the surface area of the cell divided by the number of receptors:

$$\frac{1.26 \times 10^{-9} \text{ m}^2}{10^3 \text{ receptors}} = 1.26 \times 10^{-12} \text{ m}^2 \text{ per receptor}$$

Assuming that each maximum area is a circle, then the diameter of the circle gives the maximum distance between neighboring receptors:

$$\tfrac{1}{4}\pi D^2 = 1.26 \times 10^{-12} \text{ m}^2$$

$$D = \left(\frac{1.26 \times 10^{-12}}{0.785}\right)^{1/2} = 1.27 \times 10^{-6} \text{ m}$$

CHAPTER 2

Protein Structure and Function

Proteins are macromolecules that play central roles in all the processes of life. Stryer begins Chapter 2 with a description of the chemical properties of amino acids, the building blocks of proteins. It is essential that you learn the names, symbols, and properties of the twenty common amino acids at this point as they will recur throughout the text in connection with protein structures, enzymatic mechanisms, metabolism, protein synthesis, and the regulation of gene expression. It is also important to review the behavior of weak acids and bases, either in the Appendix to Chapter 2 or in an introductory chemistry text. Following the discussion of amino acids, Stryer turns to peptides and to the linear sequences of amino acid residues in proteins. Next, he describes the folding of these linear polymers into the specific three-dimensional structures of proteins. You should note that the majority of functional proteins exist in water and that their structures are stabilized by the forces and interactions you learned about in Chapter 1. Stryer concludes this chapter with examples of how protein conformations determine their functions.

LEARNING OBJECTIVES

When you have mastered this chapter, you should be able to complete the following objectives.

Introduction (Stryer pages 15–16)

1. List examples of *protein* functions.

Amino Acids (Stryer pages 16–21)

2. Draw the structure of an *amino acid* and indicate the following features, which are common to all amino acids: *functional groups, side chains, ionic forms, and isomeric forms.*

3. Classify each of the twenty amino acids according to the side chain on the α-carbon as *aliphatic, aromatic, sulfur-containing, aliphatic hydroxyl, basic, acidic,* or *amide derivative.*

4. Give the name and one-letter and three-letter *symbol* of each amino acid. Describe each amino acid in terms of *size, charge, hydrogen-bonding capacity, chemical reactivity,* and *hydrophilic* or *hydrophobic* nature.

5. Define *pH* and *pK.* Use these concepts to predict the *ionization state* of any given amino acid or its side chain in a protein.

Amino Acid Sequences of Proteins (Stryer pages 22–25)

6. Draw a *peptide bond* and describe its *conformation* and its role in *polypeptide* sequences. Indicate the *N-* and *C-terminal residues* in *peptides.*

7. Define *main chain, side chains,* and *disulfide bonds* in polypeptides. Give the range of *molecular weights* of proteins.

8. Explain the origin and significance of the unique *amino acid sequences* of proteins.

9. List examples of the *modification* and *cleavage* of proteins that expand their functional roles.

Three-Dimensional Structure of Proteins (Stryer pages 25–37)

10. Explain the importance of protein conformation to *biological activity.*

11. Differentiate between the two major *periodic structures* of proteins: the *α helix* and the *β pleated sheet.* Describe the patterns of hydrogen bonding, the shapes, and the dimensions of these structures.

12. Describe *α helical coiled coils* in specialized proteins and the role of *β-turns* in the structure of common proteins.

13. List the types of interactions among amino acid side chains that stabilize the *three-dimensional structures* of proteins.

14. Using *myoglobin* as an example, describe the main characteristics of a native folded protein structure.

15. Describe the *primary, secondary, supersecondary, tertiary,* and *quaternary structures* of proteins. Describe *domains.*

16. Using *ribonuclease* as an example, describe the evidence that the

information needed to specify the three-dimensional structure of a protein is contained in its amino acid sequence.

17. Summarize the possible sequence of steps in *protein folding*.

18. Explain the significance of *Ramachandran plots* and the use of amino acid sequences and frequencies in *predicting* protein conformations.

Specific Binding by Proteins (Stryer pages 37–39)

19. Explain the ability of proteins to interact with diverse molecules.

20. Give examples of protein interactions in enzymatic reactions, the transmission of signals, and the conversion of one form of energy into another.

SELF-TEST

Introduction

1. Match the proteins in the left column with the appropriate function in the right column.

(a) Hemoglobin _____

(b) Ribonuclease _____

(c) Acetylcholine receptor _____

(d) Myosin _____

(e) Collagen _____

(f) γ-Globulins _____

(g) Nerve growth factor _____

(1) Enzymatic catalysis
(2) Transport
(3) Generation and transmission of nerve impulses
(4) Immune protection
(5) Coordinated motion
(6) Control of growth
(7) Mechanical support

2. Explain how proteins can have the diverse functions they exhibit.

_____

_____

Amino Acids

3. Examine the four amino acids given below.

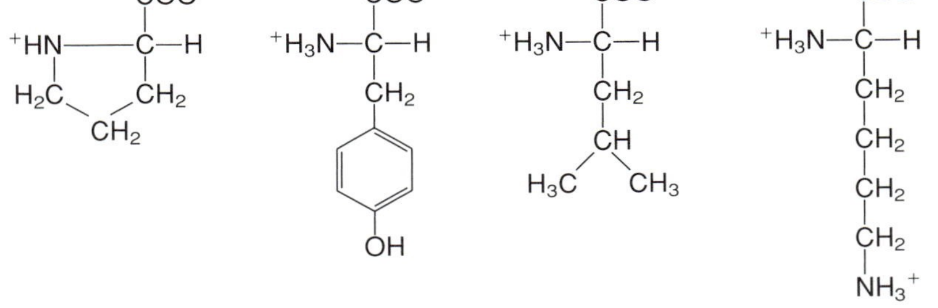

A   B   C   D

Indicate which of these amino acids are associated with the following properties.

(a) Aliphatic side chain _____

(b) Basic side chain _____

(c) Three ionizable groups _____

(d) Charge of +1 at pH 7.0 _____

(e) pK ~10 in proteins _____

(f) Secondary amino group _____

(g) Designated by the symbol K _____

(h) In the same class as phenylalanine _____

(i) Most hydrophobic of the four _____

(j) Side chain capable of forming hydrogen bonds _____

(k) May cross-link polypeptide chains of some specialized proteins _____

(l) Name the four amino acids:

A _____

B _____

C _____

D _____

(m) Name the other amino acids of the same class as D:

_____

_____

4. Draw the structure of cysteine at pH 1.

5. Match the amino acids in the left column with the appropriate side chain types in the right column.

(a) Lys _____   (1) Nonpolar aliphatic
                    (2) Nonpolar aromatic
(b) Glu _____   (3) Basic
(c) Leu _____   (4) Acidic
                    (5) Sulfur containing
(d) Cys _____   (6) Hydroxyl containing

(e) Trp _____

(f) Ser _____

6. Which of the following amino acids have side chains that are negatively charged under physiological conditions (i.e., near pH 7)?

(a) Asp
(b) His
(c) Trp
(d) Glu
(e) Cys

7. Why does histidine act as a buffer at pH 6.0? What can you say about the buffering capacity of histidine at pH 7.6?

_____

_____

8. Where stereoisomers of biomolecules are possible, only one is usually found in most organisms; for example, only the L-amino acids occur in proteins. What problems would occur if, for example, the amino acids in the body proteins of herbivores were in the L-isomer form whereas those in a large number of the plants they fed upon were in the D-isomer form?

_____

_____

Amino Acid Sequences of Proteins

9. How many different dipeptides can be made from the twenty L-amino acids? What is the minimum and the maximum number of pK values for any dipeptide?

_____

10. For the pentapeptide Glu-Met-Arg-Thr-Gly,

(a) name the carboxyl-terminal residue: _____

(b) give the number of charged groups at pH 7: _____

(c) give the net charge at pH 1: _____

(d) write the sequence using one-letter symbols: _____

(e) draw the peptide bond between the Thr and Gly residues, including both side chains:

11. If a polypeptide has 400 amino acid residues, what is its approximate molecular weight?

    (a) 11,000 daltons
    (b) 22,000 daltons
    (c) 44,000 daltons
    (d) 88,000 daltons

12. Which amino acid can stabilize protein structures by forming covalent cross-links between polypeptide chains?

    (a) Met
    (b) Ser
    (c) Glu
    (d) Gly
    (e) Cys

13. Several amino acids can be modified after the synthesis of a polypeptide chain to enhance the functional capabilities of the protein. Match the type of modifying group in the left column with the appropriate amino acid residues in the right column.

    (a) Phosphate _____
    (b) Hydroxyl _____
    (c) γ-Carboxyl _____
    (d) Acetyl _____

    (1) Glu
    (2) Thr
    (3) Pro
    (4) Ser
    (5) N-terminal
    (6) Tyr

Three-Dimensional Structure of Proteins

14. Which of the following statements about the peptide bond are *true?*

    (a) The peptide bond is planar because of the partial double bond character of the bond between the carboxyl carbon and the nitrogen.
    (b) There is relative freedom of rotation of the bond between the carboxyl carbon and the nitrogen.
    (c) The hydrogen that is bonded to the nitrogen atom is *trans* to the oxygen of the carboxyl.
    (d) There is no freedom of rotation around the bond between the α-carbon and the carboxyl carbon.

15. The α-helix structure

    (a) is maintained by hydrogen bonding between amino acid side chains.
    (b) makes up about the same percentage of all proteins.
    (c) can serve a mechanical role by forming stiff bundles of fibers in some proteins.
    (d) is stabilized by hydrogen bonds between backbone amino groups and carboxyl groups of polypeptide chains.
    (e) includes all twenty amino acids at equal frequencies.

16. Which of the following properties are common to α-helical and β-pleated-sheet structures in proteins?

    (a) Rod shape
    (b) Hydrogen bonds between main chain CO and NH groups
    (c) Axial distance between adjacent amino acids of 3.5 Å
    (d) Variable numbers of participating amino acid residues

17. Explain why α-helix and β-pleated-sheet structures are often found in the interior of water-soluble proteins.

_____

_____

18. Which of the following amino acids may alter the direction of polypeptide chains and interrupt α helices?

    (a) Phe
    (b) Cys
    (c) Trp
    (d) His
    (e) Pro

19. Which of the following amino acid residues are likely to be found on the inside of a water-soluble protein?

    (a) Val
    (b) His
    (c) Ile
    (d) Arg
    (e) Asp

20. Match the levels of protein structures in the left column with the appropriate descriptions in the right column.

    (a) Primary _____

    (b) Secondary _____

    (c) Supersecondary _____

    (d) Tertiary _____

    (e) Quaternary _____

    (1) Association of protein sub-units
    (2) Aggregate of α-helical and β-sheet structures
    (3) Linear amino acid sequence
    (4) Spatial arrangement of amino acids that are near each other in the linear sequence
    (5) Necessary for the catalytic activity of an enzyme

21. The hydrophobic interactions of the aliphatic side chains of amino acids help stabilize the folded structure of proteins. What purpose is served by the variety of sizes and shapes of aliphatic side chains?

_____

_____

22. Which of the following statements are *true*?

    (a) Ribonuclease (RNase) can be treated with urea and reducing agents to produce a random coil.
    (b) If one oxidizes random-coil RNase in urea, it quickly regains its enzymatic activity.
    (c) If one removes the urea and oxidizes RNase slowly, it will re-nature and regain its enzymatic activity.
    (d) Although renatured RNase has enzymatic activity, it can be readily distinguished from native RNase.

23. When most proteins are exposed to acidic pH (e.g., pH 2), they lose biological activity. Explain why.

_____

_____

24. A Ramachandran plot

(a) represents the sterically allowed conformations of a polypeptide backbone.
(b) gives the frequency of occurrence of amino acids in β-sheet structures.
(c) predicts α-helical structures from given amino acid sequences.
(d) shows the x-ray diffraction pattern of a protein.

Specific Binding by Proteins

25. Which of the following amino acids has a lone electron pair at one of the ring nitrogens that makes it a potential ligand in binding the iron atoms in hemoglobin?

(a) Lys
(b) Trp
(c) His
(d) Pro
(e) Arg

26. Explain how a protein may act as a molecular switch.

_____

_____

PROBLEMS

1. The hydration of carbon dioxide by carbonic anhydrase requires the participation of a zinc ion. Why would you not expect to observe enhancement of bicarbonate formation in an aqueous solution that contains only CO_2 and Zn^{2+}?

2. Suppose you are studying the conformation of a monomeric protein that has an unusually high proportion of aromatic amino acid residues throughout the length of the polypeptide chain. Compared to a protein containing many glycine residues, what would you observe for allowed ranges of ϕ and ψ? What can you say about the relative α-helical content in each of the two types of proteins?

3. Glycophorin A is a glycoprotein that extends across the red blood cell membrane. The portion of the polypeptide that extends across the membrane bilayer contains nineteen amino acid residues and is folded into an α helix. What is the width of the bilayer that could be spanned by this helix? The interior of the bilayer includes long acyl chains that are nonpolar. Which of the twenty L-amino acids would you expect to find among those in the portion of the polypeptide that traverses the bilayer?

4. Iodoacetate reacts with cysteine side chains in proteins to form *S*-carboxymethyl derivatives. Why is it a good idea to treat a protein first with β-mercaptoethanol before carrying out carboxymethylation of cysteine residues? Why might treatment with urea also be useful?

5. The net charge of a polypeptide at a particular pH can be determined by considering the p*K* value for each ionizable group in the protein. For a linear polypeptide composed of ten amino acids, how many α-carboxyl and α-amino groups must be considered?

6. For the formation of a polypeptide composed of twenty amino acids, how many water molecules must be removed when the peptide bonds are formed? Although the hydrolysis of a peptide bond is energetically favored, the bond is very stable in solution. Why?

7. In a particular enzyme, an alanine residue is located in a cleft where the substrate binds. A mutation that changes this residue to a glycine has no effect on activity; however, another mutation, which changes the alanine to a glutamate residue, leads to a complete loss of activity. Provide a brief explanation for these observations.

8. Before Anfinsen carried out his work on refolding in ribonuclease, some scientists argued that directions for folding are given to the protein during its biosynthesis. How did Anfinsen's experiments contradict that argument?

9. Many types of proteins can be isolated only in quantities that are too small for the direct determination of a primary amino acid sequence. Recent advances in gene cloning and amplification allow for relatively easy analysis of the gene coding for a particular protein. Why would an analysis of the gene provide information about the protein's primary sequences? Suppose that two research groups, one in New York and the other in Los Angeles, are both analyzing the same protein from the same type of human cell. Why would you not be surprised if they publish exactly the same primary amino acid sequence for the protein?

10. Wool and hair are elastic; both are α-keratins, which contain long polypeptide chains composed of α helices twisted about each other to form cable-like assemblies with cross-links involving Cys residues. Silk, on the other hand, is rigid and resists stretching; it is composed primarily of antiparallel β pleated sheets, which are often stacked and interlocked. Briefly explain these observations in terms of the characteristics of the secondary structures of these proteins.

11. What is the molarity of pure water? Show that a change in the concentration of water by ionization does not appreciably affect the molarity of the solution.

12. When sufficient H^+ is added to lower the pH by one unit, what is the corresponding increase in hydrogen ion concentration?

13. You have a solution of HCl that has a pH of 2.1. What is the concentration of HCl needed to make this solution?

14. The charged form of the imidazole ring of histidine is believed to participate in a reaction catalyzed by an enzyme. At pH 7.0, what is the probability that the imidazole ring will be charged?

15. Calculate the pH at which a solution of cysteine would have no net charge.

Introduction

1. (a) 2 (b) 1 (c) 3 (d) 5 (e) 7 (f) 4 (g) 6. You will learn later that myosin, a major protein of muscle, also has an enzymatic activity.

2. Proteins are made up of twenty amino acids arranged in unique sequences that are folded to form specific three-dimensional structures. Therefore, there is a great variety of distinct protein conformations that are especially adapted for diverse biological functions.

Amino Acids

3. (a) C (b) D (c) B, D (d) D (e) B, D (f) A (g) D (h) B (i) C (j) B, D (k) D (l) A is proline, B is tyrosine, C is leucine, and D is lysine. (m) Histidine and arginine (basic amino acids)

4. See the structure of cysteine in the margin. At pH 1, all the ionizable groups are protonated.

5. (a) 3 (b) 4 (c) 1 (d) 5 (e) 2 (f) 6

6. a, d

7. Histidine acts as a buffer at pH 6.0 because this is the pK of the imidazole group. At pH 7.6, histidine is a poor buffer because no one ionizing group is partially protonated and therefore capable of donating or accepting protons without markedly changing the pH.

8. All metabolic reactions in an organism are catalyzed by enzymes that are generally specific for either the D- or the L-isomer form of a substrate. If an animal (a herbivore in this case) is to be able to digest the protein from a plant and build its own protein from the resulting amino acids, both the animal and the plant must make their proteins from amino acids having the same configuration.

Amino Acid Sequences of Proteins

9. The twenty L-amino acids can form $20 \times 20 = 400$ dipeptides. The minimum number of pK values for any dipeptide is two; the maximum is four.

10. (a) Glycine
    (b) 4
    (c) +2, contributed by the N-terminal amino group and the arginine residue.
    (d) E-M-R-T-G
    (e) See the structure of the peptide bond in the margin.

11. c

12. e

13. (a) 2, 4, 6 (b) 3 (c) 1 (d) 5

Three-Dimensional Structure of Proteins

14. a, c

15. c, d

COOH
|
^+H_3N—C—H
|
CH_2
|
SH

Cysteine

H   O   H   O
| || | ||
wwN—C—C—N—C—C—O$^-$
| | |
H—C—OH H   H
|
CH_3

Peptide bond

16. b, d

17. In both α-helix and β-sheet structures, the polar peptide bonds of the main chain are involved in internal hydrogen bonding, thereby eliminating potential hydrogen-bond formation with water. Overall the secondary structures are less polar than the corresponding linear amino acid sequences.

18. e

19. a, c. Specific charged and polar amino acid residues may be found inside some proteins, in active sites, but nonfunctional residues of this type are usually located on the surface of proteins.

20. (a) 3 (b) 4 (c) 2 (d) 5 (e) 1, 5

21. The different sizes and shapes of aliphatic side chains fill the space in the interior of proteins and promote contacts that increase van der Waals interactions.

22. a, c

23. A low pH (pH 2) will cause the protonation of all ionizable side chains and will change the charge distribution on the protein; furthermore, it will impart a large net positive charge to the protein. The resulting repulsion of adjacent positive charges and the disruption of salt bridges often cause unfolding of the protein and loss of biological activity.

24. a

Specific Binding by Proteins

25. c

26. In proteins, particularly those that contain subunits, the specific binding of a molecule or ion at one site can elicit conformational changes that are propagated over significant distances to produce a change in another site, often in another subunit. The conformational changes in the second site then result in a change of its biological activity.

ANSWERS TO PROBLEMS

1. In carbonic anhydrase, the zinc atom is located in a deep pocket in coordination with three histidine side chains. Water bound to the fourth coordination position is deprotonated to form hydroxide ion, which is stabilized at pH 7, where OH^- does not usually exist. Other nearby sites on the enzyme bind CO_2, which then reacts with OH^- to form bicarbonate. Zinc ion in solution would be less likely to be found in a tetrahedral configuration with other ligands, and even if hydroxyl ion were generated, its effective concentration in solution would be low. Carbon dioxide would seldom be in proximity for the reactions with OH^- to occur.

2. The larger the side chain, the more likely it is that steric hindrance will restrict the range of ϕ and ψ. Since aromatic side chains, like the

phenyl group of tyrosine, are much bulkier than the hydrogen atom of glycine, the range of ϕ and ψ values will be lower. As Table 2-3 in the text shows, the relative levels of glycine and the aromatic amino acids in α-helical regions are low; hence, one would expect both types of proteins to have relatively low α-helical content. Presumably the large side chains of aromatic residues interfere with establishment of the regular structure of the α helix. Although there seems to be ample room for glycine residues to fit within an α-helical structure, its frequency in such chains is almost as low as that of proline, a fact that is not yet understood.

3. Since each residue in the α helix is 1.5 Å from its neighbor, the length of the chain that spans the membrane bilayer is 19×1.5 Å $= 28.5$ Å, which is also the width of the membrane. One would expect to find nonpolar amino acid residues in the polypeptide portion associated with the membrane bilayer. These would include Ala, Val, Leu, Ile, Met, and Phe. The actual sequence of the buried chain is

$$\text{I-T-L-I-I-F-G-V-M-A-G-V-I-G-T-I-L-L-I}$$

4. Many proteins contain disulfides (cystine), which must be reduced by thiol reagents to sulfhydryls (cysteine) before they can react with iodoacetate. Urea promotes the unfolding of the protein, providing exposure of otherwise buried disulfides to β-mercaptoethanol.

5. Only the N-terminal α-amino group and the C-terminal α-carboxyl group will undergo ionization. The internal groups will be joined by peptide bonds and are not ionizable.

6. For a peptide of n residues, $n - 1$ water molecules must be removed. A significant activation energy barrier makes peptide bonds kinetically stable.

7. Both alanine and glycine are neutral polar residues with small side chains, whereas the side chain of glutamate is acidic and bulkier than that of alanine. Either feature of the glutamate R-group could lead to the loss of activity by altering the protein conformation or by interfering with the binding of the substrate.

8. The fact that ribonuclease folded in vitro to yield full activity indicated that the biosynthetic machinery is not required to direct the folding process for this protein.

9. Because the sequence of DNA specifies, through a complementary sequence of RNA, the amino acid sequence of a protein, knowledge about any one of the three types of sequences yields information about the other two. One would also expect the coding sequence for a particular protein to be the same among members of the same species, allowing for an occasional rare mutation. For that reason, the published primary amino acid sequences are likely to be the same.

10. When the α helices in wool are stretched, intrahelix hydrogen bonds are broken as are some of the interhelix disulfide bridges; maximum stretching yields an extended β-sheet structure. The Cys crosslinks provide some resistance to stretch and help pull the α helices back to their original positions. In silk, the β sheets are already maximally stretched to form hydrogen bonds. Each β pleated sheet resists stretching, but since the contacts between the sheets primarily involve van der Waals forces, the sheets are somewhat flexible.

11. The molarity of water equals the number of moles of water per liter. A liter of water weighs 1000 grams, and its molecular weight is 18, so the molarity of water is

$$M = \frac{1000}{18} = 55.6$$

At 25°C, K_w is 1.0×10^{-14}; at neutrality, the concentration of both hydrogen and hydroxyl ions is each equal to 10^{-7} M. Thus, the actual concentration of H_2O is $(55.6 - 0.0000007)$ M; the difference is so small that it can be disregarded.

12. Because pH values are based on a logarithmic scale, every unit change in pH means a tenfold change in hydrogen ion concentration. When pH = 2.0, $[H^+] = 10^{-2}$ M; when pH = 3.0, $[H^+] = 10^{-3}$ M.

13. Assume that HCl in solution is completely ionized to H^+ and Cl^-. Then find the concentration of H^+, which equals the concentration of Cl^-.

$$pH = -\log[H^+] = 2.1$$
$$[H^+] = 10^{-2.1}$$
$$= 10^{0.9} \times 10^{-3}$$
$$= 7.94 \times 10^{-3} \text{ M}$$

Thus, $[H^+] = [Cl^-] = [HCl] = 7.94 \times 10^{-3}$ M

14. Use the Henderson-Hasselbalch equation to calculate the concentration of histidine, whose imidazole ring is ionized at neutral pH. The value of pK for the ring is 6.0.

$$pH = pK + \log\frac{[His]}{[His^+]}$$
$$7.0 = 6.0 + \log\frac{[His]}{[His^+]}$$
$$\log\frac{[His]}{[His^+]} = 1.0$$
$$\frac{[His]}{[His^+]} = 10$$

At pH 7.0, the ratio of uncharged histidine to charged histidine is 10:1, making the probability that the side chain is charged only 9%.

15. To see which form of cysteine has no net charge, examine all the possible forms, beginning with the one that is most protonated:

COOH		COO⁻		COO⁻		COO⁻
$^+H_3N-\overset{\displaystyle |}{\underset{\displaystyle |}{C}}-H$	$\xrightarrow{+ OH^-}$	$^+H_3N-\overset{\displaystyle |}{\underset{\displaystyle |}{C}}-H$	$\xrightarrow{+ OH^-}$	$^+H_3N-\overset{\displaystyle |}{\underset{\displaystyle |}{C}}-H$	$\xrightarrow{+ OH^-}$	$H_2N-\overset{\displaystyle |}{\underset{\displaystyle |}{C}}-H$
CH_2		CH_2		CH_2		CH_2
SH		SH		S⁻		S⁻
Net +1		0		−1		−2
charge						

The pH of the cysteine solution at which the amino acid has no net charge will be that point at which there are equal amounts of the compound with a single positive charge and a single negative charge. This is, in effect, the average of the two corresponding pK values (see p. 42 of the text), one for the α-carboxyl and the other for the α-amino group. Thus, $(1.8 + 10.8)/2 = 12.6/2 = 6.3$. This value is also known as the isoelectric point.

Exploring Proteins

Chapter 3 is an extension of Chapter 2. In Chapter 3, Stryer introduces the most important methods used to investigate proteins. Many of these methods were essential in discovering the principles of protein structure presented in Chapter 2. They also constitute the backbone of modern biochemical research and underlie current developments in biotechnology. First, Stryer describes methods for the analysis and purification of proteins; he then describes methods of sequencing the amino acids in proteins. Next, Stryer explains the use of x-ray crystallography in the determination of the three-dimensional structures of proteins. He concludes the chapter with a discussion of antibodies as highly specific reagents and a description of methods used in the synthesis of peptides.

LEARNING OBJECTIVES

When you have mastered this chapter, you should be able to complete the following objectives.

Separation and Purification of Proteins (Stryer pages 44–50)

1. Describe the principle of *electrophoresis* and its application in the separation of proteins.

2. Explain the determination of *protein mass* by *sodium-dodecyl-sulfate-polyacrylamide gel electrophoresis*.

3. Define the *isoelectric point (pI)* of a protein and describe the *isoelectric focusing* separation method.

4. List the properties of proteins that can be used to accomplish their *separation* and *purification*, and correlate them with the appropriate methods: *gel-filtration chromatography, dialysis, salting-out, ion-exchange chromatography,* and *affinity chromatography*. Describe the basic principles of each of these methods.

5. Compare the factors that affect the *velocity of migration* of a particle during *ultracentrifugation* and during *electrophoresis*.

6. Define the *sedimentation coefficient s*. Note the range of *S values* for biomolecules and cells.

7. Relate the *sedimentation velocity* of a particle to the *hydrodynamic properties* of the particle and to the density of the solution. The hydrodynamic properties of a particle include its *size, shape,* and *density*.

8. Describe *zonal centrifugation* and *sedimentation equilibrium* and explain their applications to the study of proteins.

Determination of Amino Acid Sequences (Stryer pages 50–59)

9. Outline the steps in the determination of the *amino acid composition* and the *amino-terminal residue* of a peptide.

10. Describe the sequential *Edman degradation method* and the automated determination of the amino acid sequences of peptides.

11. List the most common reagents for the *specific cleavage* of proteins. Explain the use of *overlap peptides*.

12. Describe the additional steps that must be used for sequencing *disulfide-linked* polypeptides and *oligomeric* proteins.

13. Explain how *recombinant DNA technology* is used to determine the amino acid sequences of *nascent* proteins. Note the differences between a nascent protein and a protein that has undergone *posttranslational modifications*.

14. Give examples of the important information that amino acid sequences can provide.

X-Ray Crystallography (Stryer pages 59–62)

15. Describe the essential components of the *x-ray crystallographic analysis* of a protein, and give the basic physical principles underlying this technique.

16. Note that fundamental insights are gained by x-ray crystallography into the structure, interactions, and evolution of proteins.

Use of Specific Antibodies (Stryer pages 62–64)

17. Define the terms *antibody, antigen, antigenic determinant (epitope),* and *immunoglobulin G.*

18. Contrast *polyclonal antibodies* and *monoclonal antibodies* and describe their preparation.

19. Outline the various methods that use specific antibodies in the analysis or localization of proteins.

Synthesis of Peptides (Stryer pages 64–67)

20. List the most important uses of *synthetic peptides.*

21. Describe the steps of the *solid-phase method* for the synthesis of peptides.

SELF-TEST

Separation and Purification of Proteins

1. The following five proteins, which are listed with their molecular weights and isoelectric points, were separated by SDS-polyacrylamide gel electrophoresis. Give the order of their migration from the top (the point of sample application) to the bottom of the gel.

	Molecular weight (daltons)	pI
(a) α-Antitrypsin	45,000	5.4
(b) Cytochrome c	13,400	10.6
(c) Myoglobin	17,000	7.0
(d) Serum albumin	69,000	4.8
(e) Transferrin	90,000	5.9

Top _____ Bottom

2. Which of the following statements are *not* true?

(a) The pI is the pH value at which a protein has no charges.
(b) At a pH value equal to its pI, a protein will not move in the electric field of an electrophoresis experiment.
(c) An acidic protein will have a pI greater than 7.
(d) A basic protein will have a pI greater than 7.

3. If the five proteins in Question 1 were separated in an isoelectric-focusing experiment, what would be their distribution between the positive (+) and negative (−) ends of the gel? Indicate the high and low pH ends.

cathode (−) _____ (+) anode

4. SDS-polyacrylamide gel electrophoresis and the isoelectric-focusing method for the separation of proteins have which of the following characteristics in common? Both

(a) separate native proteins.

(b) make use of an electrical field.
(c) separate proteins according to their mass.
(d) require a pH gradient.
(e) are carried out on supporting gel matrices.

5. Before high performance liquid chromatography (HPLC) methods were devised for the separation and analysis of small peptides, electrophoresis on a paper support was frequently used. Separation was effected on the basis of the charge on a peptide at different pH values. Predict the direction of migration for the following peptides at the given pH values. Use C for migration toward the cathode, the negative pole; A for migration toward the anode, the positive pole; and O if the peptide remains stationary.

	pH			
	2.0	4.0	6.0	11.0
(a) Lys-Gly-Ala-Gly				
(b) Lys-Gly-Ala-Glu				
(c) His-Gly-Ala-Glu				
(d) Glu-Gly-Ala-Glu				
(e) Gln-Gly-Ala-Lys				

6. Match the properties of proteins listed in the left column with the appropriate separation or purification methods listed in the right column.

(a) Size _____
(b) Charge _____
(c) Specific binding _____
(d) Solubility _____

(1) Electrophoresis
(2) Gel filtration
(3) Salting-out
(4) Immunoprecipitation
(5) Isoelectric focusing
(6) Affinity chromatography
(7) Ion-exchange chromatography
(8) Zonal ultracentrifugation

7. Which of the methods listed in Question 6 are sufficiently specific to isolate a particular protein from a complex mixture of proteins in one step? Explain why.

_____

_____

8. The sedimentation velocity of a protein in a centrifuge *does not* depend on the

(a) density of the solution.
(b) density of the protein.
(c) charge on the protein.
(d) shape of the protein.
(e) mass of the protein.

9. The molecular weight of a protein can be determined by SDS-polyacrylamide gel electrophoresis or by sedimentation equilibrium. Which method would you use to determine the molecular weight of a

protein containing four subunits, each consisting of two polypeptide chains cross-linked by two disulfide bridges? Explain your answer.

_____

_____

Determination of Amino Acid Sequences

10. Which of the following are useful in identifying the amino-terminal residue of a protein?

    (a) Cyanogen bromide
    (b) Fluorodinitrobenzene
    (c) Performic acid
    (d) Dabsyl chloride
    (e) Phenyl isothiocyanate

11. Which of the following statements concerning the Edman degradation method are *true?*

    (a) Phenyl isothiocyanate is coupled to the amino-terminal residue.
    (b) Under mildly acidic conditions, the modified peptide is cleaved into a cyclic derivative of the terminal amino acids and a shortened peptide (minus the first amino acid).
    (c) Once the PTH amino acid is separated from the original peptide, a new cycle of sequential degradation can begin.
    (d) If a protein has a blocked amino-terminal residue (as does *N*-formyl methionine, for example), it cannot react with phenyl isothiocyanate.

12. When sequencing proteins, one tries to generate overlapping peptides by causing specific cleavages. Which of the following statements about the cleavages caused by particular chemicals or enzymes are *true?*

    (a) Cyanogen bromide cleaves at the carboxyl side of threonine.
    (b) Trypsin cleaves at the carboxyl side of Lys and Arg.
    (c) Chymotrypsin cleaves at the carboxyl side of aromatic and bulky amino acids.
    (d) Trypsin cleaves at the amino side of Lys and Arg.
    (e) Chymotrypsin cleaves at the carboxyl side of aspartate and glutamate.

13. What treatments could you apply to the following hemoglobin fragment to determine the amino-terminal residue and to obtain two sets of peptides with overlaps so that the complete amino acid sequence can be established? Give the sequences of the overlap peptides obtained.

    Val-Leu-Ser-Pro-Ala-Lys-Thr-Asn-Val-Lys-Ala-Ala-Trp-Gly-Lys-Val-Gly-Ala-His-Ala-Gly-Glu-Tyr-Gly-Ala-Glu-Ala-Thr-Glu

_____

_____

_____

14. The following reagents are often used in protein chemistry:

    (1) CNBr   (6) Dabsyl chloride
    (2) Urea   (7) 6 N HCl

(3) β-Mercaptoethanol   (8) Ninhydrin
(4) Trypsin   (9) Phenyl isothiocyanate
(5) Dicyclohexylcarbodiimide   (10) Chymotrypsin

Which of these reagents are best suited for the following tasks?

(a) Determination of the amino acid sequence of a small peptide:

_____

(b) Identification of the amino-terminal residue of a peptide (of which you have less than 10^{-7} grams): _____

(c) Reversible denaturation of a protein devoid of disulfide bonds: _____

(d) Hydrolysis of peptide bonds on the carboxyl side of aromatic residues: _____

(e) Cleavage of peptide bonds on the carboxyl side of methionine:

_____

(f) Hydrolysis of peptide bonds on the carboxyl side of lysine and arginine residues: _____

(g) Reversible denaturation of a protein that contains disulfide bonds (two reagents are needed): _____

(h) Activation of carboxyl groups during peptide synthesis:

_____

(i) Determination of the amino acid composition of a small peptide: _____

15. Which of the following techniques are used to locate disulfide bonds in a protein?

    (a) The protein is first reduced and carboxymethylated.
    (b) The protein is cleaved by acid hydrolysis.
    (c) The protein is specifically cleaved under conditions that keep the disulfide bonds intact.
    (d) The peptides are separated by SDS-polyacrylamide gel electrophoresis.
    (e) The peptides are separated by two-dimensional electrophoresis with an intervening performic acid treatment.

16. In spite of the convenience of recombinant DNA techniques for determining the amino acid sequences of proteins, chemical analyses of amino acid sequences are frequently required. Explain why.

_____

_____

17. Which of the following are important reasons for determining the amino acid sequences of proteins?

    (a) Knowledge of amino acid sequences helps elucidate the molecular basis of biological activity.
    (b) Alteration of an amino acid sequence may cause abnormal functioning and disease.
    (c) Amino acid sequences provide insights to evolutionary pathways.

(d) The three-dimensional structure of a protein can be predicted from its amino acid sequence.

(e) Amino acid sequences provide information about the destination and processing of some proteins.

X-Ray Crystallography

18. Which of the following statements concerning x-ray crystallography is *not* true?

(a) Only crystallized proteins can be analyzed.
(b) The x-ray beam is scattered by the protein sample.
(c) All atoms scatter x-rays equally.
(d) The basic experimental data are spots of different intensities and positions on x-ray film.
(e) The electron-density maps are obtained by applying the Fourier synthesis to the spot intensities.

19. How can x-ray crystallography provide information about the interaction of an enzyme with its substrate?

_____

_____

Use of Specific Antibodies

20. Match the terms in the left column with the appropriate items from the right column.

(a) Antigens _____

(b) Antigenic determinants _____

(c) Polyclonal antibodies _____

(d) Monoclonal antibodies _____

(1) Immunoglobulins
(2) Foreign proteins, polysaccharides, or nucleic acids
(3) Antibodies produced by hybridoma cells
(4) Groups recognized by antibodies
(5) Heterogeneous antibodies
(6) Homogeneous antibodies
(7) Antibodies produced by injecting an animal with a foreign substance
(8) Epitopes

21. The method used to localize a specific protein in an intact cell is

(a) Western blotting.
(b) solid-phase immunoassay.
(c) enzyme-linked immunosorbent assay.
(d) immunoelectron microscopy.

22. Explain why immunoassays are especially useful for detecting and quantitating small amounts of a substance in a complex mixture.

_____

_____

Synthesis of Peptides

23. The amino acid sequence of a protein is known and strong antigenic determinants have been predicted from the sequence; however, you do not have enough of the pure protein to prepare antibodies. How could you circumvent this problem?

_____

_____

24. Which of the following is commonly used as a protecting group during peptide synthesis?

   (a) *tert*-Butyloxycarbonyl
   (b) Dicyclohexylcarbodiimide
   (c) Dicyclohexylurea
   (d) Hydrogen fluoride
   (e) Phenyl isothiocyanate

PROBLEMS

1. A glutamine residue that is the amino-terminal residue of a peptide often undergoes spontaneous cyclization to form a heterocyclic ring; the cyclization is accompanied by the release of ammonium ion. Diagram the structure of the ring, showing it linked to an adjacent amino acid residue. How would the formation of the ring affect attempts to use the Edman procedure for sequence analysis? A similar heterocyclic ring is formed during the biosynthesis of proline; in this case, *glutamate* is the precursor. Can you propose a pathway for the synthesis of proline from glutamate?

2. Mass spectrometry is often used for the sequence analysis of peptides from two to twenty amino acids in length. The procedure requires only microgram quantities of protein and is very sensitive; cationic fragments are identified by their charge to mass ratio. In one procedure, peptides are treated with triethylamine and then with acetic anhydride. What will such a procedure do to amino groups? Next, the modified peptide is incubated with a strong base and then with methyl iodide. What groups will be methylated? Which two amino acids cannot be distinguished by mass spectrometry?

3. Many of the methods described in Chapter 3 are used to purify enzymes in their native state. Why would the use of SDS-polyacrylamide gel electrophoresis be unlikely to lead to the successful purification of an active enzyme? What experiments would you conduct to determine whether salting-out with ammonium sulfate would be useful in enzyme purification?

4. A peptide composed of twelve amino acids does not react with dabsyl chloride or with phenyl isothiocyanate. Cleavage with cyanogen bromide yields a peptide with a carboxyl-terminal homoserine lactone residue, which is readily hydrolyzed, in turn yielding a peptide whose sequence is determined by the Edman procedure to be

E-H-F-W-D-D-G-G-A-V-L

Cleavage with staphylococcal protease yields an equivalent of aspartate and two peptides. Use of the Edman procedure gives the following sequences for these peptides:

G-G-A-V-L-M-E   and   H-F-W-D

Why does the untreated peptide fail to react with dabsyl chloride or with phenyl isothiocyanate?

5. Most enzymes are relatively stable in the pH range from 5.5 to 8.5 or 9.0. Proteins with few histidine residues exhibit relatively stable affinities for ion exchangers over that pH range, whereas proteins with significant percentages of histidine residues vary greatly in their affinity for ion exchangers over the same pH range. Why? Suppose you have an active protein in a crude extract that, at various pH ranges, adsorbs neither to cation nor to anion exchangers. Why should you continue to investigate the use of ion exchangers as a means of purifying the protein?

6. A laboratory group wishes to prepare a monoclonal antibody that can be used to react with a specific viral-coat protein in a Western-blotting procedure. Why would it be a good idea to treat the viral-coat protein with SDS before attempting to elicit monoclonal antibodies?

ANSWERS TO SELF-TEST

Separation and Purification of Proteins

1. Top <u>e d a c b</u> Bottom

2. a and c. Regardless of the pH, a protein is never devoid of charges; at the pI, the sum of all the charges is zero.

3. High pH − <u>b c e a d</u> + Low pH

4. b, e

5.

	pH			
	2.0	4.0	6.0	11.0
(a)	C	C	C	A
(b)	C	C	O	A
(c)	C	C	A	A
(d)	C	O	A	A
(e)	C	C	C	A

For example, peptide b carries a net charge of $+1.5$ at pH 2.0 (Lys side chain, $+1$; α-amino group, $+1$; Glu side chain, 0; and terminal carboxyl, -0.5, since the pH coincides with its pK value). At pH 4.0, the net charge is $+0.5$; the Glu side chain is half ionized (-0.5), but the terminal carboxyl is almost completely ionized (-1). At pH 6.0, the net charge is 0 due to a $+2$ charge contributed by the Lys residue and a -2 charge contributed by the Glu residue. At pH 11.0, the α-amino group is deprotonated (charge of 0) and the Lys side chain is half-protonated (charge of +0.5); thus, the net charge is -1.5. The same answer for peptide b can be given graphically:

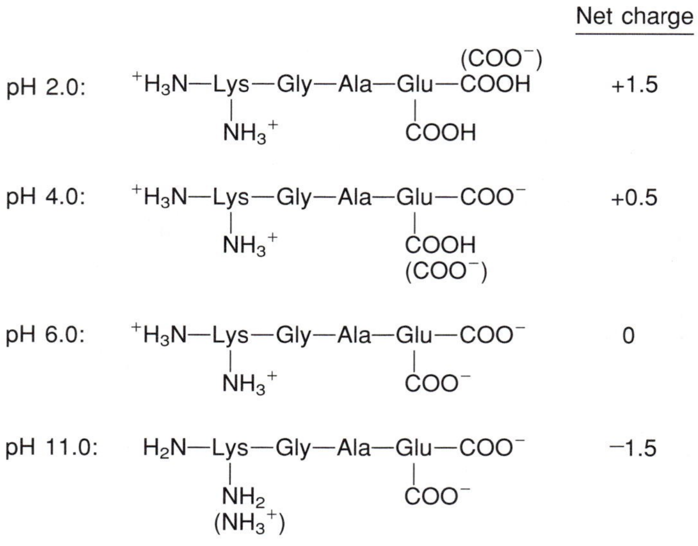

		Net charge
pH 2.0:	^+H_3N—Lys—Gly—Ala—Glu—COOH with NH$_3^+$ on Lys, (COO$^-$)/COOH on Glu	+1.5
pH 4.0:	^+H_3N—Lys—Gly—Ala—Glu—COO$^-$ with NH$_3^+$ on Lys, COOH (COO$^-$) on Glu	+0.5
pH 6.0:	^+H_3N—Lys—Gly—Ala—Glu—COO$^-$ with NH$_3^+$ on Lys, COO$^-$ on Glu	0
pH 11.0:	H_2N—Lys—Gly—Ala—Glu—COO$^-$ with NH$_2$ (NH$_3^+$) on Lys, COO$^-$ on Glu	-1.5

6. (a) 1, 2, 8 (b) 1, 3, 5, 7 (c) 4, 6 (d) 3. Note that some electrophoresis methods, such as SDS-polyacrylamide gel electrophoresis, separate proteins according to their size, although their migration on the gel is due to the overall charge. Salting-out may depend on the charge of the protein.

7. Immunoprecipitation and affinity chromatography depend on the specific interaction of a protein with an antibody and a binding substance, respectively; therefore, these methods should be sufficiently specific to isolate a particular protein. Isoelectric focusing can also be quite specific, but it is not practical for very complex mixtures.

8. c

9. Determinations of mass by SDS-polyacrylamide gel electrophoresis are carried out on proteins that have been denatured by the detergent in a reducing medium; the reducing agent in the medium disrupts disulfide bonds. Therefore, to determine the molecular weight of a native protein containing subunits with disulfide bridges, you must use the sedimentation equilibrium method. Another nondenaturing method, gel-filtration chromatography, can be used to obtain approximate native molecular weights.

Determination of Amino Acid Sequences

10. b, d, e

11. a, b, c, d

12. b, c

13. The amino-terminal residue of the hemoglobin fragment can be determined by labeling it with fluorodinitrobenzene or dabsyl chloride or by analyzing the intact fragment by the Edman degradation method; this shows that the amino-terminal residue is Val. Trypsin digestion, separation of peptides, and Edman degradation give

Val-Leu-Ser-Pro-Ala-Lys

Thr-Asn-Val-Lys

Ala-Ala-Trp-Gly-Lys

Val-Gly-Ala-His-Ala-Gly-Glu-Tyr-Gly-Ala-Glu-Ala-Thr-Glu

Chymotrypsin digestion, separation of peptides, and Edman degradation give

Val-Leu-Ser-Pro-Ala-Lys-Thr-Asn-Val-Lys-Ala-Ala-Trp

Gly-Lys-Val-Gly-Ala-His-Ala-Gly-Glu-Tyr

Gly-Ala-Glu-Ala-Thr-Glu

14. (a) 9 (b) 6, 7, 9 (c) 2 (d) 10 (e) 1 (f) 4 (g) 2, 3 (h) 5 (i) 7, 8

15. c and e. The performic acid oxidizes the disulfide bonds to SO_3^- groups and releases new peptides.

16. The amino acid sequence derived from the DNA sequence is that of the nascent polypeptide chain prior to any posttranslational modifications. Since the function of a protein depends on its mature structure, it is often necessary to analyze the protein chemically to find out the changes that have occurred in its sequence after translation.

17. a, b, c, e. Answer d may be correct in some cases where homologous proteins are compared in terms of amino acid sequences and three-dimensional structures.

X-Ray Crystallography

18. c

19. If an enzyme can be crystallized with and without its substrate and the three-dimensional structures of both are obtained using x-ray crystallography, the difference between the two structures should reveal how the substrate fits in its binding site and which atoms and bonds are involved in the interaction.

Use of Specific Antibodies

20. (a) 2 (b) 4, 8 (c) 1, 5, 7 (d) 1, 3, 6

21. d

22. Because the interaction of an antibody with its antigen is highly specific, recognition and binding can occur in the presence of many other substances. If the antibody is coupled to a radioactive or fluorescent group, then a sensitive method is available for the detection and quantitation of the antigen-antibody complex.

Synthesis of Peptides

23. You could synthesize peptides containing the putative antigenic determinants, couple these peptides to an antigenic macromolecule, and prepare antibodies against the synthetic peptides. If the same antigenic determinants are present in the protein of interest and are not occluded by the structure of the protein, then the antibodies prepared against the synthetic peptides should also react with the protein.

24. a

1. Glutamine cyclizes to form pyrrolidone carboxylic acid:

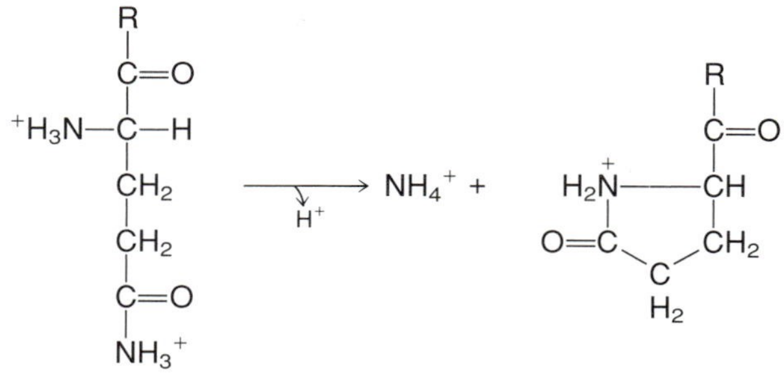

N-terminal glutamine residue → NH$_4^+$ + **Pyrrolidone carboxylate residue**

The Edman procedure begins with the reaction of phenyl isothiocyanate with the terminal α-amino group of the peptide. In the pyrrolidine ring, that group is not available. Therefore, the cyclized residue must be removed enzymatically before the Edman procedure can be used. During the biosynthesis of proline, glutamate undergoes reduction to form glutamate-γ-semialdehyde; this compound cyclizes, with the loss of water, to form Δ-pyrroline-5-carboxylate, which is then reduced to form proline.

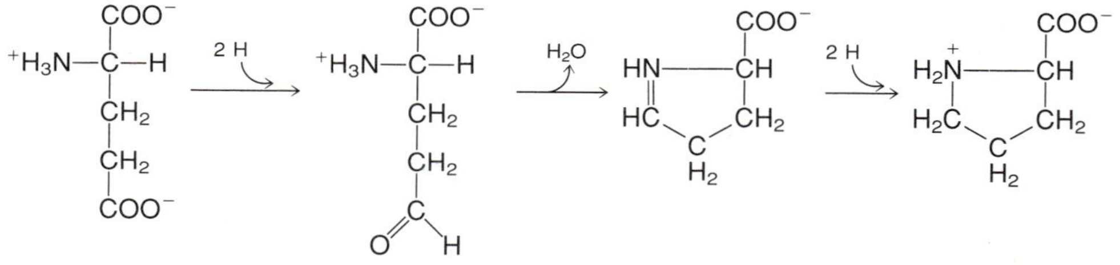

Glutamate → **Glutamic-γ-semialdehyde** → **Δ'-Pyrroline-5-carboxylate** → **Proline**

2. Treatment with triethylamine and then with acetic anhydride will yield acetylated amino groups. The strong base removes protons from amino, carboxyl, and hydroxyl groups. These groups would then be methylated by methyl iodide. Leucine and isoleucine have identical molecular weights, so they cannot be distinguished by mass spectrometry.

3. SDS disrupts nearly all noncovalent interactions in a native protein, so the renaturation of a purified protein, which is necessary to restore enzyme activity, could be difficult or impossible. You should therefore conduct small-scale, pilot tests to determine whether enzyme activity would be lost upon SDS denaturation. Similarly, when salting-out with ammonium sulfate is considered, pilot experiments should be conducted. In many instances, concentrations of ammonium sulfate can be chosen such that the active enzyme remains in solution while other proteins are precipitated, thereby affording easy and rapid purification.

4. The sequences of the peptides produced by the two cleavage methods are circular permutations of each other. Thus, the peptide is circular, so it has no free α-amino group that can react with dabsyl chloride or with phenyl isothiocyanate.

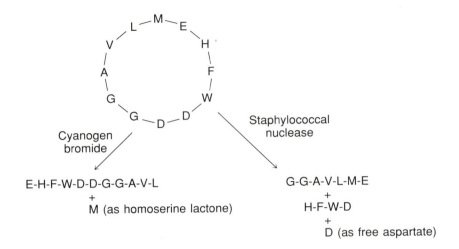

5. Because the pK for the imidazole group of histidine is approximately 6.5, the charge on the residue can vary considerably over the pH range from 5.5 to 9.0. The greater the number of histidine residues in the protein, the greater the protein charge varies when pH changes. Although the desired protein may not be adsorbed by either ion exchanger, other proteins in the extract may be adsorbed; this increases the specific activity (grams of enzyme per grams of total protein) and thereby helps to purify the protein of interest.

6. Samples for assay by Western blotting are separated by electrophoresis in SDS before blotting and antibody staining, so the reacting proteins are denatured. The use of an SDS-denatured antigen to generate the monoclonal antibody response to the viral-coat protein could assure that similar specificities are achieved in the test.

CHAPTER 4

DNA and RNA:
Molecules of Heredity

Chapters 2 and 3 introduced you to proteins. Stryer now turns to a second class of macromolecules, the nucleic acids. First, he describes the structures of the nucleotide building blocks of DNA and the phosphodiester bond that links them together. He then discusses the experimental bases for the conclusion that genes are DNA. Following this, the Watson-Crick DNA double helix is presented, an overview of how DNA is replicated is given, and DNA polymerase is introduced. Stryer concludes the chapter by describing viruses in which the genetic material is not duplex DNA but rather is single-stranded DNA or RNA. He describes the way in which these viruses replicate through double-stranded nucleic acid intermediates formed by specific base pairing. The replication of one of the RNA viruses is described to illustrate the reversal of the usual direction of the flow of information from DNA to RNA.

When you have mastered this chapter, you should be able to complete the following objectives.

DNA Composition (Stryer pages 72–73)

1. Recognize the structural components of *DNA*, that is, the *nitrogenous bases*, the *sugar*, and the *phosphate group*. Recognize the various conventions used to represent these components and the structure of DNA.

2. Distinguish among *purines*, *pyrimidines*, *ribonucleosides*, *deoxyribonucleosides*, *ribonucleotides*, and *deoxyribonucleotides*. Recognize the *deoxyadenosine*, *deoxycytidine*, *deoxyguanosine*, and *deoxythymidine* constituents of DNA, and describe the *phosphodiester bond* that joins them together to form DNA. Relate the direction of writing a DNA sequence to the *polarity of the DNA chain*.

Genes Are DNA (Stryer pages 73–76)

3. Describe the major experiments that indicated DNA is the genetic material.

DNA Structure (Stryer pages 76–78, 82–84)

4. List the important features of the *Watson-Crick DNA double helix*. Relate the pairing of *adenine* with *thymine* and *cytosine* with *guanine* to the *duplex structure* of DNA and to the replication of the helix. Explain the determinants of the specific base pairs in DNA.

5. Compare the lengths of DNA molecules in viruses, bacteria, and eucaryotes.

6. Describe *supercoiling* and state its biological consequences.

DNA Replication (Stryer pages 78–82, 84–85)

7. Describe the *Meselson-Stahl experiment* and relate it to *semiconservative replication*. Define the *melting temperature* (T_m) for DNA and relate it to the separation of the strands of duplex DNA. Describe *annealing*.

8. List the substrates and the important enzymatic properties of *DNA polymerase I* as they relate to replication. Distinguish between a *primer* and a *template* and describe their functions.

RNA and Single-Stranded DNA Viruses (Stryer pages 85–88)

9. Describe the data that led to the conclusion that bacteriophage ϕX174 contains single-stranded DNA. Explain how single-stranded DNA replicates via a *replicative form*.

10. Contrast the composition and structure of *RNA* and DNA. Distinguish between *uracil* and thymine and between *ribose* and *deoxyribose*.

11. Outline the experiments with tobacco mosaic virus (TMV) that led to the conclusion that RNA can function as a gene. Describe the replication of TMV including the function of *RNA-directed RNA polymerase*.

12. Relate the catalytic activity of *reverse transcriptase (an RNA-directed DNA polymerase)* to the replication of *retroviruses*.

SELF-TEST

DNA Composition

1.

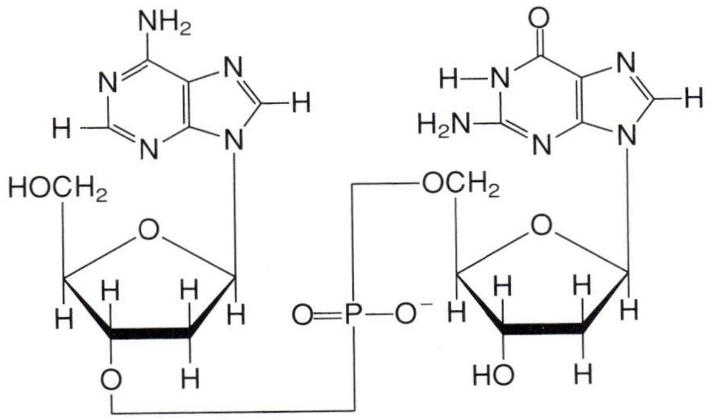

A

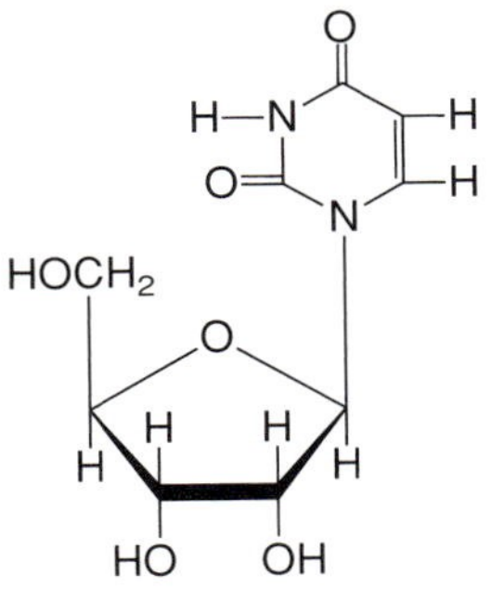

B

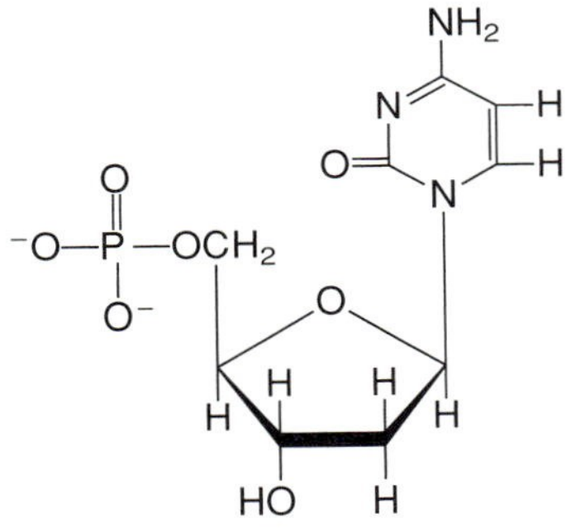

C

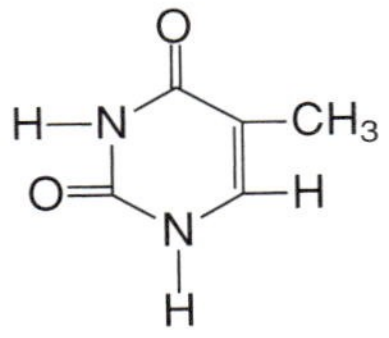

D

Which of the preceding structures

(a) contains ribose? _____

(b) contains deoxyribose? _____

(c) contains a purine? _____

(d) contains a pyrimidine? _____

(e) contains guanine? _____

(f) contains a phosphate monoester? _____

(g) contains a phosphodiester? _____

(h) is a nucleoside? _____

(i) is a nucleotide? _____

(j) would be found in RNA? _____

(k) would be found in DNA? _____

Genes Are DNA

2. The transforming principle of pneumococcus found by Avery, MacLeod, and McCarty was

(a) DNA.
(b) mRNA.
(c) polysaccharide capsular material.
(d) protein.

3. The transforming principle of pneumococcus had which of the following properties?

(a) Its activity was not affected by ribonuclease.
(b) Elemental chemical analysis of it was consistent with that of DNA.
(c) Elemental chemical analysis of it was consistent with that of RNA.
(d) Its activity was affected by deoxyribonuclease.
(e) Its activity was affected by protease.

4. Why did the radioisotopes ^{32}P and ^{35}S specifically label the DNA and the protein coat, respectively, of bacteriophage T2 in the Hershey-Chase experiment?

_____

_____

DNA Structure

5. Which of the following are characteristics of the Watson-Crick DNA double helix?

(a) The two polynucleotide chains are coiled about a common axis.
(b) Hydrogen bonds between A and C and between G and T hold the two chains together.
(c) The helix makes one complete turn every 34 Å because each base pair is rotated by 36° with respect to adjacent base pairs and is separated by 3.4 Å from them along the helix axis.
(d) The purines and pyrimidines are on the inside of the helix and the phosphodiester backbones are on the outside.
(e) Base composition analyses of DNA duplexes isolated from many organisms have shown that the amounts of A and T are equal as are the amounts of G and C.
(f) The sequence in one strand of the helix varies independently of that in the other strand.

6. If a region of one strand of a Watson-Crick DNA double helix has the sequence ACGTAACC, what is the sequence of the complementary region of the other strand?

_____

_____

7. DNA polymerase I activity requires

(a) a template.
(b) a primer with a free 5'-hydroxyl group.
(c) dATP, dCTP, dGTP, and dTTP.
(d) ATP.
(e) Mg^{2+}.

8. Derive the polarity of the synthesis of a DNA strand by DNA polymerase I from the mechanism for the formation of the phosphodiester bond.

_____

_____

_____

9. You are provided with a long, single-stranded DNA molecule having a base composition of C = 24.1%, G = 18.5%, T = 24.6%, and A = 32.8%; DNA polymerase I; [α-^{32}P]dATP (dATP with the innermost phosphate labeled), dCTP, dGTP, and dTTP; a short primer that is complementary to the single-stranded DNA; and a buffer solution with Mg^{2+}. What is the base composition of the radiolabeled product DNA after the completion of one round of synthesis?

_____

10. You are given two solutions containing different purified DNAs. One is from the bacterium *P. aeruqinosa* and has a GC composition of 68%, whereas the other is from a mammal and has a GC composition of 42.5%.

(a) You measure the absorbance of light of each solution as a function of increasing temperature. Which solution has the higher T_m value and why?

_____

_____

(b) After melting the two solutions, mixing them together, and allowing them to cool, what would you expect to happen?

_____

_____

(c) Would appreciable amounts of bacterial DNA be found associated in a helix with mammalian DNA? Explain.

_____

_____

_____

11. Outline the basic process by which a Watson-Crick duplex replicates to give two identical daughter duplexes. Explain the reasons for the accuracy of the process.

The DNA in a bacterium is uniformly labeled with ^{15}N, and the organism shifted to a growth medium containing ^{14}N-labeled DNA precursors. After two generations of growth, the DNA is isolated and is subjected to density-gradient equilibrium sedimentation. What proportion of light-density DNA to intermediate-density DNA would you expect to find?

_____

_____

_____

_____

12. Haploid human DNA has 2.9×10^6 kilobases (a kilobase, abbreviated kb, is 1000 bases or base pairs). What is the length of human DNA in centimeters?

_____

13. Purified duplex DNA molecules can be

(a) linear.
(b) circular and supercoiled.
(c) linear and supercoiled.
(d) circular and relaxed.

RNA and Single-Stranded DNA Viruses

14. For the viruses in the left column, indicate the appropriate characteristics from the right column.

(a) Tobacco mosaic virus _____

(b) Bacteriophage T7 _____

(c) Rous sarcoma virus _____

(1) Linear genome
(2) Genome contains U rather than T
(3) Single-stranded nucleic acid genome
(4) DNA intermediates are involved in replication
(5) Uses RNA-directed RNA polymerase to replicate
(6) Uses RNA-directed DNA polymerase to replicate

15. Describe how a single-stranded DNA virus could replicate by incorporating semiconservative replication into the process.

_____

_____

_____

_____

16. From the following nucleic acids, select those that appear during the infection of a cell with a retrovirus and place them in the order in which genetic information flows during the process.

_____

(a) Double-stranded DNA-RNA helix in the cell
(b) Single-stranded RNA in the virus
(c) Single-stranded RNA in the cell
(d) Double-stranded DNA in the cell
(e) Double-stranded RNA in the virus
(f) Double-stranded RNA in the cell

1. The genome of the mammalian virus SV40 is a circular DNA double helix containing 5243 base pairs. When a solution containing intact DNA molecules is heated, one observes an increase in the absorbance of light at 260 nm. When the solution is then cooled slowly, a decrease in absorbance is observed. If one or more breaks are made in the sugar-phosphate backbones of the SV40 double-stranded circles, heating causes a similar hyperchromic effect. However, when the solution of nicked molecules is cooled, the reduction in absorbance is much slower than that observed in the solution containing intact molecules. Why do the two types of molecules behave differently when they are cooled after heating?

2. A number of factors influence the behavior of a linear, double-stranded DNA molecule in a 0.25 M sodium chloride solution. Considering this, explain each of the following observations:

(a) The T_m increases in proportion to length of the molecule.
(b) As the concentration of sodium chloride decreases, the T_m decreases.
(c) Renaturation of single strands to form double strands occurs more rapidly when the DNA concentration is increased.
(d) The T_m is reduced when urea is added to the solution.

3. (a) Many proteins that interact with double-stranded DNA bind to specific sequences in the molecule. Why is it unlikely that these enzymes operate by sensing differences in the diameter of the helix?
(b) What other features of the double-stranded helix might be recognized by the protein?

4. You have a double-stranded linear DNA molecule, the appropriate primers, all the enzymes required for DNA replication, four ^{32}P-labeled nucleoside triphosphates, Mg^{2+} ion, and the equipment needed to detect newly synthesized radioactive DNA. Why is this system not sufficient to distinguish between conservative and semiconservative replication of the DNA molecule?

5. Certain deoxyribonuclease enzymes cleave any sequence of single-stranded DNA to yield nucleoside monophosphates; these enzymes do not hydrolyze base-paired DNA sequences. What products would you expect when you incubate a solution containing a single-strand specific deoxyribonuclease and the following oligonucleotide?

5′-ApGpTpCpGpTpApTpCpCpTpCpTpApCpGpApCpT-3′

6. Formaldehyde reacts with amino groups to form hydroxymethyl derivatives. Would you expect formaldehyde to react with bases in

DNA? Suppose you have a solution that contains separated complementary strands of DNA. How would the addition of formaldehyde to the solution affect reassociation of the strands?

7. When double-stranded DNA is placed in a solution containing tritiated water, hydrogens associated with the bases readily exchange with protons in the solution. The greater the percentage of AT base pairs in the DNA, the greater the rate of exchange. Why?

8. At the same time that many experiments were suggesting that DNA in chromosomes is very long and continuous, it was established that DNA polymerase I adds deoxyribonucleotides to the 3'-hydroxyl terminus of a primer chain and that a DNA template is essential. Why was a great deal of attention then given to determining whether chromosomal DNA contains breaks in the sugar-phosphate backbone?

9. The value of the T_m for DNA in degrees Celsius can be calculated using the formula

$$T_m = 69.3 + 0.41(GC)$$

where GC is the mole percentage of guanine plus cytosine.

   (a) A sample of DNA from *E. coli* contains 50 mole percent GC. At what temperature would you expect this DNA molecule to melt?
   (b) The melting curves for most naturally occurring DNA molecules reveal that the T_m is normally greater than 65°C. Why is this important for most organisms?

10. During early studies of the denaturation of double-stranded DNA, it was not known whether the two strands unwind and completely separate from each other. Suppose that you have double-stranded DNA in which one strand is labeled with ^{14}N and the other is labeled with ^{15}N. If density-gradient equilibrium sedimentation can be used to distinguish between both double- and single-stranded molecules of different densities, how can you determine whether DNA strands separate completely during denaturation?

11. Under strongly acidic conditions, several atoms of DNA bases are protonated; these include the N-1 and N-7 of adenine, the N-7 of guanine, the N-3 of cytosine, and the O-4 of thymine. Relate these observations to the stability of double-stranded DNA at low pH.

12. Sol Spiegelman found that some types of single-stranded RNA can associate with single-stranded DNA to form double-stranded molecules. What is the most important condition that must be satisfied in order to allow the formation of these hybrid molecules?

13. Many cells can synthesize deoxyuridine 5'-triphosphate (dUTP). Can dUTP be utilized as a substrate for DNA polymerase I? If so, with which base will uracil pair in newly replicated DNA?

14. The DNA of bacteriophage λ is a linear double-stranded molecule that has complementary single-stranded ends. These molecules can form closed circular molecules when two "cohesive" ends on the same molecule join, and they can form linear dimers, trimers, or longer molecules when sites on different molecules are joined.

   (a) What conditions should be chosen to insure that λ phage DNA molecules form closed circular monomers?
   (b) Under certain conditions, λ phage DNA molecules are infec-

tive. When a very low concentration of λ phage DNA is incubated with DNA polymerase I and the four deoxyribonucleoside triphosphates, the infectious activity of λ phage DNA is destroyed. Brief treatment of λ phage DNA with bacterial exonuclease III, an enzyme that removes 5'-mononucleotides from the 3'-ends of double-stranded DNA molecules and about 400 nucleotides, also destroys infectivity, but subsequent treatment of the DNA with DNA polymerase I and nucleotide substrates can restore infectivity. Describe more completely the structure of λ phage DNA, and provide an interpretation of the action of the two enzymes on the molecule.

15. The isolation of viral DNA from animal cells that have been infected with adenovirus yields linear double-stranded molecules that, when denatured and allowed to reassociate under conditions favoring intramolecular annealing, form single-stranded circles. Although the circles can be visualized using the electron microscope, resolution is not sufficient to visualize the ends of the molecule. Other analyses of the single-stranded molecule show that each end has a sequence that allows the structure shown in Figure 4.1 to be formed.

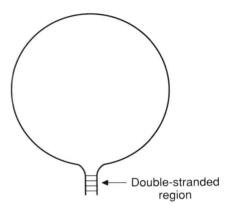

Figure 4-1
A single-strand circle formed by intramolecular annealing of adenovirus DNA.

(a) Suppose that the base sequence at one end of the single-stranded molecule is 5'-ACTACGTA . . .'. What is the corresponding sequence at the other end? Show how these sequences would allow full-length, double-stranded, linear molecules to be formed.

(b) An alternate suggestion for the formation of the single-stranded molecules was also proposed; it is shown in Figure 4.2. Why is this proposed pairing scheme unlikely?

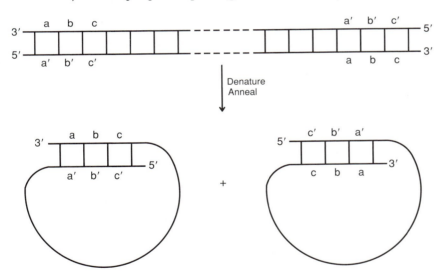

Figure 4-2
Another proposal for formation of single-stranded molecules of adenovirus DNA.

ANSWERS TO SELF-TEST

DNA Composition

1. (a) B (b) A, C (c) A (d) B, C, D (e) A (f) C (g) A (h) B (i) C; strictly speaking, A is called a dinucleotide, not a nucleotide (j) B (k) A, C, D

Genes Are DNA

2. a

3. a, b, d

4. The radioisotope ^{32}P labeled the DNA because there is one phosphorus atom per phosphodiester bond linking the deoxynucleosides of DNA to one another. The radioisotope ^{35}S labeled the protein coat because sulfur is a constituent of the amino acids cysteine and methionine in the protein coat. There is no sulfur in DNA and no phosphorus in the protein coat of bacteriophage T2.

DNA Structure

5. a, c, d, and e. Answer b is not correct because A pairs with T and G pairs with C. Answer f is not correct because the sequence of one strand determines the sequence of the other by base pairing.

6. GGTTACGT. The convention for indicating polarity is that the 5′-end of the sequence is written to the left. The two chains of the Watson-Crick double helix are antiparallel, so the correct complementary sequence is *not* TGCATTGG.

DNA Replication

7. a, c, and e. Answer b is not correct because, although the enzyme requires a primer, the state of its 5′-end is irrelevant since dNMP residues are added to the 3′-end. Answer d is not correct because the enzyme uses dNTP and not NTP molecules, where N means A, C, G, T, or U.

8. The 3′-hydroxyl of the terminal nucleotide of the primer makes a nucleophilic attack on the innermost phosphorus atom of the incoming dNTP that is appropriate for Watson-Crick base pairing to the template strand to form the phosphodiester bond. As a result, a dNMP residue is added onto the 3′-end of the primer with the concomitant release of PP$_i$ and the chain grows in the 5′→3′ direction.

9. After the completion of one round of synthesis, the template strand will have directed the polymerization of a complement in which C = 18.5%, G = 24.1%, T = 32.8%, and A = 24.6%. Since the primer is short with respect to the template, its contribution to the composition of the product strand can be neglected.

10. (a) The bacterial DNA solution has the higher T_m because it has the higher GC content and is therefore more stable to the thermal-induced separation of its strands.
    (b) The complementary DNA strands from each species will anneal to form Watson-Crick double helices as the solution cools.
    (c) No; each strand will find its partner because the perfect match between the linear arrays of the bases of complementary strands is far more stable than the mostly imperfect matches in duplexes composed of one strand of bacterial and one strand of mammalian DNA would be.

11. When replication occurs, the two strands of the Watson-Crick double helix must separate so that each can serve as a template for the synthesis of its complement. Since the two strands are complementary to one another, each bears a definite sequence relationship to the other. When one strand acts as a template, it directs the synthesis of its com-

plement. The product of the synthesis of the complement of each template strand is therefore a duplex molecule that is identical to the starting duplex. The process is accurate because of the specificity of base pairing and because the protein apparatus that catalyzes the replication can remove mismatched bases.

After two generations, you should expect to find equal amounts of light-density DNA, in which both strands of each duplex were synthesized from ^{14}N precursors, and intermediate-density DNA, in which each duplex consists of a heavy ^{15}N strand paired with a light ^{14}N strand.

12. 98.6 cm. The math is as follows:

$$(2.9 \times 10^6 \text{ kb}) \times 10^3 \text{ bases/kb} \times 3.4 \text{ Å/base} \times 10^{-8} \text{ cm/Å} = 98.6 \text{ cm}$$

13. a, b, and d. Answer d is correct because, if at least one discontinuity exists in the phosphodiester backbone of either chain of a circular duplex molecule, the chains are free to rotate about one another to assume the relaxed circular form. Answer c is incorrect because supercoiling requires closed circular molecules. In a linear molecule, the ends of each strand are not constrained with respect to rotation about the helical axis; therefore, the molecule cannot be supercoiled.

RNA and Single-Stranded DNA Viruses

14. (a) 1, 2, 3, 5
    (b) 1, 4
    (c) 1, 2, 3, 4, 6

15. The single-stranded DNA penetrates the cell, where it is converted enzymatically to a duplex replicative form through Watson-Crick base pairing. The replicative form is then reproduced by a mechanism similar to that used for the semiconservative replication of the duplex chromosome of the host cell. Finally, after this stage of replication, the mechanism shifts to one in which the replicative form serves as a template to produce copies of the single-stranded DNA found in the mature virus.

16. Starting with the single-stranded RNA in the virus and ending with the single-stranded progeny viruses, the order in which genetic information flows during the infection of a cell with a retrovirus is b, c, a, d, c, b. Retroviruses use the enzyme reverse transcriptase to convert their single-stranded genomes into a DNA-RNA replicative form that is subsequently converted into a duplex DNA replicative form prior to insertion into the host chromosome and ultimate reconversion into the single-stranded viral RNA.

ANSWERS TO PROBLEMS

1. When the intact double-stranded circular DNA molecule is heated in solution, its base pairs are disrupted and an increase in the absorbance of light at 260 nm is observed. However, the two resulting single-stranded circles are so tangled about each other that they remain closely associated. When the molecules are cooled, the interlocked strands move relative to each other until their base sequences are properly

aligned and a double-stranded molecule is reformed. This molecule absorbs less light at 260 nm than does the pair of denatured single strands. Breaks in one or both strands of a double-stranded DNA molecule allow them to separate completely from each other during denaturation. In order to form a double-stranded molecule, the separate strands collide randomly until at least a small number of correct base pairs is formed; then the remaining bases pairs form to generate a double-stranded molecule. Reassociation of a pair of separate strands in solution is slower than that of a pair of interlocked circles, so a corresponding difference in the reduction of absorbance will be observed.

2. (a) The longer the DNA molecule, the larger the number of base pairs it contains. As a result, more thermal energy is required to disrupt the helical structure of the longer DNA molecule. Experiments show that such a relationship is true for molecules up to 4000 base pairs in length.

   (b) Sodium ions neutralize the negative charges of the phosphate groups in both strands. As the concentration of NaCl decreases, repulsion among the negatively charged phosphate groups increases, making it easier to separate the two strands. The tendency for the strands to separate more easily means that dissociation occurs at a lower temperature, which means that the T_m of the molecule is lower.

   (c) The reassociation of single strands begins when a short sequence of bases in one strand forms hydrogen bonds with a complementary sequence in another. Once a short stretch of base pairs is formed, reassociation to form the longer double-stranded molecule occurs rapidly. The higher the concentration of DNA, the greater the number of complementary sequences there are in the solution, and thus the quicker the complementary sequences will find and pair with each other.

   (d) Urea in the solution disrupts hydrogen bonds. Because hydrogen bonds are partly responsible for the stability of the double helix, the disruption of these bonds makes the structure more sensitive to denaturation by thermal energy and thereby reduces the T_m. In addition to hydrogen bonding, the tendency of bases to stack also contributes significantly to the stability of the helix. Base stacking minimizes the contact of the relatively insoluble bases with water, and it also allows the sugar-phosphate chain to be located on the outside of the helix, where it can be highly solvated. Urea may also cause destabilization of the helix by allowing bases to associate more readily with water.

3. (a) The four base pairs found in the DNA double helix are almost identical in size and shape, so the diameter of the double helix is essentially uniform all along its length. It is therefore unlikely that a protein can identify a specific sequence by sensing differences in the diameter of the helix.

   (b) Proteins that interact with specific sequences might do so by forming hydrogen bonds with the bases; in some cases, it would be necessary for the double strand to undergo local unwinding or melting in order for the bases to form hydrogen bonds with a protein. However, hydrogen-bond donor and acceptors are also found in the grooves of the helix. A protein could also bind to a specific location on DNA by forming hydrogen bonds with a particular group of atoms in one of the grooves of the helix.

4. Although the system described could yield ^{32}P-labeled daughter DNA molecules, chemical methods cannot distinguish DNA in which both strands are radioactively labeled from DNA in which one strand is labeled and one strand is unlabeled. In their experiments, Meselson and Stahl used a physical technique, density gradient equilibrium sedimentation, to separate the labeled molecules according to their content of ^{14}N and ^{15}N.

5. In solution, the oligonucleotide forms an interchain double-stranded molecule with flush ends and a small single-stranded loop containing the sequence 5′-pTpCpCpTpCp-3′. The deoxyribonuclease hydrolyzes the nucleotides in this single-stranded region to form nucleoside monophosphates, leaving a small double-stranded linear molecule containing seven base pairs.

6. Formaldehyde could react with the amino groups on the C-6 carbon of adenine, the C-2 of guanine, and the C-4 of cytosine to form hydroxymethyl derivatives. Because these derivatives cannot form hydrogen bonds with complementary bases, formaldehyde-treated single strands would reassociate to a lesser extent than would untreated single strands. The actual sites of the reaction of formaldehyde with DNA are not precisely known; these sites may also include the ring nitrogen atoms in pyrimidines.

7. This experiment suggests that the hydrogen bonds of base-paired regions of double-stranded DNA may undergo reversible dissociation to form single-stranded regions, often known as *bubbles*. The transient disruption of these hydrogen bonds allows the exchange of protons with the tritiated water. AT pairs open more easily than GC pairs. Thus, the greater the percentage of AT pairs, the greater the rate of proton exchange.

8. A continuous double-stranded DNA molecule has only two 3′-OH groups available for the initiation of DNA synthesis by DNA polymerase I; because each is located at an opposing end of the molecule, no template sequence is available. In order to construct a relatively simple mechanism for chromosomal replication, one could postulate that the enzyme initiates DNA replication at a number of breaks along the chromosome, with each of the breaks offering the 3′-OH group required for the initiation of the new DNA strand. The template required for replication would then be located on the strand opposite the break, thus ensuring that DNA synthesis could continue. It is now well-established that DNA in chromosomes is very long and continuous. The fact that there are no breaks in the molecule makes the mechanism of replication more complex. It involves a number of enzyme activities, as well as the use of RNA to prime the synthesis of DNA. For details, see page 672 of Stryer's text.

9. (a) The expected melting temperature for *E. coli* DNA containing 50% GC base pairs is

$$T_m = 69.3 + 0.41(GC)$$
$$= 69.3 + 0.41(50)$$
$$= 69.3 + 20.5$$
$$= 89.8°C$$

(b) Most organisms live at temperatures that are considerably lower than 65°C. Because both the transmission and expression of genetic information depends on the integrity of the double-stranded DNA molecule, it is important that the molecule not be disrupted by thermal energy.

10. First, you must determine the temperature at which the hydrogen bonds are disrupted and single strands are formed. You can do this by heating the double-stranded DNA to various temperatures and measuring the extent of hyperchromicity. Once the DNA has been melted, centrifuge the sample using the density-gradient equilibrium sedimentation technique to attempt to separate the ^{14}N-labeled DNA strands from the ^{15}N-labeled DNA strands, which will be the denser of the two. If you are successful, this would suggest that the strands separate completely during thermal denaturation.

11. The protonation of the N-1 and N-7 of adenine, the N-7 of guanine, the N-3 of cytosine, and the O-4 of thymine makes hydrogen bonding at these locations impossible. Therefore, at low pH, where proton concentrations are high and protonation is likely to occur, double-stranded DNA is less stable than at higher pH values.

12. The association of a molecule of RNA with a molecule of DNA to form a hybrid molecule depends primarily on the two molecules having complementary sequences of bases. The formation of hydrogen bonds between complementary bases will allow the formation of a double helix composed of RNA and DNA.

13. The deoxyribonucleotide dUTP can be used as a substrate for DNA polymerase I during DNA replication because the structure of uracil is very similar to that of thymine. When incorporated into a double-stranded DNA polymer, uracil pairs with adenine, as does thymine. For a discussion of the reasons uracil is not normally incorporated into DNA, refer to page 679 of the text.

14. (a) To insure that λ phage DNA molecules form closed circular monomers, the concentration of λ phage DNA should be relatively low so that the intrachain formation of hydrogen bonds is favored. At higher concentrations, the probability of interchain joining is enhanced. At very low DNA concentrations, the molecules will remain as linear monomers.

(b) The most reasonable model for the structure of the λ phage DNA molecule is a double-stranded molecule having single-stranded protrusions at the 5′-ends, as illustrated in the margin. The 3′-ends have hydroxyl groups, which allow them to serve as primers for DNA replication catalyzed by DNA polymerase I. This enzyme fills in the single-stranded regions of the molecule, perhaps producing a molecule with flush ends. Such a molecule no longer has cohesive ends, which could be required for infectivity.

Molecules treated with exonuclease III have a longer single-stranded sequence at either end; they may not be infective because the newly exposed bases may not be complementary to each other, which would mean that the ends could no longer be joined. When the exonuclease-treated DNA is treated with DNA polymerase I, the single-stranded regions are filled in until the molecule has protruding single strands that are approximately the same length as those in the native molecule.

Hence, the molecule once again becomes infective. Further treatment of the molecule with DNA polymerase I will once again produce a molecule that has flush ends and is no longer infective.

15. (a) The sequence at the other end of the single-stranded molecule must be composed of complementary bases. It must therefore be

$$\ldots \text{TACGTAGT-3}'$$

The structure of the full-length, double-stranded, linear molecule would be

5'-ACTACGTA ———————— TACGTAGT-3'

3'-TGATGCAT ———————— ATGCATCA-5'

Each single strand has a pair of inverted repeats.

(b) The formation of double-stranded helical segments depends upon hydrogen bond formation between bases in nucleotide chains that are *antiparallel,* as follows:

$$5' \ldots \text{PuPyPuPyPu} \ldots 3'$$

$$3' \ldots \text{PyPuPyPuPy} \ldots 5'$$

where Py = pyrimidine and Pu = purine. When the suggested structure is labeled using this scheme, as in Figure 4.3, it can be seen that it would require the formation of base pairs in *parallel* chains, and such pairing cannot take place.

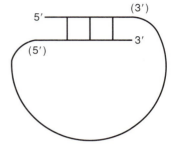

Figure 4-3
Base pairing between parallel nucleotide chains, required to form the circular structures shown in Figure 4.2, is not known to occur in DNA.

CHAPTER 5

Flow of Genetic Information

In Chapter 4, Stryer introduced you to DNA and RNA as the storage materials of genetic information. In this chapter, he begins a discussion of how this information is expressed in the proteins of a cell. Stryer starts with descriptions of the basic structures of RNA and the kinds of RNA and an explanation of their central roles in the overall flow of genetic information in the cell. He then presents the functions of messenger RNA, transfer RNA, and ribosomal RNA in protein synthesis, along with a description of the enzyme that synthesizes all cellular RNA. The experiments that elucidated the genetic code and its general features are described. The colinear relationship between the sequences of nucleotides in the DNA and the amino acids of the encoded protein is compared in procaryotes and in eucaryotes, in which some genes are interrupted by noncoding sequences. Stryer next describes the mechanism by which the noncoding sequences are removed from the initial transcript to form messenger RNA and the biological consequences of such splicing. He concludes the chapter with a discussion of catalytically active RNA and the implications of its existence with respect to hypotheses of molecular evolution.

When you have mastered this chapter, you should be able to complete the following objectives.

Introduction (Stryer pages 91–93)

1. State the role of DNA in protein synthesis. Outline the flow of genetic information during gene expression.

2. Define the terms *transcription* and *translation* and relate these processes to the *flow of genetic information.*

3. Describe the structure of the *phosphodiester bond* that links the *ribonucleosides* of *RNA* to one another. Contrast the composition and structures of RNA and DNA.

4. Name the *three major classes of RNA* found in *E. coli* and outline their functions. Compare their sizes and their relative proportions.

Messenger RNA and Transcription (Stryer pages 93–96)

5. Identify the primary cellular locations of RNA and protein synthesis in eucaryotes and relate these locations to the concept of *messenger RNA (mRNA)*. List the properties of mRNA and describe its role in the flow of genetic information.

6. Describe the experiments that indicated mRNA was an informational intermediate in the conversion of the genetic information in DNA into proteins.

7. Describe *ribosomes* and explain their role in *gene expression*.

8. Describe the technique of *hybridization* and explain how it was used to show the relationship of mRNA, *transfer RNA (tRNA),* and *ribosomal RNA (rRNA)* to their *DNA templates*.

RNA Polymerases and Transcription (Stryer pages 96–98)

9. List the substrates, products, and important enzymatic properties of *RNA polymerases (DNA-dependent RNA polymerases)*. Explain the roles of the *DNA template,* the *promoter,* and the *terminator* in transcription.

10. Describe the *transcription* of duplex DNA to form single-stranded RNA and relate this phenomenon to the *conservative* nature of RNA synthesis.

11. Compare DNA-dependent RNA polymerases and *RNA-dependent RNA polymerases.*

The Genetic Code and Protein Synthesis (Stryer pages 99–109)

12. Describe the role of tRNA as the *adaptor molecule* between mRNA and amino acids during protein synthesis. Outline how specific amino acids are covalently attached to specific tRNA molecules. Explain the relationship of the *codon* and *anticodon* to the specific interaction between mRNA and tRNA.

13. Explain what the *genetic code* is and list its major characteristics. Define the terms *degenerate, ambiguous, synonyms, triplet,* and *nonoverlap-*

ping as they apply to the genetic code. Recognize the *termination codons* and the *initiation codon*.

14. Recount the historical development of the concepts of information flow and name the major contributors.

15. Using the genetic code, predict the sequence of amino acids encoded by any template DNA or mRNA sequence.

16. Discuss the universality of the genetic code.

Introns and Splicing (Stryer pages 109–113)

17. Describe the *colinear relationship* between the sequence of DNA in a bacterial gene and the sequence of the amino acids in the protein it encodes. Relate this concept to the *intervening sequences* or *introns* that frequently occur in eucaryotic genes.

18. Recount a hypothesis relating *exons* and *functional domains* to the generation and evolution of protein diversity. Differentiate between *genetic recombination* and *RNA splicing* in this process.

Catalytic RNA and Molecular Evolution (Stryer pages 113–114)

19. List the properties of the self-splicing precursor ribosomal RNA *(ribozyme)* in *Tetrahymena*. Explain how the existence of *catalytically active RNA* molecules fits into a hypothesis concerning the evolution of molecules prior to the existence of DNA and proteins.

SELF-TEST

Introduction

1. Transcription is directly involved in which of the following possible steps in the flow of genetic information.

   (a) DNA to RNA
   (b) RNA to DNA
   (c) DNA to DNA
   (d) RNA to protein
   (e) Protein to RNA

2. Translation is involved in which of the following possible steps in the flow of genetic information?

   (a) DNA to RNA
   (b) RNA to DNA
   (c) DNA to DNA
   (d) RNA to protein
   (e) Protein to RNA

3. Answer the following questions about RNA.

   (a) What is the name of the bond joining the ribonucleoside components of RNA to one another? _____

   (b) Is this bond between the 2′- or the 3′-hydroxyl group of one ribose and the 5′-hydroxyl of the next? _____

(c) Intramolecular base pairs form what kinds of structures in RNA molecules?

_____

(d) What bases pair with one another in RNA?

_____

(e) What are the three major classes of RNA in a cell and which is most abundant?

_____

Messenger RNA and Transcription

4. After the infection of _E. coli_ with T2 bacteriophage, which of the following events occur?

   (a) New proteins are synthesized.
   (b) New RNA is synthesized.
   (c) New ribosomes are synthesized.
   (d) New RNA becomes associated with new ribosomes.
   (e) New RNA becomes associated with preexisting ribosomes.

5. The following are parts of the eucaryotic cell:

   (1) Lysosomes   (3) Plasma membranes

   (2) Nucleus   (4) Cytoplasm

   (a) In which part does the synthesis of RNA usually occur? _____

   (b) In which part are most proteins synthesized? _____

6. If you have samples of pure RNA and duplex DNA, how can you tell whether they have any complementary nucleotide sequences?

_____

_____

7. If all the RNA referred to in Question 6 turns out to have sequences that were complementary to the DNA, will its percentage of G and C be identical to that of the DNA? Explain.

_____

_____

8. If samples of each of the three major classes of RNA found in a cell were hybridized to denatured DNA from the same cell and the RNA-DNA hybrids were filtered through a nitrocellulose membrane, which of the classes would be retained on the filter?

   (a) mRNA
   (b) rRNA
   (c) tRNA

RNA Polymerases and Transcription

9. Which of the following are required for the DNA-dependent RNA polymerase reaction to produce a unique RNA transcript?

(a) ATP   (g) RNA
(b) CTP   (h) Mg^{2+}
(c) GTP   (i) Promoter sequence
(d) dTTP   (j) Operator sequence
(e) UTP   (k) Terminator sequence
(f) DNA

10. What is the sequence of the mRNA that will be synthesized from a template strand of DNA having the following sequence:

. . . ACGTTACCTAGTTGC . . .

_____

11. Describe the mechanism of chain growth in RNA synthesis. What is the polarity of synthesis and how is it related to the polarity of the template strand of DNA?

_____

_____

The Genetic Code and Protein Synthesis

12. Which of the following are characteristics or functions of tRNA?

(a) It contains a codon.
(b) It contains an anticodon.
(c) It can become covalently attached to an amino acid.
(d) It interacts with mRNA during transcription.
(e) It can have any of a number of different sequences.
(f) It serves as an adaptor between the information in mRNA and an individual amino acid.

13. What is the minimum number of contiguous nucleotides in mRNA that can serve as a codon? Explain.

_____

_____

_____

14. What is the sequence of the polypeptide that would be encoded by the DNA sequence given in Question 10? Assume that the reading frame starts with the given nucleotide. The genetic code is given on page 107 of Stryer.

_____

15. The following is a partial list of mRNA codons and the amino acids they encode:

AGU = serine   AGC = serine

AAU = asparagine   AAC = asparagine

AUG = methionine   AUA = isoleucine

Based on this list, which of the following statements are correct?

(a) The genetic code is degenerate.
(b) The alteration of a single nucleotide in the DNA directing the

synthesis of these codons could lead to the substitution of a serine for an asparagine in a polypeptide.

(c) The alteration of a single nucleotide in the DNA directing the synthesis of these codons would necessarily lead to an amino acid substitution in the encoded polypeptide.

(d) A tRNA with the anticodon ACU would be bound by a ribosome in the presence of one of these codons.

16. Explain why mitochondria can use a genetic code that is different from the standard code used in the nucleus.

_____

_____

Introns and Splicing

17. Explain how genetic techniques and amino acid sequence analyses were used to show the colinear relationship of a procaryotic gene and the protein it encodes.

_____

_____

_____

18. Answer the following questions about what was discovered when the sequence of the gene for the β chain of hemoglobin was determined.

(a) What was the major finding when the sequence of the gene and the amino acid sequence of the β chain were compared?

_____

_____

(b) What did hybridization between the partially denatured DNA and the mRNA for β-globin show?

_____

_____

(c) What must happen to the primary transcript from the β-globin gene before it can serve as an mRNA for protein synthesis?

_____

_____

19. How might the fact that some exons encode discrete functional domains in proteins be related to the evolution of new proteins?

_____

_____

Catalytic RNA and Molecular Evolution

20. What would you predict would happen if the gene for the precursor of rRNA found in *Tetrahymena*, which contains an intron, were in-

troduced into *E. coli* and converted into a primary transcript by the DNA-dependent RNA polymerase of *E. coli*?

1. Certain DNA endonucleases degrade double-stranded DNA to yield mononucleotides and dinucleotides, but these enzymes do not degrade those sequences to which other proteins are bound.

    (a) How can you use a DNA endonuclease and RNA polymerase to locate a promoter site?

    (b) Why should this process be performed in the absence of ribonucleotide triphosphates?

2. The amino acid at position 102 in the primary sequence of a bacterial enzyme is valine, and the corresponding codon in the mRNA sequence for the enzyme is GUU. Suppose a mutation that alters the codon to GCU has no effect on the activity of the enzyme, but another mutation that changes the codon to GAU completely inactivates the enzyme. Briefly explain these observations.

3. It is essential for spliceosomes to remove introns precisely, that is, between the terminal nucleotide of an intron and the first nucleotide of an exon. To see why, suppose that the sequence at the junction between an intron and an exon is

$$\underline{\text{. . . UUAG}}\Big|\underline{\text{GCUAACGG . . .}}$$
$$\text{Intron} \qquad \text{Exon}$$

Suppose further that a spliceosome occasionally cleaves the RNA transcript between the C and U residues in the exon sequence to yield the following two molecules:

$$\text{. . . UUAGGC} \qquad \text{UAACGG . . .}$$

What would be the consequence of this cleavage?

4. Although nearly all the proteins synthesized by a bacterial cell after it has been infected with T2 bacteriophage are determined by the viral genome, some bacterial proteins are also required for successful infection. What bacterial enzyme is needed to initiate viral infection when T2 DNA first enters the cell?

5. (a) The genome of bacteriophage ϕX174 is a single strand of DNA containing 5386 nucleotides. If only one AUG in the genome were used as an initiation signal, how many amino acids could be encoded by the genome? If the average molecular weight of an amino acid is 112, what is the maximum molecular weight of protein encoded by the genome?

    (b) Studies have shown that the ϕX174 genome can encode a larger number of proteins than expected. One reason for this increased encoding capacity is that some of the genes overlap each other. For example, the coding sequence for gene B is located entirely within the sequence that codes for gene A. However, the amino acid sequences of the two proteins specified by these genes are entirely different. How is this possible?

6. In contrast to DNA polymerase, RNA polymerase has no nuclease capability to excise mismatched nucleotides. Suggest why the two enzymes are different in this respect.

7. The genome of bacteriophage G4 is a small, single-stranded circle of DNA. Replication of the circle is initiated when an RNA polymerase, a product of the *dnaG* gene of *E. coli*, synthesizes a small segment of RNA that binds to a unique sequence on the G4 chromosome. Initiation of G4 DNA synthesis does not occur in bacterial *dnaG* gene mutants, which have an inactive RNA polymerase. Suggest a function for the small segment of RNA.

8. (a) In *E. coli*, a tRNA that carries tyrosine is composed of 85 nucleotides. However, transcription of the gene that codes for tyrosine tRNA yields an RNA molecule consisting of 350 nucleotides. At least three ribonuclease enzymes cooperate in removing a 41-base segment on the 5′ side of the tRNA sequence and a 224-base segment that extends from the 3′ terminus of the tRNA sequence. The tRNA sequence in the primary transcript is continuous, and no nucleotides are removed from that part of the transcript during processing. How does this type of RNA processing differ from the splicing operations described in Chapter 5 of the text?

   (b) Another primary transcript that is synthesized in bacteria contains 6500 nucleotides, including sequences for the 23S, 16S, and 5S RNA molecules found in ribosomes. This primary transcript has sequences on either side of the set of rRNA sequences, as well as "spacer" sequences between each of them. Suggest a reason for the synthesis of a transcript containing all three rRNA sequences.

9. As discussed on page 107 of the text, the codons UAA, UAG, and UGA are signals for chain termination in protein synthesis because none of these codons are read by tRNA molecules. These codons are normally found at the ends of coding sequences for proteins. However, single-base mutations in certain codons can also cause premature termination of the protein chain.

   (a) Which codons can be converted to the chain-termination codon UAA by a single base change?

   (b) Suppose a mutation creates a UAA codon that is three codons away from the 3′ end of the normal mRNA coding sequence. Why might you assume that the prematurely terminated protein could still be functional?

   (c) Revertants of chain-termination mutants include those in which a single base substitution changes a termination codon to one that can again be read by a tRNA molecule. For example, a UAG codon can mutate to UCG. What amino acid would then be found at the corresponding position in the protein?

   (d) Other revertants retain the original termination codon at the premature termination site, but an amino acid is inserted at the corresponding site in the protein so that the protein has the same length as the nonmutant protein would have. These revertants are due to another mutation in which the anticodon of a tRNA molecule is altered so that the tRNA molecule can read a termination codon. These tRNA molecules are called *suppressor tRNAs* because they suppress the effect of a chain-termination mutation. Suppose you have a chain-termination mutation that is due to the presence of a UAG codon in the normal coding sequence. If the effect of the UAG codon is suppressed by a tRNA mutation, which amino acids could be found at the site corresponding to the premature termination signal? Assume that a single base change occurs in each case.

10. Polynucleotide phosphorylase was used in the laboratory to synthesize polyribonucleotides that were useful in determining the genetic code. Why is it unlikely that this enzyme synthesizes RNA in the cell? Suggest how the cell uses this enzyme.

11. (a) How did Spiegelman's experiments with T2 RNA suggest that bacterial transcription takes place on only one of the two DNA strands at any location on the chromosomes?
    (b) Later the principle of transcription on only one strand of DNA was established more firmly by studies with DNA from the virus SP8, which infects the bacterium *Bacillus subtilis*. Because the two complementary strands of SP8 DNA have very different base compositions, they can be easily separated by density centrifugation. How could you use these separated strands to learn more about the transcription of SP8 DNA?

ANSWERS TO SELF-TEST

Introduction

1. a and b. Answer b is correct because the reverse transcription of RNA sequences into DNA sequences occurs during the replication of retroviruses (see Stryer, Chapter 4).

2. d

3. (a) The bond is called the phosphodiester bond.
   (b) The bond joins the 3'-hydroxyl to the 5'-hydroxyl to form a 3'→5' phosphodiester bond.
   (c) Hairpin loops are formed when the RNA chain folds back upon itself and base pairs become hydrogen bonded.
   (d) A pairs with U, and G paris with C; G can also pair with U, but the bond is weaker than that of the GC base pair.
   (e) The three major classes of RNA found in a cell are mRNA, rRNA, and tRNA; the most abundant is rRNA.

Messenger RNA and Transcription

4. a, b, e

5. (a) 2 (b) 4

6. You could sequence the RNA and DNA and then compare the specific sequences of each to see if the two are complementary; this method provides definitive evidence of identity. An easier but less pre-

cise way would be to use the hybridization technique. You would mix the samples, heat the mixture to melt the double-stranded DNA, slowly cool the solution, and then examine it to see if it contains any double-stranded DNA-RNA hybrids. If it does, this would indicate that the RNA and DNA sequences are complementary.

7. Not necessarily; RNA synthesis is asymmetric, and only one strand of any region of the DNA serves as a template. This can lead to RNA with a GC composition different from that of the duplex DNA.

8. a, b, c

RNA Polymerases and Transcription

9. a, b, c, e, f, h, i, and k. Answer f is correct because DNA is needed to serve as the template. Answers k and i are correct because the promoter and terminator sequences are needed to specify the precise start and stop points, respectively, for the transcription.

10. The mRNA sequence will be . . . GCAACUAGGUAACGU . . . , written in the 5′→3′ direction.

11. The 3′-hydroxyl terminus of the growing RNA chain makes a nucleophilic attack on the α-phosphate (the innermost phosphate) of the ribonucleoside triphosphate that has been selected by the template strand of the duplex DNA. RNA polymerase catalyzes the reaction. A ribonucleoside monophosphate residue is added to the chain as a result, and the chain has grown in the 5′→3′ direction; that is, the chain has grown at its 3′ end. As with all Watson-Crick base pairing, the stands are antiparallel; that is, the RNA chain is assembled in the 3′→5′ direction on the template strand of the DNA.

The Genetic Code and Protein Synthesis

12. b, c, e, and f. Answer d is incorrect because the interaction of tRNA with mRNA takes place during translation, not transcription.

13. Three contiguous nucleotides is the minimum that can serve as a codon. There are four kinds of nucleotides in mRNA. A codon consisting of only two nucleotides (either of which could be any of the four possible nucleotides) allows only 16 possible combinations (4 × 4 = 16). This would not be sufficient to specify all 20 of the amino acids. A codon consisting of three nucleotides, however, allows 64 possible combinations (4 × 4 × 4 = 64), more than enough to specify the 20 amino acids.

14. The sequence of the polypeptide would be Ala-Thr-Arg. The reading frame is set by the nucleotide at the 5′ end of the mRNA transcript; the fourth codon of the mRNA transcript is UAA, which is a translation termination codon.

15. a, b, and d. Answer a is correct because both AGU and AGC specify serine; since more than one codon can specify some amino acids, the genetic code is said to be degenerate. Answer b is correct because the alteration of a single nucleotide in the DNA could change a codon on the mRNA transcript from AGU, which specifies serine, to AAU, which specifies asparagine. Answer d is correct because the anticodon ACU would base-pair with the codon AGU. Answer c is *not* correct because the alteration of a single nucleotide in the DNA could result in another

codon that specifies the same amino acid; for example, a codon changed from AGU to AGC would still specify serine.

16. Mitochondria can use a genetic code that differs from the standard code because mitochondrial DNA encodes a distinct set of tRNAs that are matched to the genetic code used in their mRNAs.

Introns and Splicing

17. The mutations in a given gene of *E. coli* were mapped by recombination analysis. The proteins encoded by the wild-type and the mutant genes were then sequenced, and the location and nature of the amino acid substitution for each mutation was identified. It was seen that the order of the mutations on the genetic map was the same as the order of the corresponding changes in the amino acid sequence of the polypeptide produced by the gene; this established that genes and their polypeptide products are colinear.

18. (a) The number of nucleotides in the gene was much greater than three times the number of amino acids in the protein. There were two stretches of extra nucleotides between the exon sequences that encode the amino acids in the β-chain.

    (b) The mRNA hydridized to the DNA under conditions where DNA-RNA hybrids are more stable than DNA-DNA hybrids, but there were sections of duplex DNA between the hybrid regions. This indicated that there are intron sequences in the DNA that have no corresponding sequences in the mRNA. (See Figures 5-18 and 5-19 on p. 110 of Stryer.)

    (c) The intervening sequences in the nascent or primary transcript, which are complementary to the template strand of the DNA of the gene but do not encode amino acids in the protein, must be removed by splicing to generate the mRNA that functions in translation.

19. The shuffling of exons that encode discrete functional domains, such as catalytic sites, binding sites, or structural elements, preserves the functional units but allows them to interact in new ways, thereby generating new kinds of proteins.

Catalytic RNA and Molecular Evolution

20. Because the sequence of ribonucleotides is catalytically active and can splice itself to remove its own intron, it would do so in the *E. coli* that produced the transcript.

ANSWERS TO PROBLEMS

1. (a) To locate a promoter site, you would first incubate the double-stranded DNA with RNA polymerase; the RNA polymerase will bind to the promoter site. Next, you would add the DNA endonuclease, which will degrade the DNA that is not protected by RNA polymerase. Electrophoresis can be used to determine the size of the protected fragments of DNA, and the

base sequence can be determined using methods discussed in Chapter 6 of the text.

(b) Ribonucleotide triphosphates are substrates for the transcribing activity of RNA polymerase. If present when this process is performed, they will allow the polymerase molecule to move from the promoter site to the site on the template where transcription begins as well as onwards as transcription progresses. As a result, the promoter site would no longer be protected from endonuclease degradation.

2. The GCU codon in the first mutation corresponds to a substitution of alanine for valine at position 102. Although the side chain of alanine is smaller than that of valine, both are aliphatic amino acids, so the alteration in the structure of the enzyme does not affect the enzyme activity. The GAU codon of the second mutation specifies the substitution of aspartate for valine at position 102. This substitution could have a detrimental effect because the side-chain carboxyl group of aspartate has a negative charge at neutral pH. The charged group could disrupt the native conformation of the enzyme, thereby inactivating it.

3. If the spliceosome cleaves the transcript within the normal exon sequence as shown, the exon coding sequence will be altered because of the loss of two bases. Instead of beginning with the codon GCU in the normal exon the reading frame will begin with the codon UAA in the altered exon. This codon is, in fact, a termination signal for protein synthesis, which means that the translation of the polypeptide specified by the spliced messenger RNA will be terminated prematurely, indeed, before it even starts.

4. Because only T2 DNA enters the cell, the synthesis of viral proteins cannot begin until T2 messenger RNA has become available. Transcription of the T2 DNA must therefore be carried out by bacterial RNA polymerase using ribonucleotide substrates synthesized by the bacterial cell.

5. (a) Three consecutive bases are required to encode an amino acid, so up to 1794 amino acids could be specified by the ϕX174 genome. The molecular weight of this much protein would be approximately 201,000.

(b) Overlapping genes can yield proteins with different primary amino acid sequences only if each of the protein coding sequences is read in a different reading frame. The critical feature for the establishment of the proper reading frame is the site of the AUG initiation signal. As an example, consider the following mRNA sequence, which contains two AUG codons that are in overlapping but different reading frames:

When an initiator tRNA binds to the first AUG codon, reading frame 1 is established; similarly, reading frame 2 is established when an initiator tRNA binds to the second AUG codon. The polypeptides specified by the two different mRNA sequences will necessarily have different amino acid sequences.

6. DNA polymerase is responsible for the duplication of chromosomes, which are the repositories of the genetic information that is

passed on to progeny cells. Any error that occurs in the copying of a DNA template will be transmitted not only to the duplicated chromosome but also to all the messenger RNA molecules transcribed from the miscopied DNA template. Therefore, it is very important that DNA polymerase be able to correct errors that occur due to mismatched nucleotides. RNA polymerase can make many copies of mRNA, but these molecules have relatively brief lives in the cell, and very few are passed on to progeny cells. Occasional errors in transcription can result in the production of defective proteins, but it appears that the cell can tolerate such errors provided that not too many occur.

7. The small RNA molecule serves as the primer for the initiation of DNA synthesis by DNA polymerase. As described in Chapter 4 of the text, DNA polymerase requires a primer nucleotide with a 3'-hydroxyl terminus along with a template. Studies show that either a deoxyribonucleotide or a ribonucleotide can serve as a primer. DNA polymerase cannot synthesize a primer sequence on a closed DNA circle, but the RNA polymerase produced by the *dnaG* gene can and does synthesize a short primer sequence, which is then extended as DNA is synthesized by DNA polymerase. When replication has extended around the circle and the 5' terminus of the RNA primer is reached, the primer is hydrolyzed and the small gap is filled in with a DNA sequence. Bacterial *dnaG* mutants cannot support G4 infection because they cannot synthesize the RNA primer.

8. (a) The processing of the primary tRNA transcript removes RNA on either side of the uninterrupted tRNA sequence of 85 nucleotides, whereas splicing operations remove RNA sequences that are located within regions that code for a continuous polypeptide; these sequences must be removed to ensure that the protein specified by the RNA will have the correct amino acid sequence.

   (b) The most obvious reason for transcribing all three rRNA sequences simultaneously is that it ensures that an equal number of ribosomal RNA molecules will be available for the assembly of ribosomes. In addition, only one promoter, rather than three, is required for rRNA synthesis.

9. (a) The protein-encoding codons that could be mutated to UAA by a single base change are CAA, GAA, AAA, UCA, UUA, UAU, and UAC.

   (b) Chain termination near the 3' end of the normal coding sequence could allow the prematurely terminated protein to be functional because most of the polypeptide sequence would be intact. Removing a few amino acids from the C-terminal end of many, but not all, proteins does not appreciably affect their normal function.

   (c) When a UAG codon reverts to UCG, a serine residue will be found at the site on the protein that corresponds to the chain-termination site; it will be linked by a peptide bond to the next amino acid in the polypeptide.

   (d) To determine which amino acids would be carried by suppressor tRNAs to a UAG codon site, you should identify all those tRNA molecules having an anticodon that, by a single base change, can read a UAG codon. Each such tRNA molecule would suppress the premature termination of the chain by inserting an amino acid at the corresponding site in the protein.

The amino acids that could be found at the site (along with their codons are Glu (GAG), Gln (CAG), Leu (UUG), Lys (AAG), Ser (UCG), Trp (UGG), and Tyr (UAC and UAU). When a cell contains a suppressor tRNA, proteins whose mRNA sequence normally ends with a single stop codon may not be terminated. Although the extension of such proteins could be lethal, most cells are able to tolerate suppression. One explanation is that other proteins involved in chain termination may recognize a stop codon even though a tRNA that reads the codon is present. More work is needed to develop a full understanding of the reasons for toleration of suppression.

10. Equilibrium for the reaction catalyzed by polynucleotide phosphorylase lies toward the direction of RNA degradation rather than synthesis. High concentrations of ribonucleotides are required to achieve the net synthesis of RNA; and it is likely that the concentrations of ribonucleotides in the cell are not sufficient to drive net polynucleotide synthesis. Also, polynucleotide phosphorylase does not use a template, so the polyribonucleotides it synthesizes contain random sequences, which makes them of no value for protein synthesis. The cell may use polynucleotide phosphorylase as a degrading enzyme in conjunction with other nucleases that regulate the lifetimes of RNA molecules, including mRNA, which in bacteria are relatively short.

11. (a) If transcription occurred simultaneously on both DNA strands, the complementary RNA molecules synthesized from the two templates would be double-stranded. Spiegelman found no evidence of such molecules in his studies.

   (b) To establish whether one or both strands of SP8 DNA are used for transcription, you can carry out hybridization experiments with the separate strands using radioactive RNA synthesized during the infection of *Bacillus* with SP8. The results show that such RNA hybridizes to only one of the two strands, which means that only one of the two strands of the DNA of the SP8 virus is transcribed. In most other organisms, parts of both strands are used for transcription; SP8 virus is exceptional in that only one strand is used for all mRNA synthesis.

CHAPTER 6

Exploring Genes: Analyzing, Constructing, and Cloning DNA

The nature of hereditary material and the flow of information from DNA to protein via RNA were outlined in Chapters 4 and 5. Review these chapters in preparation for studying Chapter 6.

In this chapter, Stryer presents the methods and techniques that are used to analyze and manipulate DNA. He begins with a description of restriction endonucleases and their products—specific DNA restriction fragments. He then presents two major ways of determining the sequence of DNA and an automated method for synthesizing DNA by chemical means. After this introduction, Stryer describes how genes can be constructed, amplified through cloning, and expressed to yield protein products. A more detailed description of restriction enzymes and DNA ligase follows; these enzymes make possible the precise production and joining of DNA fragments. Next, the self-replicating carriers of the target genes are discussed, and three of the most widely used of these vectors are described. The problems of locating specific genes in the genome and of inserting and expressing foreign genes in eucaryotes are considered. The special role in recombinant DNA technology of cDNA, which is produced from mRNA, is discussed. Stryer then describes how site-specific mutagenesis can be used with cloned genes to produce proteins having any desired amino acid at any position. Finally, he describes how these powerful new biochemical technologies make possible previously unimaginable manipulations of living organisms for the benefit of humankind.

LEARNING OBJECTIVES

When you have mastered this chapter, you should be able to complete the following objectives.

Introduction (Stryer pages 117–118)

1. Define *recombinant DNA*.

Restriction Enzymes and Restriction Fragments (Stryer pages 118–120)

2. Describe the reaction of *restriction enzymes* and the sequence characteristics of the sites they recognize.

3. Explain why *gel electrophoresis of DNA* is essential to recombinant DNA technology. Describe how DNA molecules can be detected in gels.

4. Contrast the *Southern, Northern,* and *Western blotting* techniques.

DNA Sequencing (Stryer pages 120–123)

5. Outline *DNA sequencing* by *specific chemical cleavage* (the *Maxam-Gilbert method*). Describe the function of *polynucleotide kinase* in this process.

6. Outline DNA sequencing by the *controlled interruption of enzymatic replication* (the *Sanger dideoxy* method).

7. Compare and contrast the Maxam-Gilbert and Sanger dideoxy DNA sequencing methodologies.

Automated Chemical DNA Synthesis (Stryer pages 123–124)

8. Describe how *oligonucleotides* are chemically synthesized.

9. List the common uses of oligonucleotides.

Genome Construction and Cloning (Stryer pages 124–131)

10. List the desired characteristics of a *vector*. Outline the three major steps in *cloning* a DNA molecule.

11. Name the substrates of *DNA ligase*, and describe the reaction catalyzed by it. Describe the termini of the DNA fragments joined by DNA ligase.

12. Explain how *linkers* are used in recombinant DNA technology.

13. Name some common vectors used in procaryotes and eucaryotes and compare their properties and relative merits.

14. Distinguish between *screening* and *selection*. Explain *insertional inactivation*.

15. Describe what a *probe* is and explain the biochemical basis of its specificity. Describe how probes are obtained.

16. Explain how to design a probe by converting an amino acid sequence into a nucleotide sequence using the genetic code.

17. Outline *chromosome walking* and state the kinds of information that it reveals.

18. Define *cDNA*. Distinguish between *genomic* and *cDNA libraries*.

19. Describe the reactions catalyzed by *reverse transcriptase* and *S1 nuclease*. Outline the process for converting the information in mRNA into duplex DNA.

20. Explain how the presence of *introns* in some eucaryotic genes prevents the expression of these genes in procaryotes.

21. List the properties of *expression vectors*.

22. Contrast screening for recombinant cells using *nucleic acid probes* versus *immunochemical methods*.

23. List the methods by which DNA can be incorporated into plant and animal cells. Define *transgenic organism* and give an example.

Protein Engineering and Site-Specific Mutagenesis (Stryer pages 136–137)

24. Outline the process of *oligonucleotide-directed mutagenesis*, including the detection of the mutant progeny. Define *stringency* as it relates to *nucleic acid hybridization*.

25. Describe the process of protein engineering and explain its value. List four kinds of mutational changes that can be engineered into genes. Define *chimeric protein*.

26. List some of the actual and potential uses of recombinant DNA technology.

27. Reflect on the implications of recombinant DNA technology with respect to moral and social values.

SELF-TEST

Restriction Enzymes and Restriction Fragments

1. Which of the following portions of a longer duplex DNA segment are likely to be the recognition sequences of a restriction enzyme?

    (a) 5'-AGTC-3'   (c) 5'-ACCT-3'
         3'-TCAG-5'   3'-TGGA-5'
    (b) 5'-ATCG-3'   (d) 5'-ACGT-3'
         3'-TAGC-5'   3'-TGCA-5'

2. Which of the following reagents would be useful for visualizing DNA restriction fragments that have been separated by electrophoresis in an agarose gel and remain in the wet gel?

    (a) $^{32}P_i$
    (b) $[\alpha\text{-}^{32}P]ATP$
    (c) Diphenylamine
    (d) Ethidium bromide
    (e) DNA polymerase
    (f) Polynucleotide kinase

3. Which blotting technique is used for the detection of DNA that has been separated from a mixture of DNA restriction fragments by elec-

trophoresis through an agarose gel and then transferred onto a nitro-cellulose sheet?

(a) Eastern blotting
(b) Northern blotting
(c) Southern blotting
(d) Western blotting

DNA Sequencing

4. Which of the following reagents would be useful for labeling the oligonucleotide d(GGAATTCC)?

(a) $[\gamma\text{-}^{32}P]ATP$
(b) $^{32}P_i$
(c) DNA-dependent RNA polymerase
(d) Polynucleotide kinase
(e) DNA ligase

5. In producing fragments for sequencing, which of the following reagents is used to modify the purines in a DNA molecule so that they become alkali labile?

(a) Dimethylsulfate
(b) Hydrazine
(c) NaCl
(d) Piperidine
(e) Phosphite triester

6. Complete the following statements about the Sanger dideoxy method of DNA sequencing.

(a) The incorporation of a ddNMP onto a growing DNA chain stops the reaction because

_____

_____

(b) The fragments are usually labeled with ^{32}P by

_____

_____

(c) A "universal" primer may be used when sequencing any insert cloned into M13 because

_____

_____

(d) It is preferable to label the oligonucleotide primer with a fluorescent rather than radioactive group because

_____

_____

Automated Chemical DNA Synthesis

7. Which of the following statements are *correct*? Chemically synthesized oligonucleotides can be used

(a) to synthesize genes.
(b) to construct linkers.
(c) to introduce mutations into cloned DNA.
(d) as primers for sequencing DNA.
(e) as probes for hybridization.

8. Which of the following statements are *correct?* The successful chemical synthesis of oligonucleotides requires

(a) high yields at each condensation step.
(b) the protection of groups not intended for reaction.
(c) a single treatment for the removal of all blocking groups.
(d) methods for the removal of the blocking groups that do not rupture phosphodiester bonds.
(e) a computer-controlled, automated "gene machine."

Genome Construction and Cloning

9. State whether each of the following is or is not a desired characteristic of vectors and explain why or why not.

(a) Autonomous replication _____

_____

(b) Unique restriction sites _____

_____

(c) Genes that confer antibiotic resistance _____

_____

(d) Small size _____

_____

(e) Circularity _____

_____

10. You have been supplied with the oligonucleotide d(GGAATTCC) and an isolated and purified DNA restriction fragment that has been excised from a longer DNA molecule with a restriction endonuclease that produces blunt ends. Which of the following reagents would you need to tailor the ends of the fragment so it could be inserted into an expression vector at a unique EcoRI cloning site?

(a) DNA polymerase
(b) All four dNTPs
(c) EcoRI restriction endonuclease
(d) ATP
(e) DNA ligase
(f) Polynucleotide kinase

What would happen if the restriction fragment has an internal EcoRI site?

_____

_____

11. Inserting a long DNA fragment into the middle of a vector gene

that specifies an enzyme that hydrolyzes an antibiotic and incorporating the altered vector into a bacterium

    (a) leads to drug resistance transfer.
    (b) is called insertional inactivation.
    (c) renders the cell sensitive to that antibiotic.
    (d) can be used to identify bacteria that contain the vector with the DNA fragment.
    (e) is a method of destroying pathogenic bacteria.

12. Briefly describe genomic and cDNA libraries.

_____

_____

_____

13. Which of the following partial amino acid sequences from a protein whose gene you wish to clone would be most useful in designing an oligonucleotide probe to screen a cDNA library?

    (a) Met-Leu-Arg-Leu
    (b) Met-Trp-Cys-Trp

Explain why.

_____

_____

Expression of Recombinant DNA

14. Explain how the presence of introns in eucaryotic genes complicates the production of the protein products they encode when expression is attempted in bacteria. How can this problem be circumvented?

_____

_____

_____

15. Which of these reagents would be required to perform an immunochemical screen of a population of bacteria for the presence of a particular cloned gene if you have the pure protein encoded by the gene?

    (a) $[\gamma\text{-}^{32}P]$ATP
    (b) Polynucleotide kinase
    (c) DNA polymerase
    (d) All four dNTPs
    (e) A radioactive antibody to the protein encoded by the cloned gene

16. The gene for a eucaryotic polypeptide hormone was isolated, cloned, sequenced, and overexpressed in a bacterium. After the polypeptide was purified from the bacterium, it failed to function when it was subjected to a bioassay in the organism from which the gene was isolated. Speculate why the recombinant DNA product was inactive.

_____

_____

_____

17. Which of the following statements are *correct?* Oligonucleotide-directed site-specific mutagenesis

(a) depends upon having an oligonucleotide with a sequence completely different from that of the target gene.
(b) can be used to produce deletion, insertion, and point mutations.
(c) is a good method for identifying the functional domains of an enzyme.
(d) is useful for determining the involvement of a particular amino acid in the catalytic mechanism of an enzyme.
(e) can involve the use of the same oligonucleotide to produce the mutant and to detect it.

18. Outline the steps necessary to synthesize a gene. Be explicit about the information and reagents, including enzymes, you would need.

_____

_____

_____

_____

PROBLEMS

1. You are studying a newly isolated bacterial restriction enzyme that cleaves double-stranded circles of pBR322 once to yield unit-length, linear, double-stranded molecules. After these molecules are denatured and are allowed to reanneal, all of the double-stranded molecules are unit-length linears. In another experiment, an enzyme that cleaves double-stranded DNA at random sites is used at low concentration to cleave intact pBR322 molecules approximately once per molecule, again yielding unit-length, double-stranded linears. Denaturation and renaturation yield a high percentage of double-stranded circles with a single, randomly located nick in each strand. How do these experiments show that the new restriction enzyme cleaves pBR322 DNA at a single specific site?

2. Before the development of methods for the analysis and manipulation of genes, many attempts were made to transform both procaryotic and eucaryotic cells with DNA. Although the experiments with the pneumococcus bacterium described in Chapter 4 were successful, most others were not. Suggest why these early efforts to transform cells were largely unsuccessful.

3. Pseudogenes are composed of nonfunctional DNA sequences that are related by sequence similarity to actively expressed genes. Some researchers have proposed that pseudogenes are copies of functional genes that have been inactivated during genome evolution. Suggest several ways that such genes could have become nonfunctional. Suppose you clone a number of closely related sequences, any of which may code for a particular protein. How can you tell which of the sequences is the functional gene, that is, which of the sequences codes for the protein?

4. Both the Maxam-Gilbert and Sanger dideoxy methods for determining DNA sequence are limited in that a stretch of only 500 or fewer bases can be analyzed at one time. Suppose you wish to sequence a mutant of the simian virus SV40, which contains 5000 base pairs. You decide to use restriction fragments of the DNA for sequencing, and you find an enzyme that makes a sufficient number of cuts to give fragments of 275 nucleotides or less. Why might it be a good idea also to sequence another set of fragments cleaved by another restriction enzyme?

5. The denaturation and reassociation of complementary DNA strands can be used as a tool for genetic analysis. Heating double-stranded DNA in a dilute solution of sodium chloride or increasing the pH of the solution above 11 causes dissociation of the complementary strands; when the solution of single-stranded molecules is cooled or when the pH is lowered, the complementary strands will reanneal as complementary base pairs reform. What causes base pairs to dissociate at pH 11 or higher? Both double- and single-stranded DNA molecules can be visualized using electron microscopy in a technique called heteroduplex analysis. Suppose that two types of double-stranded molecules, one type containing the sequence for a single gene and the other type containing the same sequence as well as an insertion of nonhomologous DNA, are mixed and used in a reannealing experiment. If the two types of molecules undergo denaturation and reannealing, what types of molecules would you expect to see?

6. Bacterial chromosome deletions of more than 50 base pairs can be detected by electron microscopy, using heteroduplex analysis as described in Problem 5. When a heteroduplex is formed between a single-stranded DNA molecule from a deletion strain and a single-stranded molecule from a nondeletion, or wild-type, strain, a single-stranded loop will be visible at the location of the deletion. Suppose you are studying a bacterial mutation, which appears to be a deletion of about 200 base pairs, located at a unique site on the bacterial chromosome, which contains over 3000 genes. Why would it be a good idea to clone DNA containing the site of the deletion, as well as the corresponding site in the wild-type strain, in order to study the deletion using heteroduplex analysis?

7. You wish to clone a yeast gene in lambda phage. Why is it desirable to cleave both the yeast DNA and the lambda phase DNA with the same restriction enzyme?

8. Suppose you are studying the structure of a protein that contains a proline residue, and you wish to determine whether the substitution of a glycine residue will change the conformation of the polypeptide. You have cloned the gene for the protein, and you know the sequence of the protein. Using site-specific mutagenesis, what alterations would you make in the gene sequence in order to replace proline with glycine?

9. Cleavage of a double-stranded DNA fragment that contains 500 bases with restriction enzyme A yields two unique fragments, one 100 bases and the other 400 bases in length. Cleavage of the DNA fragment with restriction enzyme B yields three fragments, two containing 150 nucleotides and one containing 200 nucleotides. When the 500-base fragment is incubated with both enzymes (this is called the double-digest technique), two fragments 100 bases in length and two 150 bases in length are found. Diagram the 500-base fragment, showing the

cleavage sites of both enzymes. Now suppose you also have a double-stranded DNA fragment that is identical to the original fragment, except that the first 75 base pairs at the left end are deleted. How can this fragment help you construct a cleavage map for the two enzymes?

ANSWERS TO SELF-TEST

1. d. It has twofold rotational symmetry; that is, the top strand, 5′-ACGT-3′, has the same sequence as the bottom strand. Many restriction enzymes recognize and cut such palindromic sequences.

2. d. Ethidium bromide would intercalate into the DNA duplex, and its quantum yield of fluorescence would consequently increase. Upon UV irradiation, it will fluoresce with an intense orange color wherever DNA is present.

3. c

4. a and d. The reaction of the oligonucleotide with these reagents would yield [5′-^{32}P]d(pGAATTCC).

5. a. Dimethylsulfate methylates guanine at N-7 and adenine at N-3. Hydrazine reacts with the pyrimidines, and piperidine is the base that catalyzes the cleavage of the phosphodiester backbone at apurinic (a site lacking a purine) or apyrimidinic sites.

6. (a) . . . the newly synthesized ddNMP terminus lacks the requisite 3′-hydroxyl onto which the next dNMP residue would add if the normal dNTP were present in the enzymatic elongation reaction mixture.
   (b) . . . using an [α-^{32}P]dNTP in the reaction mixture so that DNA polymerase will incorporate the [5′-^{32}P]dNMP portion of the nucleotide into the growing DNA chain.
   (c) . . . a single primer sequence that is complementary to a region adjacent to the site of insertion can be used to sequence any DNA segment cloned into that site (see Stryer, p. 130).
   (d) . . . it avoids the use of radioisotopes and could allow the automated detection of the terminated primers.

7. All are correct.

8. a, b, and d. Answer c is not correct because the blocking groups must be removed differentially; for example, the dimethoxytrityl group must be removed from the 5′-hydroxyl (so that the next condensation with an incoming nucleotide can occur) without removing the blocking groups on the exocyclic amines of the bases. Answer e is not correct because the synthesis can be carried out manually, albeit doing so is slower and more laborious.

9. Answers a, b, c, and d are desired characteristics. Autonomous replication allows amplification of the vector in the absence of extensive cell growth. Unique restriction sites allow the cutting of vectors at single, specific sites for the insertion of the foreign DNA. Antibiotic resistance allows for the selection of those bacteria that carry the vector or for insertional inactivation. Small size allows the insertion of long pieces of foreign DNA without interfering with the introduction of the recom-

binant molecule into the host bacterium. Answer e is not a desired characteristic because vectors do not have to be circular to function effectively; an example is lambda phage.

10. c, d, e, and f. The oligonucleotide must have a 5′-phosphate group to serve as a substrate for DNA ligase. Therefore, ATP and polynucleotide kinase would be used, as would DNA ligase and ATP, to allow the joining of the duplex form of the palindromic oligonucleotide to the blunt-ended fragment. Finally, the fragment with the linker covalently joined to it would be cut with EcoRI endonuclease to produce cohesive ends that match those of the cut vector. It is assumed that the fragment itself lacks EcoRI sites because, if one or more internal EcoRI sites were present, the fragment would be cut when it is treated with the enzyme to generate the cohesive ends.

11. b, c, and d. Insertion of the DNA fragment into the gene disrupts the gene's production of the enzyme that confers drug resistance. A cell containing the altered vector is therefore sensitive to the antibiotic. When the vector contains a gene that confers resistance to a second drug, insertional inactivation can be incorporated into a selection scheme for isolating cells that contain vectors having the inserted fragment. Cells that remain resistant to the second antibiotic while sensitive to the first probably contain the vector with the foreign DNA.

12. A genomic library is composed of a collection of clones, each of which contains a fragment of DNA from the target organism. The entire collection should contain all the sequences present in the genome of the target organism. A cDNA library is composed of a collection of clones that contain the sequences present in the mRNA of the target organism from which the mRNA was isolated. A cDNA library contains far fewer sequences than does a genomic library because only a small fraction of the genome is being transcribed into mRNA at any given time. The content of a cDNA library depends upon the type of cell and the period of development, the environmental influences, and so forth of the cells from which the mRNA was isolated.

13. b. This is the better choice for reverse translation into a DNA sequence because it contains fewer amino acid residues having multiple codons; Trp and Met have one codon each and Cys has two. Thus, for the (b) sequence, Met-Trp-Cys-Trp, there are $1 \times 1 \times 2 \times 1 = 2$ different dodecameric oligonucleotide coding sequences. In contrast, Leu and Arg each have six codons, so for the (a) sequence, Met-Leu-Arg-Leu, there are $1 \times 6 \times 6 \times 6 = 216$ different coding sequences. Therefore, the probe for (b) would be simpler to construct and would be more likely to give unambiguous results.

14. In eucaryotes the introns are removed from the primary transcript by processing to produce the mRNA that is translated. Procaryotes lack the machinery to perform this processing; consequently, the translation product of the primary transcript would not be functional because it would contain amino acid sequences specified by the intron sequences. The problem can be circumvented by using the cDNA prepared from mRNA from the gene encoding the protein; the cDNA will contain only the sequences present in the processed RNA; that is, the intron sequences will have been removed.

15. e. An immunochemical screen could be performed by adding the radioactive antibody to lysed bacterial colonies and then examining the population by autoradiography to see if the antigen of the cloned gene—that is, the product of the cloned gene—is present.

16. Omitting such an obvious explanation as the destruction of the polypeptide during the bioassay, it is possible that the polypeptide might not have undergone some posttranslational modification that is required for it to function. For example, the polypeptide might need to be acetylated, methylated, or trimmed at the N-terminus, or it might need to have a carbohydrate group attached to it. The bacterium in which it was produced would be unlikely to contain the enzymatic machinery necessary to carry out these modifications, or if it did, it might lack the ability to recognize the eucaryotic signals that direct these modifications. It is also possible that the bacterium might have contained a peptidase or protease that inactivated the polypeptide without destroying its antigenic properties.

17. b, d, and e. If the sequence of the oligonucleotide were completely different from the sequence of the target gene, it could not hybridize to the target gene and serve as a primer for DNA polymerase even under low-stringency hybridization conditions. Functional domains could be better identified by deletion mutagenesis, in which relatively large regions of the gene would be systematically removed and the resulting functional consequences would be tested. Although oligonucleotide-directed mutagenesis can be used to make deletions, it is not the method of choice for an initial survey to find functional domains, since only a single, precisely defined deletion is produced with each oligonucleotide. For exploratory deletion analysis, nucleases are used to generate populations of deleted sequences for functional testing. Oligonucleotide-directed mutagenesis is better suited for changing specific regions when one wishes to test a specific model or hypothesis regarding the function of one or a few amino acids. At the correct conditions of hybridization stringency, the mutagenizing oligonucleotide will form a more stable hybrid with the newly produced mutant sequence than with the original unmodified sequence because it will form a perfect complement. Thus, it can be used to differentiate the mutant and the original sequences from one another.

18. You would need to know the sequence of the gene you wish to synthesize. This could be derived from the amino acid sequence of the protein the gene encodes by reverse translation using the genetic code. You would also need to know what restriction sites you wish to build into the synthetic sequence to allow for the cloning of the synthetic product. You would have to decide upon the individual sequences of the oligonucleotides that compose both strands of the gene. These sequences would be based upon the final desired sequence, the individual lengths (30 to 80 nucleotides long) that can be easily synthesized and purified, and the requirement for overlapping ends that will be necessary to allow unique joinings of the cohesive ends of the partially duplex segments. Self-complementary oligomers would be mixed together to form duplex fragments with cohesive ends. DNA ligase and ATP would be added to join these together to form the whole duplex. If appropriate ends have been designed into the synthesis, the product can then be ligated into a vector for cloning.

1. Cleavage of a circular molecule at one specific site, followed by denaturation, will yield single-stranded DNA molecules with a specific end-to-end base sequence; that is, the molecules have base sequences that are perfectly complementary. Such molecules will anneal to form double-stranded linears, rather than circles. Random cleavage of the original intact molecules yields double-stranded linears with a variety of end-to-end (or permuted) sequences. Denaturation and renaturation allow the random association of these linears, which results in the formation of double-stranded linears with overlapping, complementary ends. Such molecules then form circles as their overlapping ends anneal.

2. During the early years of such experiments, few ways were available to determine what happened to the DNA during transformation attempts. Among the reasons that these transformation attempts were not successful were the failure of the cells to take up the DNA, the rapid degradation of the DNA inside the cell (restriction enzymes are a good example of a cause of this particular problem), the lack of accurate transcription or translation, and the inability of the host cells to replicate and maintain the foreign DNA as they divided.

3. Among the ways that a gene could be inactivated are the insertion of a stop codon in the sequence, which would prevent the complete translation of the protein; a mutation in the promoter region of the gene, which would prevent proper transcription; and other mutations that could prevent proper splicing or processing. To distinguish a functional gene from a pseudogene, you would have to determine the sequence of the protein and then compare it with the coding sequence for each of the gene sequences. These types of analyses remind us that protein sequencing remains a very necessary tool in molecular biology.

4. Whenever one attempts to locate all the sites cleaved by a particular enzyme, there is always a risk that very small fragments generated by the cleavages may not be detected. Determining the sequences of fragments that extend across the junctions of the original set of fragments serves as a check on the overall assignment of sequence.

5. At high pH, protons dissociate from some of the bases, making them unable to participate in base pairing. One example is guanine, for which the pK for the proton on N-1 is 9.2. Removal of the hydrogen at this location disrupts the ability of guanine to pair with cytosine.

If you mix the two types of double-stranded molecules, you would expect to see linear molecules that are double-stranded all along their length as well as some molecules that are only partially double-stranded. These partially double-stranded molecules will contain a single-strand loop that denotes the position of the insertion; they are formed between one strand of the molecule containing the normal gene and one strand of the molecule containing the insertion.

6. Even if you were able to isolate intact, unbroken bacterial chromosomes, formation of intact heteroduplex molecules between the deletion and wild-type DNAs is difficult because the very long single strands become entangled as they pair with each other, making them impossible to analyze by electron microscopy. In addition, the time required

for complete reassociation of the strands is very long. Generating shorter, randomly cleaved DNA fragments for heteroduplex analysis permits faster reassociation and easier analysis, but since the deletion is located at a single unique site in the chromosome, the probability of finding the desired molecule among the mixture of many heteroduplex molecules is rather low. Cloning DNA molecules containing the deletion or its corresponding wild-type sequence allows you to carry out reannealing experiments that yield a high concentration of heteroduplex molecules with the loop characteristic of deletion mutations.

7. In order to insert the yeast gene into the lambda phage vector, you must have complementary base pairs on the ends of each duplex in order for them to join. Because each restriction enzyme cleaves at a unique and constant site, the yeast and lambda phage molecules will have complementary ends if both have been cleaved with the same enzyme. Of course, you must also make sure that the sites of cleavage are in appropriate places so that the gene is intact when it is inserted into the vector.

8. The RNA codon for proline is 5′-CCX-3′ and the codon for glycine is 5′-GGX-3′, where X is any ribonucleotide. Suppose you determine that the proper codon for your protein is 5′-CCC-3′. Using the scheme outlined in Figure 6-33 of the text, you would prepare an oligonucleotide primer that is complementary to the region of the gene that specifies the proline residue, except that it would contain the DNA sequence 5′-CCC-3′ instead of 5′-GGG-3′. Elongation of the primer using DNA polymerase, followed by closure and replication, will yield some progeny plasmids that will express a protein with a glycine substitution.

9. The 500-base fragment has one site that is cleaved by enzyme A. This yields two fragments with two possible patterns:

Enzyme B cleaves the 500-base molecule twice, so there are three possible cleavage patterns:

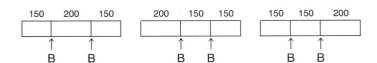

Since we cannot distinguish between ends of the molecule by this type of analysis, let us arbitrarily assume that enzyme A cuts the molecule 100 nucleotides from the left end. We can then superimpose the possible cleavage patterns for enzyme B:

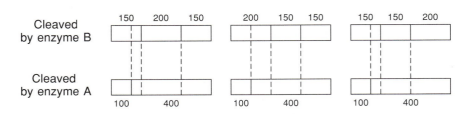

Only one of the patterns for enzyme B, that in which a cut occurs 200 bases from the left end, yields the results obtained when the fragment is incubated with both enzymes. The other patterns would yield at least one 50-base fragment. The correct pattern is therefore:

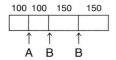

Because we still cannot distinguish between the right- and left-hand ends of the molecule, an alternative cleavage pattern can also be constructed from the analysis outlined above:

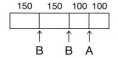

The deletion fragment serves as a marker for the left-hand end of the molecule, and using both enzymes in the double-digest technique allows us to establish which cleavage pattern is correct. For example, if the cleavage pattern shown below on the left is correct, the cleaved deletion molecule will yield four fragments, including one only 25 nucleotides in length. Alternatively, the pattern shown below on the right means that double digestion of the deletion fragment will again yield four fragments, but the smallest will have a length of 75 nucleotides.

The complexity of cleavage patterns (also known as restriction maps) increases greatly when additional cleavages are involved. Often the best ways to use the double-digest technique is to isolate the fragments generated by one enzyme and then digest each of them with the other. This allows you to determine the location of different cleavage sites within a particular fragment. In restriction mapping, as in genetic mapping, it is important to remember that the sum of the fragment lengths generated by one enzyme must equal the sum of the fragment lengths generated by the other.

CHAPTER 7

Oxygen-Transporting Proteins: Myoglobin and Hemoglobin

The principles of protein structure and the methods used to investigate proteins were presented in Chapters 2 and 3. In this chapter, Stryer uses these concepts to describe the structure-function relationships of two very important and well-studied oxygen-carrying proteins: myoglobin and hemoglobin. First, Stryer describes the structures of myoglobin and hemoglobin; he then discusses the oxygen binding properties of hemoglobin in some detail. Hemoglobin exemplifies the allosteric proteins, which are modulated in their activity by specific effectors. Finally, Stryer describes how mutations in the hemoglobin gene can produce defective hemoglobin and cause molecular disease.

When you have mastered this chapter, you should be able to complete the following objectives.

Myoglobin Structure and Heme Properties (Stryer pages 143–150)

1. Explain the physiological roles of *myoglobin* and *hemoglobin* in vertebrates.

2. List the major features of the structure of myoglobin as revealed by *x-ray crystallography.*

3. Describe the structure of the *heme prosthetic group* and its properties in the free form and when bound to the *globins.*

4. Relate the structure of the *myoglobin gene* to the functional domains of myoglobin.

Hemoglobin Structure (Stryer pages 150–153)

5. Compare the *subunit* compositions of *adult* and *fetal hemoglobins.*

6. Compare the three-dimensional structures of myoglobin and hemoglobin.

7. Describe the significance of the *invariant amino acid residues* in the amino acid sequences of hemoglobins from different species.

Oxygen Binding and the Allosteric Properties of Hemoglobin (Stryer pages 154–163)

8. Contrast the *oxygen binding* properties of myoglobin and hemoglobin.

9. Explain the significance of the differences in *oxygen dissociation curves,* in which the *fractional saturation* (Y) of the oxygen-binding sites is plotted as a function of the *partial pressure of oxygen* (pO_2), for myoglobin and hemoglobin.

10. Starting with the equilibrium expression for oxygen binding by myoglobin, derive the equation

$$Y = \frac{pO_2}{pO_2 + P_{50}}$$

11. Relate the empirical expression for the fractional saturation of hemoglobin

$$Y = \frac{(pO_2)^n}{(pO_2)^n + (P_{50})^n}$$

with the equation

$$\frac{Y}{1 - Y} = \left(\frac{pO_2}{P_{50}}\right)^n$$

and with the *Hill plot.* Explain the significance of the *Hill coefficient.*

12. Describe the *Bohr effect,* that is, the effects of CO_2 and H^+ on the binding of oxygen by hemoglobin.

13. Explain the effect of *2,3-bisphosphoglycerate (BPG)*, also known as *2,3-diphosphoglycerate (DPG)*, on the affinity of hemoglobin for oxygen.

14. Describe the major structural differences between the *oxygenated* and *deoxygenated* forms of the hemoglobin molecule.

15. Describe the structural bases for the effects of H^+, CO_2, and BPG on the binding of oxygen by hemoglobin.

16. Summarize the general properties of an *allosteric protein*, as exemplified by hemoglobin.

Sickle-Cell Anemia and Other Molecular Diseases (Stryer pages 163–171)

17. Describe *sickle-cell anemia* as a genetically transmitted *molecular disease*.

18. Contrast the biochemical and structural properties of deoxygenated *sickle-cell hemoglobin (Hb S)* with those of *hemoglobin A (Hb A)*.

19. Correlate the clinical observations of sickle-cell anemia and the *sickle-cell trait* to the molecular defect in the hemoglobin.

20. Explain how deoxyhemoglobin S forms *fibrous precipitates*.

21. Correlate the genetics of sickle-cell anemia with the geographical distribution of the disease.

22. Describe the detection of the sickle-cell gene by *endonuclease digestion* of DNA and *Southern blotting*.

23. Explain the possible functional consequences of *amino acid substitutions* in *mutant hemoglobins*.

24. Define *thalassemias*, and indicate the steps in hemoglobin synthesis where defects can occur.

SELF-TEST

Myoglobin Structure and Heme Properties

1. Match the forms of myoglobin in the left column with their corresponding properties from the right column.

(a) Ferrimyoglobin _____

(b) Oxymyoglobin _____

(c) Deoxymyoglobin _____

(1) Iron in the +2 oxidation state
(2) Iron in the +3 oxidation state
(3) Oxygen bound to the sixth coordination position of iron
(4) Empty sixth coordination position
(5) Water bound to the sixth coordination position
(6) Histidine at the fifth coordination position

2. Which of the following statements about heme structure are *true?*

(a) Heme contains a tetrapyrrole ring with four methyl, four vinyl, and four propionate side chains.
(b) The iron atom in heme may be present in the ferrous or the ferric state.
(c) The iron in free heme is mostly present in the ferrous state.
(d) The iron atom is coplanar with the tetrapyrrole ring in deoxymyoglobin.
(e) The axial coordination positions of heme are occupied by tyrosine residues in myoglobin.

3. Which of the following are structural features of myoglobin?

(a) Seventy-five percent of the amino acids are in α helices.
(b) All of the prolines are found within α helices.
(c) The interior of the molecule consists mainly of nonpolar residues.
(d) The interior of the molecule is hollow so that it can accommodate many molecules of water.
(e) The heme group is bound on the surface of the protein.

4. Which of the following are properties of the heme binding site in myoglobin?

(a) The proximal histidine is involved in binding iron.
(b) Oxygenation involves an insertion of O_2 in a coordination complex with the iron of heme.
(c) Sequestering the heme in a nonpolar environment is important in preventing the oxidation of ferrous iron.
(d) The heme binds CO with the Fe, C, and O atoms in a linear array.
(e) The distal histidine is involved in binding iron.

5. The heme prosthetic group found in hemoglobin and myoglobin is also found in cytochromes b and c, which are proteins that transfer electrons in the electron transport chain. How can the same prosthetic group serve such different functions as oxygen binding and electron transport?

_____

_____

6. The role of the distal histidine in myoglobin is to

(a) bind O_2.
(b) prevent tight CO_2 binding.
(c) prevent tight CO binding.
(d) bind to the fifth coordination position of Fe.
(e) help keep Mb in the ferrous state.

Hemoglobin Structure

7. Hemoglobin is a tetrameric protein consisting of two α and two β polypeptide subunits. The structures of the α and β subunits are remarkably similar to that of myoglobin. However, at a number of positions, hydrophilic residues in myoglobin have been replaced by hydrophobic residues in hemoglobin.

(a) How can this observation be reconciled with the generalization that hydrophobic residues fold into the interior of proteins?

_____

_____

(b) In this regard, what can you say about the nature of the interactions that determine the quaternary structure of hemoglobin?

_____

_____

8. Much of the information we have concerning hemoglobin structure comes from the study of crystals. Which of the following lines of evidence support the notion that crystallized hemoglobin has a structure similar to that of hemoglobin in solution?

(a) The amino acid sequence of crystallized hemoglobin is the same as that found in hemoglobin in solution.
(b) The visible absorption spectrum of crystallized hemoglobin is virtually the same as that of hemoglobin in solution.
(c) Crystallized hemoglobin is functionally active.
(d) The α helix content of crystallized hemoglobin is similar to that found in hemoglobin in solution.

9. What are the meanings of the codes that are used to designate the positions of the invariant amino acid residues in hemoglobins, as in glycine B6 or tyrosine HC2, for example?

_____

_____

10. There are nine positions in the amino acid sequences of the hemoglobins of diverse species that have invariant amino acid residues. On the basis of the properties of amino acids, which you learned in Chapter 2, match the invariant amino acids in the left column with the appropriate role from the right column.

(a) Glycine B6 _____

(b) Histidine F8 _____

(c) Leucine F4 _____

(d) Phenylalanine CD1 _____

(e) Proline C2 _____

(f) Tyrosine HC2 _____

(1) terminates a helix.
(2) contacts the heme (forms part of the nonpolar heme pocket).
(3) allows the close approach of the B and E helices.
(4) crosslinks two helices (H and F) with a hydrogen bond.
(5) binds to the heme iron.

Oxygen Binding and the Allosteric Properties of Hemoglobin

11. Hemoglobin differs from myoglobin in that

(a) hemoglobin is multimeric whereas myoglobin is monomeric.
(b) hemoglobin binds O_2 more tightly than does myoglobin at any given O_2 concentration.
(c) hemoglobin binds CO_2 more effectively than does myoglobin.
(d) the Hill coefficient for O_2 binding is smaller for hemoglobin than it is for myoglobin.

(e) the binding of O_2 by hemoglobin depends on the concentrations of CO_2, H^+, and BPG whereas the binding of O_2 by myoglobin does not.

12. Which of the following statements are *false?*

(a) The oxygen dissociation curve of myoglobin is sigmoidal, whereas that of hemoglobin is hyperbolic.
(b) The affinity of hemoglobin for O_2 is regulated by organic phosphates, whereas the affinity of myoglobin for O_2 is not.
(c) Hemoglobin has a higher affinity for O_2 than does myoglobin.
(d) The affinity of both myoglobin and hemoglobin for O_2 is independent of pH.

13. Several oxygen dissociation curves are shown in Figure 7-1. Assuming that curve 3 corresponds to isolated hemoglobin placed in a solution containing physiological concentrations of CO_2 and BPG at a pH of 7.0, indicate which of the curves reflects the following changes in conditions.

(a) Decreased CO_2 concentration   _____

(b) Increased BPG concentration   _____

(c) Increased pH   _____

(d) Dissociation of hemoglobin into subunits _____

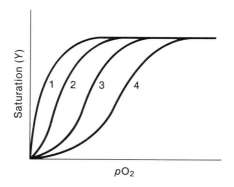

Figure 7-1
Oxygen dissociation curves.

14. Which of the following statements concerning the Bohr effect are *true?*

(a) Lowering the pH shifts the oxygen dissociation curve of hemoglobin to the right.
(b) The acidic environment of an exercising muscle allows hemoglobin to bind O_2 more strongly.
(c) The affinity of hemoglobin for O_2 is diminished by high concentrations of CO_2.
(d) In the lung, the presence of higher concentrations of H^+ and CO_2 allows hemoglobin to become oxygenated.
(e) In the lung, the presence of higher concentrations of O_2 promotes the release of CO_2 and H^+.

15. Match the parameters in the left column with the appropriate definitions from the right column.

(a) Hill coefficient (n) _____

(b) $(pO_2/P_{50})^n$ _____

(c) P_{50} _____

(d) Y _____

(1) The fractional occupancy of the oxygen-binding sites
(2) The pO_2 at half-saturation of the hemes with oxygen
(3) The ratio of oxyheme to deoxyheme
(4) The cooperativity of oxygen binding

16. BPG (2,3-bisphosphoglycerate) plays a role in high-altitude adaptation. What is the effect of an increase in the level of BPG on the amount of oxygen transported to muscle in a person living at 10,000 feet as compared to a person living at sea level? Assume that the increase in BPG level has shifted P_{50} from 26 torr to 35 torr and that alveolar pO_2 is 67 torr and pO_2 in muscle capillaries is 20 torr at 10,000 feet. At sea

level the alveolar pO_2 is 100 torr and pO_2 in muscle capillaries is still 20 torr. Assume that the Hill coefficient is 2.8.

_____

_____

_____

17. Explain why fetal hemoglobin has a higher affinity for oxygen than does maternal hemoglobin, and why this is a necessary adaptation.

_____

_____

18. The oxygen dissociation curve for hemoglobin reflects allosteric effects that result from the interaction of hemoglobin with O_2, CO_2, H^+, and BPG. Which of the following structural changes occur in the hemoglobin molecule when O_2, CO_2, H^+, or BPG bind?

  (a) The binding of O_2 pulls the iron into the plane of the heme and causes a change in the interaction of all four globin subunits, mediated through His F8.
  (b) BPG binds at a single site between the four globin subunits in deoxyhemoglobin and stabilizes the deoxyhemoglobin form by cross-linking the β subunits.
  (c) The deoxy form of hemoglobin has a greater affinity for H^+ because the molecular environment of His and the α-NH_2 groups of the α chains changes, rendering these groups less acidic when O_2 is released.
  (d) The binding of CO_2 stabilizes the oxy form of hemoglobin.

19. The structure of deoxyhemoglobin is stabilized by each of the following interactions *except* for

  (a) BPG binding
  (b) salt bridges between acidic and basic side chains.
  (c) coordination of the hemes with the distal histidine.
  (d) hydrophobic interactions.
  (e) salt bridges involving N-terminal carbamates.

20. In the transition of hemoglobin from the oxy to the deoxy form, a glutamate residue is brought to the vicinity of His 146. This increases the affinity of this histidine for protons. Explain why.

_____

_____

Sickle-Cell Anemia and Other Molecular Diseases

21. Which of the following observations indicate that hemoglobin S *differs* from hemoglobin A by the substitution of Val for Glu in position 6 of the β chain?

  (a) Porphyrin isolated from hemoglobin S has one more valine group than does porphyrin isolated from hemoglobin A.
  (b) All the tryptic peptides of hemoglobin S are identical to those of hemoglobin A.
  (c) One altered peptide of hemoglobin S migrates toward the cath-

ode ($-$) more than does the corresponding peptide of hemoglobin A.

(d) The isoelectric point of hemoglobin S is slightly lower than that of hemoglobin A.

22. The N-terminal tryptic peptides from the β chains of Hb A and Hb S are as follows:

Hb A: Val-His-Leu-Thr-Pro-Glu-Glu-Lys

Hb S: Val-His-Leu-Thr-Pro-Val-Glu-Lys

(a) Would these peptides separate from one another in an electric field at pH 7.0?

_____

(b) What is the approximate net charge on each peptide at pH 7.0?

_____

_____

23. Which of the following statements are *true*? Hemoglobin S forms fibrous precipitates

(a) because the valine at position 6 of the β chain forms a sticky hydrophobic patch on the surface of the protein
(b) that are reversible upon oxygenation.
(c) that distort the shape of red cells.
(d) only in the deoxy form.
(e) only in homozygotes.

24. The first step in detecting the mutation in sickle-cell hemoglobin is to cleave the DNA with a specific restriction endonuclease. Which of the following would be appropriate subsequent steps in the analysis?

(a) The DNA fragments are separated by gel electrophoresis and are identified with a specific radiolabeled DNA probe.
(b) The DNA fragments are separated by gel filtration and are detected by their light absorption.
(c) The DNA fragments are separated by gel electrophoresis, and the abnormal fragment is identified by its altered mobility with respect to that of the normal fragment.
(d) The DNA fragments are separated by gel electrophoresis and are identified with a fluorescent antibody.

25. A mutant hemoglobin gene has been isolated and shown to encode an abnormal α chain by endonuclease digestion. The electrophoretic mobility of the mutant hemoglobin and its affinity for O_2 are essentially the same as for hemoglobin A, however. With regard to these observations, which of the following statements are *correct*?

(a) The mutation is most likely at the active site.
(b) There has been a substitution of a basic amino acid residue for an acidic one.
(c) There has been a substitution of an acidic amino acid residue for a neutral one.
(d) There has been a substitution that involves a homologous residue in terms of charge.
(e) The mutation is most likely on the surface of the hemoglobin.

26. The molecular defects in thalassemias and in abnormal hemoglobins, such as Hb S, can be similar in which of the following ways?

    (a) The molecular defect occurs outside of the DNA encoding the protein.
    (b) The molecular defect results in a stop codon.
    (c) The molecular defect results in the substitution of one amino acid for another.
    (d) The molecular defect occurs within the structural genes.
    (e) The molecular defect results in a frameshift mutation.

PROBLEMS

1. An effective respiratory carrier must be able to pick up oxygen from the lungs and deliver it to peripheral tissues. Oxygen dissociation curves for substances A and B are shown in Figure 7-2. What would be the disadvantage of each of these substances as a respiratory carrier? Where would the curve for an effective carrier appear in the figure?

2. Glutathione is a tripeptide consisting of glutamic acid, cysteine, and glycine; it is abundant in human erythrocytes as well as in many types of tissue. Reduced glutathione (GSH) acts as a reducing agent in tissues because its sidechain —SH group can be readily oxidized to form disulfide bonds. Unless glutathione is maintained substantially in its reduced form, as opposed to its oxidized form, in erythrocytes, they lose their ability to transport oxygen effectively. Why do you think this is so?

3. Although one expects proline to occur primarily in nonhelical regions of proteins, three of the four proline residues (C2, F3, and G1) of sperm-whale myoglobin occur within regions designated as helical. How can this be the case? (Refer to Stryer, Figure 7-6, p. 146.)

4. The interhelical amino acid residues of sperm-whale myoglobin are mostly charged and polar. How does this contribute to the stability of the protein?

5. One molecule of 2,3-bisphosphoglycerate binds to one molecule of hemoglobin in a central cavity of the hemoglobin molecule. Is the interaction between BPG and hemoglobin stronger or weaker than it would be if BPG bound to the surface of the protein instead? Explain your answer.

6. The partial pressure of oxygen in the capillaries of active muscles is approximately 20 torrs.

    (a) Calculate the fractional saturation of myoglobin and hemoglobin under these conditions.
    (b) How do the values you have calculated in part (a) relate to the physiological role of myoglobin in muscle?

7. The partial pressure of oxygen in the venous blood of a human at rest at sea level is approximately 40 torrs.

    (a) Calculate the fractional saturation of hemoglobin under these conditions.

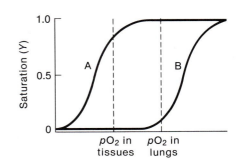

Figure 7-2
Oxygen dissociation curves for substances A and B.

(b) What percentage of the oxygen originally bound to hemoglobin in the alveoli remains unreleased?

(c) Is the residual oxygen bound to hemoglobin under these conditions of any benefit? Explain.

8. Predict whether each of the following manipulations will increase or decrease the tendency of Hb S molecules to polymerize in vitro. Give a brief rationale for each answer.

(a) An increase in temperature

(b) An increase in the partial pressure of oxygen

(c) Stripping the Hb S molecules of BPG

(d) An increase in the pH

9. One avenue of approach to therapy for sickle-cell anemia involves finding a way of turning on the synthesis of Hb F in afflicted adults. Briefly explain why such a manipulation might be beneficial.

10. Another approach to the management of sickle-cell anemia involves a search for osmotically active agents that would expand the volume of erythrocytes.

(a) Give a brief rationale for this approach.

(b) Suppose that the volume of an erythrocyte is increased by 10%. Calculate the rate of Hb S polymer information in the enlarged cell as a fraction of its rate in a normal-sized cell.

11. An abnormal hemoglobin, Hb Hammersmith, is characterized by the substitution of Ser for Phe at position 42 of the β chain.

(a) Can Hb Hammersmith be separated from Hb A by electrophoresis at pH 6.0? Why or why not?

(b) Give the nucleotide change(s) (on RNA) that most likely produced this mutant. (See the genetic code, Stryer, p. 107.)

12. Prenatal diagnosis of sickle-cell anemia can be carried out by the treatment of fetal DNA with the restriction endonuclease MstII, which recognizes the sequence CCTNAGG, where N is any nucleotide. (See Stryer, p. 169.) Digestion of the β^S gene with MstII yields a 1.3-kb fragment, whereas digestion of the β^A gene yields a 1.1-kb fragment.

(a) Using your knowledge of the amino acid substitution that occurs in Hb S, give the identity of nucleotide N on both the β^A and β^S genes. Also identify the mutational change at the DNA level that leads to the amino acid substitution. Explain your answer. (Refer to the genetic code, Stryer, p. 107.)

(b) Give the identity of the amino acid at position 5 in the β chain using the information provided here and the genetic code (Stryer, p. 107).

ANSWERS TO SELF-TEST

Myoglobin Structure and Heme Properties

1. (a) 2, 5, 6 (b) 1, 3, 6 (c) 1, 4, 6

2. b

3. a, c

4. a, b, c

5. The oxidation state and the binding properties of heme vary markedly with its environment; therefore, the different amino acid residue environments of these two classes of proteins change the function of the heme.

6. c, e

Hemoglobin Structure

7. (a) Hydrophobic patches occur on the surface of the hemoglobin subunits where the α and β chains fit together. As a result, these patches will be found in the interior of the multimeric protein.
   (b) Hydrophobic interactions play an important role in stabilizing the tetrameric subunit structure of hemoglobin.

8. b, c, and d. Regarding answer b, the visible absorption spectrum of heme varies with its oxidation state and its environment. The fact that both crystallized hemoglobin and hemoglobin in solution have the same spectra is indicative of a similar structure of the heme pocket.

9. Glycine B6 means that the glycine residue is in the sixth position of the B helix; tyrosine HC2 designates the second position of the carboxyl-terminal segment. In general, the helical segments are designated A, B, C, D, E, F, G, H, and the interhelical regions, AB, BC, etc.; NA denotes the region between the amino-terminal residue and the first helix (A), and HC denotes the segment between the last helix (H) and the carboxyl-terminal residue.

10. (a) 3 (b) 5 (c) 2 (d) 2 (e) 1 (f) 4

Oxygen Binding and the Allosteric Properties of Hemoglobin

11. a, c, e

12. a, c, d

13. (a) 2 (b) 4 (c) 2 (d) 1

14. a, c, e

15. (a) 4 (b) 3 (c) 2 (d) 1

16. Using the equation

$$Y = \frac{(pO_2)^n}{(pO_2)^n + (P_{50})^n}$$

the increased level of BPG at 10,000 feet changes P_{50} and yields a ΔY (the difference in the fractional saturation of hemoglobin with oxygen in the lungs versus that in the tissues) of 0.69. Compare this with a ΔY of 0.65 at sea level. Thus, the increase in BPG results in a similar delivery of oxygen to the tissues even when alveolar pO_2 is significantly decreased. In general, the larger the value of ΔY, the higher the delivery of O_2 to the tissues.

17. Fetal hemoglobin is composed of different subunits than adult hemoglobin and binds BPG less strongly. As a result, the affinity of fetal hemoglobin for oxygen is higher, and the fetus can extract the O_2 that is transported in maternal blood.

18. a, b, c

19. c

20. The pK values of ionizable groups are sensitive to their environment. The change in the environment of His 146 in deoxyhemoglobin increases its affinity for protons as a result of the electrostatic attraction between the negative charge of the glutamate and the proton.

Sickle-Cell Anemia and Other Molecular Diseases

21. c

22. (a) Yes, the peptides will separate.
    (b) The net charge for Hb S is between -0.9 and -1, and that for Hb A is between 0 and $+0.1$. Since the pK of His is approximately 6.0, this residue will be partially ionized with a charge of less than $+0.1$ at pH 7.0, which leads to the nonintegral answers.

23. a, b, c, d

24. a, c

25. d, e

26. d. Answers a, b, and e only apply to thalassemias, whereas answer c only applies to abnormal hemoglobins.

ANSWERS TO PROBLEMS

1. Substance A would never unload oxygen to peripheral tissues. Substance B would never load oxygen in the lungs. An effective carrier would have an oxygen dissociation curve between those depicted for substance A and substance B. It would be relatively saturated with oxygen in the lungs and relatively unsaturated in the peripheral tissues.

2. Reduced glutathione helps keep the iron of hemoglobin in the ferrous ($+2$) valence state. When the iron of heme is oxidized to the ferric ($+3$) state to form methemoglobin, it can no longer combine reversibly with oxygen.

3. The α helix has 3.6 residues per turn. The three proline residues that occur within helical regions of sperm-whale myoglobin are all in the first three positions of a helical run, where the helix is not maximally stabilized by hydrogen bonding. Therefore, some local distortion in bond angles can be accommodated.

4. In the tertiary structure of sperm-whale myoglobin, the charged amino acid residues between the helical runs are on the exterior of the protein, where they are stabilized by hydration.

5. BPG binds to hemoglobin by electrostatic interactions. These interactions between the negatively charged phosphates of BPG and the positively charged residues of hemoglobin are much stronger in the interior, hydrophobic environment than they would be on the surface, where water would compete and weaken the interaction by binding both to BPG and the positively charged residues. Remember that the force of electrostatic interactions, given by Coulomb's Law (Stryer,

p. 7), is inversely proportional to the dielectric constant of the medium. The dielectric constant in the interior of a protein may be as low as 2. Hence, electrostatic interactions there are much more stable than those on the surface, where the dielectric constant is approximately 80.

6. (a) The applicable relationship for determining the fractional saturation of myoglobin is equation 5 on page 155 of Stryer:

$$Y_{Mb} = \frac{pO_2}{pO_2 + P_{50}}$$

$$= \frac{20}{20 + 1}$$

$$= 0.952$$

For the fractional saturation of hemoglobin, equation 7 applies:

$$Y_{Hb} = \frac{(pO_2)^n}{(pO_2)^n + (P_{50})^n}$$

$$= \frac{(20)^{2.8}}{(20)^{2.8} + (26)^{2.8}}$$

$$= 0.324$$

   (b) Since hemoglobin has a smaller fractional saturation than myoglobin does under these conditions, hemoglobin will unload oxygen to myoglobin, as it must if myoglobin is to serve as an oxygen store in muscle.

7. (a) The applicable relationship for calculating the fractional saturation of hemoglobin is equation 7 (see the answer to Problem 6):

$$Y_{Hb} = \frac{(40)^{2.8}}{(40)^{2.8} + (26)^{2.8}} = 0.770$$

   (b) The percentage of oxygen in the alveoli that remains unreleased is $(0.77/0.98) \times 100 = 79\%$.

   (c) The remaining oxygen constitutes a reserve; it can be released to extremely active tissues where the partial pressure of oxygen is very low.

8. (a) An increase in temperature will favor the polymerization of Hb S. The interaction between Hb molecules in polymer formation is hydrophobic in nature. Hydrophobic interactions have negative temperature coefficients; that is, they become more stable with increasing temperature.

   (b) An increase in the partial pressure of oxygen will inhibit polymerization because only deoxy hemoglobin S polymerizes.

   (c) BPG stabilizes deoxyhemoglobin S. Since only the deoxy form polymerizes, the removal of BPG from Hb S would inhibit polymer formation.

   (d) Increasing the pH (decreasing the acidity) stabilizes oxyhemoglobin S. Since only the deoxy form polymerizes, this would inhibit polymer formation.

9. Hb F is devoid of β chains, having gamma chains instead, so it does not polymerize. If adults with sickle-cell anemia could synthesize Hb F. each erythrocyte would contain a mixture of Hb S and Hb F, which would reduce the degree of the polymerization of Hb S.

10. (a) The rate of polymerization of Hb S is proportional to the tenth power of its concentration. Therefore, increasing the cell volume would decrease the Hb S concentration, which would slow the rate of Hb S polymerization.

(b) If the cell has been expanded by 10%, the Hb S concentration has been decreased to $\frac{10}{11}$, or 90.9% of its original value. The rate of polymerization under these conditions is $(0.909)^{10} = 0.386$ of the rate in an unexpanded cell.

11. (a) No, electrophoresis at pH 6.0 will not separate Hb Hammersmith from Hb A. Both Ser and Phe have uncharged sidechains. Thus, there is no difference in net charge of the two hemoglobins at pH 6.0.

(b) A C was substituted for a U. Either UUU was changed to UCU or UUC was changed to UCC. Other possibilities would involve simultaneous changes in two positions and are therefore less likely.

12. (a) Nucleotide N must be C on both the β^A and β^S genes. In Hb A, Glu is present at position 6; it is encoded by GAG (on mRNA). Therefore, the sequence CTC must be present on the informational strand of β^A DNA. The mutation that leads to the substitution of Val for Glu at position 6 is T $\rightarrow$ A.

(b) The amino acid at position 6 in the β chain is Pro, which is encoded by CCU on RNA (or AGG on DNA).

CHAPTER 8

Introduction to Enzymes

Enzymes catalyze almost all biochemical reactions, and they are intimately involved in the transformations of one form of energy into another. Stryer begins this chapter with an introduction to the catalytic power, specificity, and regulation of enzymes; he then reviews some fundamental thermodynamic principles that are essential for the understanding of whether or not chemical reactions can occur. In the second part of the chapter, enzyme catalysis, enzyme kinetics, and inhibitors of enzymatic reactions are discussed. This chapter draws on your knowledge of protein structure (Chapter 2) and the interactions between biomolecules (Chapters 1 and 7). It sets the stage for the majority of the remaining chapters of this text that deal with biochemical reactions.

When you have mastered this chapter, you should be able to complete the following objectives.

Introduction (Stryer pages 177–180)

1. Explain why *enzymes* are the major and most versatile *biological catalysts*.

2. List the different ways in which the *catalytic activity* of enzymes may be *regulated*.

Thermodynamic Principles (Stryer pages 180–183)

3. State the *first* and *second laws of thermodynamics*. Define the *entropy (S)*, *enthalpy (H)*, and *free energy (G)* of a *system*, and give their mathematical relationship.

4. Describe how ΔG can be used to predict whether a reaction can occur spontaneously.

5. Write the equation for the ΔG of a chemical reaction. Define the *standard free-energy change* (ΔG°); define $\Delta G'$ and $\Delta G^{\circ\prime}$.

6. Derive the relationship between $\Delta G^{\circ\prime}$ and the *equilibrium constant* (K'_{eq}) of a reaction. Relate each tenfold change in K'_{eq} to the change in $\Delta G^{\circ\prime}$ in kilocalories per mole.

Enzyme Catalysis (Stryer pages 183–187)

7. Explain why enzymes do not alter the *equilibrium* of chemical reactions but only change the *rates* of chemical reactions.

8. Define the *transition state* and the *free energy of activation* $(\Delta G^{\ddagger})$, and describe the effect of enzymes on $\Delta G^{\ddagger}$.

9. Describe the formation of *enzyme-substrate (ES) complexes*, and discuss their properties.

10. Summarize the key features of the *active sites* of enzymes.

Enzyme Kinetics and Enzyme Inhibitors (Stryer pages 187–197)

11. Outline the *Michaelis-Menten model of enzyme kinetics* and define all the terms.

12. Reproduce the derivation of the *Michaelis-Menten equation* in the text. Relate the Michaelis-Menten equation to experimentally derived plots of *velocity (V)* versus *substrate concentration* [S].

13. Define V_{max} and K_M and explain how these parameters can be obtained from a plot of V versus [S] or a plot of $1/V$ versus $1/[S]$ (*a Lineweaver-Burk plot*).

14. Explain the significance of V_{max}, K_M, k_3, and k_{cat}/K_M.

15. Describe the different kinds of *enzyme inhibitors*.

16. Contrast the *kinetics of allosteric enzymes* with simple Michaelis-Menten kinetics; note their similarity to the oxygen-dissociation kinetics of hemoglobin and myoglobin (Chapter 7).

17. Describe the effects of *competitive* and *noncompetitive inhibitors* on the kinetics of enzyme reactions. Contrast reversible and irreversible inhibitors.

18. Using the examples of ethanol in ethylene glycol poisoning and penicillin in microbial infection, explain the importance of enzyme inhibitors in medicine.

SELF-TEST

Introduction

1. Which of the following is *not* a general property of enzymes?

   (a) Enzymes are almost exclusively proteins.
   (b) Enzymes have great catalytic power.
   (c) Enzymes bind substrates specifically.
   (d) Enzymes use only hydrophobic interactions in binding substrates.
   (e) The catalytic activity of enzymes can be regulated.

2. Which of the following are mechanisms for the regulation of the catalytic activity of enzymes?

   (a) The binding of regulatory proteins
   (b) The covalent modification of tryptophan residues
   (c) The proteolytic cleavage of an inactive enzyme precursor
   (d) The binding of regulatory peptides via disulfide bonds

3. List some of the molecular mechanisms by which the catalytic activity of enzymes is controlled. What is the common feature of these mechanisms?

   _____

   _____

   _____

Thermodynamic Principles

4. Which of the following statements is *correct?* The entropy of a reaction refers to

   (a) the heat given off by the reaction.
   (b) the tendency of the system to move toward maximal randomness.
   (c) the energy of the transition state.
   (d) the effect of temperature on the rate of the reaction.

5. Explain why the thermodynamic parameters ΔE and ΔS cannot be used to predict in which direction a reaction will proceed.

   _____

   _____

   _____

6. If the standard free-energy change ($\Delta G°$) for a reaction is zero, which of the following statements about the reaction are *true*?

    (a) The entropy ($\Delta S°$) of the reaction is zero.
    (b) The enthalpy ($\Delta H°$) of the reaction is zero.
    (c) The equilibrium constant for the reaction is 1.0.
    (d) The reaction is at equilibrium.
    (e) The concentrations of the reactants and products are all 1 M at equilibrium.

7. The enzyme triose phosphate isomerase catalyzes the following reaction:

$$\text{Dihydroxyacetone phosphate} \underset{k_2}{\overset{k_1}{\rightleftharpoons}} \text{glyceraldehyde 3-phosphate}$$

The $\Delta G°'$ for this reaction is 1.83 kcal/mol. In light of this information, which of the following statements are *correct?*

    (a) The reaction would proceed spontaneously from left to right under standard conditions.
    (b) The rate of the reaction in the reverse direction is higher than the rate in the forward direction at equilibrium.
    (c) The equilibrium constant under standard conditions favors the synthesis of the compound on the left, dihydroxyacetone phosphate.
    (d) The data given are sufficient to calculate the equilibrium constant of the reaction.
    (e) The data given are sufficient to calculate the left-to-right rate constant (k_1).

8. Phosphorylase catalyzes the reaction

$$\text{Glycogen}_n + \text{phosphate} \rightleftharpoons \text{glucose 1-phosphate} + \text{glycogen}_{n-1}$$

$$K'_{eq} = \frac{[\text{glucose 1-phosphate}][\text{glycogen}_{n-1}]}{[\text{phosphate}][\text{glycogen}_n]} = 0.088$$

Based on these data, which of the following statements are *correct?*

    (a) Because phosphorylase *degrades* glycogen in cellular metabolism, there is a paradox in that the equilibrium constant favors synthesis.
    (b) The $\Delta G°'$ for this reaction at 25°C is 1.44 kcal/mol.
    (c) The phosphorolytic cleavage of glycogen$_n$ to yield glucose 1-phosphate and glycogen$_{n-1}$ requires an input of energy.
    (d) If the ratio of phosphate to glucose 1-phosphate in cells is high enough, phosphorylase can degrade glycogen.

9. The hydrolysis of glucose 6-phosphate to give glucose and phosphate has a $\Delta G°' = -3.3$ kcal/mol. The reaction takes place at 25°C. Initially, the concentration of glucose 6-phosphate is 10^{-5} M, that of glucose is 10^{-1} M, and that of phosphate is 10^{-1} M. Which of the following statements pertaining to this reaction are *correct?*

    (a) The equilibrium constant for the reaction is 267.
    (b) The equilibrium constant cannot be calculated because standard conditions do not prevail initially.
    (c) The $\Delta G'$ for this reaction under the initial conditions is -0.78 kcal/mol.

(d) Under the initial conditions, the synthesis of glucose 6-phosphate will take place rather than hydrolysis.
(e) Under standard conditions, the hydrolysis of glucose 6-phosphate will proceed spontaneously.

Enzyme Catalysis

10. Enzyme catalysis of a chemical reaction

(a) decreases $\Delta G'$ so that the reaction can proceed spontaneously.
(b) increases the energy of the transition state.
(c) does not change $\Delta G^{\circ\prime}$, but it changes the ratio of products to reactants.
(d) decreases the entropy of the reaction.
(e) increases the forward and reverse reaction rates.

11. Which of the following statements regarding an enzyme-substrate complex is *not true?*

(a) The heat stability of an enzyme frequently changes upon the binding of a substrate.
(b) At sufficiently high concentrations of substrate, the catalytic sites of the enzyme become filled and the reaction rate reaches a maximum.
(c) An enzyme-substrate complex cannot be isolated because of the constant turnover of the substrate.
(d) Enzyme-substrate complexes can be visualized by x-ray crystallography or electron microscopy.
(e) Spectroscopic changes in the substrate or the enzyme can be used to detect the formation of an enzyme-substrate complex.

12. Why is there a high degree of stereospecificity in the interaction of enzymes with their substrates?

_____

_____

_____

13. Explain why the forces that bind a substrate at the active site of an enzyme are usually weak.

_____

_____

_____

Enzyme Kinetics and Enzyme Inhibitors

14. Which of the following statements regarding simple Michaelis-Menten enzyme kinetics are *correct?*

(a) The maximal velocity, V_{max}, is related to the maximal number of substrate molecules that can be "turned over" in unit time by a molecule of enzyme.
(b) K_M is expressed in terms of a reaction velocity (e.g., $mol\ s^{-1}$).
(c) K_M is the dissociation constant of the enzyme-substrate complex.

(d) K_M is the concentration of substrate required to achieve one-half of V_{max}.

(e) K_M is the concentration of substrate required to convert one-half of the total enzyme into the enzyme-substrate complex.

15. Explain the relationship between K_M and the dissociation constant of the enzyme-substrate complex, K_{ES}.

_____

_____

_____

16. Note the similarity between the Michaelis-Menten equation

$$\frac{V}{V_{max}} = \frac{[S]}{[S] + K_M}$$

and the oxygen dissociation equation for myoglobin

$$Y = \frac{pO_2}{pO_2 + P_{50}}$$

Explain the relationship between the two equations.

_____

_____

_____

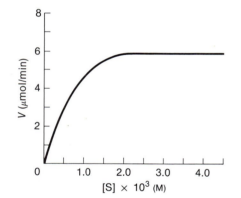

Figure 8-1
Plot of reaction velocity versus substrate concentration.

17. From the plot of velocity versus substrate concentration shown in Figure 8-1, obtain the following parameters. The amount of enzyme in the reaction mixture is 10^{-3} μmol.

(a) K_M _____

(b) V_{max} _____

(c) k_3/K_M _____

(d) Turnover number _____

18. What is the significance of k_3/K_M or k_{cat}/K_M?

_____

_____

_____

19. The turnover number for chymotrypsin is 100 s^{-1} and for DNA polymerase it is 15 s^{-1}. This means that

(a) chymotrypsin binds its substrate with higher affinity than does DNA polymerase.
(b) the velocity of the chymotrypsin reaction is always greater than that of the DNA polymerase reaction.
(c) the velocity of the chymotrypsin reaction at a particular enzyme concentration and saturating substrate levels is lower than that of the DNA polymerase reaction under the same conditions.
(d) the velocities of the reactions catalyzed by both enzymes at saturating substrate levels could be made equal if 6.7 times more DNA polymerase than chymotrypsin were used.

20. Which of the following statements about the different types of enzyme inhibition are *correct?*

    (a) Competitive inhibition is seen when a substrate competes with an enzyme for binding to an inhibitor protein.

    (b) Competitive inhibition is seen when the substrate and the inhibitor compete for the active site on the enzyme.

    (c) Noncompetitive inhibition of an enzyme cannot be overcome by adding large amounts of substrate.

    (d) Competitive inhibitors are often similar in chemical structure to the substrates of the inhibited enzyme.

    (e) Noncompetitive inhibitors often bind to the enzyme irreversibly.

21. If the K_M of an enzyme for its substrate remains constant as the concentration of the inhibitor increases, what can be said about the mode of inhibition?

_____

_____

_____

22. The kinetic data for an enzymatic reaction in the presence and absence of inhibitors are plotted in Figure 8-2. Identify the curve that corresponds to each of the following.

    (a) No inhibitor   _____

    (b) Noncompetitive inhibitor   _____

    (c) Competitive inhibitor   _____

    (d) Mixed inhibitor   _____

23. Draw approximate Lineweaver-Burk plots for each of the inhibitor types in Question 22.

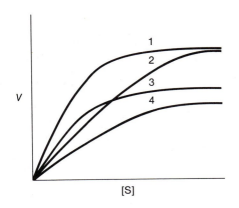

Figure 8-2
Effects of inhibitors on a plot of V versus [S].

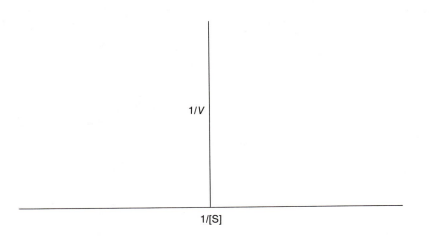

24. The inhibition of bacterial cell-wall synthesis by penicillin is a classic example of a medically significant inhibition of an enzymatic reaction. Which of the following statements about the inhibition of glycopeptide transpeptidase by penicillin is *true?*

    (a) The inhibition is noncompetitive.

    (b) Penicillin binds irreversibly to an allosteric site of the enzyme.

(c) Penicillin inhibits bacterial cell-wall synthesis by incorrectly cross-linking the peptides of the proteoglycan.

(d) The penicilloyl-enzyme intermediate may be dissociated by high concentrations of D-alanine.

(e) Penicillin resembles acyl-D-Ala-D-Ala, one of the substrates of the transpeptidase.

PROBLEMS

1. The change in internal energy, ΔE, for the transition from ice to water at 0°C is approximately +1.1 kcal/mol. Yet the transition occurs spontaneously. Briefly explain why.

2. Calculate the values for $\Delta G°'$ that correspond to the following values of K'_{eq}. Assume that the temperature is 25°C.

   (a) 1.5×10^4
   (b) 1.5
   (c) 0.15
   (d) 1.5×10^{-4}

3. Calculate the values for K'_{eq} that correspond to the following values of $\Delta G°'$. Assume that the temperature is 25°C.

   (a) -10 kcal/mol
   (b) -1 kcal/mol
   (c) $+1$ kcal/mol
   (d) $+10$ kcal/mol

4. The enzyme hexokinase catalyzes the following reaction:

$$\text{Glucose} + \text{ATP} \rightleftharpoons \text{glucose 6-phosphate} + \text{ADP}$$

For this reaction, $\Delta G°' = -4.0$ kcal/mol.

(a) Calculate the change in free energy, $\Delta G'$, for this reaction under typical intracellular conditions using the following concentrations: glucose, 55 mM; ATP, 5.0 mM; ADP, 1.0 mM; and glucose 6-phosphate, 0.1 mM. Assume that the temperature is 25°C.

(b) In the typical cell, is the reaction catalyzed by hexokinase close to equilibrium or far from equilibrium? Explain.

5. The enzyme aldolase catalyzes the following reaction:

$$\text{Fructose 1,6-bisphosphate} \rightleftharpoons$$
$$\text{dihydroxyacetone phosphate} + \text{glyceraldehyde 3-phosphate}$$

For this reaction, $\Delta G°' = +5.7$ kcal/mol.

(a) Calculate the change in free energy, $\Delta G'$, for this reaction under typical intracellular conditions using the following concentrations: fructose 1,6-bisphosphate, 0.15 mM; dihydroxyacetone phosphate, 4.3×10^{-6} M; and glyceraldehyde 3-phosphate, 9.6×10^{-5} M. Assume that the temperature is 25°C.

(b) Explain why the aldolase reaction occurs in cells in the direction written despite the fact that it has a positive free-energy change under standard conditions.

6. Stryer makes the statement (p. 184) that a decrease of 1.36 kcal/mol in the free energy of activation of an enzyme-catalyzed reaction has the effect of increasing the rate of conversion of substrate to product by a factor of 10. What effect should a decrease of 1.36 kcal/mol in the free energy of activation have on the reverse reaction, the conversion of product to substrate? Explain.

7. What is the ratio of [S] to K_M when the velocity of an enzyme-catalyzed reaction is 80% of V_{max}?

8. The simple Michaelis-Menten model (equation 14, Stryer, p. 187) applies only to the initial velocity of an enzyme-catalyzed reaction, that is, to the velocity when no appreciable amount of product has accumulated. What feature of the model is consistent with this constraint? Explain.

9. Two first-order rate constants, k_2 and k_3, and one second-order rate constant, k_1, define K_M by the relationship

$$K_M = \frac{k_2 + k_3}{k_1}$$

By substituting the appropriate units for the rate constants in this expression, show that K_M must be expressed in terms of concentration.

10. Suppose that two tissues, tissue A and tissue B, are assayed for the activity of enzyme X. The activity of enzyme X, expressed as the number of moles of substrate converted to product per gram of tissue, is found to be five times greater in tissue A than in tissue B under a variety of circumstances. What is the simplest explanation for this observation?

11. Sketch the appropriate plots on the following axes. Assume that simple Michaelis-Menten kinetics apply. (See Stryer, pp. 187–189.)

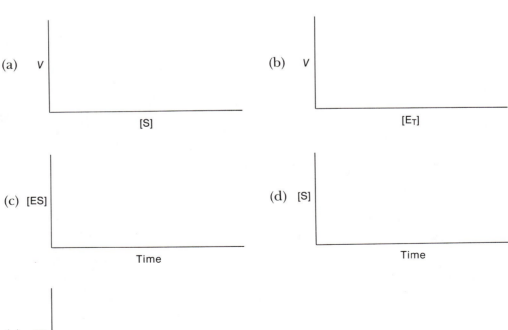

(a) v [S]

(b) v [E_T]

(c) [ES]   Time

(d) [S]   Time

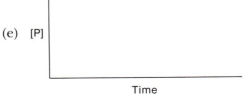

(e) [P]   Time

[S] (mM)	V (mmol ml^{-1} min^{-1})
0.1	3.33
0.2	5.00
0.5	7.14
0.8	8.00
1.0	8.33
2.0	9.09

[S] (mM)	V (mmol ml^{-1} min^{-1})	
	Without X	With X
0.2	5.0	3.0
0.4	7.5	5.0
0.8	10.0	7.5
1.0	10.7	8.3
2.0	12.5	10.7
4.0	13.6	12.5

[S] (mM)	V (mmol ml^{-1} min^{-1})	
	Without Y	With Y
0.2	5.0	2.0
0.4	7.5	3.0
0.8	10.0	4.0
1.0	10.7	4.3
2.0	12.5	5.0
4.0	13.6	5.5

12. Suppose that the data shown in the margin are obtained for an enzyme-catalyzed reaction.
   (a) From a double-reciprocal plot of the data, determine K_M and V_{max}.
   (b) Assuming that the enzyme present in the system had a concentration of 10^{-6} M, calculate its turnover number.

13. Suppose that the data shown in the margin are obtained for an enzyme-catalyzed reaction in the presence and absence of inhibitor X.
   (a) Using double-reciprocal plots of the data, determine the type of inhibition that has occurred.
   (b) Does inhibitor X combine with E, with ES, or with both? Explain.
   (c) Calculate the inhibitor constant, K_i, for substance X, assuming that the final concentration of X in the reaction mixture was 0.2 mM.

14. Suppose that the data shown in the margin are obtained for an enzyme-catalyzed reaction in the presence and absence of inhibitor Y.
   (a) Using double-reciprocal plots of the data, determine the type of inhibition that has occurred.
   (b) Does inhibitor Y combine with E, with ES, or with both? Explain.
   (c) Calculate the inhibitor constant, K_i, for substance Y, assuming that the final concentration of Y in the reaction mixture was 0.3 mM.

15. Suppose that a modifier Q is added to an enzyme-catalyzed reaction with the results depicted in Figure 8-3. What role does Q have? Does it combine with E, with ES, or with both E and ES?

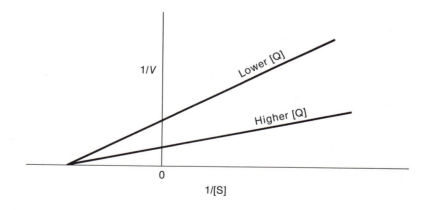

Figure 8-3
Effects of modifier Q on a plot of V versus 1/[S].

ANSWERS TO SELF-TEST

Introduction

1. d

2. a, c

3. Molecular mechanisms that control the catalytic activity of enzymes include feedback inhibition, the binding of regulatory proteins, covalent modification, and proteolytic cleavage. These mechanisms of enzyme regulation depend on conformational changes that occur at the site of modulation and are transmitted to the active site of the enzyme. Consequently, both enzyme-substrate interactions and catalytic activity are affected.

Thermodynamic Principles

4. b

5. The thermodynamic parameters ΔE and ΔS for a chemical reaction are not easily measured. Also, both depend on changes that occur not only in the system under study but also in the surroundings. Intrinsically unfavorable reactions can take place if a change in the surroundings compensates for an increase in the energy (positive ΔE) or a decrease in the entropy (negative ΔS) of the system.

6. c, e

7. c, d

8. All the statements are correct.

(a) Resolution of the paradox comes from the fact that the ratio of phosphate to glucose 1-phosphate in cells is usually high enough that phosphorolysis of glycogen is favored.

(b) Using K'_{eq}, one can calculate the $\Delta G^{\circ\prime}$ for the phosphorylase reaction:

$$\Delta G^{\circ\prime} = -2.303\ RT \log_{10} K'_{eq}$$

$$= -2.303 \times 1.98\ \frac{cal}{mol \cdot {}^{\circ}K} \times 298^{\circ}K \times \log_{10}(0.088)$$

$$= -1360\ cal/mol \times -1.055$$

$$= 1440\ cal/mol = 1.44\ kcal/mol$$

(c) In part (b) the $\Delta G^{\circ\prime}$ for the phosphorylase reaction of $+1.44$ kcal/mol was calculated; therefore, energy is consumed rather than released by this reaction.

(d) In cells, the ratio of phosphate to glucose 1-phosphate is so large that phosphorylase is mainly involved with glycogen degradation.

9. a, d, e
(a) Correct.

$$\Delta G^{\circ\prime} = -2.303\ RT \log_{10} K'_{eq}$$

$$-3.3\ kcal/mol = -1.36\ kcal/mol \times \log_{10} K'_{eq}$$

$$\log_{10} K'_{eq} = 2.426$$

$$K'_{eq} = 267$$

(b) Incorrect. K'_{eq} is a constant; it is independent of the initial concentrations.

(c) Incorrect.

$$\Delta G' = \Delta G^{\circ\prime} + 2.303\, RT \log_{10} \frac{[\text{glucose}][\text{phosphate}]}{[\text{glucose 6-phosphate}]}$$

$$= -3.3\ \text{kcal/mol} + \left(1.36\ \text{kcal/mol} \times \log_{10} \frac{10^{-1} \times 10^{-1}}{10^{-5}}\right)$$

$$= -3.3\ \text{kcal/mol} + \left(1.36\ \text{kcal/mol} \times \log_{10} \frac{10^{-2}}{10^{-5}}\right)$$

$$= -3.3\ \text{kcal/mol} + (1.36\ \text{kcal/mol} \times 3)$$

$$= +0.78\ \text{kcal/mol}$$

(d) Correct. Under the initial conditions, $\Delta G'$ is positive; therefore, the reaction will proceed towards the formation of glucose 6-phosphate.

(e) Correct. The negative $\Delta G^{\circ\prime}$ value (at standard conditions) indicates that the reaction will proceed spontaneously towards the hydrolysis of glucose 6-phosphate.

Enzyme Catalysis

10. e

11. c. In reactions requiring two substrates, an enzyme-substrate complex of one of the substrates can be isolated if the complex is very stable.

12. The formation of an enzyme-substrate complex involves a close, complementary fitting of the atoms of the amino acid residues that make up the active site of the enzyme with the atoms of the substrate. Since stereoisomers have different spatial arrangements of their atoms, only a single stereoisomer of the substrate usually fits in the active site.

13. The enzyme-substrate and enzyme-product complexes have to be reversible so that catalysis may proceed; therefore, weak forces are involved in the binding of substrates to enzymes.

Enzyme Kinetics and Enzyme Inhibitors

14. a, d, e

15. K_M can be equal to K_{ES} when the rate constant $k_3 \ll k_2$. Since $K_M = (k_2 + k_3)/k_1$, when k_3 is negligible relative to k_2, K_M becomes equal to k_2/k_1, which is the dissociation constant of the enzyme-substrate complex.

16. These equations are related because they express the occupancy of saturable binding sites as a function of either O_2 or substrate concentration. The fraction of active sites filled, V/V_{max}, is analogous to Y, the degree of myoglobin saturation with oxygen; [S] and pO_2 are the concentrations of substrate and O_2, respectively; and K_M and P_{50}, are substrate or O_2 concentrations at half-maximal saturation.

17. (a) $K_M = 5 \times 10^{-4}$ M. K_M is obtained from Figure 8-1; it is equal to [S] at $\frac{1}{2}V_{max}$. Note that the units of [S] are mM. The factor 10^3 is used to multiply the actual concentrations. For example,

$$[S] \times 10^3 = 2.0\ \text{M}$$

$$[S] = 2.0 \times 10^{-3}\ \text{M}$$

(b) $V_{max} = 6\ \mu\text{mol/min}$. V_{max} is obtained from Figure 8-1; it is the maximum velocity.

(c) $k_3/K_M = 2 \times 10^5 \text{ s}^{-1} \text{ M}^{-1}$. In order to calculate this ratio, k_3 must be known. Since $V_{max} = k_3[E_T]$, $k_3 = V_{max}/[E_T]$. Thus

$$k_3 = \frac{6\ \mu\text{mol/min}}{10^{-3}\ \mu\text{mol}}$$

$$= 6 \times 10^3 \text{ min}^{-1}$$

$$= 100 \text{ s}^{-1}$$

Using K_M from part (a),

$$\frac{k_3}{K_M} = \frac{100 \text{ s}^{-1}}{5 \times 10^{-4} \text{ M}}$$

(d) The turnover number is 100 s^{-1}, equal to k_3, which was calculated in part (c).

18. The ratios k_3/K_M and k_{cat}/K_M allow one to estimate the catalytic efficiency of an enzyme. The upper limit for k_3/K_M, 10^8 to 10^9 $\text{M}^{-1} \text{s}^{-1}$, is set by the rate of diffusion of the substrate in the solution, which limits the rate at which it encounters the enzyme. If an enzyme has a k_3/K_M in this range, its catalytic velocity is restricted only by the rate at which the substrate can reach the enzyme, which means that the enzymatic catalysis has attained kinetic perfection.

19. d. $V_{max} = k_3[E_T]$; thus, if 6.7 times more DNA polymerase than chymotrypsin is used, V_{max} for both enzymes is the same:

$$100 \text{ s}^{-1} = 6.7 \times 15 \text{ s}^{-1}$$

20. b, c, d

21. The inhibition is noncompetitive since the proportion of bound substrate remains the same as the concentration of the inhibitor increases.

22. (a) 1 (b) 3 (c) 2 (d) 4

23. See Figure 8-4. Plots 1 and 2 have the same $1/V$ intercept; plots 1 and 3 have the same $1/[S]$ intercept; and plots 1 and 4 have different $1/V$ and $1/[S]$ intercepts.

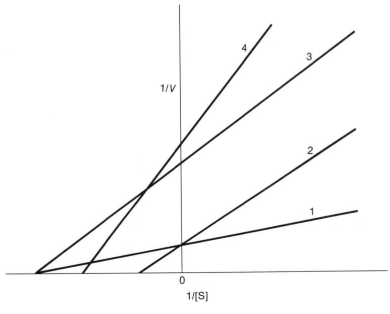

Figure 8-4
Lineweaver-Burk plots for competitive (2), noncompetitive (3), and mixed (4) inhibition, relative to the enzymatic reaction in the absence of inhibitors (1).

24. e

1. The increase in entropy must be sufficient to offset the increase in internal energy, which leads to a negative change in the free energy. We may envision this as follows: as ice melts, it loses its ordered structure and its entropy therefore increases.

2. The values for $\Delta G^{\circ\prime}$ are found by substituting the values for K'_{eq} into equation 11 on page 182 of Stryer.

   (a) $\Delta G^{\circ\prime} = -2.303 \, RT \log_{10} K'_{eq}$

   $= -2.303 \times 1.98 \times 10^{-3} \times 298 \times \log_{10} (1.5 \times 10^4)$

   $= -5.7$ kcal/mol

   (b) -0.24 kcal/mol
   (c) $+1.1$ kcal/mol
   (d) $+5.2$ kcal/mol

3. Equation 13 on page 183 of Stryer is used to find the answers.

   (a) $K'_{eq} = 10^{-\Delta G^{\circ\prime}/1.36}$

   $= 10^{-(-10/1.36)}$

   $= 2.3 \times 10^7$

   (b) 5.4
   (c) 0.18
   (d) 4.4×10^{-8}

4. (a) The applicable relationship is equation 6 on page 182 of Stryer:

   $$\Delta G' = \Delta G^{\circ\prime} + 2.303 \, RT \log_{10} \frac{[C][D]}{[A][B]}$$

   $$= \Delta G^{\circ\prime} + 2.303 \, RT \log_{10} \frac{[\text{glucose 6-phosphate}][\text{ADP}]}{[\text{glucose}][\text{ATP}]}$$

   $= -4.0$ kcal/mol

   $+ \left(2.303 \times 1.98 \times 10^{-3} \times 298 \times \log_{10} \frac{[0.1 \times 10^{-3}][1.0 \times 10^{-3}]}{[55 \times 10^{-3}][5 \times 10^{-3}]} \right)$

   $= -4.0$ kcal/mol -4.7 kcal/mol

   $= -8.7$ kcal/mol

   (b) The large negative value for $\Delta G'$ means that the reaction is far from equilibrium in the typical cell and thus has a strong thermodynamic drive to go in the direction of product formation. (Remember that at equilibrium, $\Delta G' = 0$.)

5. (a) The applicable relationship is again equation 6 on page 182 of Stryer:

$$\Delta G' = \Delta G^{\circ\prime} + 2.303\ RT\ \log_{10} \frac{[C][D]}{[A][B]}$$

$$= \Delta G^{\circ\prime} + 2.303\ RT\ \log_{10} \frac{[DHAP][G3P]}{[FBP]}$$

$$= +5.7\ \text{kcal/mol} + (2.303 \times 1.98 \times 10^{-3} \times 298$$

$$\times \log_{10} \frac{(4.3 \times 10^{-6}) \times (9.6 \times 10^{-5})}{0.15 \times 10^{-3}}$$

$$= +5.7\ \text{kcal/mol} - 7.6\ \text{kcal/mol}$$

$$= -1.9\ \text{kcal/mol}$$

(b) The reaction occurs in the direction written because of the effects of the concentrations on the free-energy change. The concentration term in the equation is much smaller than 1.0, which is its value under standard conditions.

6. The rate of the reverse reaction must also increase by a factor of 10. Enzymes do not alter the equilibria of processes; they affect the rate at which equilibrium is attained. Since the equilibrium constant, K_{eq}, is the quotient of the rate constants for the forward and reverse reactions, both rate constants must be altered by the same factor. If the rate of the forward reaction is increased by a factor of 10, the rate of the reverse reaction must increase by the same factor

7. Start with the Michaelis-Menten equation, equation 28 on page 189 of Stryer:

$$V = V_{max} \frac{[S]}{[S] + K_M}$$

Substituting $0.8\ V_{max}$ for V yields

$$0.8\ V_{max} = V_{max} \frac{[S]}{[S] + K_M}$$

$$0.8[S] + 0.8K_M = [S]$$

$$0.8K_M = 0.2[S]$$

$$[S] = 4\ K_M$$

$$\frac{[S]}{K_M} = 4$$

Thus a substrate concentration four times as great as the Michaelis constant yields a velocity that is 80% of maximal velocity.

8. Equation 14 shows the k_3 step as being irreversible. This is true initially because P and E cannot recombine to give ES at an appreciable rate if little P is present. Note that this says nothing about the relative magnitudes of k_3 and the reverse rate constant, k_4:

$$\text{E} + \text{S} \underset{k_2}{\overset{k_1}{\rightleftharpoons}} \text{ES} \underset{k_4}{\overset{k_3}{\rightleftharpoons}} \text{E} + \text{P}$$

The reverse constant, k_4, may actually be quite large compared to k_3, but the reverse reaction still will not occur when little product is present since the rate of the k_4 step depends on the concentrations of P and E as well as on the magnitude of k_4.

9. The first-order rate constants have the dimensions t^{-1}, whereas the second-order constant has the dimension $\text{conc}^{-1}\, t^{-1}$. Thus,

$$K_{\mathrm{M}} = \frac{k_2 + k_3}{k_1}$$

$$= \frac{t^{-1} + t^{-1}}{\text{conc}^{-1}\, t^{-1}}$$

$$= \text{conc}$$

10. For the activity of enzyme X to be five times greater in tissue A than in tissue B, tissue A must have five times the amount of enzyme X as does tissue B. Enzyme activity is directly proportional to enzyme concentration.

11. The sketches should resemble the following:

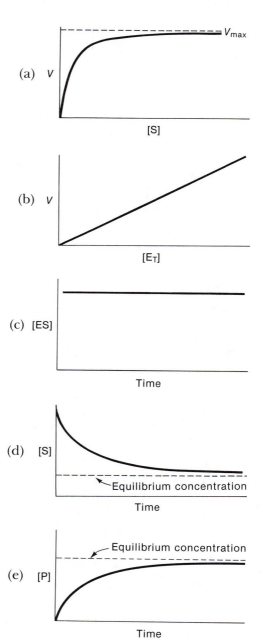

12. (a) See the graph at the right, Figure 8-5. $V_{max} = 1/0.1 = 10$ mmol ml^{-1} min^{-1}.

$$\text{Slope} = \frac{0.3 - 0.1}{10} = 0.02$$

$$\text{Slope} = \frac{K_M}{V_{max}}$$

$$K_M = 0.02 \times 10 = 0.2 \text{ mM}$$

(b) The turnover number is equal to the rate constant k_3 in equation 33 on page 190 of Stryer. Rearrangement of the equation gives

$$k_3 = \frac{V_{max}}{[E_T]}$$

$$= \frac{10 \text{ mmol ml}^{-1} \text{ min}^{-1}}{10^{-6} \text{ mol liter}^{-1}}$$

$$= \frac{10 \text{ mol liter}^{-1} \text{ min}^{-1}}{10^{-6} \text{ mol liter}^{-1}}$$

$$= 10^7 \text{ min}^{-1} \qquad \text{or} \qquad 1.7 \times 10^5 \text{ s}^{-1}$$

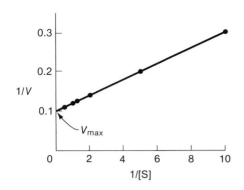

Figure 8-5
A double-reciprocal plot of data for Problem 12.

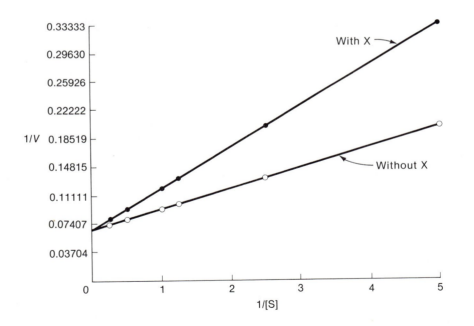

Figure 8-6
A double-reciprocal plot of data for Problem 13 showing the effects of an inhibitor X.

13. (a) See Figure 8-6. The double-reciprocal plots intersect on the y axis, so the inhibition was competitive.
   (b) The inhibitor combines only with E, the free enzyme. A competitive inhibitor cannot combine with ES because the inhibitor and the substrate compete for the same binding site on the enzyme.

(c) An inhibitor increases the slope of a double-reciprocal plot by a factor of $1 + [I]/K_i$:

$$\text{Slope}_{\text{inhib}} = \text{slope}_{\text{uninhib}} \left(1 + \frac{[I]}{K_i}\right)$$

The slope with X is

$$\text{Slope}_{\text{inhib}} = \frac{0.333 - 0.067}{5} = 0.0532$$

The slope without X is

$$\text{Slope}_{\text{uninhib}} = \frac{0.200 - 0.067}{5} = 0.0266$$

Substituting in these values yields

$$0.0532 = 0.0266 \left(1 + \frac{0.2 \text{ mM}}{K_i}\right)$$

$$K_i = 0.2 \text{ mM}$$

14. (a) See Figure 8-7. The inhibition was noncompetitive as indicated by the fact that the double-reciprocal plots intersect to the left of the y axis.

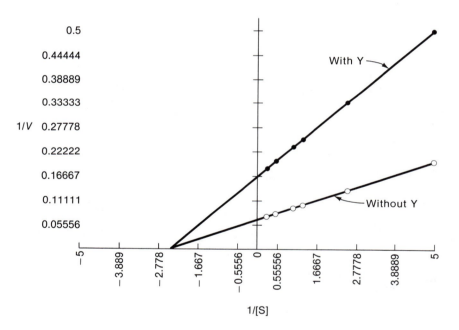

Figure 8-7
A double-reciprocal plot of data for Problem 14 showing the effects of an inhibitor Y.

(b) A noncompetitive inhibitor combines at a site other than the substrate binding site. Thus, it may combine both with E and ES. In the case illustrated, the inhibitor has equal affinity for E and ES, which is shown by the fact that the plots intersect on the x axis.

(c) Again, the slope increases by a factor of $1 + [I]/K_i$ in the presence of an inhibitor.

$$\text{Slope}_{\text{inhib}} = \text{slope}_{\text{uninhib}} \left(1 + \frac{[I]}{K_i}\right)$$

The slope with Y is

$$\text{Slope}_{\text{inhib}} = \frac{0.500}{5.0 - (-2.5)} = 0.0266$$

The slope without Y is

$$\text{Slope}_{\text{uninhib}} = \frac{0.200}{5.0 - (-2.5)} = 0.0666$$

Substituting in these values yields

$$0.0666 = 0.0266 \left(1 + \frac{0.3 \text{ mM}}{K_{\text{i}}} \right)$$

$$K_{\text{i}} = 0.2 \text{ mM}$$

15. Q increases the rate of reaction, so it is an activator, or perhaps a second substrate. It combines with both E and ES.

CHAPTER 9

Mechanisms of Enzyme Action

In the previous chapter, you learned that the catalytic action of enzymes is based on their ability to stabilize the transition states of chemical reactions. In this chapter, Stryer describes in detail the structures, active-site configurations, binding of substrates, and catalytic mechanisms of four well-understood enzymes: lysozyme, ribonuclease, carboxypeptidase A, and chymotrypsin. Using these specific examples, some fundamental principles of enzyme catalysis are illustrated: specific binding of substrates, induced fit of the enzyme-substrate complex, general acid-base catalysis by active-site residues, formation and stabilization of transition states, and reversibility of the catalytic steps. Since the interactions of enzymes with substrates depend on the chemical properties of amino acid residues and on protein structure in general, a review of Chapter 2 may be helpful at this time.

LEARNING OBJECTIVES

When you have mastered this chapter, you should be able to complete the following objectives.

Lysozyme (Stryer pages 201–210)

1. Describe the *polysaccharide* and *oligosaccharide substrates* of *lysozyme*. (The nomenclature, structures, and properties of carbohydrates are described in Chapter 14 of Stryer.)

2. Explain how *x-ray crystallography* may be used to find the *active site* of an enzyme.

3. Summarize the predictions made about the *binding site* and the *catalytic groups* of lysozyme on the basis of x-ray data.

4. Outline the *catalytic mechanism* of lysozyme. Explain the roles of *general acid catalysis* and the formation and stabilization of a *carbonium ion intermediate* in the process.

5. List the experiments that support the mode of substrate binding and the mechanism of catalysis proposed for lysozyme.

Ribonuclease and RNA Enzymes (Stryer pages 211–215)

6. Describe the *substrate specificity* of *ribonuclease*.

7. Summarize the chemical studies of ribonuclease and the x-ray crystallographic evidence that implicate *histidine* residues in catalysis.

8. Describe the catalytic mechanism proposed for ribonuclease. Explain the general acid-base catalysis by the histidine residues and the formation and stabilization of a *pentacovalent transition state* of phosphorus.

9. Discuss the discovery that RNA molecules can act as enzymes.

Carboxypeptidase A (Stryer pages 215–220)

10. State the substrate specificity of *carboxypeptidase A*.

11. Describe the x-ray crystallographic information about the *zinc ion* binding site and the substrate binding site of carboxypeptidase A.

12. Describe the structural changes that occur at the active site of carboxypeptidase A upon the binding of substrate *(induced fit)*.

13. Summarize the involvement of the zinc ion, glutamate 270, and arginine 127 in the catalytic mechanism of carboxypeptidase A. Note especially the induction of *electronic strain* in the substrate by the zinc ion.

14. Outline the use of *site-specific mutagenesis* in elucidating enzyme mechanisms.

Chymotrypsin and Other Proteolytic Enzymes (Stryer pages 220–229)

15. Describe the substrate specificity and the physiological role of *chymotrypsin*.

16. Outline the catalytic mechanism of chymotrypsin and explain the role of the *acyl-enzyme intermediate.*

17. Give the experimental evidence that implicates reactive serine and histidine residues in the catalytic mechanism of chymotrypsin. Describe the role of the serine-histidine-aspartate *catalytic triad.*

18. Describe the formation and stabilization of the transient *tetrahedral intermediate* produced from the planar amide group of the substrate.

19. Compare other *serine proteases,* for example, *trypsin, elastase,* and *subtilisin,* with chymotrypsin in terms of their substrate specificities, structures, catalytic mechanisms, and *evolution.*

20. List the four different classes of proteolytic enzymes and summarize their properties.

21. Explain how *catalytically active antibodies* may be generated using *transition-state analogs* as antigens.

SELF-TEST

Lysozyme

1. X-ray crystallographic determination of the structure of water-soluble enzymes, such as lysozyme and chymotrypsin,

   (a) shows compact, almost globular structures.
   (b) immediately reveals the catalytic groups.
   (c) indicates variable amounts of secondary structure.
   (d) can be readily performed with the substrate bound to the active site.
   (e) can reveal evolutionary relationships with other enzymes.

2. Given the oligosaccharide NAM-NAG-NAM-NAG-NAM-NAG and the sugar binding sites A, B, C, D, E, and F on lysozyme, identify which of the following statements is *true.*

   (a) Any segments of this oligosaccharide having more than three sugars would bind to lysozyme with the same affinity as the hexamer.
   (b) The hexamer NAG-NAM-NAG-NAM-NAG-NAM would have the same affinity for lysozyme as the given hexamer.
   (c) The hexamer in (b) would bind to lysozyme with a higher affinity than the given hexamer.
   (d) The given hexamer would be cleaved between sugars 4 and 5 from the NAM end.

3. A variety of chemical experiments support the catalytic mechanism of lysozyme proposed on the basis of x-ray crystallographic data. Among them were the following:

   (1) The use of ^{18}O enriched H_2O in the hydrolytic reaction

   (2) The dependence of the catalytic rate on pH, with the optimum rate occurring at pH 5

   (3) The cleavage pattern of hexa-NAG into tetra-NAG and di-NAG

(4) The selective chemical modification of aspartic and glutamic amino acid residues

(5) The binding of a lactone analog of tetra-NAG with high affinity

Indicate which of these experiments led to the following conclusions:

(a) Hydrolysis of the glycosidic bond occurs between sugar residues D and E. _____

(b) Acidic amino acid residues participate in catalysis. _____

(c) The oxygen of the cleaved glycosidic bond remains with sugar residue E. _____

(d) Distortion of the D sugar residue is important in catalysis. _____

4. Which of the following is a catalytic amino acid residue of lysozyme?

(a) Tryptophan 62
(b) Serine 57
(c) Histidine 24
(d) Glutamic acid 35
(e) Lysine 74

5. A pH-enzyme activity curve is shown in Figure 9-1. Which of the following pairs of amino acids would be likely candidates as catalytic groups? (See Stryer, p. 42 for the pK values of amino acid residues.)

(a) Glutamic acid and lysine
(b) Aspartic acid and histidine
(c) Histidine and cysteine
(d) Histidine and histidine
(e) Histidine and lysine

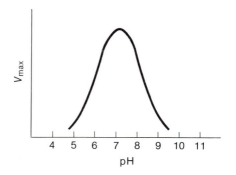

Figure 9-1
pH dependence of V_{max}.

Ribonuclease and RNA Enzymes

6. Which of the following are *true* statements regarding the ribonuclease reaction?

(a) Ribonuclease cleavage of RNA produces new 3′-OH and 5′-phosphate ends.
(b) Single-stranded DNA can bind to ribonuclease.
(c) A 3′,5′-cyclic phosphate intermediate is formed during the reaction.
(d) A pocket adjacent to the catalytic groups of ribonuclease specifically binds pyrimidine bases.
(e) Several main-chain NH and CO groups of ribonuclease form hydrogen bonds with the phosphate backbone of RNA.

7. Give a reason why genetic information is encoded in DNA rather than in RNA.

8. In the transition state during RNA hydrolysis by ribonuclease, the reactive phosphorus atom adopts a pentacovalent trigonal bipyramid geometry. The oxygen atoms at the apices of the pyramid correspond to

    (a) H_2O and $3'$-O
    (b) $2'$-O and $5'$-O
    (c) H_2O and $5'$-O
    (d) $2'$-O and $3'$-O
    (e) OH^- and $2'$-O

What is one consequence of this transition state geometry?

_____

9. Why is histidine a particularly versatile amino acid residue in terms of its involvement in enzymatic reaction mechanisms?

_____

_____

10. Which of the following is *not* a common property of enzymes (proteins) and ribozymes (RNA enzymes)?

    (a) Their active sites may contain large hydrophobic pockets.
    (b) They usually bind substrates via multiple weak interactions.
    (c) Both show susceptibility to competitive inhibition.
    (d) They exhibit saturation kinetics at high substrate concentrations.
    (e) Both have a high degree of substrate specificity.

11. What are two main structural features of an enzyme that determine its substrate specificity?

_____

_____

Carboxypeptidase A

12. The binding of substrate to carboxypeptidase A *does not* involve which of the following interactions?

    (a) The carbonyl oxygen of the susceptible peptide bond is coordinated to the zinc ion.
    (b) The terminal carboxylate group of the substrate forms an electrostatic bond with an arginine residue.
    (c) The terminal carboxylate group of the substrate forms a hydrogen bond with a tyrosine residue.
    (d) The NH group of the susceptible peptide bond forms a hydrogen bond with a glutamate residue.
    (e) The aromatic side chain of the carboxyl-terminal residue of the substrate binds to a hydrophobic pocket near the zinc ion.

13. Why does carboxypeptidase A have a positive zinc ion in its active site instead of a positively charged amino acid residue such as arginine or lysine?

_____

_____

14. Upon the binding of substrate to carboxypeptidase A, the active site undergoes a major structural rearrangement. Which of the following is a consequence of this induced-fit rearrangement of carboxypeptidase A?

(a) The zinc ion, which was in solution, now binds to the active site.
(b) Water molecules are displaced from the active site.
(c) The electrostatic and hydrogen-bond interactions between the enzyme and substrate become weaker.
(d) The active site cavity can now accommodate an internal peptide bond of a protein substrate.

15. Propose an experiment to determine whether a particular amino acid residue is essential for the activity of an enzyme.

_____

_____

Chymotrypsin and Other Proteolytic Enzymes

16. Which of the following experimental observations provide evidence of the formation of an acyl-enzyme intermediate during the chymotrypsin reaction?

(a) There is a biphasic release of p-nitrophenol during the hydrolysis of p-nitrophenyl acetate.
(b) The active serine can be specifically labeled with organic fluorophosphates.
(c) The pH dependence of the catalytic rate is bell shaped, with a maximum at pH 8.
(d) An acyl-enzyme covalent complex can be isolated at low pH.

17. Which of the following enzymes can be irreversibly inactivated with diisopropylphosphofluoridate (DIPF)?

(a) Carboxypeptidase A
(b) Trypsin
(c) Lysozyme
(d) Subtilisin
(e) Thrombin

18. Three essential amino acid residues in the active site of chymotrypsin form a catalytic triad. Which of the following are roles of these residues in catalysis?

(a) The histidine residue facilitates the reaction by acting as an acid-base catalyst.
(b) The aspartate residue orients the substrate properly for reaction.
(c) The serine residue acts as a nucleophile during the reaction.
(d) The serine residue acts as an electrophile during the reaction.
(e) The aspartate residue initiates the deacylation step by a nucleophilic attack on the carbonyl carbon of the acyl intermediate.

19. Although chymotrypsin is a proteolytic enzyme, it is quite resistant to self-hydrolysis. Explain why.

_____

_____

20. The three enzymes trypsin, elastase, and chymotrypsin

    (a) evolved from a common ancestor.

    (b) have major similarities in their amino acid sequences and three-dimensional structures.

    (c) catalyze the same reaction: the cleavage of a peptide bond.

    (d) catalyze reactions that proceed through a covalent intermediate.

    (e) have structural differences at their active sites.

21. Match the enzymes and the catalytic or functional groups in the right column with the main proteolytic enzyme classes in the left column.

    (a) Zinc proteases _____

    (b) Serine proteases _____

    (c) Thiol proteases _____

    (d) Acid proteases _____

    (1) Papain
    (2) Pepsin
    (3) Elastase
    (4) Thermolysin
    (5) Histidine
    (6) Cysteine
    (7) Aspartate
    (8) Arginine

22. Suppose that you want to prepare an antibody with the catalytic activity of lysozyme. What type of antigen would you use?

_____

_____

23. For each enzyme in the left column, indicate the appropriate transition state or chemical entity in the right column that has a postulated involvement in its catalytic mechanism.

    (a) Lysozyme _____

    (b) Ribonuclease _____

    (c) Carboxypeptidase A _____

    (d) Chymotrypsin _____

    (1) Mixed anhydride
    (2) Oxianion
    (3) Pentacovalent phosphorus
    (4) Carbonium ion
    (5) Tetrahedral peptide bond

PROBLEMS

1. The peptide hydrogen of residue 59 and the carbonyl oxygen of residue 107 of the polypeptide backbone of lysozyme form hydrogen bonds with residue C of tri-*N*-acetylglucosamine (tri-NAG) (see Stryer, p. 205). Would you expect these two residues to be found in helical regions? Explain.

2. What limitations does the use of tri-NAG have for mapping the interactions that bind substrate to lysozyme? What advantage does its use have despite these limitations? Explain.

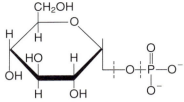

Glucose 1-phosphate

(Dashed lines indicate
possible sites of cleavage).

3. Glucose 1-phosphate (shown in the margin) may be cleaved to glucose and inorganic phosphate by the enzyme alkaline phosphatase:

$$\text{Glucose 1-phosphate} + H_2O \xrightleftharpoons[\text{phosphatase}]{\text{Alkaline}} \text{glucose} + \text{phosphate}$$

In principle, the phosphoryl group could be removed by scission of either a P–O or a C–O bond. These two possibilities may be distinguished experimentally by the use of H_2O labeled with ^{18}O as a substrate.

    (a) Which product would be labeled with ^{18}O if the P–O bond is broken? Explain.

    (b) Which product would be labeled with ^{18}O if the C–O bond is broken? Explain.

    (c) The ^{18}O is found to be incorporated exclusively into inorganic phosphate. Which bond must therefore have been broken?

4. The proposed catalytic mechanism for lysozyme presented by Stryer on page 207 requires that the side-chain group of aspartic acid 52 be in the ionized —COO⁻ form and that of glutamic acid 35 be in the un-ionized —COOH form.

    (a) What percentage of Asp side-chain groups would be in the ionized —COO⁻ form at pH 5.0, the pH optimum for the hydrolysis of chitin by lysozyme? Use the Henderson-Hasselbalch equation (see Stryer, p. 41). Assume that the pK of the side-chain carboxyl group is 3.9.

    (b) Using the same method as in part (a), calculate the percentage of Glu side-chain groups that would be in the protonated —COOH form at pH 5.0. Assume a pK of 4.3.

    (c) In view of your answer to part (b), explain how it is possible for Glu 35 in lysozyme to be present in the un-ionized form.

5. When a polysaccharide chain is cleaved by lysozyme, a C–O bond at the C-1 of sugar residue D of the substrate is broken, as indicated by the isotopic evidence presented by Stryer on page 206. Briefly explain how the cleavage of this bond rather than the C–O bond at the C-4 of sugar residue E is consistent with the notion that a carbonium ion intermediate is involved in the mechanism. See Figure 9-2.

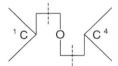

Figure 9-2
Possible cleavage sites for lysozyme.

6. An old theory of enzyme catalysis involved the notion that the distortion of substrate molecules upon binding to enzymes contributed to catalysis. For a time, this theory fell into disfavor because there was little evidence to back it up. In what respects is evidence for the theory provided by studies of the lysozyme reaction? Explain.

7. Predict the cleavage products that would be obtained when the following nucleotide is cleaved with pancreatic ribonuclease:

$$\text{pApUpGpUpCpGpGp}$$

(See Stryer, p. 212 for the mechanism of ribonuclease and p. 73 for the conventions used in abbreviating the structure of nucleotides.)

8. Suppose that the nucleotide UpGp were cleaved with pancreatic ribonuclease in the presence of H_2O labeled with ^{18}O. Give the products that would be formed, and indicate which product would contain ^{18}O. Explain your answer.

9. Trypsin, chymotrypsin, and carboxypeptidase A fail to cleave pep-

tide bonds involving proline. Trypsin, for example, will not cleave at Lys-Pro. Why do you think this is the case?

10. Place slash marks at the sites where you would expect chymotrypsin to cleave the following peptide:

Lys-Gly-Phe-Thr-Tyr-Pro-Asn-Trp-Ser-Tyr-Phe

11. A competitive inhibitor of chymotrypsin, β-phenylpropionate, protects histidine 57 against alkylation by the affinity-labeling reagent TPCK (see Stryer, p. 223). Noncompetitive inhibitors do not provide such protection. Explain.

12. Many enzymes can be protected against thermal denaturation during purification procedures by the addition of substrate. Propose an explanation for this phenomenon.

13. Transmission at many synapses in the central nervous system, including those between motor nerves and muscles, is mediated by the neurotransmitter acetylcholine. Acetylcholine is released from vesicles at the nerve endings; it diffuses across synapses to the postsynaptic membrane, where it triggers a response and is then cleaved to acetate and choline by the enzyme acetylcholinesterase:

$$\text{Acetylcholine} + H_2O \xrightleftharpoons{\text{Acetylcholinesterase}} \text{acetate} + \text{choline}$$

(a) Acetylcholinesterase is inactivated by DIPF (see Stryer, p. 213) on a mole-for-mole basis, just as chymotrypsin is. (Because of its potent effects on acetylcholinesterase, DIPF is also called nerve gas.) What deduction may be made about the mechanism of acetylcholinesterase?

(b) Give a two-step reaction mechanism for acetylcholinesterase involving an acyl-enzyme intermediate.

14. The enzyme rhodanese is abundant in mammalian liver and kidney tissue. It can catalyze the transfer of a sulfur atom from thiosulfate to cyanide to yield sulfite and thiocyanate:

$$S_2O_3{}^{2-} + CN^- \rightleftharpoons SO_3{}^- + SCN^-$$

Although this reaction is valuable for studying the mechanism of rhodanese catalysis in the laboratory, it probably does not occur to a significant extent in tissues. (Cyanide, after all, is a deadly poison.) Rather, some sulfur acceptor other than cyanide serves as the major substrate in the cell.

(a) The given reaction involves an interaction between ions with like charges. One effect that an enzyme could have would be to divide the reaction into steps so that no such unfavorable interaction would occur. It was found experimentally that when enzyme was incubated with thiosulfate labeled with ^{35}S, a stable enzyme-sulfur compound could be isolated. Give a two-step reaction mechanism for rhodanese that involves such an enzyme-sulfur intermediate.

(b) One might expect the negatively charged thiosulfate to bind to the enzyme electrostatically. In an effort to demonstrate such an interaction, neutral salts like NaCl were added to the enzyme assay system to determine whether they had any effect on thiosulfate binding. It was found that the addition of such salts

$$CH_3\overset{\overset{\textstyle O}{\|}}{C}OCH_2CH_2-\overset{\overset{\textstyle CH_3}{|}}{\underset{\underset{\textstyle CH_3}{|}}{N}}{}^{+}-CH_3$$

Acetylcholine

decreased the ability of thiosulfate to bind? Does this support the hypothesis of electrostatic interaction? Explain.

ANSWERS TO SELF-TEST

Lysozyme

1. a, c, e

2. c. Because site C cannot bind NAM, the given oligosaccharide would bind to sites A–E or B–F with the approximate affinity of a pentamer.

3. (a) 3 (b) 2, 4 (c) 1 (d) 5

4. d

5. c

Ribonuclease and RNA Enzymes

6. b, d

7. RNA can be cleaved by mildly basic solutions. This spontaneous hydrolysis of RNA proceeds via a 2', 3'-cyclic phosphate intermediate. Since DNA lacks a 2'-hydroxyl group, it is much more stable to hydrolysis than is RNA, so it is a safer repository for genetic information.

8. b. This geometry is compatible with an in-line mechanism of nucleophilic attack.

9. Histidine residues may act as nucleophiles as well as general acids or bases. Because their pK values are approximately 6 to 7, histidine residues can be protonated or deprotonated at physiological pH; they are therefore able either to donate or to accept protons.

10. a

11. Binding groups must be present on the enzyme that can interact with the substrate to position it properly for a productive reaction. In addition, the enzyme must have catalytic residues that react with the specific chemical bond of the substrate. Both of these interactions determine the substrate specificity of an enzyme.

Carboxypeptidase A

12. d

13. Carboxypeptidase A accelerates catalysis by inducing electronic strain in its substrate. For inducing electronic strain in the substrate a small, highly charged ion is more effective than a bulky, basic amino acid side chain. In addition, the zinc ion binds and perhaps activates the water molecule that participates in the cleavage of the peptide bond.

14. b

15. Site specific mutagenesis could be used to change the particular amino acid residue into a different amino acid residue. Alternatively, specific chemical modification of the residue could be attempted. If

either of these treatments make the enzyme inactive, then the particular amino acid residue could be essential for the activity of the enzyme. However, it is also possible that the replacement of the amino acid or its chemical modification inactivate the enzyme by altering its structure. Additional experiments should be performed to establish that the structure of the enzyme has not been changed.

Chymotrypsin and Other Proteolytic Enzymes

16. a, d

17. b, d, e

18. a, c

19. Chymotrypsin specifically cleaves peptide bonds that are adjacent to aromatic amino acid residues or bulky, hydrophobic residues such as methionine. Since these types of residues are often buried in the interior of proteins, including chymotrypsin, the self-hydrolysis of native, folded chymotrypsin is very inefficient. In fact, during digestion, chymotrypsin acts most effectively on partially degraded and denatured proteins.

20. a, b, c, d, e

21. (a) 4, 8 (b) 3, 5, 7 (c) 1, 5, 6 (d) 2, 7

22. To prepare an antibody with the catalytic activity of lysozyme, a transition-state analog of the substrate should be used as the antigen. The D-ring lactone analog of tetra-NAG would be a likely choice.

23. (a) 4 (b) 3 (c) 1 (d) 2, 5

ANSWERS TO PROBLEMS

1. No. If these residues were in helical regions of lysozyme, the peptide hydrogen and the carbonyl oxygen would be involved in the hydrogen bonds that maintain the α helix. They would therefore be unavailable for hydrogen bonding with the substrate.

2. Cell-wall polysaccharides are alternating copolymers of *N*-acetylglucosamine (NAG) and *N*-acetylmuramate (NAM). One would therefore expect that groups on NAM are also important in binding to lysozyme. In addition, residues D, E, and F of the substrate contribute to binding (see Stryer, Figure 9-16, p. 209). The major advantage of using tri-NAG for mapping the interactions that bind substrate to lysozyme is that it is not cleaved at an appreciable rate. Tri-NAG therefore remains stably bound in the cleft of lysozyme, where it can be detected by the difference Fourier method (see Stryer, p. 204).

3. (a) If the P—O bond is broken, the label would be incorporated into inorganic phosphate.
   (b) If the C—O bond is broken, the label would be incorporated into glucose.
   (c) Since the label is incorporated exclusively into inorganic phosphate, the P—O bond must have been broken. Such cleavage is typical of reactions involving phosphatases and is therefore called *phosphatase-type cleavage*.

4. (a) For Asp,

$$pH = pK + \log \frac{[A^-]}{[HA]}$$

$$5.0 = 3.9 + \log \frac{[COO^-]}{[COOH]}$$

$$1.1 = \log \frac{[COO^-]}{[COOH]}$$

$$12.6 = \frac{[COO^-]}{[COOH]}$$

The percentage of groups in the —COO$^-$ form is

$$\frac{12.6}{13.6} \times 100 = 92.6\%$$

(b) For Glu,

$$pH = pK + \log \frac{[A^-]}{[HA]}$$

$$5.0 = 4.3 + \log \frac{[COO^-]}{[COOH]}$$

$$0.7 = \log \frac{[COO^-]}{[COOH]}$$

$$5 = \frac{[COO^-]}{[COOH]}$$

The percentage of groups in the —COOH form is

$$\frac{1}{6} \times 100 = 16.7\%$$

(c) Because Glu 35 is present in a hydrophobic region of lysozyme, the attraction between H$^+$ and —COO$^-$ will be significantly greater than when Glu is free in solution, and the pK for the group will be considerably greater than 4.3. As a result, the percentage of —COOH groups will be greater than calculated in part (b) of this problem. (See Stryer, p. 7, for a discussion of the effects of dielectric constants on ionic equilibria.)

5. Carbonium ions are planar, having sp^2 hybridization. A carbonium ion can only be formed if the C–O bond involving the C-1 of residue D is broken. If the C–O bond involving the C-4 of residue E were cleaved, a planar intermediate could not be involved.

6. Evidence for the theory comes from the experimental observation that hexa-NAG fits best into the active-site cleft of lysozyme when sugar residue D is distorted into a half-chair conformation. The half-chair conformation facilitates the formation of a planar carbonium ion involving the C-1 of sugar residue D. (See Stryer, p. 209, and Problem 5 here.) This is substantiated by experiments showing that sugar residue D contributes negatively to the standard free energy of binding of the substrate. (See Stryer, Figure 9-16, p. 209.)

7. The points of cleavage are indicated by the arrows:

$$pApUpGpUpCpGpGp$$

The products would be pApUp, GpUp, Cp, and GpGp.

8. The products would be Up and Gp. The ^{18}O would appear in Up. The bond that is broken is that between P and the 5'-O atom of G. See Figure 9-3.

9. Because of its ring structure, proline cannot be accommodated in the substrate binding sites of trypsin, chymotrypsin, and carboxypeptidase A. Therefore, these proteases do not cleave peptide bonds involving proline.

10. Chymotrypsin would produce the following four fragments:

Lys-Gly-Phe   Thr-Tyr-Pro-Asn-Trp   Ser-Tyr   Phe

The Tyr-Pro bond is not cleaved. (See Problem 9.)

11. The competitive inhibitor β-phenylpropionate and the affinity-labeling reagent TPCK both bind at the same site on the enzyme, namely, the substrate binding site, so the binding of the competitive inhibitor would prevent the binding of TPCK. Noncompetitive inhibitors combine with chymotrypsin at a site on the enzyme away from the substrate-binding site. Therefore, noncompetitive inhibitors do not prevent the binding of TPCK to chymotrypsin.

12. When the substrate occupies a cleft in the enzyme, the weak bonds that it forms with groups on the enzyme help to stabilize the tertiary structure of the enzyme. This helps to protect the enzyme against thermal denaturation.

13. (a) Ser is involved. DIFP specifically combines with serine.
    (b) Step 1:

$$E\text{-Ser-OH} + acetylcholine \rightleftharpoons E\text{-Ser-O}\overset{\overset{\textstyle O}{\|}}{C}CH_3 + choline$$

Step 2:

$$E\text{-Ser-O}\overset{\overset{\textstyle O}{\|}}{C}CH_3 + H_2O \rightleftharpoons E\text{-Ser-OH} + acetate$$

14. (a) Step 1:

$$S_2O_3^{2-} + E \rightleftharpoons E\text{-S} + SO_3^-$$

Step 2:

$$E\text{-S} + CN^- \rightleftharpoons E + SCN^-$$

(b) Yes, it does support the hypothesis of electrostatic interaction. If an interaction between a negatively charged thiosulfate and a positively charged functional group on the enzyme occurred, the addition of a neutral salt would weaken that interaction by competition.

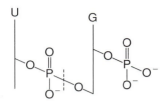

Figure 9-3
Bond cleaved by pancreatic ribonuclease.

CHAPTER 10

Control of Enzymatic Activity

The theme of Chapter 10 is the regulation of enzyme activity. Three of the major types of regulatory mechanisms are discussed in detail: allosteric control, proteolytic activation, and stimulation or inhibition by control proteins. Stryer uses specific examples to illustrate the general structure-function relationships involved in these control mechanisms. To illuminate allosteric control, Stryer discusses *E. coli* aspartate transcarbamoylase, a well-understood enzyme. You were introduced to allosteric interactions in Chapter 7, where oxygen binding by hemoglobin and its regulation by pH, CO_2, and 2,3-bisphosphoglycerate were described. At the end of the section on allosteric control, Stryer presents two current theoretical models of allosteric regulation. Next, Stryer turns to the activation of digestive enzymes. He describes the proteolytic steps and conformational rearrangements that produce the active forms of chymotrypsin, trypsin, and pepsin from their inactive zymogens. The mechanisms of action of the digestive enzymes were presented in Chapter 9. Stryer concludes Chapter 10 with a discussion of the blood clotting cascade—the series of proteolytic activations of clotting factors that lead to the formation of fibrin clots. Several specific stimulating and inhibiting proteins are described in connection with the proteolytic enzymes.

When you have mastered this chapter, you should be able to complete the following objectives.

Introduction (Stryer pages 233–234)

1. List the four major regulatory mechanisms that control enzyme activity, and give examples of each.

Allosteric Interactions in Aspartate Transcarbamoylase (Stryer pages 234–239)

2. Describe the reaction catalyzed by *aspartate transcarbamoylase (ATCase)*, the regulation of ATCase by *CTP* and *ATP*, and the biological significance of this regulation.

3. Describe the subunit composition of ATCase. Explain the effects of subunit dissociation on the *allosteric behavior* of the enzyme.

4. Describe the three-dimensional structure of ATCase in terms of the subunit arrangement and the major features of its active site as revealed by the binding of *N-(phosphonacetyl)-L-aspartate (PALA)*.

5. Outline the structural effects of the binding of PALA, CTP, or ATP on ATCase.

Models for Allosteric Interactions (Stryer pages 239–243)

6. Give the major assumption of the *concerted model* for *allosteric interactions*.

7. Note the effects of substrates and *allosteric activators* or *inhibitors* on the $R \rightleftharpoons T$ conformational equilibrium.

8. Define *homotropic* and *heterotropic (positive* or *negative)* allosteric effects.

9. Give the major assumptions of the *sequential model* for allosteric interactions; contrast the assumptions for and the predictions from the concerted and sequential models.

10. Describe the experimental evidence for a concerted allosteric transition upon the binding of substrate analogs to ATCase.

Proteolytic Activation of Digestive Enzymes (Stryer pages 243–248)

11. Define *zymogen (proenzyme)* and *proprotein*. Give examples of enzymes and proteins that are derived from zymogens and proproteins.

12. Outline the pathway for the secretion of *pancreatic zymogens*.

13. Describe the activation of *chymotrypsinogen* and the general structural changes that lead to the formation of the active site.

14. Describe the activation of *trypsinogen* and *pepsinogen;* compare the conformational changes that occur during the activation of the three major pancreatic zymogens.

15. Summarize the enzymes and conditions required for the activation of all the *digestive enzymes*.

16. Explain how *trypsin* is inhibited by the *pancreatic trypsin inhibitor*.

17. Describe the role of α_1-*antitrypsin* in the inactivation of *elastase*. Relate it to the clinical condition called *emphysema*.

The Cascade of Zymogen Activations in Blood Clotting (Stryer pages 248–255)

18. Explain the biological significance of the *enzymatic cascade in clotting*.

19. Contrast the structure and properties of *fibrinogen* with those of *fibrin*.

20. Describe the stabilization of the *fibrin clot* by *Factor XIII$_a$*, an active *transamidase*. Explain the role of *thrombin* in the activation of fibrinogen and *Factor XIII*.

21. Compare the properties of thrombin with those of the *pancreatic serine proteases*.

22. Discuss the requirement for *vitamin K* in the synthesis of *prothrombin*. Outline the mechanism of prothrombin activation.

23. Describe the genetic defect in *hemophilia*. Explain how recombinant DNA technology has been used to produce human *Factor VIII (antihemophilic factor)*.

24. Summarize the roles of the various clotting factors in the *intrinsic*, *extrinsic*, and *common pathways* of clotting. List the factors that are serine proteases and those that contain γ-*carboxyglutamate domains*.

Inhibition of Clotting Factors (Stryer pages 255–256)

26. Describe the general mechanisms for the control of clotting.

27. Explain the specific role of *antithrombin III* in blocking the serine proteases in the clotting cascade. Note the effect of *heparin* on antithrombin III.

28. Describe the *lysis* of fibrin clots by *plasmin* and the activation of *plasminogen* by *tissue-type plasminogen activator (TPA)*.

SELF-TEST

Allosteric Interactions in Aspartate Transcarbamoylase

1. The dependence of the reaction velocity on the substrate concentration for an allosteric enzyme is shown in Figure 10-1 as curve A. A shift to curve B could be caused by the

    (a) addition of an irreversible inhibitor.
    (b) addition of an allosteric activator.
    (c) addition of an allosteric inhibitor.
    (d) dissociation of the enzyme into subunits.

2. In *E. coli*, ATCase is inhibited by CTP and is activated by ATP. Explain the biological significance of these effects.

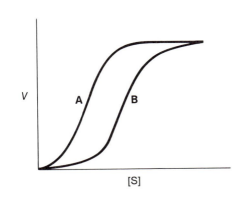

Figure 10-1
Allosteric enzyme kinetics.

3. Which of the following statements regarding the structure of ATCase in *E. coli* is *incorrect*?

(a) ATCase consists of two kinds of subunits and a total of twelve polypeptide chains.
(b) Desensitization by mercurials dissociates each ATCase into three r_2 and two c_3 subunits.
(c) ATCase has a threefold axis of symmetry and a large inner cavity.
(d) The active sites of ATCase are located at the interface between c and r subunits.
(e) The separate subunits r_2 and c_3 retain their respective ligand-binding capacitites.

4. Which of the following methods can provide information about the subunit dissociation of ATCase or the structural changes that occur when ATCase binds a substrate analog?

(a) X-ray crystallography
(b) Western blotting
(c) Sedimentation velocity ultracentrifugation
(d) SDS-polyacrylamide gel electrophoresis
(e) Gel-filtration chromatography

5. Explain why the reagent *N*-(phosphonacetyl)-L-aspartate (PALA) has been especially useful in the investigation of the properties of ATCase.

_____

_____

Models for Allosteric Interactions

6. If T (tense) and R (relaxed) are the two possible conformational states of the subunits of an allosteric enzyme in the concerted model, which of the following statements are true?

(a) Enzyme that is saturated with substrate is in the TT form.
(b) Allosteric activators preferentially bind to subunits in the R state.
(c) Subunits in the R and T states can coexist in a single molecule at low substrate concentrations.
(d) Allosteric inhibitors increase the proportion of enzyme molecules in the TT form relative to the RR form.

7. The allosteric effect of CTP on ATCase is called

(a) homotropic activation.
(b) homotropic inhibition.
(c) heterotropic activation.
(d) heterotropic inhibition.

8. Match the assumptions or predictions in the right column with the appropriate model for allosteric interactions in the left column.

(a) Concerted model _____
(b) Sequential model _____

(1) The R form is induced by substrate binding.
(2) The R and T forms are in equilibrium.
(3) Only RR and TT states exist.

(4) RR, RT, and TT states exist.

(5) The binding of substrate by one subunit always increases the binding affinity of the other.

(6) The binding of substrate by one subunit may increase or decrease the binding affinity of the other.

9. Propose an experiment to show whether the cooperative homotropic effects in hemoglobin are concerted or sequential.

_____

_____

Proteolytic Activation of Digestive Enzymes

10. Which of the following proteins are synthesized and secreted as zymogens or proproteins?

(a) Pepsin
(b) Lysozyme
(c) Collagen
(d) Enteropeptidase
(e) Thrombin

11. The pancreas is the source of the proteolytic enzyme trypsin. Which of the following are reasons trypsin does not digest the tissue in which it is produced?

(a) It is synthesized in the form of an inactive precursor that requires activation.

(b) It is stored in zymogen granules that are enclosed by a membrane.

(c) It is active only at the pH of the intestine, not at the pH of the pancreatic cells.

(d) It requires a specific noncatalytic modifier protein in order to become active.

12. Activation of chymotrypsinogen requires

(a) the cleavage of at least two peptide bonds by trypsin.

(b) structural rearrangements that complete the formation of the substrate cavity and the oxyanion hole.

(c) major structural rearrangements of the entire protein molecule.

(d) the concerted proteolytic action of trypsin and pepsin to give α-chymotrypsin.

13. Match the zymogens in the left column with the enzymes that participate directly in their activation or with their mechanism of activation, which are listed in the right column.

(a) Chymotrypsinogen _____

(b) Pepsinogen _____

(c) Trypsinogen _____

(d) Proelastase _____

(e) Procarboxypeptidase _____

(1) Intramolecular hydrolysis
(2) Trypsin
(3) Enteropeptidase
(4) Carboxypeptidase

14. Explain why the new carboxyl-terminal residues of the polypeptide chains produced during the activation of pancreatic zymogens are usually Arg or Lys.

_____

_____

15. The inactivation of trypsin by pancreatic trypsin inhibitor involves

    (a) an allosteric inhibition.
    (b) the covalent binding of a phosphate to the active site serine.
    (c) the facilitated self-digestion of the enzyme.
    (d) denaturation at the alkaline pH of the duodenum.
    (e) the nearly irreversible binding of the protein inhibitor at the active site.

16. Explain the effects of an α_1-antitrypsin deficiency on the lungs.

_____

_____

The Cascade of Zymogen Activations in Blood Clotting

17. Match fibrinogen and fibrin with the appropriate properties in the right column.

    (a) Fibrinogen _____   (1) is soluble in blood.
                                  (2) is insoluble in blood.
    (b) Fibrin _____   (3) forms ordered fibrous arrays.
                                  (4) contains α-helical coiled coils.
                                  (5) contains tryosine-O-sulfate groups.
                                  (6) has the lower pI of the two.
                                  (7) may be cross-linked by transamidase.

18. Which of the following amino acid residues are involved in the transamidase reaction catalyzed by Factor XIII_a.

    (a) Asparagine
    (b) Glutamine
    (c) Glutamate
    (d) Lysine
    (e) Arginine

19. Thrombin and trypsin are both serine proteases that are capable of cleaving the peptide bond on the carboxyl side of arginine; thrombin, however, is specific for Arg-Gly bonds. Describe briefly the similarities and differences in the active sites of these two enzymes.

_____

_____

20. Which of the following statements about prothrombin are _incorrect_?

    (a) It requires vitamin K for its synthesis.
    (b) It can be converted to thrombin by the decarboxylation of γ-carboxyglutamate residues.
    (c) It is activated by Factor IX_a and Factor VIII.

(d) It is anchored to platelet phosphlipid membranes through Ca^{2+} bridges.

(e) It is part of the common pathway of clotting.

21. Explain the role of the γ-carboxyglutamate residues found in clotting factors.

_____

_____

22. Examine the clotting pathways shown in Figure 10-2.

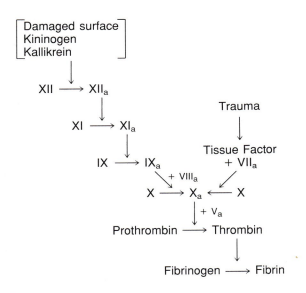

Figure 10-2
Enzymatic cascade in blood clotting.

(a) List the serine proteases:

_____

(b) Give the factors that are vitamin K dependent:

_____

(c) List the zymogens in the common pathway:

_____

(d) List the activating proteins:

_____

(e) Which is the antihemophilic factor? _____

(f) Which pathway is the intrinsic pathway?

_____

Inhibition of Clotting Factors

23. Which of the following mechanisms is _not_ involved in the control of the clotting process?

(a) The specific inhibition of fibrin formation by antielastase

(b) The degradation of factors V_a and VIII_a by protein C, which is in turn switched on by thrombin

(c) The dilution of clotting factors in the blood and their removal by the liver.

(d) The specific inhibition of thrombin by antithrombin III

24. Explain the effects of each of the following substances on blood coagulation or clot dissolution:

(a) Heparin: _____

_____

(b) Dicumarol: _____

_____

(c) Tissue-type plasminogen activator: _____

_____

_____

25. Which of the following statements about plasmin are *true?*

(a) It is a serine protease.
(b) It diffuses into clots.
(c) It cleaves fibrin at connector rod regions.
(d) It is inactivated by α_1 antitrypsin.
(e) It contains a "kringle" region in its structure for binding to clots.

PROBLEMS

1. (a) Plot the fractional saturation, Y, as a function of substrate concentration, [S], for a dimeric allosteric enzyme when L has each of the following values: 10^4, 10^3, and 0. Use the following substrate concentrations in each case: 0.0001 M, 0.0002 M, 0.0004 M, 0.001 M, 0.002 M, and 0.004 M. Assume that the Monod-Wyman-Changeux (MWC) model applies and that substrate does not bind to the T state. Use a value of 10^{-5} M for K_R. (See Stryer, p. 240.)
   (b) Describe in words the effect of decreasing values for L.

2. Show algebraically that when the allosteric constant, L, is equal to 0, the relationship between the substrate concentration and the velocity for an allosteric enzyme has the form of the Michaelis-Menten equation. Assume that the enzyme is dimeric, that the Monod-Wyman-Changeux model applies, and that the substrate binds only to the R form. (See Stryer, p. 240 and p. 189. Assume that equation 8 on p. 240 applies.)

3. What would be the kinetic consequences if a substrate were to have exactly equal affinities for the R form and the T form of an allosteric enzyme? (See the discussion of the Monod-Wyman-Changeux model on pp. 239–240 of Stryer.)

4. Figure 10-3 shows time courses for the activation of two zymogens, I and II. Which of the time courses most resembles that of the activation of trypsinogen, and which corresponds to the activation of chymotrypsinogen? Explain.

5. What is the significance of the fact that amino acid residues 2, 3, 4, and 5 of trypsinogen are all aspartate?

6. Trypsin has thirteen lysine and two arginine residues in its primary structure. Why does trypsin not cleave itself into sixteen smaller peptides?

7. Although thrombin has many properties in common with trypsin, the conversion of prothrombin to thrombin is not autocatalytic whereas the conversion of trypsinogen to trypsin is autocatalytic. Why is the conversion of prothrombin to thrombin not autocatalytic?

8. A patient suffering from a very rare blood disorder has shown a propensity for thrombosis since late childhood. Normal levels of plasminogen were detected in the plasma of the patient by immunoreactive techniques. Give two possible mechanisms that might account for the defect in this patient.

9. Because many clotting factors are present in blood in small concentrations, direct chemical measurements often cannot be used to determine whether the factors are within normal concentration ranges or are deficient. Once a deficiency has been established, however, plasma from the affected individual can be used to screen for the presence of the deficiency in other individuals. A rare deficiency in Factor XII leads to a prolongation of clotting time. Assuming that you have plasma from an individual in which this deficiency has been established, design a test that might help determine whether another individual has a Factor XII deficiency.

10. Aspartate transcarbamoylase catalyzes the first step in the biosynthetic pathway leading to the synthesis of cytidine triphosphate (CTP). (See Stryer, p. 235.) CTP serves as an allosteric inhibitor of aspartate transcarbamoylase that shuts off the biosynthetic pathway when the cell has an ample supply of CTP. Although the first step in a pathway may often be the principal regulatory step, such is not always the case. Figure 10-4 shows a hypothetical degradative metabolic pathway in which step 3 is the principal regulatory step. In this pathway what advantage does regulation at step 3 have over regulation at step 1 or 2?

11. In general, regulatory enzymes catalyze reactions that are irreversible in cells, that is, reactions that are far from equilibrium. Why must this be the case?

12. Amplification cascades, such as the one involved in blood clotting, are important in a number of regulatory processes. Figure 10-5 shows a hypothetical cascade involving conversions between inactive and active forms of enzymes. Active enzyme A serves as a catalyst for the activation of enzyme B. Active B in turn activates C, and so forth. Assume that each enzyme in the pathway has a turnover number of 10^3. How many molecules of enzyme D will be activated per unit time when one molecule of active enzyme A is produced per unit time?

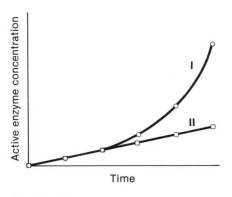

Figure 10-3
Time courses for the activation of two zymogens, I and II.

Figure 10-4
A hypothetical metabolic pathway.

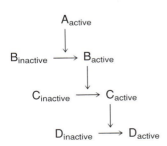

Figure 10-5
A hypothetical regulatory cascade.

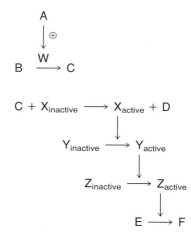

Figure 10-6
Another hypothetical regulatory
cascade.

13. Another hypothetical scheme for an amplification cascade is shown in Figure 10-6. The letters W, X, Y, and Z represent enzymes, whereas the letters A, B, C, D, E, and F represent metabolites. Compound A is an allosteric activator of enzyme W, which converts B to C. Compound C in turn combines stoichiometrically with the inactive form of enzyme X to give active X and compound D. Active enzyme X activates enzyme Y, which in turn activates Z. Enzyme Z catalyzes the conversion of compound E to compound F. Assuming that each enzyme in the scheme has a turnover number of 10^3, calculate the number of molecules of E that are converted to F per unit time when one molecule of A is added to the system.

ANSWERS TO SELF-TEST

Allosteric Interactions in Aspartate Transcarbamoylase

1. c

2. The activation of ATCase by ATP occurs when metabolic energy is available for DNA replication and the synthesis of pyrimidine nucleotides. Feedback inhibition by CTP prevents the overproduction of pyrimidine nucleotides and the waste of precursors.

3. d

4. a, c, e

5. PALA is a bisubstrate analog; that is, it resembles a combination of both substrates, and it is a transition-state analog for the carbamoyl phosphate–aspartate complex during catalysis by ATCase. X-ray diffraction analysis of ATCase with bound PALA has revealed the location of the active site and interactions that occur within it. In addition, comparisons of structures with and without PALA have indicated the large structural changes that ATCase undergoes upon binding substrates.

Models for Allosteric Interactions

6. b, d

7. d

8. (a) 2, 3, 5 (b) 1, 4, 6

9. Both the pH and the fractional occupancy (Y) of hemes by O_2 in hemoglobin could be determined as functions of the partial pressure of O_2. Because of the structural consequences of the Bohr effect, the pH would indicate the fraction of hemoglobin molecules in the R state (f_R). If the change in f_R precedes the change in Y, then the change from the T state to the R state is concerted.

10. a, c, e

11. a, b

12. b

13. (a) 2 (b) 1 (c) 3, 2 (d) 2 (e) 2

14. Because trypsin is the common activator of the pancreatic zymogens, its specificity for Arg-X and Lys-X peptide bonds will produce Arg and Lys carboxyl-terminal residues.

15. e

16. α_1-Antitrypsin specifically binds to and inactivates elastase. When the activity of elastase is not effectively controlled, it can destroy the elastic proteins of the alveoli of the lungs, leading to emphysema.

The Cascade of Zymogen Activations in Blood Clotting

17. (a) 1, 4, 5, 6 (b) 2, 3, 4, 7. Fibrinogen has a lower pI than fibrin because fibrinogen contains many more negatively charged amino acid residues than does fibrin.

18. b, d

19. Because both thrombin and trypsin are serine proteases, they both have an oxyanion hole and a catalytic triad at the active site. Also, the substrate-specificity sites of both have a similar, negatively charged pocket capable of binding Arg. However, thrombin probably has just enough space to accommodate a Gly residue next to the Arg binding site in contrast to trypsin which has no restrictions as to the amino acid residue that can be accommodated at the corresponding position.

20. b, c

21. The γ-carboxyglutamate residues are effective chelators of Ca^{2+}. This Ca^{2+} is the electrostatic anchor that binds the protein to a phospholipid membrane, thereby bringing interdependent clotting factors into close proximity.

22. (a) Kallikrein, XII_a, XI_a, IX_a, VII_a, X_a, and thrombin
    (b) Prothrombin, VII, IX, and X
    (c) X, prothrombin, and fibrinogen
    (d) VIII_a, V_a, and tissue factor
    (e) VIII
    (f) The intrinsic pathway starts with a damaged surface + kininogen + kallikrein and ends with IX_a + VIII_a.

Inhibition of Clotting Factors

23. a

24. (a) Heparin enhances the inhibitory action of antithrombin III.
    (b) Dicumarol is a vitamin K analog, and as such, it interferes with the synthesis of the factors that contain γ-carboxyglutamate residues.
    (c) Tissue-type plasminogen activator facilitates clot dissolution by converting plasminogen into plasmin directly on the clot.

25. a, b, c

ANSWERS TO PROBLEMS

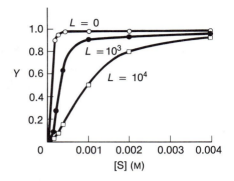

Figure 10-7
The fractional saturation, Y, as a function of substrate concentration, [S].

1. (a) The applicable equation is equation 7 on page 240 of Stryer:

$$Y = \left(\frac{[S]}{K_R}\right)\frac{1 + [S]/K_R}{L + (1 + [S]/K_R)^2}$$

Substitution of the given values results in the plots shown in Figure 10-7.

   (b) As the value of L decreases, the plot becomes less sigmoidal. When $L = O$, the plot is hyperbolic.

2. The velocity is given by equation 8 on page 240 of Stryer:

$$V = YV_{max}$$

The fractional saturation Y, is given by equation 7 on page 240:

$$Y = \left(\frac{[S]}{K_R}\right)\frac{1 + [S]/K_R}{L + (1 + [S]/K_R)^2}$$

Substituting equation 7 into equation 8 gives

$$V = V_{max}\left(\frac{[S]}{K_R}\right)\frac{1 + [S]/K_R}{L + (1 + [S]/K_R)^2}$$

When $L = 0$, this equation becomes

$$V = V_{max}\left(\frac{[S]}{K_R}\right)\frac{1 + [S]/K_R}{(1 + [S]/K_R)^2}$$

$$= V_{max}\left(\frac{[S]}{K_R}\right)\frac{1}{(1 + [S]/K_R)}$$

$$= V_{max}\frac{[S]/K_R}{(1 + [S]/K_R)}$$

Multiplying the numerator and denominator by $1/K_R$ gives

$$V = \frac{V_{max}[S]}{[S] + K_R}$$

This equation has the same form as the Michaelis-Menten equation (equation 28 on p. 189 of Stryer). The only difference is that K_M is replaced by K_R.

3. If a substrate were to have equal affinities for the R and T forms, the forms would be indistinguishable kinetically, and the system would behave as if all the enzyme were present in a single form. Thus, Michaelis-Menten kinetics would apply, and a plot of the reaction velocity versus the substrate concentration would be hyperbolic.

4. Curve I corresponds to the activation of trypsinogen, a process that is autocatalytic. As the process occurs, trypsin is produced, which can then cleave yet more trypsinogen. Curve II corresponds to the activation of chymotrypsinogen. The activation of chymotrypsinogen is not autocatalytic. Rather, tryspin catalyzes the conversion of chymotrypsinogen to active π-chymotrypsin. Therefore, its time course is initially linear.

5. The conversion of trypsinogen to trypsin involves the removal of an N-terminal hexapeptide (see Stryer, p. 246). The run of polar aspartate residues at positions 2 through 5 of trypsinogen insures that the N-terminus of the polypeptide is at the surface of the zymogen in a nonhelical region where it may be cleaved by either enteropeptidase or by trypsin.

6. The lysine and arginine residues must be partially buried and inaccessible to the active site of trypsin.

7. Thrombin specifically cleaves Arg-Gly bonds. The two bonds that are broken when prothrombin is converted to thrombin are Arg-Thr and Arg-Ile. Therefore, the conversion cannot be autocatalytic.

8. Because the patient has a tendency to form thrombi (blood clots), a defect in fibrinolysis is likely. The fact that normal levels of plasminogen are present suggests either (1) that there is a defect in the conversion of plasminogen to plasmin or (2) that the plasmin that is produced is enzymatically inactive and fails to lyse fibrinogen. In the actual case cited, the second possibility turned out to be the correct one; however, the data presented in the problem are not sufficient for discrimination between the two alternatives.

9. Prepare two samples of blood from the individual to be tested. Add normal plasma to one sample and Factor XII deficient plasma to another. If clotting time is restored to normal in both samples, Factor XII deficiency is probably not involved. If the addition of normal plasma restores normal clotting time but the addition of Factor XII deficient plasma does not, then a Factor XII deficiency must be suspected.

10. The pathway shown in Figure 10-4 is branched. If regulation were to occur at step 1 only, there would be no control over the production of X from B. If only step 2 were regulated, there would be no regulation over the production of X from A. Regulation at step 3 provides control of the amount of X produced from both A and B. In branched pathways, the principal regulatory step is usually after the branch point.

11. Suppose that a reaction is at equilibrium. If the enzyme catalyzing that reaction were made more active, nothing would happen. The reaction would still be at equilibrium. If, on the other hand, the reaction is displaced far from equilibrium and the enzyme catalyzing the reaction is made more active, more product will be produced. Thus, a regulatory enzyme must catalyze an irreversible step if it is to increase the flux rate through a pathway when it is allosterically activated.

12. One molecule of active A will lead to the activation of 10^9 molecules of enzyme D per unit time. Active A will produce 10^3 molecules of active B. Each of the 10^3 molecules of active B will activate 10^3 molecules of C per unit time. Since there are 10^3 molecules of B, this gives a total of 10^6 molecules of active C. Similar reasoning leads to the answer of 10^9 molecules of active D.

13. One molecule of A leads to the production of 10^{12} molecules of F per unit time. There are 10^3 molecules of C produced when one molecule of A is added to the system. For each molecule of C, there is one molecule of active enzyme X produced since the interaction of C with X is stoichiometric and not catalytic. Since there are 10^3 molecules of C, there will be 10^3 molecules of active X, leading to 10^6 molecules of active Y, 10^9 molecules of active Z, and 10^{12} molecules of F.

CHAPTER 11

Connective-Tissue Proteins

In this chapter, Stryer describes proteins that serve structural purposes. He begins with collagen, the most abundant family of proteins in mammalian tissues, stressing the basic structural features of collagen and explaining how they are modified to serve diverse structural functions in mature and developing tissues. In the last part of the chapter Stryer briefly describes the structures, properties, and biological roles of three other extracellular macromolecules—elastin, the proteoglycans, and fibronectin. Chapters 2 and 10 contain important background material for this chapter because of its emphasis on protein structure and the parallels that exist between the maturation of collagen and the formation of fibrin clots.

LEARNING OBJECTIVES

When you have mastered this chapter, you should be able to complete the following objectives.

The Basic Structural Features of Collagen (Stryer pages 261–268)

1. Describe the subunit composition of *tropocollagen* in different types of *collagen*.

2. List the salient characteristics of the primary structure of collagen.

3. Describe the *hydroxylation* of *proline* and *lysine* residues in collagen, and note the *glycosylation* of *hydroxylysine* residues.

4. Describe the *triple-stranded collagen helix* and contrast it with the α-helix. Explain the role of *glycine* residues in the triple helix.

5. Explain the significance of the *melting temperature*, T_m, of tropocollagen and the *shrinkage temperature*, T_s, of collagen fibers.

6. Discuss the biochemical basis of *scurvy*.

Maturation of Collagen (Stryer pages 268–274)

7. Describe the extracellular conversion of *procollagen* into tropocollagen.

8. Describe the assembly of tropocollagen in *collagen fibers*.

9. Explain the nature and the role of the *cross-links* in collagen fibers

10. Outline the steps leading from the synthesis of polypeptide chains to the formation of mature collagen fibers. Compare the overall process with the formation of fibrin clots.

11. Describe the structures of *collagen genes*, and relate them to the functional domains of the mature protein. Explain the diversity of the gene structures.

12. Explain the role of *collagenases* in the degradation of collagen.

Elastin (Stryer pages 274–275)

13. Describe the biological role of *elastin* and its tissue localization. Compare it with collagen.

14. Summarize the structural characteristics of elastin: its amino acid composition, the nature of the cross-links between its polypeptide chains, and its secondary structure.

Proteoglycans (Stryer pages 275–277)

15. Describe the roles of *proteoglycans* in the extracellular medium of connective tissue and in the joints.

16. List the major kinds of *glycosaminoglycans* and give their basic sugar composition.

17. Discuss the general structural organization of the proteoglycan from cartilage.

Fibronectin (Stryer pages 277–278)

18. Describe the domain structure of *fibronectin*.

19. Explain the role of fibronectin in blood clotting and in cell migration.

20. Outline the organization of the *fibronectin gene* and explain how different fibronectin domains are expressed.

The Basic Structural Features of Collagen

1. Which of the following are properties of the primary structure of collagen?

    (a) One-third of the amino acids are hydroxyprolines.
    (b) Hydroxyproline and hydroxylysine residues are present.
    (c) The sequence glycine-proline-hydroxylysine recurs frequently.
    (d) Proline residues are more abundant in collagen than in most other proteins.
    (e) The sugars glucose and galactose are often found attached to serine residues.

2. The hydroxylation of proline and lysine in tropocollagen polypeptide chains *does not* require which of the following?

    (a) O_2
    (b) Specific dioxygenases
    (c) Ascorbate
    (d) Pyridoxal phosphate
    (e) α-Ketoglutarate

3. Match the collagen helix and the α helix with their corresponding features from the right column.

    (a) Collagen helix _____
    (b) α helix _____

    (1) Single polypeptide
    (2) Three polypeptides
    (3) Rise per residue of 2.9 Å
    (4) Rise per residue of 1.5 Å
    (5) Hydrogen bonds parallel to the helix axis
    (6) Hydrogen bonds perpendicular to the helix axis
    (7) Side chains that participate in hydrogen bonds

4. The collagen triple-helix structure is characterized by extensive sequences of $(Gly-X-Pro)_n$ or $(Gly-X-Hyp)_n$, where X is any amino acid.

    (a) Why must every third residue be glycine?

    _____

    _____

    (b) What are the principal bonds that hold the three helices of the superhelix together?

    _____

    _____

5. Why does hydroxylation increase the stability of the collagen triple helix?

    (a) It promotes hydrogen bonding with water.
    (b) It increases hydrogen bonding between polypeptide chains.

(c) It expands the helix and allows the glycine residues to fit better in the interior.

(d) It decreases the melting temperature of tropocollagen.

(e) It neutralizes the charge on lysine residues.

Maturation of Collagen

6. If 2 moles of the α1 chain of tropocollagen are mixed with 1 mole of the α2 chain in physiological buffer, they do not self-assemble to form tropocollagen efficiently. Why not? How is this related to the type VII Ehlers-Danlos syndrome?

_____ _____

_____

7. Bone formation involves which of the following?

(a) Tropocollagen precipitation by Ca^{2+}

(b) Preformed collagen fibers

(c) Nucleation of calcium phosphate crystals at regular intervals

(d) Nucleation of calcium phosphate crystals at random sites on the fibers

(e) Disulfide bond formation between tropocollagen units

8. The formation of the intermolecular cross-links in collagen fibers involves which of the following?

(a) Hydroxylysine residues

(b) Hydroxyproline residues

(c) Lysine residues

(d) Lysyl hydroxylase

(e) Lysyl oxidase

9. Starving people have sometimes eaten certain legumes that have appreciable quantities of β-aminopropionitrile. How do you think the prolonged ingestion of these foods would affect the collagen of these people?

_____

_____

10. Match the selected steps in the synthesis and maturation of collagen in the left column with the appropriate items in the right column.

(a) Hydroxylation and glycosylation _____

(b) Hydrolysis of propeptides _____

(c) Cross-linking of collagen fibers _____

(1) Intracellular
(2) Extracellular
(3) Procollagen peptidases
(4) Lysyl oxidase
(5) Prolyl hydroxylase
(6) Galactosyl transferase
(7) Lathyrism
(8) Scurvy
(9) Type VII Ehlers-Danlos syndrome

11. Which of the following statements about collagen genes is *incorrect*?

(a) All the collagen genes in an organism are identical; the diversity of collagens arises only from post-translational modifications.

(b) Each gene contains about 50 exons.

(c) Each gene likely evolved by duplication of an ancestral 54-bp exon.

(d) Each gene encodes repeated Gly-X-Y sequences that correspond to the helical regions of collagen.

12. Which of the following statements about the properties shared by collagen and fibrin are *not* correct?

(a) Both are synthesized as proproteins.

(b) Both are stabilized by the same type of covalent cross-links.

(c) Both can be enzymatically digested by specific proteases.

(d) Both are water-insoluble supramolecular assemblies.

(e) Both have unusual repeating amino acid sequences.

13. Which of the following statements about tissue collagenases are correct?

(a) They are synthesized as proenzymes.

(b) They have the same specificity as thrombin.

(c) They can cleave through a collagen triple helix.

(d) They are serine proteases.

(e) They are active in tissues that are undergoing rapid reorganization.

14. Vertebrate collagenases are metalloproteases containing Zn^{2+}. Suggest a possible role for Zn^{2+} in the active site of these enzymes, using carboxypeptidase as a model.

_____

_____

Elastin

15. Which of the following are properties of elastin?

(a) High "stretchability"

(b) High hydroxylysine content

(c) High aliphatic side-chain amino acid content

(d) Cross-linking through complex lysine derivatives

(e) Organization into triple helices

16. Considering the properties of collagen and elastin, which would you expect to find in large amounts in the following tissues? Use *A* for collagen, *B* for elastin, and *C* for neither.

(a) Tendon _____

(b) Liver _____

(c) Ligament _____

(d) Aorta _____

(e) Bone _____

17. The cross-linking of elastin by lysyl oxidase involves which of the following?

(a) Two lysines to give lysinonorleucine cross-links.

(b) Two lysines to give bislysine cross-links.

(c) Two hydroxylysines and one lysine to give hydroxypyridinium cross-links.

(d) Four lysines to give desmosine cross-links.

18. Suggest how the structural features of elastin may be linked to its ability to stretch and to return to its original size and shape.

_____

_____

_____

Proteoglycans

19. Which of the following statements about glycosaminoglycans are *true*?

    (a) They contain derivatives of either glucosamine or galactosamine.
    (b) They constitute 5% of the weight of proteoglycans.
    (c) They contain positively charged substituent groups.
    (d) They bind substantial amounts of water and cations.
    (e) They have repeating units of four sugar groups.

20. Which of the following statements about the proteoglycan from cartilage are *true*?

    (a) It contains heparin sulfate.
    (b) It contains chondroitin sulfate.
    (c) It has a core protein with hyaluronate filaments bound to it.
    (d) It contains two kinds of proteins: a core protein and a link protein.
    (e) It has a mass of about 2×10^6 daltons.

Fibronectin

21. Fibronectin may have binding domains for which of the following?

    (a) Elastin
    (b) Collagen
    (c) Heparin
    (d) Thrombin
    (e) Fibrin

22. In contrast to the fibronectin secreted by fibroblasts, the fibronectin secreted by liver cells lacks a cell binding domain. Explain the biochemical basis of this difference.

_____

_____

PROBLEMS

1. How many stereoisomers may exist for 4-hydroxyproline? How many are found in proteins? Explain your answer. (Refer to Figure 11-2, p. 263 of Stryer.)

2. Would you expect prolyl hydroxylase to hydroxylate free proline? Why or why not?

3. What role do you think that iron likely plays in the hydroxylation of proline and lysine? Explain your answer with reference to the role of iron in hemoglobin and myoglobin.

4. What agent has the same role in human red blood cells that ascorbate has in fibroblasts?

5. Suppose that a portion of a nascent polypeptide chain of collagen has the following amino acid sequence:

-Gly-Pro-Lys-Gly-Pro-Pro-Gly-Ala-Ser-Gly-Lys-Asn-
  1   2   3   4   5   6   7   8   9 10 11 12

(a) Which of the proline residues might subsequently be converted to 4-hydroxyproline? Explain.
(b) Which might be converted to 3-hydroxyproline? Explain.
(c) Which lysine residue might subsequently be hydroxylated? Explain.
(d) Which of the amino acid residues might subsequently have a carbohydrate unit attached?

6. Explain why glycine is the most important amino acid residue that contributes to the interhelical hydrogen bonds that help to stabilize the structure of the collagen triple helix.

7. Suppose that a mutant gene leads to the synthesis of an aberrant $\alpha1(I)$ gene of collagen, and that the aberrant $\alpha1(I)$ gene, designated $\alpha1(I)^M$, is dominant to the normal allele, $\alpha1(I)^+$. If twice the number of normal collagen $\alpha1(I)$ chains as abnormal $\alpha1(I)$ chains are produced, what fraction of procollagen molecules would be spoiled and what fraction would be normal? What about when normal and abnormal $\alpha1(I)$ chains are produced in equal amounts? (See Stryer, p. 266.)

8. Studies of the melting temperature of synthetic polypeptides that form a collagen-like triple helix show that the helix is markedly stabilized by the hydroxylation of proline (see Stryer, p. 267). Why do you think this is the case?

9. Suppose that four kinds of synthetic helical collagen were prepared containing the repeating triplet sequences that follow. Predict the relative thermal stability of helices containing each of the four sequences, giving the most stable first.

(a) $[Gly-Pro-Y]_n$
(b) $[Gly-Pro-Hyp]_n$
(c) $[Gly-X-Y]_n$
(d) $[Gly-Pro-Pro]_n$

10. Indicate with slash marks the sites where the following peptide will be cleaved by collagenase from *Clostridium histolyticum*.

Gly-Ala-Ser-Gly-Pro-Met-Gly-Pro-Arg-Gly-Pro-Hyp

11. Heparin is frequently used as an anticoagulant. Would you also expect keratan sulfate to be effective in that role? Explain.

12. Arrange the following in order of decreasing axial distance between the α-carbon atoms of adjacent amino acid residues: (a) α helix, (b) β pleated sheet, (c) collagen triple helix.

13. Nascent collagen peptides treated with a chemical reducing agent soon after being biosynthesized form helices less readily than do untreated collagen peptides. Propose an explanation for this observation.

14. The rate of collagen turnover in the body is frequently assessed clinically by measuring the concentration of hydroxyproline in the urine.

  (a) Give two reasons why hydroxyproline is a better choice than proline for these purposes.
  (b) Why might this procedure tend to underestimate the rate of collagen turnover?

ANSWERS TO SELF-TEST

The Basic Structural Unit of Collagen

1. b, d

2. d

3. (a) 2, 3, 6, 7 (b) 1, 4, 5

4. (a) Every third residue in each of the three helices falls in the interior of the superhelix, where there is no room for an amino acid residue larger than glycine, which has a single hydrogen atom on the α-carbon.
   (b) Each polypeptide folds into a helical structure, designated the type II *trans* helix. The amide hydrogens and the carbonyl oxygens of each peptide bond extend perpendicularly toward the helix axis and form hydrogen bonds with complementary groups on the adjacent helices.

5. b

Maturation of Collagen

6. The natural precursor of tropocollagen is procollagen. It is composed of two pro-α1 chains and one pro-α2 chain. Amino acid sequences at the ends of these chains specify their proper alignment, thereby allowing the triple helix to form efficiently. When these regions are absent, as in the α1 and α2 chains of tropocollagen, then the assembly of the triple helix is inefficient.

   In the type VII Ehlers-Danlos syndrome, the pro-α chains of procollagen are not efficiently cleaved to form their mature counterparts, and thus the formation of collagen fibers is impaired.

7. b, c

8. a, c, e

9. β-Aminopropionitrile is an irreversible inhibitor of lysyl oxidase, which converts lysine side chains in collagen to the aldehyde forms involved in cross-linking. Individuals ingesting the substance would have collagen that is insufficiently cross-linked and unstable. Thus, prolonged ingestion of β-aminopropionitrile would result in extensible skin as well as hypermobile joints, and in addition may lead to defective

skeletal development and increased urinary excretion of hydroxyproline peptides.

10. (a) 1, 5, 6, 8 (b) 2, 3, 9 (c) 2, 4, 7

11. a

12. b, e

13. a, c, e

14. By analogy with carboxypeptidase, another Zn^{2+} protease, the probable role of the Zn^{2+} in vertebrate collagenases is to contribute to the electronic strain in the substrate and thereby accelerate the hydrolysis of the peptide bond. (See Stryer, p. 219.)

Elastin

15. a, c, d

16. (a) A (b) C (c) B, A (d) B, A (e) A. The ligaments and the aorta contain large amounts of collagen, as well as elastin.

17. a, d

18. Elastin contains helical arrays of β-turns, known as the β-spiral, which could stretch. The lysinonorleucine and desmosine cross-links, which occur outside of the β-spiral regions, would allow the elastin fibers to return to their original size and shape.

Proteoglycans

19. a, d

20. b, d, e

Fibronectin

21. b, c, e

22. Fibronectin is encoded by a single gene, but different types of cells splice the nascent RNA transcript differently to express specific binding domains encoded in the exons. The splicing pattern in liver cells produces plasma fibronectin, which lacks a binding domain for integrin, a cell-surface protein. In contrast, fibroblast fibronectin has a binding domain for integrin.

ANSWERS TO PROBLEMS

1. Hydroxyproline has two asymmetric carbon atoms, at positions 2 and 4 of the ring; it therefore has four possible stereoisomers. Only one of the four is found in proteins, however. The α-carbon atom must have the L configuration, as is true of the α-carbon atoms of all amino acyl or imino acyl residues found in proteins. The carbon atom at position 4 must have the configuration indicated in Figure 11-3 on page 263 of Stryer. Notice that the hydroxylation of the proline residue is stereospecific and that the resulting —OH group extends in front of the ring.

2. Prolyl hydroxylase hydroxylates the prolyl residues of polypeptides. It does not work on free proline.

3. Iron likely binds the molecular oxygen that is used in these reactions. The clue to this role is the fact that iron must be maintained in the reduced (ferrous) state by ascorbate in order to be effective, just as the iron of hemoglobin and myoglobin must be maintained in the ferrous state to bind reversibly with oxygen (see Stryer, p. 144).

4. The agent that maintains iron in reduced form in human erythrocytes is reduced glutathione. (See the answer to Problem 2, Chapter 7, p. 89.)

5. (a) The proline at position 6 may subsequently be converted to 4-hydroxyproline. Proline residues must be on the amino side of a glycine residue in order to be converted to 4-hydroxyproline.

   (b) The proline at positions 2 and 5 may subsequently be converted to 3-hydroxyproline. Proline residues must be on the carboxyl side of a glycine residue in order to be converted to 3-hydroxyproline.

   (c) The lysine at position 3 may be subsequently hydroxylated. To be converted to hydroxylysine, a lysine residue must be on the amino side of a glycine residue.

   (d) Carbohydrate units attach to the 3-hydroxylysine that is formed at position 11 of the polypeptide.

6. Glycine is the most abundant amino acid residue that is capable of serving as a hydrogen bond donor. Every third residue of collagen is glycine. Remember that proline, which lacks a peptide hydrogen, cannot serve as a hydrogen bond donor.

7. The fraction of spoiled collagen is $\frac{5}{9}$, or 55.6%; and the fraction of normal collagen is $\frac{4}{9}$, or 44.4%. Two-thirds of the $\alpha1(I)$ chains are normal and one-third are mutant. Since procollagen contains two $\alpha1(I)$ chains, the possible combinations together with their frequencies are given by the expression:

$$[\tfrac{2}{3}\alpha1(I)^+ + \tfrac{1}{3}\alpha1(I)^M]^2 =$$
$$\tfrac{4}{9}\alpha1(I)^+\alpha1(I)^+ + \tfrac{2}{9}\alpha1(I)^+\alpha1(I)^M + \tfrac{2}{9}\alpha1(I)M+\alpha1(I)^+ + \tfrac{1}{9}\alpha1(I)^M\alpha1(I)^M$$

Type $\alpha1(I)^+\alpha1(I)^+$ will be normal. Types $\alpha1(I)^+\alpha1(I)^M$ and $\alpha1(I)^M\alpha1(I)^M$ will be mutant.

When normal and abnormal chains are produced in equal amounts, the fraction of spoiled collagen is $\frac{3}{4}$, or 75%. In individuals carrying one normal and one mutant gene, the normal product is sometimes produced in greater amounts to overcome the effect of the mutant.

8. The hydroxyl group of hydroxyproline adds another hydrogen-bond forming species to the structure of collagen, which allows for the formation of more hydrogen bonds. The —OH group is both a hydrogen bond donor and a hydrogen bond acceptor.

9. The order of relative stability is b > d > a > c. Hyp stabilizes the collagen helix more than Pro does, and Pro stabilizes it more than the amino acids X and Y do.

10. The peptide will be cleaved by collagenase as follows:

Gly-Ala-Ser | Gly-Pro-Met | Gly-Pro-Arg | Gly-Pro-Hyp

11. Heparin functions as an anticoagulant because its negatively charged groups (two per sugar residue) are effective chelators of calcium. Keratan sulfate has a lower density of negative charges (one per two sugar residues); hence, it is not effective as an anticoagulant.

12. The correct order is b, c, a. The greatest axial distance between adjacent α-carbon atoms is in the β pleated sheet, in which the polypeptide chain is almost fully extended, a distance of 0.35 nm. Adjacent α-carbon atoms in a collagen triple helix are 0.29 nm apart, whereas those in the α helix are 0.15 nm apart.

13. Nascent collagen contains cysteine residues in the carboxyl-terminal propeptide that form interchain disulfide bonds that are thought to be important in promoting the formation of the helix. The addition of a reducing agent prevents the formation of these disulfide bonds and thereby leads to a reduced rate of helix formation.

14. (a) Hydroxyproline is the better choice because (1) hydroxyproline is not found to any extent in any other protein, and (2) hydroxyproline is not recycled for the biosynthesis of new collagen. Rather, proline is incorporated into the polypeptides and is then converted to hydroxyproline.

    (b) The method tends to underestimate the rate of collagen turnover because some hydroxyproline is metabolized to other end products rather than being directly excreted.

CHAPTER 12

Introduction to
Biological Membranes

In this chapter, Stryer describes the composition, structural organization, and general functions of biological membranes. After listing the common features of biological membranes, Stryer introduces a new class of biomolecules, the lipids, in their roles as membrane components. The structures of membrane lipids and their organization into bilayers in water are described first. Next, Stryer turns to membrane proteins, the major functional constituents of biological membranes. He describes the arrangement of proteins and lipids in membranes and stresses the fluid nature of membranes. Stryer then discusses the erythrocyte membrane in some detail. He concludes the chapter with some comments on high-resolution analyses of the structures of a few membrane proteins. The most useful background material for this chapter can be found in the first three chapters: intermolecular forces in Chapter 1, protein structure in Chapter 2, and the methods used in the investigation of proteins in Chapter 3.

LEARNING OBJECTIVES

When you have mastered this chapter, you should be able to complete the following objectives.

Introduction (Stryer pages 283–284)

1. List the functions of *membranes*.

2. Describe the common features of biological membranes.

Membrane Lipids (Stryer pages 284–287)

3. Define *lipid*, and list the major kinds of *membrane lipids*.

4. Recognize the structures and the constituent parts of *phospholipids* (*phosphoglycerides* and *sphingomyelin*), *glycolipids*, and *cholesterol*.

5. Describe the general properties of the *fatty acid chains* found in phospholipids and glycolipids.

6. Draw the general chemical formula of a phosphoglyceride, and recognize the most common *alcohol moieties* of phosphoglycerides (e.g., *choline, ethanolamine,* and *glycerol*).

Lipid Bilayers (Stryer pages 288–292)

7. Describe the properties of an *amphipathic molecule* and the symbol used to represent it.

8. Distinguish between a *micelle* and a *lipid bilayer*.

9. Describe the *self-assembly process* for the formation of lipid bilayers. Note the stabilizing intermolecular forces.

10. Outline the methods used to prepare *lipid vesicles (liposomes)* and *planar bilayer membranes*. Point out some applications of these systems.

11. Explain the relationship between the *permeability coefficients* of small molecules and ions and their *solubility* in a nonpolar solvent relative to their solubility in water.

Organization and Dynamics of Membranes (Stryer pages 292–301)

12. Give examples of the compositional and functional varieties of biological membranes.

13. Distinguish between *peripheral* and *integral membrane proteins*. Describe the use of *freeze-fracture electron microscopy* in visualizing integral membrane proteins.

14. Describe the evidence for the *lateral diffusion* of membrane lipids and proteins. Contrast the rates for *lateral diffusion* with those for *transverse diffusion*.

15. Describe the features of the *fluid mosaic model* of biological membranes.

16. Discuss the origin and the significance of *membrane asymmetry*.

17. Explain the roles of the fatty acid chains of membrane lipids and cholesterol in controlling the *fluidity of membranes*.

18. Note the roles and location of the *carbohydrate components* of membranes.

19. Describe the *reconstitution* of functional membrane systems from purified components and explain the role of *detergents* in this process.

The Red-Cell Membrane Proteins and Other Membrane Proteins (Stryer pages 301–309)

20. Discuss the advantages of *erythrocytes* in studies of membranes. Give examples of the membrane proteins found in erythrocytes.

21. Describe the structure of *glycophorin A* and the properties of the *anion channel*. Explain how *transmembrane α helices* can be predicted from *hydropathy plots*.

22. Explain the role of *spectrin* in the formation of the *membrane skeleton* of an erythrocyte.

23. Discuss the use of electron-microscopic and x-ray analyses in obtaining high-resolution images of the structures of *bacteriorhodopsin* and the *photosynthetic reaction center*.

SELF-TEST

Introduction

1. Identify the functions in the right column that are associated with lipids or proteins, the major components of cell membranes.

(a) Lipids _____

(b) Proteins _____

(1) are specific receptors for external stimuli.
(2) form the permeability barrier.
(3) act as pumps and gates.
(4) determine the fluidity of membranes.
(5) carry out concerted enzymatic reactions.

2. Which of the following statements about biological membranes is *not* true?

(a) They contain carbohydrates that are covalently bound to proteins and lipids.
(b) They are very large, sheetlike structures with closed boundaries.
(c) They have symmetric structures because of the symmetric nature of lipid bilayers.
(d) They can be regarded as two-dimensional solutions of oriented proteins and lipids.
(e) They contain specific proteins that mediate their distinctive functions.

Membrane Lipids

3. Which of the following are membrane lipids?

   (a) Cholesterol

   (b) Glycerol

   (c) Phosphoglycerides

   (d) Choline

   (e) Cerebrosides

4. After examining the structural formulas of the four lipids in Figure 12-1, answer the following questions.

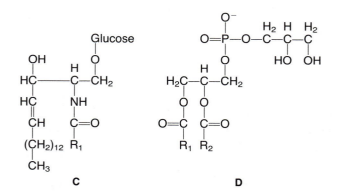

Figure 12-1
Membrane lipids. R_1 and R_2 represent hydrocarbon chains.

   (a) Which are phosphoglycerides?   _____

   (b) Which is a glycolipid?   _____

   (c) Which contain sphingosine?   _____

   (d) Which contain choline?   _____

   (e) Which contain glycerol?   _____

   (f) Name the lipids: A _____

                                        B _____

                                        C _____

                                        D _____

4. (g) Draw the structures of the components—that is, the possible hydrolysis products—of lipid D. Name the components and indicate how many moles of each are present per mole of lipid.

5. Calculate the molecular weight of a phosphatidyl choline with a palmitic acid esterified to the C-1 and an oleic acid esterified to the C-2 carbon of the glycerol backbone. Calculate the percentage of the molecular weight of the lipid that is due to the fatty acid chains.

Lipid Bilayers

6. The phosphoinositol portion of the phosphatidyl inositol molecule is called which of the following?

(a) The amphipathic moiety
(b) The hydrophobic moiety
(c) The hydrophilic moiety
(d) The micelle
(e) The polar head group

7. Which of the following are *not* features that a micelle and a lipid bilayer have in common?

(a) Both assemble spontaneously in water.
(b) Both are made up of amphipathic molecules.
(c) Both are very large, sheetlike structures.
(d) Both have the thickness of two constituent molecules in one of their dimensions.
(e) Both are stabilized by hydrophobic interactions, van der Waals forces, hydrogen bonds, and electrostatic interactions.

8. A triglyceride is a glycerol derivative that is similar to a phosphoglyceride, except that all three of its glycerol OH groups are esterified to fatty acid chains. Would you expect a triglyceride to form a lipid bilayer? Explain.

9. What is the volume of the inner water compartment of a liposome that has a diameter of 500 Å and a bilayer that is 40 Å thick?

    (a) 5.6×10^5 Å^3
    (b) 7.3×10^6 Å^3
    (c) 3.9×10^7 Å^3
    (d) 7.3×10^7 Å^3
    (e) 3.9×10^8 Å^3

10. What features of liposomes make them potentially useful as a delivery system for transporting water-soluble drugs to target cells. Suggest how one could prepare a liposome that is specific for a particular type of cell.

_____

_____

_____

11. Arrange the following in the order of decreasing permeability through a lipid bilayer

_____

    (a) Isoleucine
    (b) Tyrosine
    (c) H_2O
    (d) Na^+
    (e) O_2

Organization and Dynamics of Membranes

12. Show which of the properties listed on the right are characteristics of peripheral membrane proteins and which are characteristics of integral membrane proteins.

    (a) Peripheral _____

    (b) Integral _____

    (1) require detergents or organic-solvent treatment for dissociation from the membrane.
    (2) require mild salt or pH treatment for dissociation from the membrane.
    (3) appear in freeze-fracture planes as globular particles.
    (4) bind to the surface of membranes.
    (5) have transmembrane domains.

13. Which of the following statements about the diffusion of lipids and proteins in membranes is *not* true?

    (a) Many membrane proteins can diffuse rapidly in the plane of the membrane.
    (b) In general, lipids show a faster lateral diffusion than do proteins.
    (c) Membrane proteins do not diffuse across membranes at measurable rates.
    (d) Lipids diffuse across and in the plane of the membrane at equal rates.

14. Why do membrane proteins not diffuse, that is, flip-flop, across membranes?

_____

_____

15. In a membrane, an integral membrane protein diffuses laterally an average distance of 4×10^{-6} m in one minute, whereas a phospholipid molecule diffuses an average distance of $2 \ \mu$m in one second.

(a) Calculate the ratio of the diffusion rates for these two membrane components.

_____

(b) Explain the reason for the difference between the two rates.

_____

_____

16. Which of the following statements about the asymmetry of membranes are _true_?

(a) It is absolute for glycoproteins.
(b) It is absolute for phospholipids but is only partial for glycolipids.
(c) It arises during biosynthesis.
(d) It is structural but not functional.

17. If phosphoglyceride A has a higher T_m than phosphoglyceride B, which of the following differences between A and B may exist? (In each case only one parameter—either chain length or double bonds—is compared.)

(a) A has shorter fatty acid chains than B.
(b) A has longer fatty acid chains than B.
(c) A has more unsaturated fatty acid chains than B.
(d) A has more saturated fatty acid chains than B.
(e) A has _trans_ unsaturated fatty acid chains whereas B has _cis_ unsaturated fatty acid chains.

18. Explain the role of cholesterol in cell membranes.

_____

_____

19. Which of the following statements about the carbohydrate components of eucaryotic cell membranes are _true_?

(a) They make up approximately one-third of the weight of the membrane.
(b) They are exposed to the interior of the cell.
(c) They can be recognized by lectins.
(d) They bind to membrane proteins and lipids via hydrogen bonds.
(e) They contribute to and help maintain the asymmetry of the membranes.

20. During the solubilization of membranes, the purification of integral membrane proteins, and the reconstitution of membranes, gentler

detergents, such as octyl glucoside, are used in preference to sodium dodecyl sulfate (SDS). Explain why.

_____

_____

The Red-Cell Membrane Proteins and Other Membrane Proteins

21. Several properties of glycophorin A are listed in the left column. Match these properties with the methods listed in the right column that can be used to demonstrate them.

(a) It is an integral membrane protein _____

(b) It has one transmembrane α-helical domain of 19 amino acids _____

(c) It has a cytoplasmic C-terminus _____

(d) It is a glycoprotein _____

(e) It has a molecular weight lower than that of spectrin _____

(1) SDS polyacrylamide-gel electrophoresis
(2) Lectin binding
(3) Hydropathy plot
(4) Extraction with detergent
(5) PAS staining
(6) Binding of ferritin-labeled specific antibodies

22. Which of the following statements about the anion channel (the band 3 protein) is *incorrect*?

(a) It allows HCO_3^- to be exchanged for Cl^-.
(b) It comprises about a third of the total membrane protein.
(c) Its C-terminal half contains the single transmembrane helical domain.
(d) It has the peripheral protein ankyrin associated with its N-terminal domain.

23. Why is an α helix the preferred structure for transmembrane protein segments?

_____

_____

24. Spectrin

(a) is an integral membrane protein.
(b) is a glycoprotein.
(c) helps to maintain the shape and integrity of erythrocytes.
(d) binds to plasma membrane components through ankyrin and the band 4.1 protein.
(e) consists of two polypeptide chains, each of which is organized into multiple, triple-stranded, α-helical coiled coils.

25. Which of the following characteristics do bacteriorhodopsin and the photosynthetic reaction center have in common?

(a) A compact, essentially globular shape
(b) The same number of transmembrane segments
(c) A similar length for each transmembrane helix
(d) Similar sizes of the hydrophilic domains
(e) A similar orientation of the transmembrane helices with respect to the plane of the membrane

1. The ability of bacteria, yeasts, and fungi to convert aliphatic hydrocarbons to carbon dioxide and water has been studied intensively over the past decade because of concerns about the effects of crude oil spills on the environment. Microorganisms cannot survive when they are placed in high concentrations of crude oil or any of its components. However, they can utilize hydrocarbons very efficiently when they are placed in a medium in which an extensive lipid-water interface is created by agitation and aeration. Why?

2. Phytol, a long-chain alcohol, appears as an ester in plant chlorophyll. When consumed as part of the diet, phytol is converted to phytanic acid.

$$H_3C-\left(\underset{\substack{| \\ CH_3}}{CH}-CH_2-CH_2-CH_2\right)_3-\underset{\substack{| \\ H}}{\overset{\substack{CH_3 \\ |}}{C}}=C-CH_2OH$$

Phytol

$$H_3C-\left(\underset{\substack{| \\ CH_3}}{CH}-CH_2-CH_2-CH_2\right)_3-\underset{\substack{CH_3 \\ |}}{CH}-CH_2-\overset{\substack{O \\ \|}}{C}_{\diagdown O^-}$$

Phytanic Acid

Individuals who cannot oxidize phytanic acid suffer from a number of neurological disorders that together are known as Refsum's disease. The symptoms may be related to the fact that phytanic acid accumulates in the membranes of nerve cells. What general effects of phytanic acid on these membranes would be observed?

3. Bacterial mutants that are unable to synthesize fatty acids will incorporate fatty acids into their membranes when fatty acids are supplied in their growth medium. Suppose that each of two cultures contains a mixture of several types of straight-chain fatty acids, some saturated and some unsaturated, ranging in chain length from ten to twenty carbon atoms. If one culture is maintained at 18°C and the other is maintained at 40°C over several generations, what differences in the composition of the cell membranes of the two cultures would you expect to observe?

4. The formation of lipid bilayers in a mixture of water and phospholipids is a rapid and spontaneous process of self-assembly. For such a process to be spontaneous, the free-energy change must be negative ($\Delta G < 0$). Given that

$$\Delta G = \Delta H - T \, \Delta S$$

explain why the free-energy change for membrane assembly is negative.

5. Given two bilayer systems, one composed of phospholipids having saturated acyl chains 20 carbons in length and the other having acyl chains of the same length but with *cis* double bonds at C-5, C-8, C-11, and C-14, compare the effect of the acyl chains on T_m for each system.

H3C
H3C
CH3
CH3
H3C
H3C
H3C—CH
CH2
CH2
HC—OH
HO—CH
HC—OH
CH2
CH2OH

Bacteriohopanetetrol

6. Hopanoids are pentacyclic molecules that are found in bacteria and in some plants. A typical bacterial hopanoid, bacteriohopanetetrol, is shown in the margin. Compare the structure of this compound with that of cholesterol. What effect would you expect a hopanoid to have on a bacterial membrane?

7. As early as 1972, it was known that many biological membranes are asymmetric with respect to the distribution of phospholipids between the inner and outer leaflets of the bilayer. Once such asymmetry is established, what factors act to preserve it?

8. As shown on page 297 of Stryer's text, the transition between the rigid and liquid states for a particular phospholipid in water can be monitored by differential scanning calorimetry, which measures heat absorbed by a sample as it is warmed.

   (a) The value of T_m for a pure sample of phosphatidyl choline that contains two twelve-carbon fatty acyl chains is $-1°C$. Values for phosphatidyl choline species with longer acyl chains increase by about 20°C for each two-carbon unit added. Why?
   (b) Suppose you have a phosphatidyl choline species that has one palmitoyl group esterified to the glycerol moeity, as well as an oleoyl group esterified at C-2 of glycerol (see page 286 of Stryer's text). How would T_m for this species compare with that of dipalmitoylphosphatidyl choline, which contains two esterified palmitoyl groups?
   (c) Suppose you have a sample of sphingomyelin that has palmitate esterified to the sphingosine backbone. Compare the T_m for this phospholipid to that of dipalmitoylphosphatidyl choline.
   (d) The transition temperature for dipalmitoylphosphatidyl ethanolamine is 63°C. Suppose you have a sample of this phospholipid in excess water at 50°C, and you add cholesterol until it constitutes about 50% of the total lipid, by weight, in the sample. What would you expect when you attempt to determine the transition temperature for the mixture?

9. At least two segments of the polypeptide chain of a particular glycoprotein span the membrane of an erythrocyte. All the sugars in the glycoprotein are O-linked.

   (a) Which amino acids might be found in the portion of the chain that is buried in the lipid bilayer, and how many amino acids could those segments contain?
   (b) Why would you expect to find serine or threonine residues in the glycoprotein?
   (c) Describe a procedure that would enable you to determine on which side of the erythrocyte membrane the sugar residues are located.

ANSWERS TO SELF-TEST

Introduction

1. (a) 2, 4 (b) 1, 3, 5

2. c

Membrane Lipids

3. a, c, e

4. (a) A, D (b) C (c) B, C (d) A, B (e) A, D (f) A is phosphatidyl choline, B is sphingomyelin, C is cerebroside, and D is phosphatidyl glycerol

(g)

Fatty acids (2 moles)	Glycerol (2 moles)	Phosphate (1 mole)

5. The total molecular weight of the molecule is 759 daltons; the fatty acid chains account for 71% of the molecular weight. The molecule contains a total of 42 carbon atoms, 8 oxygen atoms, 82 hydrogen atoms, 1 phosphorus atom, and 1 nitrogen atom, giving a molecular weight of 759. The oleoyl chain has a molecular weight of 281 daltons and the palmitoyl chain has a molecular weight of 255 daltons. The percentage of the molecular weight of the phospholipid due to the fatty acids is therefore

$$\frac{281 + 255}{759} \times 100\% = 70.6\%$$

Lipid Bilayers

6. c, e

7. c

8. No. Although a triglyceride has hydrophobic fatty acyl chains attached to a glycerol backbone, it lacks a polar head group; therefore, it is not an amphipathic molecule and is incapable of forming a bilayer.

9. c. The radius of the water compartment is

$$\frac{500 \text{ Å} - (2 \times 40 \text{ Å})}{2} = 210 \text{ Å}$$

and the volume of a sphere is $\frac{4}{3}\pi r^3$.

10. Liposomes are essentially impermeable to water-soluble molecules. Therefore, water-soluble drugs could be trapped inside the liposomes and could then be delivered into the target cells by fusing the liposomes with the cell membrane. To make a liposome specific for a particular type of cell, antibodies that have been prepared against a surface protein of the target cell could be attached to the liposome via a covalent bond with a bilayer lipid, for example, phosphatidyl ethanolamine. This would enable the liposome to recognize the target cells. Of course, strategies would also have to be developed to prevent the premature, nonspecific fusion of the liposome with other cells.

11. e, c, a, b, and d. Though not discussed in the text, a lipid bilayer is even more permeable for O_2 than for water because O_2 does not have a dipole and is a smaller molecule. Also recall that Ile is more hydrophobic than Tyr (see Stryer p. 18 and Table 12-2).

Organization and Dynamics of Membranes

12. (a) 2, 4 (b) 1, 3, 5

13. d

14. Membrane proteins are very bulky molecules that contain numerous charged amino acid residues and polar sugar groups (in the case of glycoproteins) that are highly hydrated. Such molecules do not diffuse through the hydrophobic interior of the lipid bilayer.

15. (a) Rate of protein diffusion:

$$(4 \times 10^{-6} \text{ m/min})\frac{(1 \text{ min})}{(60 \text{ s})} = 6.7 \times 10^{-8} \text{ m/s}$$

Rate of phospholipid diffusion:

$$2 \ \mu\text{m/s} = 2 \times 10^{-6} \text{ m/s}$$

Ratio of phospholipid diffusion rate to protein diffusion rate:

$$\frac{2 \times 10^{-6} \text{ m/s}}{6.7 \times 10^{-8} \text{ m/s}} = 30$$

(b) The difference in diffusion rates is due primarily to the difference in size between phospholipids, which have a molecular weight of approximately 800, and proteins, which have a molecular weight greater than 10,000. In addition, integral membrane proteins may associate with peripheral proteins, which would further decrease their lateral diffusion.

16. a, c

17. b, d, and e. *Trans* unsaturated fatty acid chains have a straighter conformation than do *cis* unsaturated chains; the packing of *trans* chains in bilayers is therefore more highly ordered, so they require higher temperatures to melt.

18. Cholesterol modulates the fluidity of membranes. By inserting between the fatty acid chains, cholesterol prevents their crystallization at temperatures below T_m and sterically blocks large motions of the fatty acid chains at temperatures above T_m. In fact, high concentrations of cholesterol abolish phase transitions of bilayers. This modulating effect of cholesterol maintains the fluidity of membranes in the range required for biological function.

19. c, e

20. Although sodium dodecyl sulfate (SDS) is a very effective detergent for solubilizing membrane components, the strong electrostatic interactions of its polar head groups with charged groups on the membrane proteins disrupt protein structure. A detergent such as octyl glucoside, which has an uncharged head group, allows the proteins to retain their three-dimensional structures while it interacts with their hydrophobic domains. Recall the use of SDS as a protein denaturant in the determination of protein molecular weights (see Chapter 3) and in the separation of membrane proteins (see Figure 12-35 in Stryer).

The Red-Cell Membrane Proteins and Other Membrane Proteins

21. (a) 4 (b) 3 (c) 6 (d) 2, 5 (e) 1

22. c

23. Transmembrane protein segments usually consist of nonpolar amino acids. The main chain CO and NH groups, however, are polar and tend to form hydrogen bonds with water. In an α helix, these

groups hydrogen bond to each other, thereby decreasing their overall polarity and facilitating the insertion of the protein segment into the lipid bilayer.

24. c, d, e

25. a, c, e

---

1. All organisms require water for many biochemical reactions, and thus organisms can live only in an aqueous environment. While a bacterial cell placed in a solution of crude oil might at least survive, it would not have enough water for growth and division. Microorganisms can best utilize crude oil or its hydrocarbon components when the microorganisms are present at a boundary layer between water and lipid. Aeration and agitation increase the effective area of such a layer. Some microorganisms that degrade hydrocarbons have a glycolipid-rich cell wall in which those compounds are soluble. After being solubilized, the compounds are transferred to the cytoplasmic membrane, where water-requiring reactions that initiate hydrocarbon degradation occur.

2. The four methyl side chains of each phytanic acid molecule interfere with the ordered association of fatty acyl chains; thus, they increase the fluidity of nerve cell membranes. This increase in fluidity could interfere with myelin function or ion transport, but the actual molecular basis for the symptoms is not yet known. Many of the symptoms of Refsum's disease can be eliminated by adapting diets that are free of phytol. Because phytol that is associated with chlorophyll is poorly absorbed, it is more important to eliminate products from cows and other ruminants, which are heavy consumers of plant tissues, from the diet. These products include dairy products as well as beef.

3. You would expect to find that the bacteria grown at the higher temperature will have incorporated a higher number of the longer fatty acids and a greater proportion of the saturated fatty acids. The membranes of bacteria grown at 18°C will have more short-chain fatty acids and more that are unsaturated. These cells select fatty acids that will remain fluid at a lower temperature in order to prevent their membranes from becoming too rigid. The cells grown at the higher temperature can select fatty acids that pack more closely. Cells in both cultures thus employ strategies designed to achieve optimal membrane fluidity.

4. Before the bilayer forms, water must arrange itself in a highly ordered way around the acyl chains of each phospholipid; as the phospholipids begin to assemble, the process is driven by the association of hydrophobic chains and the tendency of water to form a more disordered structure with itself. Thus, the entropy of the system increases as assembly occurs. There are, of course, many more water molecules than phospholipids in the system. Even though the completed bilayer is ordered, the overall system has undergone a large increase in entropy ($\Delta S > 0$) that exceeds the small change in enthalpy and results in a negative value for ΔG.

5. The higher the number of *cis* double bonds, the less ordered the bilayer structure will be and the more fluid the membrane system will

be. You would therefore expect T_m for the bilayer system containing the acyl chains with four unsaturated bonds to be much lower than that for the system containing the saturated fatty acid chains.

6. Like cholesterol, bacteriohopanetetrol is a pentacyclic molecule with a rigid, platelike, hydrophobic ring structure; it has a hydrophilic region as well, although that region is on the opposite end of the molecule when compared with cholesterol. In bacterial membranes, hopanoids may have a function similar to that of cholesterol in mammalian membranes; that is, they may moderate bacterial membrane fluidity by blocking the motion of fatty acyl chains and by preventing their crystallization.

7. Phospholipids have polar head groups, so their transfer across the hydrophobic interior of the bilayer as well as their dissociation from water at the bilayer surface would require a positive change in free energy. Without the input of free energy to make the process a spontaneous one, the transfer of the polar head group is very unlikely, so the asymmetric distribution of the phospholipids is preserved.

8. (a) The longer the acyl groups, the larger the number of noncovalent interactions that can form among the hydrocarbon chains. Higher temperatures are therefore required to disrupt the interactions of phospholipid species that have longer fatty acyl groups.

   (b) The *cis* double bond in oleate produces a bend in the hydrocarbon chain, interfering with the formation of noncovalent bonds among the acyl chains. Less heat energy is therefore required to cause a phase transition; in fact, the melting temperature for phosphatidyl choline with a palmitoyl and an oleoyl unit is −5°C, while T_m for dipalmitoylphosphatidyl choline is 41°C.

   (c) As shown in Figure 12-7 of Stryer's text, the structures of phosphatidyl choline and of sphingomyelin are very similar to each other; both contain phosphoryl choline and both have a pair of hydrocarbon chains. Given similar chain lengths in palmitoylsphingomyelin and in dipalmitoylphosphatidyl choline, you would expect that values of T_m for the two molecules are similar. Both species in fact exhibit a phase transition at 43°C.

   (d) At 50°C, you should expect cholesterol to diminish or even to abolish the transition, by preventing the close packing of the fatty acyl chains that impart rigidity to the molecular assembly. At higher temperatures, cholesterol in the mixture also prevents larger motions of fatty acyl chains, making the assembly less fluid. Studies show that in mixtures containing 30 to 35 mol % cholesterol, phase transitions are extinguished.

9. (a) You should expect to find nonpolar amino acid residues in the portions of the glycoprotein chain that are buried in the membrane. These amino acids could include many of those shown in the top part of Table 12-2 on page 304 of Stryer's text. Because the core portion of the membrane is 30 Å wide, up to 20 amino acids could be included in the buried segments, assuming that the amino acids are part of an α helix, in which the translation distance for each amino acid is 1.5 Å.

   (b) The glycoprotein has *O*-linked carbohydrate residues, and in most such proteins the sugars are attached to the side chains of serine or threonine residues. Were the sugars *N*-linked instead, you would expect to find one or more asparagine moieties in the glycoprotein.

(c) First, you would use a lectin that binds to the carbohydrate residues of the intact erythrocyte, and then you would determine the number of bound lectins per intact cell. Then, in another experiment, you would lyse the cells and repeat the lectin binding. If the carbohydrate residues are located only on the exterior surface of the cell, as is usually the case, then the number of bound lectins per cell will be the same in each of the two experiments.

CHAPTER 13

Metabolism: Basic Concepts
and Design

This chapter is an introduction to the next two parts of the text, which are devoted to metabolism. Metabolism is the collection of chemical reactions that allows cells to extract energy and reducing power from their environments, synthesize the building blocks of their macromolecules, and carry out all the other processes that are required to sustain life. Because energy is an essential concept in metabolism, Stryer begins this chapter with a review of the concept of the free-energy change in the context of coupled reactions. Thermodynamic principles were discussed in Chapter 8 and should be reviewed for a better understanding of metabolism. The most important molecules for storing and coupling energy in metabolic processes are described next, including ATP, the universal currency of energy in biological systems, and the important electron carriers, NAD^+ and FAD. The relationship of these molecules and other enzyme cofactors to the vitamins is examined. Stryer concludes this chapter with a broad outline of energy metabolism and its regulation and a discussion of nuclear magnetic resonance (NMR) spectroscopy as a noninvasive tool for investigating the metabolism of tissues.

When you have mastered this chapter, you should be able to complete the following objectives.

Introduction (Stryer pages 315–316)

1. Define *metabolism*.

2. State the significance of the *free-energy change (ΔG)* of reactions and the relationship of ΔG to $\Delta G°$, the *equilibrium constant,* and the *concentrations of reactants* and *products* of the reaction.

3. Describe the *additivity* of ΔG values for *coupled reactions* and the ability of a thermodynamically favorable reaction to drive an unfavorable one.

Adenosine Triphosphate (Stryer pages 316–320)

4. Give the structure of *adenosine triphosphate (ATP)* and describe its role as the major energy-coupling agent *(energy currency)* in metabolism.

5. Describe the structural basis for the *high phosphate group-transfer potential* of ATP and give the free energy liberated by the hydrolysis of ATP under standard and cellular conditions.

6. Explain how coupling a reaction with the hydrolysis of ATP can change the equilibrium ratio of the concentrations of the products to the concentrations of the reactants by a factor of 10^8.

7. Recognize compounds that have a *high group-transfer potential,* that is, compounds that release large amounts of free energy upon hydrolysis or oxidation.

8. Describe the *ATP-ADP cycle* of energy exchange in biological systems.

Coenzymes and Vitamins (Stryer pages 320–324)

9. Recognize the structures of *nicotinamide adenine dinucleotide (NAD⁺)* and *flavin adenine dinucleotide (FAD),* describe their reduction to *NADH* and *FADH₂*, and explain their roles in metabolism.

10. Distinguish the metabolic role of *NADPH* from that of *NADH*.

11. Note that *high-energy compounds* are kinetically stable and require enzymes for their reactions.

12. Describe the structure of *coenzyme A (CoA)* and its role as a *carrier for acetyl* or *acyl groups.*

13. List the *activated carriers* in metabolic reactions.

14. Define *coenzyme* and list the *water-soluble vitamins* that are components of coenzymes.

15. Recognize the *lipid-soluble vitamins* and state their physiological roles.

16. Describe the three major stages in the extraction of energy from foodstuffs.

17. Discuss the major mechanisms for the *regulation of metabolism.*

18. Define *energy charge* and compare it to the *phosphorylation potential.*

19. Describe how *nuclear magnetic resonance (NMR)* methods are used to study metabolism in intact tissues.

SELF-TEST

Introduction

1. Which of the following is *not* a function or purpose of metabolism?

   (a) To extract chemical energy from substances obtained from the external environment.
   (b) To form and degrade the biomolecules of the cell.
   (c) To convert exogenous foodstuffs into building blocks and precursors of macromolecules.
   (d) To equilibrate extracellular substances and the biomolecules of the cell.
   (e) To assemble the building-block molecules into macromolecules.

2. If the ΔG of the reaction A $\rightarrow$ B is -3.0 kcal/mol, which of the following statements are *correct*? (Recall that the prime symbol means that a thermodynamic parameter is measured at pH 7.0.)

   (a) The reaction will proceed spontaneously from left to right at the given conditions.
   (b) The reaction will proceed spontaneously from right to left at standard conditions.
   (c) The equilibrium constant favors the formation of B over the formation of A.
   (d) The equilibrium constant could be calculated if the initial concentrations of A and B were known.
   (e) The value of $\Delta G^{\circ\prime}$ is also negative.

3. Glucose 1-phosphate is converted to fructose 6-phosphate in two successive reactions:

   $$\text{Glucose 1-P} \longrightarrow \text{glucose 6-P} \qquad \Delta G^{\circ\prime} = -1.7 \text{ kcal/mol}$$
   $$\text{Glucose 6-P} \longrightarrow \text{fructose 6-P} \qquad \Delta G^{\circ\prime} = -0.4 \text{ kcal/mol}$$

   What is $\Delta G^{\circ\prime}$ for the overall reaction?

   (a) -2.1 kcal/mol
   (b) -1.7 kcal/mol
   (c) -1.3 kcal/mol
   (d) 1.3 kcal/mol
   (e) 2.1 kcal/mol

4. The reaction

$$\text{Phosphoenolpyruvate} + \text{ADP} + \text{H}^+ \longrightarrow \text{pyruvate} + \text{ATP}$$

has a $\Delta G^{\circ\prime} = -7.5$ kcal/mol. Calculate $\Delta G^{\circ\prime}$ for the hydrolysis of phosphoenolpyruvate.

5. Inside cells, the $\Delta G^\prime$ value for the hydrolysis of ATP to ADP + P_i is approximately -12 kcal/mol. Calculate the approximate ratio of [ATP] to [ADP][P_i] found in cells at 37°C.

    (a) 5000/1
    (b) 4000/1
    (c) 2000/1
    (d) 1000/1
    (e) 200/1

Adenosine Triphosphate

6. Which of the following statements about the structure of ATP are *correct*?

    (a) It contains three phosphoanhydride bonds.
    (b) It contains two phosphate ester bonds.
    (c) The sugar moiety is linked to the triphosphate by a phosphate ester bond.
    (d) The nitrogenous base is called adenosine.
    (e) The active form is usually a complex with Mg^{2+} or Mn^{2+}.

7. Which of the following factors contribute to the high phosphate group-transfer potential of ATP?

    (a) The greater resonance stabilization of ADP and P_i than of ATP
    (b) The increase in the electrostatic repulsion of oxygens upon hydrolysis of ATP
    (c) The interaction of the terminal phosphoryl group with the ribose group in ADP
    (d) The formation of a salt bridge between the base amino group and the negative charges of the phosphate oxygens in ATP

8. Which of the following are high-energy compounds?

    (a) Glycerol 3-phosphate
    (b) Adenosine diphosphate
    (c) Glucose 1-phosphate
    (d) Acetyl phosphate
    (e) Fructose 6-phosphate

9. ATP falls in the middle of the list of compounds having high phosphate group-transfer potentials. Explain why this is advantageous for energy coupling during metabolism.

_____

_____

10. Nucleoside phosphates can be interconverted enzymatically. Which of the following reactions are _incorrect_?

    (a) $ATP + GDP \longrightarrow ADP + GTP$
    (b) $AMP + GTP \longrightarrow ATP + GDP$
    (c) $ATP + AMP \longrightarrow ADP + ADP$
    (d) $ATP + UMP \longrightarrow ADP + UDP$
    (e) $ADP + AMP \longrightarrow ATP + adenosine$

11. Which of the following are features of the ATP-ADP cycle in biological systems?

    (a) ATP hydrolysis is used to drive reactions that require an input of free energy.
    (b) The oxidation of fuel molecules forms $ADP + P_i$ from ATP.
    (c) The oxidation of fuel molecules forms ATP from $ADP + P_i$.
    (d) Light energy drives ATP hydrolysis.

Coenzymes and Vitamins

12. Match the four cofactors in the left column with the appropriate structural features and properties from the right column.

    (a) ATP _____
    (b) FAD _____
    (c) NAD^+ _____
    (d) CoA _____

    (1) Nicotinamide ring
    (2) Adenine group
    (3) Phosphoanydride bond
    (4) Sulfur atom
    (5) Isoalloxazine ring
    (6) Ribose group
    (7) Acyl group transfer
    (8) Electron transfer
    (9) Phosphate transfer

13. ATP and NADH release large amounts of free energy upon the transfer of the phosphate group to H_2O and electrons to O_2, respectively. However, both molecules are stable in the presence of H_2O or O_2. Explain why.

_____

_____

14. During the reduction of FAD,

    (a) a flavin group is transferred.
    (b) an equivalent of a hydride ion is transferred.
    (c) the isoalloxazine ring becomes charged.
    (d) two hydrogen atoms are added to the isoalloxazine ring.
    (e) the adenine ring opens.

15. Which of the following correctly pairs a coenzyme with the group transferred by that coenzyme?

    (a) CoA, electrons
    (b) Biotin, CO_2
    (c) ATP, one carbon unit
    (d) NADPH, phosphoryl group
    (e) Thiamine pyrophosphate, acyl group

16. Which of the following water-soluble vitamins forms part of the structure of CoA?

    (a) Pantothenate
    (b) Thiamine
    (c) Riboflavin
    (d) Pyridoxine
    (e) Folate

17. Which of the vitamins in Question 16 is referred to as vitamin B_1? _____

18. Match the lipid-soluble vitamins in the left column with the appropriate biological functions in the right column.

    (a) Vitamin A _____
    (b) Vitamin D _____
    (c) Vitamin E _____
    (d) Vitamin K _____

    (1) Protection of unsaturated membrane lipids from oxidation
    (2) Carboxylation of glutamate residues of clotting factors
    (3) Participation in Ca^{2+} and phosphorus metabolism
    (4) Precursor of retinal, the light-absorbing group in visual pigments
    (5) Related to fertility in rats

Outline of Energy Metabolism

19. Which of the following statements about the third of the three stages of metabolism described by Hans Krebs for the generation of energy from foodstuffs are *correct*?

    (a) It is common to the oxidation of all fuel molecules.
    (b) It involves the breakdown of food into smaller units, such as amino acids, sugars, and fatty acids.
    (c) It releases relatively little energy compared to the second stage.
    (d) It involves the degradation of sugars, fatty acids, and amino acids into a few common metabolites.
    (e) It produces most of the ATP and CO_2 in cells.

20. Which of the following are reasons the biochemical pathway for the catabolism of a molecule is almost never the same as the pathway for the biosynthesis of that molecule?

    (a) It would be extremely difficult to regulate the pathway if it served both functions.
    (b) The free-energy change would be unfavorable in one direction.
    (c) The reactions never take place in the same cell.
    (d) Enzyme-catalyzed reactions are irreversible.
    (e) Biochemical systems are usually at equilibrium.

21. Which of the following statements about the energy charge are *correct*?

    (a) It can have a value between 0 and 1.

    (b) It is around 0.1 in energy-consuming cells, such as the muscle cells.

    (c) It can regulate the rates of reactions in energy-consuming and energy-producing pathways.

    (d) It is also called the phosphorylation potential.

    (e) It is buffered in the sense that its value is maintained within narrow limits.

22. Explain why NMR spectroscopy is a very promising tool for the investigation of metabolic events in intact tissues.

_____

_____

23. Phosphorus NMR spectroscopy of muscle cells allows the intracellular measurement of which of the following?

    (a) pH

    (b) ADP

    (c) ATP

    (d) Nucleic acids

    (e) Phosphocreatine

PROBLEMS

1. Two reactions involving L-amino acids and the values of their respective free-energy changes are as follows:

Glutamate + pyruvate $\rightleftharpoons$ α-ketoglutarate + alanine
$$\Delta G^{\circ\prime} = -0.24 \text{ kcal/mol}$$

Glutamate + oxaloacetate $\rightleftharpoons$ α-ketoglutarate + aspartate
$$\Delta G^{\circ\prime} = -1.15 \text{ kcal/mol}$$

    (a) Write the overall reaction for production of alanine and oxaloacetate from aspartate and pyruvate.

    (b) Show that under standard conditions the net formation of alanine and oxaloacetate from aspartate and pyruvate is thermodynamically unfavorable.

    (c) Suppose that at 25°C the molar concentrations of reactants and products are as follows:

$$[\text{Pyruvate}] = [\text{aspartate}] = 10^{-2} \text{ M}$$

$$[\text{Alanine}] = 10^{-4} \text{ M}$$

$$[\text{Oxaloacetate}] = 10^{-5} \text{ M}$$

    Is the spontaneous synthesis of alanine and oxaloacetate possible under these conditions? Why?

2. Phosphocreatine can be used as a phosphoryl donor for the synthesis of ATP in a reaction catalyzed by creatine kinase. Refer to page 319

of Stryer's text for the free energies of hydrolysis for ATP and creatine phosphate.

(a) Write the equation for the reaction catalyzed by creatine kinase.
(b) Calculate the value of $\Delta G°'$ for the reaction. Is the reaction energetically favorable?
(c) What effect does the enzyme have on the value of $\Delta G°'$ for the reaction?
(d) Concentrations of a number of compounds in human muscle are as follows:

$$[ATP] = 10 \text{ mM}$$

$$[ADP] = 1 \text{ mM}$$

$$[P_i] = 10 \text{ mM}$$

$$[Phosphocreatine] = 30 \text{ mM}$$

$$[Creatine] = 1 \text{ mM}$$

Calculate the change in free energy when the creatine kinase reaction occurs in human muscle at 25°C.

(e) In rapidly contracting muscle, the rate of ATP utilization is approximately 3 μmol/s/g and the concentration of ATP at rest is 4.5 μmol per gram of tissue. The concentration of phosphocreatine in resting human muscle is about 20μmol/g. How can phosphocreatine be useful during rapid muscle contraction?

3. Under standard conditions, the free energy of hydrolysis of L-glycerol phosphate is -2.2 kcal/mol, and for ATP hydrolysis it is -7.3 kcal/mol. Show that when ATP is used as a phosphoryl donor for the formation of L-glycerol phosphate, the value of the equilibrium constant is altered by a factor of over 10^5.

4. When a hexose phosphate is hydrolyzed to free hexose and inorganic phosphate, the ratio of the concentration of hexose to the concentration of hexose phosphate at equilibrium is 99 to 1. What is the free-energy change for the reaction under standard conditions?

5. The process of catabolism releases free energy, some of which is stored as ATP and some of which is lost as heat to the surroundings. Explain how these observations are consistent with the fact that catabolic pathways are essentially irreversible.

6. The equilibrium constant for the oxidative deamination of glutamate to form α-ketoglutarate is 10^{-15}.

Glutamate + NADP$^+$ $\rightleftharpoons$ α-ketoglutarate + NH$_4^+$ + NADPH + H$^+$

Although the equilibrium constant is very small, α-ketoglutarate is readily formed from glutamate in the cell. How is this possible?

7. Chemotrophs derive free energy from the oxidation of fuel molecules, such as glucose and fatty acids. Which compound, glucose or a saturated fatty acid containing 18 carbons, would yield more free energy per carbon atom when subjected to oxidation in the cell?

8. In a typical cell, the concentrations of pyridine nucleotides and flavins are relatively low compared to the number of substrate molecules that must be oxidized. What does this observation suggest about the rate of oxidation and reduction of these electron carriers?

9. Why must the reaction

$$NADH \longrightarrow NAD^+ + H^+ + 2\,e^-$$

have a negative value for $\Delta G^{\circ\prime}$?

10. In an erythrocyte, the concentration of glucose is 5.0 mM, and the volume of the cell is 1.5×10^{-16} mm^3. If the oxidation of glucose yields 686 kcal/mol, how much heat can be generated by the complete oxidation of the glucose in an erythrocyte?

11. Microorganisms use derivatives of the vitamin B series as cofactors for biochemical reactions. Many bacteria, including *E. coli,* can grow and divide in a medium containing glucose and simple inorganic salts, including nitrogen in the form of ammonia. What does this tell you about vitamin B metabolism in these microorganisms?

12. Why is it desirable for a cell to regulate the *first* reaction in a biosynthetic pathway?

13. Refer to Problem 5 on page 330 of Stryer's text. The formation of a number of other important compounds in biosynthetic reactions involves the generation of pyrophosphate and its subsequent hydrolysis to two molecules of P_i. For example, the formation of UDP-glucose from UTP and glucose-1-phosphate yields PP_i, which is then cleaved. What does this tell you about the group-transfer potential of UDP-glucose?

14. How can adenylate kinase be used to generate more ATP when it is needed in rapidly contracting muscle?

15. If the free energy of ATP hydrolysis in an erythrocyte is -12.0 kcal/mol, what is the ratio of [ATP] to [ADP][P_i] in the cell?

ANSWERS TO SELF-TEST

Introduction

1. d

2. a and d. The expression for $\Delta G^\prime$ contains two variables: $\Delta G^{\circ\prime}$ (or $K_{eq}^\prime$) and the ratio of the product concentrations to the reactant concentrations. Therefore, $\Delta G^\prime$ alone cannot provide information about $\Delta G^{\circ\prime}$ or $K_{eq}^\prime$. Answer (d) is correct because $\Delta G^{\circ\prime}$ and $K_{eq}^\prime$ can be calculated when $\Delta G^\prime$ and the reactant and product concentrations are known.

3. a

4. -14.8 kcal/mol. The overall reaction can be separated into two steps:

(1)	Phosphoenolpyruvate + $H_2O \longrightarrow$ pyruvate + P_i	$\Delta G^{\circ\prime} = ?$
(2)	ADP + P_i + $H^+ \longrightarrow$ ATP + H_2O	$\Delta G^{\circ\prime} = +7.3$ kcal/mol
	Phosphoenolpyruvate + ADP + $H^+ \longrightarrow$ pyruvate + ATP	$\Delta G^{\circ\prime} = -7.5$ kcal/mol

Therefore, $\Delta G^{\circ\prime}$ for step 1 is

$$\Delta G^{\circ\prime} = -7.5 \text{ kcal/mol} - (+7.3 \text{ kcal/mol})$$

$$= -14.8 \text{ kcal/mol}$$

5. c.

$$ATP \longrightarrow ADP + P_i \qquad \Delta G^{\circ\prime} = -7.3 \text{ kcal/mol}$$

using

$$\Delta G' = \Delta G^{\circ\prime} + 2.3 \, RT \log_{10} \frac{[ADP][P_i]}{[ATP]}$$

$$-12 \frac{\text{kcal}}{\text{mol}} = -7.3 \frac{\text{kcal}}{\text{mol}} + 2.3 \times 1.98 \frac{\text{cal}}{\text{mol} \cdot {}^{\circ}\text{K}} \times 310 \, {}^{\circ}\text{K} \log_{10} \frac{[ADP][P_i]}{[ATP]}$$

$$\frac{[ADP][P_i]}{[ATP]} = 10^{-3.33}$$

$$\frac{[ATP]}{[ADP][P_i]} = 2140$$

Adenosine Triphosphate

6. c, e

7. a

8. b, d

9. The intermediate phosphate group-transfer potential of ATP means that, although ATP hydrolysis can drive a very large number of thermodynamically unfavorable biochemical reactions in metabolic pathways, it can itself be regenerated by coupling with another reaction that releases more free energy than -7.3 kcal/mol.

10. b, e

11. a, c

Coenzymes and Vitamins

12. (a) 2, 3, 6, 9 (b) 2, 3, 5, 6, 8 (c) 1, 2, 3, 6, 8 (d) 2, 3, 4, 6, 7

13. Although the transfer reactions of the cofactors ATP and NADH have large negative free-energy changes, there are high activation-energy barriers that prevent spontaneous reactions with H_2O or O_2, respectively. In other words, cofactors with high group-transfer potentials and fuel molecules are thermodynamically unstable yet kinetically stable. Consequently, specific enzymes are required to catalyze their reactions.

14. d

15. b

16. a

17. Thiamine

18. (a) 4 (b) 3 (c) 1, 5 (d) 2

Outline of Energy Metabolism

19. a, e

20. a, b

21. a, c, e

22. NMR spectroscopy is a noninvasive technique that can provide information on the amounts and types of biological molecules that are present in living tissues. It is based on the energy absorption of naturally occurring atomic nuclei when they are placed in a magnetic field.

23. a, c, e

ANSWERS TO PROBLEMS

1. (a) The individual reactions proceed as follows:

   Glutamate + pyruvate $\longrightarrow$ α-ketoglutarate + alanine

   α-Ketoglutarate + aspartate $\longrightarrow$ glutamate + oxaloacetate

   The coupled reaction is obtained by combining the two reactions:

   Pyruvate + aspartate $\longrightarrow$ alanine + oxaloacetate.

   (b) For the synthesis of alanine and α-ketoglutarate, $\Delta G^{\circ\prime}$ is -0.24 kcal/mol, and for the synthesis of glutamate and oxaloacetate, $\Delta G^{\circ\prime} = +1.15$ kcal/mol, under standard conditions. Summing those two values gives $\Delta G^{\circ\prime} = +0.91$ kcal/mol for the overall reaction, which is therefore not thermodynamically favorable under standard conditions.

   (c) For the conditions specified,

   $$\Delta G = \Delta G^{\circ\prime} + 1.364 \log_{10} \frac{[\text{alanine}][\text{oxaloacetate}]}{[\text{pyruvate}][\text{aspartate}]} \text{ kcal/mol}$$

   $$= +0.91 \text{ kcal/mol} + 1.364 \log_{10} \frac{(10^{-4} \text{ M})(10^{-5} \text{ M})}{(10^{-2} \text{ M})(10^{-2} \text{ M})} \text{ kcal/mol}$$

   $$= +0.91 + 1.364(-5) = 0.91 - 6.8$$

   $$= -5.89 \text{ kcal/mol}$$

   The negative value for ΔG indicates that the reaction occurs spontaneously under the conditions specified.

2. (a) The reaction catalyzed by creatine kinase is

   Phosphocreatine + ADP $\rightleftharpoons$ ATP + creatine

   (b) For the hydrolysis of phosphocreatine, $\Delta G^{\circ\prime} = -10.3$ kcal/mol, and for the formation of ATP it is $+7.3$ kcal/mol under standard conditions. The free-energy change for the overall reaction is equal to the sum of the two individual values:

   $$\Delta G^{\circ\prime} = (-10.3 + 7.3) = -3.0 \text{ kcal/mol}$$

   The reaction to the right, with the net formation of ATP and creatine, is therefore energetically favorable.

   (c) Although the enzyme controls the rate at which equilibrium is attained, it has no effect on the equilibrium constant, K'_{eq}. Because $\Delta G^{\circ\prime}$ is a function of the equilibrium constant, the action of the enzyme has no effect on its value.

(d) The change in free energy under nonequilibrium conditions is related to molar concentrations of reactants and products:

$$\Delta G = -3.0 \text{ kcal/mol} + 1.364 \log_{10} \frac{[\text{ATP}][\text{creatine}]}{[\text{ADP}][\text{phosphocreatine}]} \text{ kcal/mol}$$

$$= -3.0 \text{ kcal/mol} + 1.364 \log_{10} (3.3 \times 10^{-1}) \text{ kcal/mol}$$

$$= -3.0 \text{ kcal/mol} + 1.364(-0.48) \text{ kcal/mol}$$

$$= -3.65 \text{ kcal/mol}$$

Under these conditions, the reaction proceeds toward the net formation of ATP.

(e) Given the amount of ATP in a gram of resting muscle and the rapid rate of ATP utilization once rapid muscle contraction begins, the available ATP can provide energy for less than 2 seconds, whereas the available phosphocreatine can provide energy for over 6 seconds (assuming that it can be rapidly utilized for ATP synthesis). Phosphocreatine is used as an energy source until other ATP-generating reactions can accelerate to rates sufficient to sustain muscle contraction.

3. For the synthesis of L-glycerol phosphate from glycerol and phosphate, the value of $\Delta G^{\circ\prime}$ is $+2.2$ kcal/mol. Under standard conditions,

$$\Delta G^{\circ\prime} = -1.364 \log_{10} K'_{eq}$$

$$\log_{10} K'_{eq} = \frac{+2.2}{-1.364}$$

$$= -1.612$$

$$K'_{eq} = \text{antilog} -1.612$$

$$= 2.45 \times 10^{-2}$$

The overall value of $\Delta G^{\circ\prime}$ for the formation of L-glycerol phosphate using ATP as a phosphoryl donor is equal to the sum of the free-energy values for the two individual reactions:

$$\Delta G^{\circ\prime} = (+2.2) + (-7.3) = -5.1 \text{ kcal/mol}$$

The equilibrium constant for the overall reaction is

$$K'_{eq} = \frac{-5.1}{-1.364}$$

$$= +3.739$$

$$K'_{eq} = \text{antilog } 3.739$$

$$= 5.48 \times 10^3$$

The ratio of the two equilibrium constants is 2.24×10^5.

4. The reaction is

$$\text{Hexose phosphate} + H_2O \rightleftharpoons \text{hexose} + P_i$$

The expression for the equilibrium constant is

$$K'_{eq} = \frac{[\text{hexose}][P_i]}{[\text{hexose phosphate}]} = \frac{99}{1}$$

Thus, $\log_{10} K'_{eq}$ is approximately equal to 2. Using the free-energy equation,

$$\Delta G^{\circ\prime} = -1.364(2)$$

$$= -2.728 \text{ kcal/mol}$$

5. The heat that is lost contributes to an increase in the entropy of the surroundings. A positive change in entropy means that the free energy for a catabolic process is more likely to be negative. Reactions with negative free-energy values are irreversible in that they require an input of energy to procede in the opposite direction.

6. The removal of the products, α-ketoglutarate and NADPH, for use in other metabolic reactions causes the equilibrium to be shifted toward formation of more α-ketoglutarate. In addition, increased concentrations of glutamate and $NADP^+$ can displace the equilibrium toward the direction of ketoacid synthesis. In either instance, the concentrations of the reactants or products will affect the overall direction of the biochemical reaction.

7. Of the 18 carbon atoms in a saturated fatty acid, 17 are saturated (as $-CH_2-$ groups) and are more reduced than the partially oxidized carbon atoms in glucose. In glucose, 5 of the 6 carbons are partially oxidized to the hydroxymethyl level, and the sixth is at the more oxidized aldehyde level. A greater number of electrons per carbon are available in the fatty acid, so more metabolic energy is available from it than from glucose.

8. Pyridine nucleotides, such as NADH, serve as acceptors and donors of electrons in many metabolic reactions, including those that generate energy for the cell. Because the absolute number of pyridine nucleotides in the cell is low, the cycle of oxidation and reduction for these compounds must occur rapidly for the oxidation of fuel molecules to proceed at a sufficient rate.

9. The flow of electrons from reduced pyridine nucleotides, such as NADH, provides energy that can drive the formation of ATP. For such a reaction to proceed spontaneously, the overall value of $\Delta G^{\circ\prime}$ must be negative. The value of $\Delta G^{\circ\prime}$ for ATP synthesis is positive, so a negative value would be expected for the oxidation of NADH to NAD^+.

10. The concentration of glucose in the erythrocyte is 5×10^{-3} mol per liter, or 5×10^{-6} mol per milliliter. The number of moles of glucose in an erythrocyte is

$$1.5 \times 10^{-16} \text{ mm}^3 \times 5 \times 10^{-6} \text{ mol/mm}^3 = 7.5 \times 10^{-22} \text{ mol}$$

The heat released by the oxidation of glucose is

$$7.5 \times 10^{-22} \text{ mol} \times 6.86 \times 10^5 \text{ cal/mol} = 5.2 \times 10^{-16} \text{ cal}$$

11. The fact that these bacteria use vitamin B derivatives as cofactors means that they must be able to synthesize them from the simple compounds supplied in the growth medium.

12. Regulation of the first reaction in a biosynthetic pathway ensures that the intermediates in the pathway will be synthesized only when the ultimate product is required. In this way the cell can conserve energy as well as precursors of all intermediates. Such a regulatory scheme will be found when none of the intermediates are utilized in other pathways.

13. As Stryer points out in the answer to his problem (Stryer, p. 1054), the cleavage of pyrophosphate ensures that the coupled reactions will

proceed toward the net formation of desired product; that is, the overall reaction will have a rather large negative free-energy value. The fact that PP_i is formed during the synthesis of UDP-glucose suggests that the free energy released by the coupled reactions for the formation of UDP-glucose and the hydrolysis of UTP is small. Therefore, the free energy of the hydrolysis of UDP-glucose would be similar to that of UTP. Thus, you should surmise that the group-transfer potential of glucose from UDP-glucose would be high. This is the case, as UDP-glucose serves as a donor of glucose residues for the synthesis of glycogen.

14. Adenylate kinase catalyzes the interconversion of two molecules ADP to ATP and AMP. In contracting muscle, when ATP is hydrolyzed to yield ADP and P_i, an increase in ADP concentration drives the reaction toward the net formation of ATP. In this way, more ATP is made available for contraction.

15. To determine the ratio of the concentrations of products to reactants in the cell, we begin by using the known values of the free-energy changes:

$$\Delta G' = \Delta G^{\circ\prime} + 1.364 \log_{10} Q$$

where Q is the ratio of the concentrations of products to reactants, or $[ADP][P_i]/[ATP]$.

$$-12.0 = -7.3 + 1.364 \log_{10} Q$$

$$\log_{10} Q = \frac{-4.7}{1.364} = -3.446$$

$$Q = \text{antilog} -3.446$$

$$= 3.58 \times 10^{-4}$$

The ratio of the concentration of ATP to the concentration of the products is the inverse of the value for Q.

$$\frac{[ATP]}{[ADP][P_i]} = \frac{1}{Q} = 2.8 \times 10^3$$

CHAPTER 14

Carbohydrates

Carbohydrates are one of the four major classes of biomolecules; the others are proteins, nucleic acids, and lipids, all of which were discussed in earlier chapters. In Chapter 14, Stryer describes the chemical nature of carbohydrates and summarizes their principal biological roles. First, he introduces monosaccharides, the simplest carbohydrates, and describes their chemical properties. Since these sections assume familiarity with the properties of aldehydes, ketones, alcohols, and stereoisomers, students with a limited background in organic chemistry should review these topics in any standard organic chemistry text. Next, Stryer discusses simple derivatives of monosaccharides, including sugar phosphates and disaccharides. Sugar is the common name for monosaccharides and their derivatives. You have already seen some monosaccharide derivatives in the structures of nucleic acids (Chapter 4) and cofactors (Chapter 13). Finally, Stryer discusses polysaccharides and oligosaccharides as storage and structural polymers and as components of glycoproteins. Once again, these topics are not completely new; polysaccharides were introduced in connection with proteoglycans in Chapter 11, and glycoproteins and glycolipids were described as components of biological membranes in Chapter 12.

When you have mastered this chapter, you should be able to complete the following objectives.

Introduction (Stryer pages 331–332)

1. List the main roles of *carbohydrates* in nature.

Monosaccharide Structures and Chemical Properties (Stryer pages 332–337)

2. Define *carbohydrate* and *monosaccharide* in chemical terms.

3. Relate the absolute configuration of monosaccharide D or L *stereoisomers* to those of *glyceraldehyde*.

4. Associate the following monosaccharide class names with their corresponding structures: *aldose* and *ketose; triose, tetrose, pentose, hexose,* and *heptose; pyranose* and *furanose.*

5. Distinguish among *enantiomers, diastereoisomers,* and *epimers* of monosaccharides.

6. Draw the *Fisher* (open-chain) *structures* and the most common *Haworth* (ring) *structures* of D-*glucose,* D-*fructose,* D-*galactose,* and D-*ribose.*

7. Explain how ring structures arise through the formation of *hemiacetal* and *hemiketal* bonds. Draw a ring structure given a Fisher formula.

8. Distinguish between α- and β-*anomers* of monosaccharides and explain *mutarotation.*

9. Compare the *chair, boat,* and *envelope conformations* of monosaccharides.

Simple Monosaccharide Derivatives (Stryer pages 337–341)

10. Define *O-glycosidic* and *N-glycosidic* bonds in terms of *acetal* and *ketal* bonds. Draw the bonds indicated by such symbols as α-1,6 or β-1,4. Stryer's text and other sources also use α(1→6) or β1:4 to designate the same bonds.

11. Explain the role of *O*-glycosidic bonds in the formation of monosaccharide derivatives, *disaccharides,* and *polysaccharides.*

12. Give examples of important naturally occurring sugar derivatives.

13. Draw the structures of *sucrose, lactose,* and *maltose.* Give the natural sources of these common disaccharides.

14. Explain the role of *UDP-activated sugars* in metabolism. Describe the synthesis of lactose.

15. Explain the consequences of human *lactase deficiency.*

Polysaccharides and Oligosaccharides (Stryer pages 341–346)

16. Describe the structures and biological roles of *glycogen, starch, amylose, amylopectin,* and *cellulose.*

17. Give examples of enzymes involved in the digestion of carbohydrates in humans.

18. Describe the types of bonds that link carbohydrates and proteins.

19. Name the sugars that are frequent constituents of *glycoproteins*.

20. List the biological roles of glycoproteins and of the proteins that recognize them on cell surfaces.

SELF-TEST

Introduction

1. Which are roles of carbohydrates in nature? Carbohydrates

   (a) serve as energy stores in plants and animals.
   (b) are major structural components of mammalian tissues.
   (c) are constituents of nucleic acids.
   (d) are conjugated to many proteins and lipids.
   (e) are found in the structures of all the coenzymes.

2. In the human diet, carbohydrates constitute approximately half the total caloric intake, yet only 1% of tissue weight is carbohydrate. Explain this fact.

Monosaccharide Structures and Chemical Properties

3. Examine the following five sugar structures.

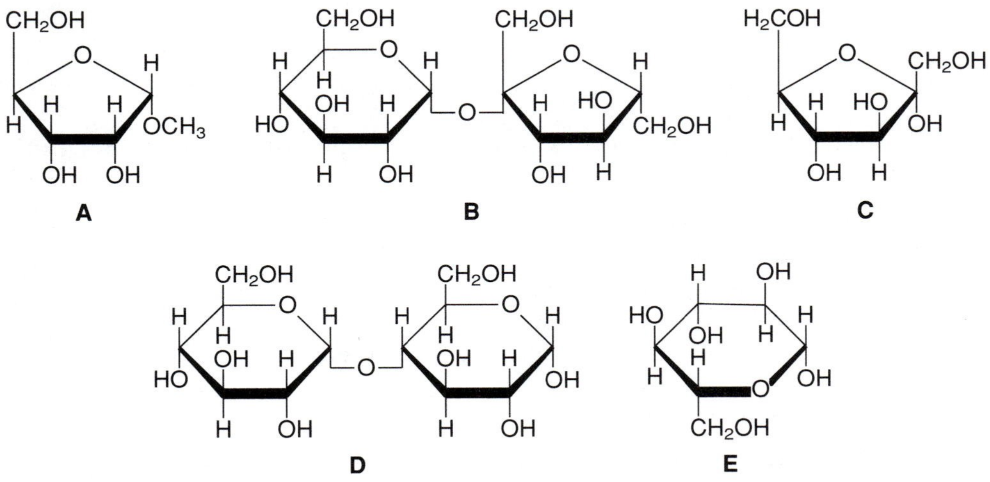

Which of these sugars

(a) contain or are pentoses? _____

(b) contain or are ketoses? _____

(c) contain the same monosaccharides? _____

Name those monosaccharides.

_____

(d) will yield different sugars after chemical or enzymatic hydrolysis of glycosidic bonds? _____

(e) are reducing sugars? _____

(f) contain a β-anomeric carbon? _____

(g) will mutarotate? _____

(h) is sucrose? _____

(i) are released upon the digestion of starch? _____

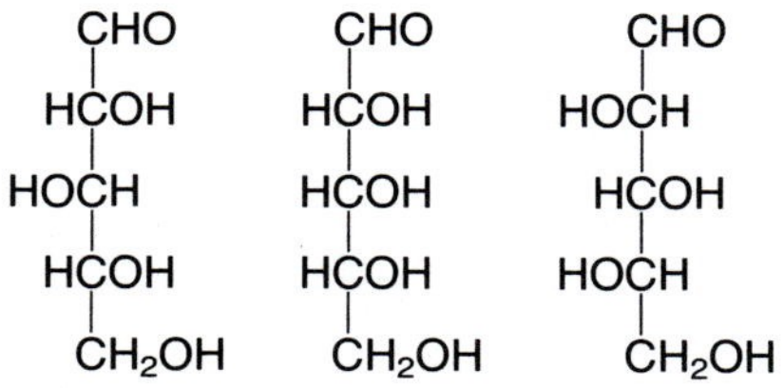

```
  CHO        CHO        CHO
 HCOH       HCOH       HOCH
 HOCH       HCOH       HCOH
 HCOH       HCOH       HOCH
 CH2OH      CH2OH      CH2OH
   A          B          C
        Aldopentoses
```

4. Consider the aldopentoses shown in the margin.

(a) Name the types of stereoisomers represented by each pair.

A and B are _____

B and C are _____

A and C are _____

(b) Name sugar B. _____

(c) Draw the α-anomeric form of the furanose Haworth ring structure for sugar A.

5. Identify the properties common to D-glucose and D-ribose. Both monosaccharides

(a) are reducing sugars.
(b) form intramolecular hemiacetal bonds.
(c) have functional groups that can form glycosidic linkages.
(d) occur in hexose form.
(e) are major constituents of glycoproteins.

Simple Monosaccharide Derivatives

6. The structure of ATP is shown at the top of page 189.

Adenosine triphosphate (ATP)

The structure of ATP

(a) contains a β-N-glycosidic linkage.
(b) contains a pyranose ring.
(c) exists in equilibrium with the open Fischer structure of the sugar.
(d) preferentially adopts a chair conformation.
(e) contains a ketose sugar.

7. Draw the structure of the disaccharide glucose-α(1→6)-galactose in the β-anomeric form.

8. How can the modifier subunit β-lactalbumin change the catalytic activity of galactosyl transferase from the synthesis of glycoprotein oligosaccharides to the synthesis of lactose?

9. Lactase deficiency is characterized by the inability to hydrolyze

(a) α-1,4-glucosidic bonds.
(b) β-1,6-galactosidic bonds.
(c) β-1,4-glucosidic bonds.
(d) β-1,4-galactosidic bonds.
(e) α-1,6-glucosidic bonds.

10. Glucose readily enters into all tissues from the blood, but only specialized tissues (liver and kidney) are capable of releasing glucose into the blood. Give two reasons why most tissues cannot supply glucose to the blood.

Polysaccharides and Oligosaccharides

11. If one carries out the partial mild-acid hydrolysis of glycogen or starch and then isolates from the product oligosaccharides all of the trisaccharides present, how many different kinds of trisaccharides would one expect to find? Disregard α- or β-anomers.

    (a) 1
    (b) 2
    (c) 3
    (d) 4
    (e) 5

12. A sample of bread gives a faint positive color with Nelson's reagent for reducing sugars. After an equivalent bread sample has been masticated, the test becomes markedly positive. Explain this result.

_____

_____

13. Why does cellulose form dense linear fibrils, whereas amylose forms open helices?

_____

_____

14. For the polysaccharides in the left column, indicate all the descriptions in the right column that are appropriate.

    (a) Amylose _____
    (b) Cellulose _____
    (c) Dextran _____
    (d) Glycogen _____
    (e) Starch _____

    (1) contains α-1,6-glucosidic bonds.
    (2) is a storage polysaccharide in yeasts and bacteria.
    (3) can be effectively digested by humans.
    (4) contains β-1,4-glucosidic bonds.
    (5) is a branched polysaccharide.
    (6) is a storage polysaccharide in humans.
    (7) is a component of starch.

15. α-Amylase

    (a) removes glucose residues sequentially from the reducing end of starch.
    (b) breaks the internal α-1,6-glycosidic bonds of starch.
    (c) breaks the internal α-1,4-glycosidic bonds of starch.
    (d) cleaves the β-1,4-glycosidic bond of lactose.
    (e) can hydrolyze cellulose in the presence of an isomerase.

16. Glycoproteins

    (a) contain oligosaccharides linked to the side chain of lysine or histidine residues.
    (b) contain oligosaccharides linked to the side chain of serine or threonine residues.
    (c) contain linear oligosaccharides with a terminal glucose residue.
    (d) bind to liver cell surface receptors that recognize sialic acid residues.
    (e) are mostly cytoplasmic proteins.

1. In alkaline solution, RNA polynucleotides are hydrolyzed to a mixture of 2'- and 3'-ribonucleotides. DNA polynucleotides, on the other hand, are not particularly sensitive to alkali.

   (a) Write out a mechanism for the hydrolysis of RNA by hydroxide ion.
   (b) Explain why DNA is not sensitive to alkaline conditions.

2. Indicate whether the following pairs of molecules are enantiomers, epimers, diastereoisomers, or anomers.

   (a) D-Xylose and D-lyxose
   (b) α-D-Galactose and β-D-galactose
   (c) D-Allose and D-talose
   (d) L-Arabinose and D-arabinose

3. L-Rhamnose is a common 6-deoxy sugar. It is derived from D-glucose. What biochemical steps would be needed for the conversion of D-glucose to L-rhamnose?

4. What is the name of the compound that is the mirror image of α-D-glucose?

5. Compound X, an aldose, is enzymatically reduced using NADPH as an electron donor, yielding D-sorbitol. This sugar alcohol is then oxidized at the C-2 position with NAD^+ as the electron acceptor; the products are NADH and a ketose, compound Y.

   (a) Name compound X and write its structure.
   (b) Will sorbitol form a furanose or pyranose ring? Why?
   (c) Name compound Y and write its structure.

6. Compare the number of dimers that can be prepared from a pair of alanine molecules and from a pair of D-galactose molecules, each of which is present as a pyranose ring. For the galactose molecules, pairs may be made using the α- or β-anomers.

7. In 1952, Morgan and Watkins showed that N-acetylgalactosamine and its α-methylglucoside inhibit the agglutination of type-A erythrocytes by type-A-specific lectins, whereas other sugars had little effect. What did this information reveal about the structure of the glycoprotein on the surface of type-A cells?

8. The steps in the conversion of glucose 6-phosphate to myoinositol 1-phosphate are shown in Figure 14-1. The enzyme inositol 1-phosphate cyclase, which catalyzes this series of steps, has as a prosthetic group a molecule of NAD^+.

   (a) Describe the role of NAD^+ in the reaction.
   (b) What type of reaction occurs when the carbocyclic ring is formed?
   (c) In solution, less than 1% of glucose 6-phosphate is present in the open-chain form, yet the enzyme quantitatively converts all of the hexose 6-phosphate to myoinositol 1-phosphate. Explain.
   (d) Which carbon atoms undergo inversion of configuration?

L-Rhamnose

D-Sorbitol

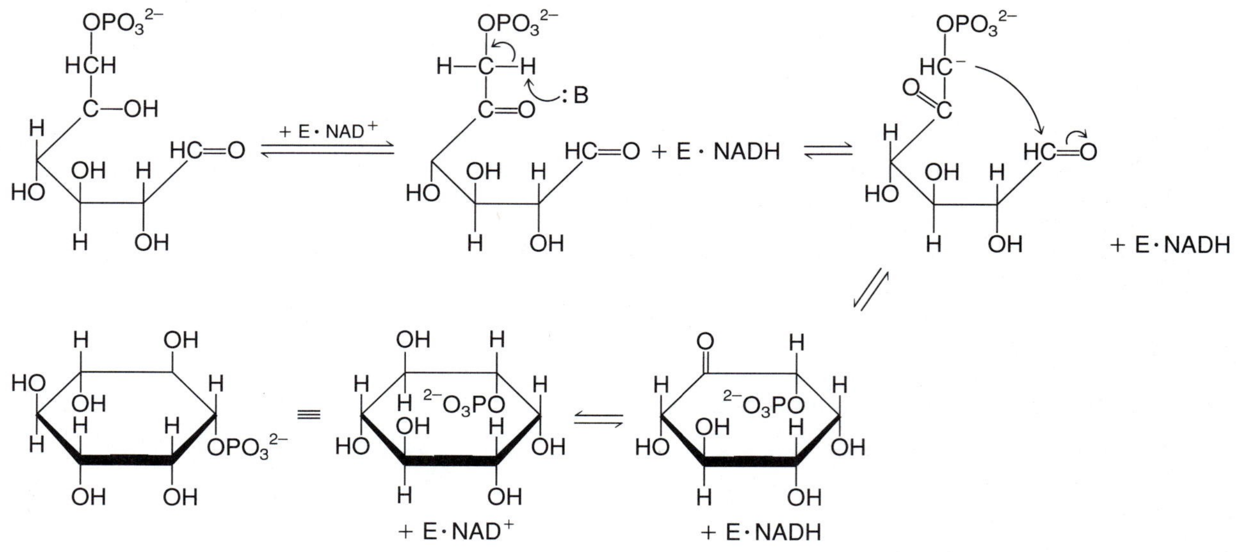

Figure 14-1
Steps in the conversion of glucose 6-phosphate to myoinositol 1-phosphate.
[From C. Walsh, *Enzymatic Reaction Mechanisms*, W. H. Freeman and Company, 1979.]

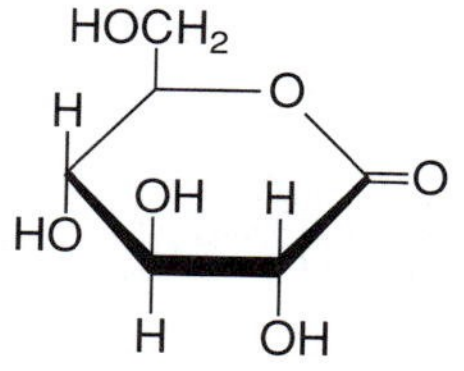

9. *Penicillium* and other fungi contain an enzyme called glucose oxidase, which converts D-glucose to the compound shown in the margin.

   (a) What is the generic name for compound formed?
   (b) Can the compound open to form a straight-chain compound? What else is required for this reaction to occur?
   (c) Why would you expect that NAD$^+$ or FAD$^+$ would be required by the enzyme?

ANSWERS TO SELF-TEST

Introduction

1. a, c, d

2. Most of the carbohydrates in the human diet are used as fuel to supply the energy requirements of the organism. Although some carbohydrate is stored in the form of glycogen, the mass stored is relatively small compared to adipose tissue and muscle mass. The carbohydrate present in nucleic acids, glycoproteins, glycolipids, and cofactors, although functionally essential, contributes relatively little to the weight of the body.

Monosaccharide Structures and Chemical Properties

3. (a) A
   (b) B, C
   (c) B and C contain fructose; B, D, and E contain or are glucose.
   (d) A, B, D
   (e) C, D, E
   (f) B and E. In structure B, the fructose ring is flipped over.

(g) C, D, E

(h) B

(i) D and E. Although E is in the β-anomer form, recall that in solution it can mutarotate.

4. (a) A and B are epimers. B and C are diastereoisomers. A and C are enantiomers.

(b) D-ribose.

(c) See the structure in the margin.

5. a, b, and c. Note that *glycosidic* refers to bonds involving any sugars; however, *glucosidic* and *galactosidic*, refer specifically to bonds involving the anomeric (reducing) carbons of glucose and galactose, respectively.

Simple Monosaccharide Derivatives

6. a

7. See the structure in the margin.

8. Interactions between protein subunits in this case between galactosyl transferase and the modifier subunit β-lactalbumin, change the structure of the active site so that glucose, instead of the *N*-acetylglucosamine residue of an oligosaccharide, is specifically bound and thus serves as the acceptor for UDP-galactose.

9. d

10. Inside cells, glucose is rapidly phosphorylated to glucose 6-phosphate. This anionic sugar phosphate is not transported through cell membranes. Furthermore, most tissues (with the exception of liver and kidney) lack the necessary enzyme (glucose 6-phosphatase) to convert glucose 6-phosphate to free glucose that can leave the cells.

Polysaccharides and Oligosaccharides

11. Both c and d are correct. Since there are two glucosidic bonds in each trisaccharide and each bond can be α(1→4) or α(1→6), the total number of possible kinds of trisaccharides is four. However, two consecutive α(1→6) bonds would be very rare in glycogen or starch; therefore, one would be more likely to find three kinds.

12. The carbohydrate in bread is mostly starch, which is a polysaccharide mixture containing D-glucose residues linked by glucosidic bonds. All the aldehyde groups in each polysaccharide, except one at the free end, are involved in acetal bonds and do not react with Nelson's reagent. During mastication, α-amylase in saliva breaks many of the internal α-1,4-glucosidic bonds and exposes reactive aldehyde groups (reducing groups).

Note: Nelson's reagent consists of copper sulfate in a hot alkaline solution; a reducing sugar, such as glucose, reduces the copper, which in turn reduces the arsenomolybdate in the reagent, producing a blue complex.

13. Both cellulose and amylose are linear polymers of D-glucose, but the glucosidic linkages of cellulose are β-1,4 whereas those of amylose are α-1,4. The different configuration at the anomeric carbons determines a different spatial orientation of consecutive glucose residues. Thus, cellulose is capable of forming a linear, hydrogen-bonded structure, whereas amylose forms an open helical structure.

Sugar A
(α-anomeric form)

Glucose-α(1→6)-galactose
(β-anomeric form)

14. (a) 3, 7 (b) 4 (c) 1, 2, 5 (d) 1, 3, 5, 6 (e) 1, 3, 5

15. c

16. b

ANSWERS TO PROBLEMS

1. (a) The mechanism is probably the same as that described for ribonuclease on pages 211–214 of the text. A 2′,3′-cyclic phosphate is an intermediate in the pathway.

(b) Hydroxide ion, instead of the enzyme histidine residue, would serve as a proton acceptor. Since the mechanism involves the formation of a cyclic 2′,-3′-phosphoryl intermediate, DNA could not undergo alkaline hydrolysis because it has a hydrogen atom instead of an hydroxyl at the 2′ position on the pentose molecule.

2. (a) D-Xylose and D-lyxose differ in configuration at a single asymmetric center; they are epimers.

(b) α-D-Galactose and β-D-galactose have differing configurations at the C-1, or anomeric, carbon; they are anomers.

(c) D-Allose and D-talose are diastereoisomers because they have opposite configurations at one or more chiral centers, but they are not complete mirror images.

(d) L-Arabinose and D-arabinose are mirror images of each other and are therefore enantiomers.

3. The conversion requires inversions of configuration at C-3, C-4, and C-5 as well as a reduction of the carbon at position 6. The sequence of reactions in vivo is rather complex and occurs while D-glucose is attached to deoxythymidine diphosphate. The steps include oxidation at C-4, dehydration and reduction at C-5 and C-6, and reduction with inversion at C-4; at some point, inversion at C-3 also occurs.

4. Although the mirror image of a D-compound is an L-compound, the mirror image of an α-compound is an α-compound. (An α-compound has a 1-hydroxyl group in the α position.) Thus, α-L-glucose is the compound that is the mirror image, or enantiomer, of α-D-glucose.

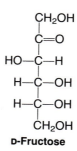

D-Fructose

5. (a) Compound X is D-glucose; it is the only aldose whose reduction will yield a hexitol with the same conformation as that of D-sorbitol.

(b) Sorbitol cannot form a hemiacetal because it has no aldehyde or ketone group. Therefore, neither type of ring can be formed by sorbitol.

(c) Compound Y is D-fructose, a ketose that is produced by the oxidation of sorbitol.

6. Only one dimer, alanylalanine, can be made from two alanine molecules linked via a peptide bond. However, the presence of several hydroxyl groups and the aldehydic function at the C-1 position of each D-galactose molecule provides an opportunity to make a larger number of dimers. Both the α and β forms of one molecule can form glycosidic linkages with the C-2, C-3, C-4, or C-6 hydroxyl groups of the other.

Recall that the C-5 position is not available because it participates in the formation of the pyranose ring. To these eight dimers can be added those dimers formed through glycosidic linkages involving the $\alpha\alpha$, $\alpha\beta$, or $\beta\beta$ configurations. Thus, eleven possible dimers exist. If one is allowed to use L forms, then the number of possible dimers increases greatly. This variety of linkages make the sugars very versatile molecules and yields many different structures that may be of utility in biology. However, this variety has also made the systematic study of the chemistry of polysaccharides very difficult.

7. The observations of Morgan and Watkins suggested that the sugar N-acetylgalactosamine in α linkage is the determinant of blood group A specificity. The galactose derivative binds to type-A lectins, occupying the sites that would otherwise bind to glycoproteins with N-acetylgalactosamine end-groups on the surfaces of type-A cells.

8. (a) First NAD^+ acts as an electron acceptor, oxidizing the C-5 carbon to the keto form. Later it acts as an electron donor, reducing the C-6 keto group to yield the inositol product. The reduction also regenerates the initial enzyme-NAD^+ complex and keeps the aldol (see b) from reopening.
   (b) An intramolecular aldol condensation involving the C-6 carbanion and the C-1 aldehyde leads to the formation of the carbocyclic ring.
   (c) An equilibrium exits among the open-chain form and the α and β anomers. As the open-chain form is consumed during the reaction, more of the open-chain form is generated from the anomers. Eventually all anomers are converted to the open-chain form, which is the substrate for the cyclase.
   (d) None of the carbons undergo inversion if one compares the carbons in the myoinositol 1-phosphate with those in the glucose 6-phosphate. The C-1 carbon is no longer an anomeric carbon because it has been reduced.

9. (a) The compound is a lactone, a cyclic ester formed between a carboxylate and an hydroxyl group.
   (b) Formation of a straight-chain compound requires hydrolysis of the ester linkage. Esters react with water in the presence of acid or base to yield carboxylic acid and alcohol groups, both part of the open-chain form of the molecule, called D-gluconic acid.
   (c) Because the C-1 atom is oxidized, an oxidized pyridine nucleotide is needed to act as an electron acceptor.

Glycolysis

Following introductions to the general principles of metabolism (Chapter 13) and the structures and properties of carbohydrates (Chapter 14), Stryer turns to the metabolism of carbohydrates via the glycolytic pathway. Glycolysis is the sequence of reactions that converts glucose into pyruvate with the concomitant trapping of the energy released as ATP. Stryer starts with a historical account of the discovery of some of the reactions of glycolysis. He then gives an overview of the key structures and types of reactions in glycolysis. Next, he describes the sequence of the reactions of glycolysis in three stages: (1) the conversion of glucose into fructose 1,6-bisphosphate; (2) the formation of triose-phosphate intermediates and the oxidation of glyceraldehyde 3-phosphate, which leads to the formation of one ATP; and (3) the conversion of 3-phosphoglycerate into pyruvate and the formation of a second ATP. Along with glucose, fructose and galactose are important dietary sugars, so their mode of entry into glycolysis is described. The regulation of glycolysis by the enzymes that catalyze the irreversible reactions in the pathway is discussed, and the fate of pyruvate is outlined. The three-dimensional arrangement of the NAD^+ binding sites of dehydrogenases, the induced-fit rearrangement of the hexokinase structure, and the mechanisms of action of several enzymes in the glycolytic pathway are described. Finally, the formation of 2,3-bisphosphoglycerate, the modulator of oxygen affinity in hemoglobin (Chapter 7), is described as a side reaction of glycolysis. In this chapter and others covering metabolism, a firm understanding of the properties of enzymes (Chapter 8), their mechanisms of action (Chapter 9), and their regulation (Chapter 10) is essential.

When you have mastered this chapter, you should be able to complete the following objectives.

Introduction (Stryer pages 349–351)

1. Define *glycolysis* and explain its role in the generation of *metabolic energy.*

2. List the alternative end-points of the glycolytic degradation of *glucose.*

3. Summarize the kinds of reactions and the types of intermediates found in glycolysis.

Stages of Glycolysis (Stryer pages 351–357)

4. Outline the three stages of glycolysis.

5. Describe the steps in the conversion of glucose to *fructose 1,6-bisphosphate,* including all the intermediates and enzymes. Note the steps where *ATP* is consumed.

6. List the reactions that convert fructose 1,6-bisphosphate, a hexose, into *glyceraldehyde 3-phosphate,* a three-carbon sugar.

7. Outline the steps of glycolysis between glyceraldehyde 3-phosphate and *pyruvate.* Recognize all the intermediates and enzymes and the co-factors that participate in the ATP-generating reactions.

8. Write the net reaction for the transformation of glucose into pyruvate and enumerate the ATP and NADH molecules formed.

Entry of Fructose and Galactose into Glycolysis (Stryer pages 357–359)

9. Describe the two pathways for the conversion of *fructose* into glyceraldehyde 3-phosphate.

10. Explain how *galactose* is transformed into *glucose 6-phosphate.* Note the role of *UDP-activated sugars* in this process.

11. Describe the biochemical defect in *galactosemia.*

Control of Glycolysis (Stryer pages 359–362)

12. List the key enzymes in the regulation of glycolysis.

13. Describe the allosteric regulation of *phosphofructokinase.*

14. Explain the role of *fructose 2,6-bisphosphate* in the regulation of phosphofructokinase. Describe the *tandem enzyme* that forms and degrades fructose 2,6-bisphosphate.

15. Discuss the regulation of *hexokinase.* Contrast the properties and physiological roles of hexokinase and *glucokinase.*

16. Define *isozyme.* Describe the regulation of the isozymes of *pyruvate kinase.*

The Fate of Pyruvate (Stryer pages 362–364)

17. Outline the reactions for the conversion of pyruvate into *ethanol, lactate,* or *acetyl CoA.*

18. Explain the role of *alcoholic fermentation* and lactate formation in the regeneration of NAD^+.

Structural Characteristics and Mechanisms of Action of Glycolytic Enzymes
(Stryer pages 364–368)

19. Describe the structure of the NAD^+-binding region common to many NAD^+-*linked dehydrogenases.*

20. Discuss the *induced-fit rearrangements* that occur in hexokinase upon glucose binding.

21. Summarize the most important features of the catalytic mechanisms of *aldolase* (*Schiff base* formation), *glyceraldehyde 3-phosphate dehydrogenase* (*thioester* formation), and pyruvate kinase (*enol → ketone* conversion).

Synthesis and Degradation of 2,3-Bisphosphoglycerate (Stryer pages 368–370)

22. Describe the formation of *2,3-bisphosphoglycerate (2,3-BPG)* from *1,3-bisphosphoglycerate* and the conversion of 2,3-BPG into *3-phosphoglycerate.*

23. Explain the role of 2,3-BPG in the interconversion of 3-phosphoglycerate and *2-phosphoglycerate.*

SELF-TEST

Introduction

1. Examine the reaction in the margin.

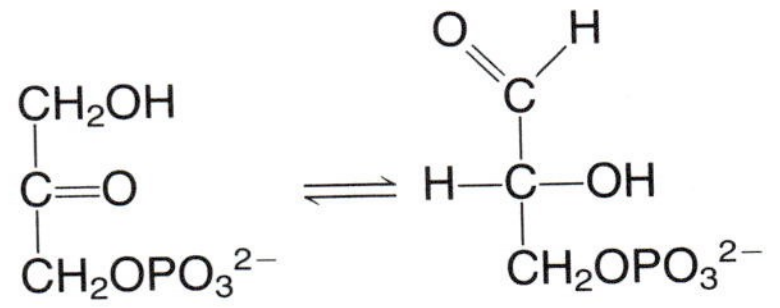

What type of a reaction is it?

   (a) Aldol cleavage
   (b) Dehydration
   (c) Phosphoryl transfer
   (d) Phosphoryl shift
   (e) Isomerization
   (f) Phosphorylation coupled to oxidation.

2. Draw the structures of

   (a) glycerate.

$$CH_2OH$$
$$C=O \rightleftharpoons$$
$$CH_2OPO_3^{2-}$$

$$\begin{array}{c} O \\ \backslash \\ C \end{array} \begin{array}{c} H \\ / \end{array}$$
$$H-C-OH$$
$$CH_2OPO_3^{2-}$$

(b) 1,3-bisphosphoglycerate; indicate the phosphate ester bond and mixed anhydride bond.

Stages of Glycolysis

3. Examine the following chemical structures of five of the intermediates of glycolysis.

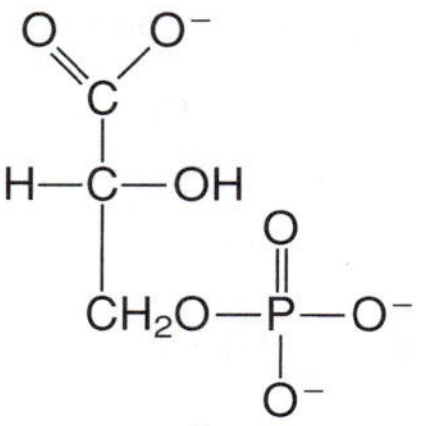

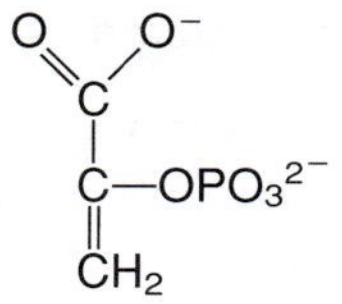

(a) Name the five intermediates:

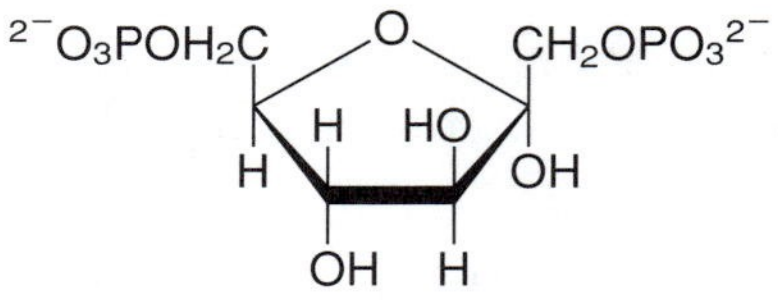

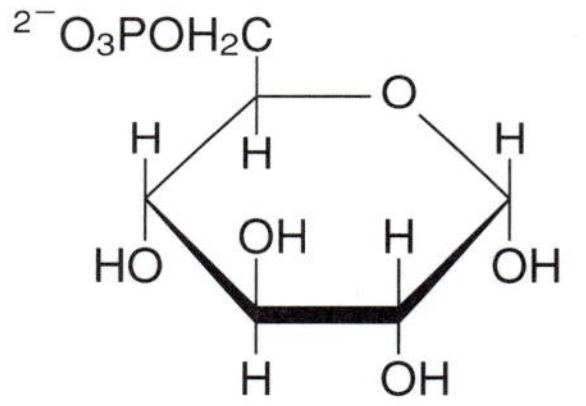

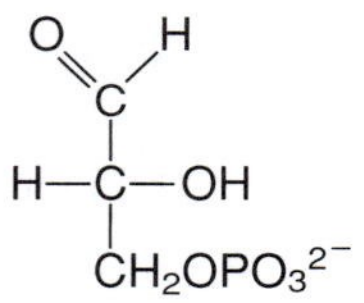

A _____

B _____

C _____

D _____

E _____

(b) Give the order in which these intermediates occur in glycolysis.

_____

(c) Name the enzyme or enzymes that produce the following intermediates:

B _____

D _____

E _____

(d) Which of these intermediates is a "high energy" compound, that is, has a large phosphate group transfer potential? _____

(e) Name the immediate precursor of intermediate A. _____

4. Hexokinase

(a) catalyzes the conversion of glucose 6-phosphate into fructose 1,6-bisphosphate.
(b) requires Ca^{2+} for activity.
(c) uses inorganic phosphate to form glucose 6-phosphate.
(d) catalyzes the transfer of a phosphoryl group to a variety of hexoses.
(e) catalyzes a phosphoryl shift reaction.

5. During the phosphoglucose isomerase reaction, the pyranose structure of glucose 6-phosphate is converted into the furanose ring structure of fructose 6-phosphate. Does this conversion require an additional enzyme? Explain.

6. The steps of glycolysis between glyceraldehyde 3-phosphate and 3-phosphoglycerate involve all of the following, *except*

(a) ATP synthesis.
(b) the utilization of P_i.
(c) the oxidation of NADH to NAD^+.
(d) the formation of 1,3-bisphosphoglycerate.
(e) catalysis by phosphoglycerate kinase.

7. The phosphofructokinase and the pyruvate kinase reactions are similar in that

(a) both generate ATP.
(b) both involve a "high energy" sugar derivative.
(c) both involve three-carbon compounds.
(d) both are essentially irreversible.
(e) both enzymes undergo induced-fit rearrangements after the binding of the substrate.

8. The reaction

$$\text{Phosphoenolpyruvate} + \text{ADP} + H^+ \longrightarrow \text{pyruvate} + \text{ATP}$$

has a $\Delta G^{\circ\prime} = -7.5$ kcal/mol and a $\Delta G' = -4.0$ kcal/mol under physiological conditions. Explain what these free-energy values reveal about this reaction.

9. If the C-1 carbon of glucose were labeled with ^{14}C, which of the carbon atoms in pyruvate would be labeled after glycolysis?

(a) The carboxylate carbon
(b) The carbonyl carbon
(c) The methyl carbon

10. Starting with fructose 6-phosphate and proceeding to pyruvate, what is the net yield of ATP molecules?

    (a) 1
    (b) 2
    (c) 3
    (d) 4
    (e) 5

Entry of Fructose and Galactose into Glycolysis

11. In the liver, the main conversion pathway from fructose to pyruvate includes the following intermediates?

    (a) Fructose 6-phosphate
    (b) Fructose 1,6-bisphosphate
    (c) Glyceraldehyde
    (d) Dihydroxyacetone phosphate
    (e) Phosphoenolpyruvate
    (f) 1,3-Bisphosphoglycerate

12. Galactose metabolism involves the following reactions:

    (1) Galactose + ATP $\longrightarrow$ Galactose 1-phosphate + ADP + H$^+$

    (2) ?

    (3) UDP-galactose $\rightleftharpoons$ UDP-glucose

    (a) Write the reaction for step 2.

    _____

    (b) Which step is defective in galactosemia? _____

    (c) Which enzymes catalyze steps 1, 2, and 3?

    _____

    (d) How is NAD$^+$ involved in the reaction in step 3?

    _____

    _____

13. Galactose is an important component of glycoproteins. Explain why withholding galactose from the diet of galactosemic patients has no effect on their synthesis of glycoproteins.

_____

_____

Control of Glycolysis

14. The essentially irreversible reactions that control the rate of glycolysis are catalyzed by which of the following enzymes?

    (a) Pyruvate kinase
    (b) Aldolase
    (c) Glyceraldehyde 3-phosphate dehydrogenase
    (d) Phosphofructokinase

(e) Hexokinase

(f) Phosphoglycerate kinase

15. Phosphofructokinase activity is enhanced by which of the following?

(a) Increased ATP concentration

(b) Increased fructose 2,6-bisphosphate concentration

(c) Decreased citrate concentration

(d) Decreased AMP concentration

(e) Increased H^+ concentration

16. When blood glucose levels are low, glucagon is secreted. Which of the following are the effects of increased glucagon levels on glycolysis and related reactions in liver?

(a) Phosphorylation of phosphofructokinase 2 and fructose bisphosphatase 2 occurs.

(b) Dephosphorylation of phosphofructokinase 2 and fructose bisphosphatase 2 occurs.

(c) Phosphofructokinase is activated.

(d) Phosphofructokinase is inhibited.

(e) Glycolysis is accelerated.

(f) Glycolysis is slowed down.

17. In which of the following is the enzyme correctly paired with its allosteric effector?

(a) Hexokinase–ATP

(b) Phosphofructokinase–glucose 6-phosphate

(c) Pyruvate kinase (L-isozyme)–alanine

(d) Phosphofructokinase–AMP

(e) Glucokinase–fructose 2,6-bisphosphate

18. Match hexokinase and glucokinase with the descriptions from the right column that are appropriate.

(a) Hexokinase _____

(b) Glucokinase _____

(1) is found in the liver.

(2) is found in nonhepatic tissues.

(3) is specific for glucose.

(4) has a broad specificity for hexoses.

(5) requires ATP for reaction.

(6) has a high K_M for glucose.

(7) is inhibited by glucose 6-phosphate.

The Fate of Pyruvate

19. Which of the following are metabolic products of pyruvate in higher organisms?

(a) Glycerol

(b) Lactic acid

(c) Acetone

(d) Acetyl CoA

(e) Ethanol

20. Write the complete reaction for the net conversion of glucose into lactate, including all the cofactors.

21. Since lactate is a "dead-end" product of metabolism in the sense that its sole fate is to be reconverted into pyruvate, what is the purpose of its formation?

_____

_____

Structural Characteristics and Mechanisms of Action of Glycolytic Enzymes

22. The characteristic three-dimensional structure of hexokinase and several other kinases includes two lobes that come together when a substrate is bound. This substrate-induced closing of the cleft

 (a) decreases the polarity of the environment at the active center.
 (b) allows for the tight binding of NAD^+.
 (c) facilitates the hydrolysis of ATP.
 (d) exposes most of the substrate on the surface of the enzyme.
 (e) rearranges the active site structure into four α helices and six parallel ß pleated sheets.

23. Match the enzymes in the left column with the appropriate reactive species or steps that occur during catalysis listed in the right column.

 (a) Aldolase _____   (1) Thioester intermediate
                                         (2) Enolpyruvate
 (b) Glyceraldehyde 3-phosphate   (3) Enolate anion
     dehydrogenase _____   (4) Removal of hydride ion
                                         (5) Protonated Schiff-base
 (c) Pyruvate kinase _____

24. For the glyceraldehyde 3-phosphate dehydrogenase reaction, explain how the oxidation of the aldehyde group ultimately gives rise to an acyl phosphate product.

_____

_____

Synthesis and Degradation of 2,3-Bisphosphoglycerate

25. 2,3-Bisphosphoglycerate

 (a) is formed from 1,3-bisphosphoglycerate.
 (b) is derived directly from glyceraldehyde 3-phosphate.
 (c) is formed by a kinase that phosphorylates 3-phosphoglycerate.
 (d) can be readily converted to 1,3-bisphosphoglycerate.
 (e) is synthesized and degraded by the enzymes of the glycolytic pathway.

26. Examine the reaction scheme in the margin.

 (a) Name the enzymes A and B. _____

 (b) The two phosphate groups on 1,3-BPG appear on 2,3-BPG after the reaction catalyzed by enzyme A. True or false? ____ Explain. _____

 (c) 2,3-BPG participates in catalytic amounts in which reaction of glycolysis? _____

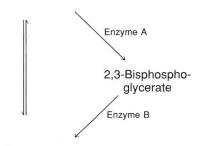

1,3-Bisphosphoglycerate

Enzyme A

2,3-Bisphospho-glycerate

Enzyme B

3-Phosphoglycerate

(d) 2,3-BPG decreases the affinity of hemoglobin for O_2. True or false? _____

Explain. _____

_____

PROBLEMS

1. Inorganic phosphate labeled with ^{32}P is added with glucose to a glycogen-free extract from liver, and the mixture is then incubated in the absence of oxygen. After a short time, 1,3-bisphosphoglycerate (1,3-BPG) is isolated from the mixture. On which carbons would you expect to find radioactive phosphate? If you allow the incubation to continue for a longer period, will you find any change in the labeling pattern? Why?

2. Mannose, the 2-epimer of glucose, and mannitol, a sugar alcohol, are widely used as dietetic sweeteners. Both compounds are only slowly transported across plasma membranes, but they can be metabolized by the liver. Propose a scheme by which mannitol and mannose can be converted into intermediates of the glycolytic pathway. You may wish to take advantage of the fact that hexokinase is relatively nonspecific. Why should such sugars be brought into glycolysis as early in the sequence as possible?

3. The value of $\Delta G^{\circ\prime}$ for the hydrolysis of sucrose to glucose and fructose is -7.30 kcal/mol. You have a solution that is 0.10 M in glucose and that contains sufficient sucrase enzyme to bring the reaction rapidly to equilibrium.

(a) What concentration of fructose would be required to yield sucrose at an equilibrium concentration of 0.01 M, at 25°C?

(b) The solubility limit for fructose is about 3.0 M. How might this limit affect your experiment?

4. Assay of hexokinase activity versus the concentration of glucose 6-phosphate yields the double-reciprocal plot shown in Figure 15-1. Explain the results shown in the figure and how the effect of glucose 6-phosphate on hexokinase is consistent with the overall control of the metabolism of glucose.

5. In 1905, Harden and Young, two English chemists, studied the fermentation of glucose using cell-free extracts of yeast. They monitored the conversion of glucose to ethanol by measuring the evolution of carbon dioxide from the reaction vessel. In one set of experiments, Harden and Young observed the evolution of CO_2 when inorganic phosphate (P_i) was added to a yeast extract containing glucose. In the graph in Figure 15-2, curve A shows what happens when no P_i is added. Curve B shows the effect of adding P_i in a separate experiment. As the evolution of CO_2 slows with time, more P_i is added to stimulate the reactions; this is shown by curve C.

(a) Why is glucose fermentation dependent upon P_i?

(b) During fermentation, what is the ratio of P_i consumed to CO_2 evolved?

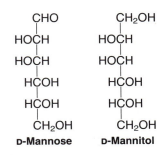

D-Mannose   **D-Mannitol**

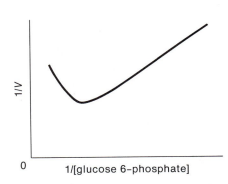

Figure 15-1
The influence of the concentration of glucose 6-phosphate on hexokinase.

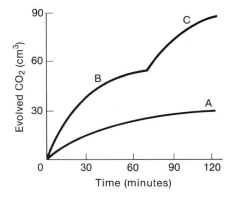

Figure 15-2
Evolution of CO_2 in the Harden-Young experiment.

(c) How does the formation of ethanol ensure that the fermentation process is in redox balance?

(d) Harden and Young found that they could recover phosphate from the reaction mixture, but it was not precipitable by magnesium citrate, as is P_i. Name at least three organic compounds that would be phosphorylated when P_i is added to the fermenting mixture.

(e) As the rate of CO_2 evolution decreased, Harden and Young found that an unusual compound accumulated in the reaction mixture. In 1907, Young identified the compound as a hexose bisphosphate. Name the compound and explain why it might accumulate when P_i becomes limiting.

(f) Later, Meyerhof showed that the addition of adenosine triphosphatase (ATPase, an enzyme that hydrolyzes ATP to yield ADP and P_i) to the reaction mixture stimulates the evolution of CO_2. Explain this result.

6. In solution, 80% of the fructose 6-phosphate is in the ß-anomeric form and 20% is in the α-anomeric form, with the half-time for anomerization being about 1.5 seconds. To determine which of the two anomers is a substrate for phosphofructokinase (PFK), Voll and his colleagues employed two model substrates (shown in the margin) that have C-2 configurations corresponding to the anomeric forms of fructose 6-phosphate.

(a) Why will neither of the substrates in the margin undergo mutarotation?

When the mannitol derivative is incubated with PFK and ATP, its rate of phosphorylation is about 80% of that for fructose 6-phosphate; K_M for fructose 6-phosphate is 0.04 mM, whereas K_M for the mannitol derivative is 0.40 mM. The glucitol derivative binds to PFK with an affinity almost equal to that of the mannitol derivative, but it is not phosphorylated by PFK; it is a competitive inhibitor of fructose 6-phosphate, with a K_i of 0.35 mM.

(b) Based on these observations, which anomer of fructose 6-phosphate is a substrate for PFK?

7. On page 368 of the text, Stryer provides the concentrations of many of the intermediates of the glycolytic pathway in erythrocytes. Given these concentrations, how much free energy is liberated when pyruvate and ATP are produced through the action of pyruvate kinase?

8. The absence of UDP-galactose-4-epimerase in the erythrocytes of a few individuals has been described; the tissues of these individuals exhibit relatively normal activity. What effect would the absence of epimerase activity have on galactose assimilation in the erythrocytes of these individuals? Would you expect to see the accumulation of galactitol in the lens? Why?

9. When the mechanism of action of UDP-galactose-4-epimerase was studied, investigators considered a 1,1-hydrogen shift or dehydration and rehydration as possible mechanisms for epimerization at C-4. However, when the enzyme was incubated with $^3H—H_2O$ and either UDP-galactose or UDP-glucose, no tritium was incorporated into either of the products. In addition, when one or the other of the substrates was incubated with epimerase in $^{18}O—H_2O$, there was no incorporation of ^{18}O into either product. Based on what you have read about the mecha-

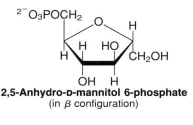

2,5-Anhydro-D-mannitol 6-phosphate
(in β configuration)

2,5-Anhydro-D-glucitol 6-phosphate
(in α configuration)

nism of the epimerase reaction, why should both sets of observations be expected?

10. Ahlfors and Mansour studied the activity of purified sheep phosphofructokinase (PFK) as a function of the concentration of ATP in experiments that were carried out at a constant concentration of fructose 6-phosphate. Typical results are shown in Figure 15-3. Explain these results, and relate them to the role of PFK in the glycolytic pathway.

11. Purified glyceraldehyde 3-phosphate dehydrogenase rapidly and irreversibly loses its catalytic activity when it is incubated with iodoacetate. Structural analysis of the acetylated protein reveals that a single cysteine residue at position 149 in the primary sequence is converted to the carboxymethyl form. In the presence of glyceraldehyde 3-phosphate, the enzyme is protected against inactivation by iodoacetate. Can you suggest a role for the cysteine residue at position 149 in the enzyme? Write out the reaction that occurs when iodoacetate reacts with the cysteine residue. What effect would you expect iodoacetate to have on glycolysis in living cells?

12. Hexokinase catalyzes the formation of glucose 6-phosphate at a maximum velocity of 2×10^{-5} mol/min, whereas V_{max} for the formation of fructose 6-phosphate is 3×10^{-5} mol/min. The value of K_M for glucose is 10^{-5} M, whereas K_M for fructose is 10^{-3} M. Suppose that in a particular cell the observed rates of phosphorylation are 1.0×10^{-8} mol/min for glucose and 1.5×10^{-5} mol/min for fructose.

(a) Estimate the concentrations of glucose and fructose in the cell.
(b) Which of these hexoses is more important in generating energy for this cell?

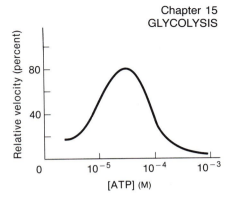

Figure 15-3
The effects of ATP concentration on sheep PFK.

ANSWERS TO SELF-TEST

Introduction

1. e

2. (a)

H—C—OH
H₂C—OH
Glycerate

(b)

H—C—OH
H₂C—OPO₃²⁻
1,3-Bisphosphoglycerate

Anhydride bond
Ester bond

Stages of Glycolysis

3. (a) A is 3-phosphoglycerate; B is fructose 1,6-bisphosphate; C is phosphoenolpyruvate; D is glucose 6-phosphate; and E is glyceraldehyde 3-phosphate.
(b) D, B, E, A, C
(c) B is produced by phosphofructokinase. D is produced by hexokinase or glucokinase. E is produced by aldolase or triose phosphate isomerase; aldolase produces E directly from fructose

1,6-bisphosphate, and triose phosphate isomerase converts dihydroxyacetone phosphate into E.
  (d) C
  (e) 1,3-Bisphosphoglycerate

4. d

5. No. The reaction catalyzed by phosphoglucose isomerase is a simple isomerization between an aldose and a ketose and involves the
open-chain structures of both sugars. Since glucose 6-phosphate and
fructose 6-phosphate are both reducing sugars, their Haworth ring
structures are in equilibrium with their open-chain forms. This equilibration is very rapid and does not require an enzyme. Note that this
isomerization reaction is of the same type as that catalyzed by triose-
phosphate isomerase.

6. c

7. d, e

8. The large negative $\Delta G^{\circ\prime}$ value indicates that equilibrium favors
product formation by a very large margin. The -4 kcal/mol value for
$\Delta G'$ means that under physiological conditions the reaction will also
proceed toward product formation essentially irreversibly. The fact
that $\Delta G'$ has a smaller negative value than $\Delta G^{\circ\prime}$ indicates that the concentrations of the products are considerably greater than the concen-
trations of the reactants under physiological conditions.

9. c

10. c

Entry of Fructose and Galactose into Glycolysis

11. c, d, e, and f. In the liver, a specific fructokinase forms fructose
1-phosphate, which is subsequently cleaved into glyceraldehyde and
dihydroxyacetone phosphate.

12. (a) Galactose 1-phosphate + UDP-glucose $\rightleftharpoons$

  glucose 1-phosphate + UDP-galactose
  (b) Step 2 is defective in galactosemia.
  (c) Galactokinase catalyzes step 1; galactose 1-phosphate uridyl
    transferase catalyzes step 2; and UDP-galactose-4-epimerase
    catalyzes step 3.
  (d) NAD^+ is tightly bound to the epimerase and is involved in
    forming a transient keto structure at C-4 that is an intermediate
    in the epimerization.

13. Because their epimerase activity is normal, galactosemic patients
are able to synthesize UDP-galactose from UDP-glucose. The UDP-
galactose is then used in the synthesis of glycoproteins.

Control of Glycolysis

14. a, d, e

15. b, c

16. a, d, f

17. c, d

18. (a) 1, 2, 4, 5, 7, (b) 1, 3, 5, 6

19. b, d

20. Glucose + 2 ADP + 2 P_i $\longrightarrow$ 2 lactate + 2 ATP + 2 H_2O

21. The reduction of pyruvate to lactate converts NADH to NAD^+, which is required in the glyceraldehyde 3-phosphate dehydrogenase reaction. This prevents glycolysis from stopping due to too low a concentration of NAD^+ and allows continued production of ATP.

Structural Characteristics and Mechanisms of Action of Glycolytic Enzymes

22. a

23. (a) 3, 5 (b) 1, 4 (c) 2

24. The oxidation of the aldehyde group by NAD^+ is an energetically favorable reaction that leads to the formation of a high-energy thioester bond between the substrate and the thiol group of a cysteine residue of the enzyme. Inorganic phosphate then attacks the thioester bond, which gives rise to an acyl-phosphate product, 1,3-bisphosphoglycerate.

Synthesis and Degradation of 2,3-Bisphosphoglycerate

25. a

26. (a) Enzyme A is bisphosphoglycerate mutase and enzyme B is 2,3-bisphosphoglycerate phosphatase.
    (b) False. The carbon atoms and the phosphate on the C-3 carbon of 2,3-BPG arise from 3-phosphoglycerate. Only the phosphate on C-2 of 2,3-BPG comes from 1,3-BPG. In fact 3-phosphoglycerate is an obligatory intermediate in the conversion of 1,3-BPG into 2,3-BPG.
    (c) The interconversion of 3-phosphoglycerate and 2-phosphoglycerate is catalyzed by phosphoglyceromutase, which requires catalytic amounts of 2,3-BPG to be active.
    (d) True. 2,3-BPG is present in high concentrations in erythrocytes; it is an allosteric effector of hemoglobin that stabilizes the R form (or deoxy form) of hemoglobin and decreases the affinity of hemoglobin for O_2.

ANSWERS TO PROBLEMS

1. After a short incubation time, labeled phosphate will be found on C-1 of 1,3-BPG. Inorganic phosphate enters the glycolytic pathway at the step catalyzed by glyceraldehyde 3-phosphate dehydrogenase:

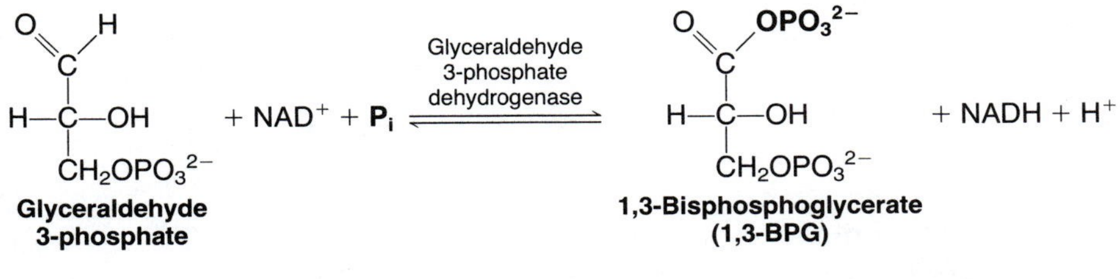

However, after a longer incubation time, the radioactive label will be found on both C-1 and C-3 of 1,3-BPG because the step subsequent to the formation of 1,3-BPG involves the phosphorylation of ADP to form ATP, which will be radioactively labeled in the γ-phosphoryl group:

$$
\begin{array}{ccc}
\underset{\text{1,3-Bisphosphoglycerate}}{
\begin{array}{c}
O \\
\parallel \\
C{-}\textbf{OPO}_3{}^{2-} \\
| \\
H{-}C{-}OH \\
| \\
CH_2OPO_3{}^{2-}
\end{array}}
& + \text{ ADP} \quad \underset{\text{kinase}}{\overset{\text{Phosphoglycerate}}{\rightleftharpoons}} \quad &
\underset{\text{3-Phosphoglycerate}}{
\begin{array}{c}
O \\
\parallel \\
C{-}O^- \\
| \\
H{-}C{-}OH \\
| \\
CH_2OPO_3{}^{2-}
\end{array}} \quad + \textbf{ ATP}
\end{array}
$$

In other glycolytic reactions, the radioactively labeled ATP can phosphorylate at C-1 of fructose 6-phosphate and C-6 of glucose (fructose 6-phosphate and glucose, respectively), both of which are equivalent to C-3 in 1,3-BPG. Thus, after prolonged incubation, both labeled inorganic phosphate *and* labeled ATP will be present in the mixture, and 1,3-BPG with a radioactive label at both C-1 and C-3 will be present in the extract. One must assume that a small amount of unlabeled ATP is available at the start to initiate hexose phosphorylation.

2. The first step is the conversion of mannitol to mannose. This requires oxidation at the C-1 of mannitol, using a dehydrogenase enzyme with NAD^+ or $NADP^+$ as an electron acceptor. One could then propose a number of schemes, using nucleotide derivatives with isomerase or epimerase activities in combination with one or more phosphorylated intermediates. An established pathway uses hexokinase and ATP for the synthesis of mannose 6-phosphate; this is then converted by mannose phosphate isomerase to form fructose 6-phosphate, an intermediate of the glycolytic pathway. Bringing such sugars into the glycolytic pathway as soon as possible means that already existing enzymes can be used to process the intermediates derived from each of a number of different sugars. Otherwise, a separate battery of enzymes would be needed to obtain energy from each of the sugars found in the diet.

3. (a) The reaction you are concerned with is

$$\text{Glucose} + \text{fructose} \longrightarrow \text{sucrose}$$

For the reaction in this direction, $\Delta G^{\circ\prime} = +7.30$ kcal/mol. The equilibrium constant K'_{eq} is

$$K'_{eq} = \frac{[\text{sucrose}]}{[\text{glucose}][\text{fructose}]}$$

At the start of the reaction, [sucrose] = 0, [glucose] = 0.10 M, and [fructose] = x. At equilibrium, [sucrose] = 0.01 M, [glucose] = 0.09 M, and [fructose] = $x - 0.01$ M. First, calculate K'_{eq} for the reaction at equilibrium:

$$\Delta G^{\circ\prime} = -1.364 \log_{10} K'_{eq}$$

$$7.3 = -1.364 \log_{10} K'_{eq}$$

$$\log_{10} K'_{eq} = \frac{7.3}{-1.364} = -5.35$$

$$K'_{eq} = \text{antilog} \,(-5.35)$$

$$= 4.46 \times 10^{-6} \text{ M}$$

Then find the concentration of fructose that satisfies the conditions at equilibrium. Assume that the unknown concentration of fructose at equilibrium ($x - 0.01$ M) is approximately equal to x.

$$K'_{eq} = 4.46 \times 10^{-6} \text{ M} = \frac{10^{-2} \text{ M}}{(0.09 \text{ M})(x)}$$

$$x = 2.49 \times 10^4 \text{ M}$$

(b) The concentration of fructose required to generate sucrose at a concentration of 0.01 M exceeds the solubility limit for fructose; it is therefore impossible to establish such conditions in solution.

4. The double-reciprocal plot shows that at relatively low concentrations of glucose 6-phosphate, velocity increases in proportion to substrate concentration. At higher concentrations of glucose 6-phosphate, the velocity of the hexokinase reaction decreases as the concentration of glucose 6-phosphate increases. In the cell, the concentration of glucose 6-phosphate rises when the levels of other intermediates in the glycolytic pathway increase. For example, when the energy requirements of the cell are being met and the levels of ATP and citrate are relatively high, the activity of phosphofructokinase decreases and the concentration of fructose 6-phosphate increases. A subsequent rise in the concentration glucose 6-phosphate occurs because it is in equilibrium with fructose 6-phosphate, and the activity of hexokinase decreases. This causes the rate of formation of glucose 6-phosphate to decrease, which allows the cell to stop utilizing glucose until it is needed for the generation of energy, NADPH, or glycogen. The reaction catalyzed by hexokinase is not the first *committed* step of the glycolytic pathway (that is, it is not the first irreversible step that is unique to glycolysis) because there are a number of possible fates of glucose 6-phosphate in the cell; for example, it can be oxidized in the pentose phosphate pathway, or it can be used as a source of glucosyl units in the synthesis of glycogen. It would therefore not be appropriate for glycolytic intermediates to regulate the synthesis of glucose 6-phosphate because the demands of other pathways would not be met by such a control system.

5. (a) Inorganic phosphate is required for one of the reactions of the glycolytic pathway: the phosphorylation by which glyceraldehyde 3-phosphate converted to 1,3-bisphosphoglycerate.

(b) The P_i/CO_2 ratio is 1.0, with one P_i being consumed for each pyruvate that undergoes decarboxylation.

(c) In most cells, the absolute concentration of NAD^+ and NADH is low. Both must undergo successive and continuous oxidation and reduction to continue to serve as donors and acceptors of electrons. In this case, NAD^+ must be constantly available for the continued activity of glyceraldehyde 3-phosphate dehydrogenase in the glycolytic pathway. The NADH generated during the oxidation of glyceraldehyde 3-phosphate is reoxidized to NAD^+ when acetaldehyde is reduced to ethanol.

(d) Initially, glyceraldehyde 3-phosphate is phosphorylated when P_i is added to the fermenting mixture. ADP is phosphorylated when 1,3-bisphosphoglycerate donates a phosphoryl group to the nucleotide and is itself converted to 3-phosphoglycerate. The ATP formed in this reaction can be used in either of two earlier reactions of glycolysis, the phosphorylations of glucose and fructose 6-phosphate.

(e) The hexose bisphosphate is fructose 1,6-bisphosphate, the only such intermediate in the glycolytic pathway. This compound accumulates when glycolytic flux is blocked at the glyceraldehyde 3-phosphate dehydrogenase step by limited P_i availability. As the phosphorylation of glucose continues, intermediates from the steps preceding the formation of 1,3-bisphosphoglycerate build up.

(f) The hydrolysis of ATP to ADP and P_i makes more P_i available for the phosphorylation of glyceraldehyde 3-phosphate. Under such conditions, glycolytic activity and CO_2 production are stimulated.

6. (a) Each of the model substrates lacks a C-2 hydroxyl group, which is required for the opening of the ring and mutarotation to the other anomeric form.

(b) The observations show that both substrates bind to phosphofructokinase (PFK) with almost equal affinities, but only the mannitol derivative, which has a β-conformation at C-2, is phosphorylated. Thus, it is likely that the β-anomer of fructose 6-phosphate is the substrate for PFK.

7. From page 357 of the text, the value of $\Delta G^{\circ\prime}$ for the reaction catalyzed by pyruvate kinase is -7.5 kcal/mol. We want to calculate $\Delta G^{\prime}$, the free energy released for the concentrations found in erythrocytes:

$$\Delta G^{\prime} = \Delta G^{\circ\prime} + 1.364 \log_{10} Q$$

where Q is the ratio of the products of the concentrations of products and substrates:

$$\Delta G^{\prime} = -7.5 + 1.364 \log_{10} \frac{[\text{pyruvate}][\text{ATP}]}{[\text{phosphoenolpyruvate}][\text{ADP}]}$$

$$= -7.5 + 1.364 \log_{10} \frac{(51)(1850)}{(23)(138)}$$

$$= -7.5 + 1.364 \log_{10} (29.73)$$

$$= -7.5 + (1.364 \times 1.472)$$

$$= -5.5 \text{ kcal/mol}$$

8. The absence of the epimerase activity in the erythrocytes means that they are unable to convert UDP-galactose to UDP-glucose. An accumulation of UDP-galactose and galactose 1-phosphate would be observed in the erythrocytes because galactose 1-phosphate uridyl transferase and galactose kinase are still active. Glucose 1-phosphate, the other product of transferase activity, can be isomerized to glucose 6-phosphate by phosphoglucomutase. Although one might expect to see some deterioration of erythrocyte function as a result of the absence of epimerase activity, this does not occur, perhaps because other tissues in the body can metabolize galactose. The ability of lens tissue to utilize galactose and its metabolites in the normal manner means that galactitol concentration in the lens are not elevated, in contrast to patients who lack galactose 1-phosphate uridyl transferase.

9. As outlined on page 359 of Stryer's text, a tightly bound NAD^+ molecule on the epimerase protein accepts the C-4 hydrogen and is converted to NADH. A 4-keto sugar nucleotide is transiently generated, which is converted to the other epimer when NADH adds a hydrogen atom to the other side of the 4-C carbon. Because the transfer of hydrogen is direct, there is no opportunity for proton exchange with the solvent when the reaction is carried out in tritiated water. Direct

hydrogen exchange also rules out dehydration and rehydration when the reaction is carried out in ^{18}O—H_2O.

10. The rate of the reaction catalyzed by PFK initially increases with the ATP concentration because ATP is a substrate for the reaction; it binds at the active site of PFK with fructose 6-phosphate and serves as a phosphoryl donor. At higher concentrations, ATP binds not only at the active site but also at the allosteric site; this alters the conformation of the enzyme and decreases the level of its activity. The effects of ATP on PFK are consistent with the role of PFK as a control element for the glycolytic pathway. When concentrations of ATP are relatively low, the activity of PFK is stimulated so that additional fructose 1,6-bisphosphate is made available for subsequent energy-generating reactions; when concentrations of ATP are higher and the demand of the cell for energy is lower, ATP inhibits PFK activity, thereby allowing glucose and other substrates to be utilized in other pathways. In many cells, ATP concentration is maintained at relatively high and constant levels, so that PFK is always subject to inhibition by ATP. Inhibition can be relieved by fructose 2,6-bisphosphate, which is synthesized when glucose is readily available. This allows cells to carry out glycolysis even when ATP levels are high, permitting the synthesis of building blocks from glucose.

11. The rapid inactivation of glyceraldehyde 3-phosphate dehydrogenase by iodoacetate and the protection against inactivation provided by glyceraldehyde 3-phosphate suggest that the cysteine residue at position 149 is unusually reactive and that it may be located near the active site. Indeed, other evidence has confirmed that the cysteine residue at position 149 is the one referred to in Stryer's description of the action of glyceraldehyde 3-phosphate dehydrogenase (p. 367 of the text). Normally the aldehyde substrate reacts with this cysteine to form a hemithioacetal. Iodoacetate reacts with the sulfhydryl group of the residue to form a carboxymethyl derivative, thereby inactivating the enzyme. This reaction is as follows:

$$E—Cys—S^- + I—CH_2—COO^- \longrightarrow E—Cys—S—CH_2—COO^- + I^-$$
Carboxymethyl enzyme
(inactive)

Glyceraldehyde 3-phosphate protects against inactivation by blocking the access of iodoacetate to the active site. Early experiments showed that iodoacetate can completely inhibit glycolysis in living cells; the irreversible inactivation of glyceraldehyde 3-phosphate dehydrogenase probably accounts for these observations.

12. (a) The values for V_{max}, K_M, and V, the measured velocity, are given for glucose and fructose. You can use the Michaelis-Menten equation by solving for [S], the substrate concentration:

$$V = \frac{V_{max}[S]}{[S] + K_M}$$

$$V_{max}[S] = V[S] + VK_M$$

$$[S](V_{max} - V) = VK_M$$

$$[S] = \frac{VK_M}{V_{max} - V}$$

Then you can use the values provided to calculate the concentrations of the two sugars in the cell. For glucose, $[S] = 5 \times 10^{-9}$ M; whereas for fructose, $[S] = 1 \times 10^{-3}$ M.

(b) The phosphorylation of a hexose such as glucose or fructose is the initial step in the oxidation of the sugar, a process in which energy in the form of ATP is generated. Fructose is more important in the provision of energy for the cell because V, the observed rate of formation, is 1500 times faster for fructose 6-phosphate than it is for glucose 6-phosphate.

CHAPTER 16

Citric Acid Cycle

The citric acid cycle, also known as the tricarboxylic acid cycle or the Krebs cycle, is the final oxidative pathway for carbohydrates, lipids, and amino acids. It is also a source of precursors for biosynthesis. Stryer begins Chapter 16 with a general description of the reactions of the citric acid cycle. He then discusses in detail the reaction mechanisms of the pyruvate dehydrogenase complex and citrate synthetase, as well as the stereospecificity of some of the reactions. In the following sections, Stryer summarizes the biosynthetic roles of the citric acid cycle and describes its relationship to the glyoxylate cycle found in bacteria and plants. He concludes the chapter with a discussion of control mechanisms, an account of the discovery of the citric acid cycle by Krebs, and some comments on the evolutionary selection of the citric acid cycle. The chapters on enzymes (Chapter 8 and 9), the introduction to metabolism (Chapter 13), and the chapter on glycolysis (Chapter 15) contain essential background material for this chapter.

LEARNING OBJECTIVES

When you have mastered this chapter, you should be able to complete the following objectives.

Introduction (Stryer pages 373–374)

1. Outline the role of the *citric acid cycle* in aerobic metabolism.

2. Locate the enzymes of the cycle in eucaryotic cells.

Reactions of the Citric Acid Cycle (Stryer pages 374–379)

3. Name all the *intermediates* of the citric acid cycle and draw their structures.

4. List the enzymatic reactions of the citric acid cycle in their appropriate sequence. Name all the enzymes.

5. Give examples of *condensation, dehydration, hydration, decarboxylation, oxidation,* and *substrate-level phosphorylation* reactions.

6. Indicate the steps of the cycle that yield CO_2, *NADH, FADH$_2$* and *GTP*.

7. Calculate the *yield of ATP* from the complete oxidation of pyruvate or of acetyl CoA.

The Pyruvate Dehydrogenase Complex, Citrate Synthase, and Stereospecific Enzymatic Reactions (Stryer pages 379–387, page 395)

8. Describe the composition of *pyruvate dehydrogenase* as a *multienzyme complex.*

9. List the *cofactors* that participate in the pyruvate dehydrogenase complex reactions and discuss the roles they play in the overall reaction.

10. Compare the pyruvate dehydrogenase complex and the *α-ketoglutarate dehydrogenase complex.*

11. Describe the consequences and the biochemical basis of *thiamine deficiency.*

12. Outline the enzymatic mechanism of *citrate synthase* (or *citrate synthetase*).

13. Explain the importance of the *induced-fit* rearrangements in citrate synthase.

14. Describe the *asymmetric reaction* of citrate. Follow the fate of each carbon atom of citrate through the cycle.

15. Define *chiral* and *prochiral* molecules. Familiarize yourself with the *RS designation of chirality.*

16. Describe the stereospecific transfer of hydrogen by the two classes of NAD^+ *dehydrogenases (A- or B-stereospecific).*

The Citric Acid Cycle as a Source of Biosynthetic Precursors and the Glyoxylate Cycle (Stryer pages 387–389)

17. Indicate the citric acid cycle intermediates that may be used as *biosynthetic precursors.*

18. Describe the role of *anaplerotic reactions* and discuss the *pyruvate carboxylase* reaction.

19. Compare the reactions of the *glyoxylate cycle* and those of the citric acid cycle. List the reactions that are unique to the glyoxylate cycle.

20. Describe the control of *isocitrate dehydrogenase* in bacteria and plants.

Control of the Pyruvate Dehydrogenase Complex and the Citric Acid Cycle
(Stryer pages 389–390)

21. Summarize the different modes of *regulation* of the pyruvate dehydrogenase complex. List the major *activators* and *inhibitors*.

22. Indicate the *control points* of the citric acid cycle and note the activators and inhibitors.

History and Evolution (Stryer pages 391–393)

23. Summarize the experiments that led to the discovery of the citric acid cycle.

24. Explain why the citric acid cycle evolved as the final oxidative pathway for fuel molecules.

SELF-TEST

Introduction

1. Which of the following statements are *correct?* The citric acid cycle

  (a) does not exist as such in plants and bacteria because its functions are performed by the glyoxylate cycle.
  (b) oxidizes acetyl CoA derived from fatty acid degradation.
  (c) produces most of the CO_2 in anaerobic organisms.
  (d) provides succinyl CoA for the synthesis of carbohydrates.
  (e) provides precursors for the synthesis of glutamic and aspartic acids.

2. If a eucaryotic cell were broken open and the subcellular organelles were separated by zonal ultracentrifugation on a sucrose gradient, in which of the following would the citric acid cycle enzymes be found?

  (a) Nucleus
  (b) Lysosomes
  (c) Golgi complex
  (d) Mitochondria
  (e) Endoplasmic reticulum

Reactions of the Citric Acid Cycle

3. Given the biochemical intermediates of the pyruvate dehydrogenase reaction and the citric acid cycle (see Figure 16-1), answer the following questions:

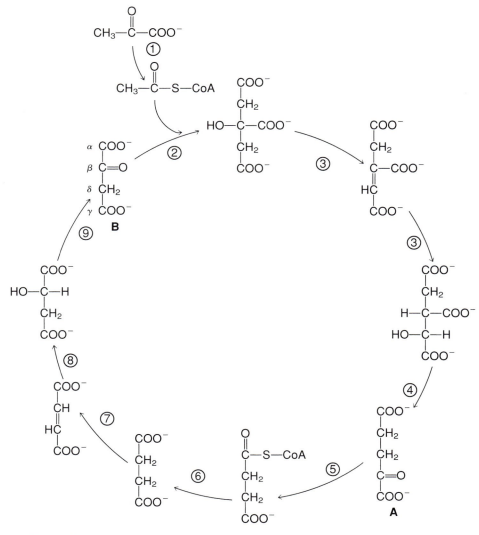

Figure 16-1
Citric acid cycle and the pyruvate dehydrogenase reaction.

(a) Name the intermediates:

A _____

B _____

(b) Draw the structure of isocitrate and show those atoms that come from acetyl CoA in bold letters.

(c) Which reaction is catalyzed by α-ketoglutarate dehydrogenase?

_____

(d) Which enzyme catalyzes step 2?

_____

(e) Which reactions are oxidations? Name the enzyme catalyzing each of them.

_____

_____

(f) At which reaction does a substrate-level phosphorylation occur? Name the enzyme and the products of this reaction.

_____

_____

(g) Which of the reactions require an FAD cofactor? Name the enzymes.

_____

_____

(h) Indicate the decarboxylation reactions and name the enzymes.

_____

_____

4. Succinate dehydrogenase

(a) is an iron-sulfur protein like aconitase.
(b) contains FAD and NAD^+ cofactors like pyruvate dehydrogenase.
(c) is an integral membrane protein unlike the other enzymes of the citric acid cycle.
(d) carries out an oxidative decarboxylation like isocitrate dehydrogenase.

5. The conversion of malate to oxaloacetate has a $\Delta G^{\circ\prime} = +7.1$ kcal/mol, yet in the citric acid cycle the reaction proceeds from malate to oxaloacetate. Explain how this is possible.

_____

_____

6. Considering the citric acid cycle steps between α-ketoglutarate and malate, how many high-energy phosphate bonds or net ATP molecules can be generated?

(a) 5
(b) 6
(c) 8
(d) 10
(e) 12

7. The standard free-energy change (in terms of net ATP production) when glucose is converted to 6 CO_2 and 6 H_2O is about how many times as great as the free-energy change when glucose is converted to two lactate molecules?

(a) 2
(b) 4
(c) 15
(d) 30

The Pyruvate Dehydrogenase Complex, Citrate Synthase, and Stereospecific Enzymatic Reactions

8. Match the cofactors of the pyruvate dehydrogenase complex in the left column with their corresponding enzyme components and with

their roles in the enzymatic steps that are listed in the right column.

(a) Coenzyme A _____

(b) NAD^+ _____

(c) Thiamine pyrophosphate _____

(d) FAD _____

(e) Lipoamide _____

(1) Pyruvate dehydrogenase component

(2) Dihydrolipoyl dehydrogenase

(3) Dihydrolipoyl transacetylase

(4) Oxidizes the hydroxyethyl group

(5) Decarboxylates pyruvate

(6) Oxidizes dihydrolipoamide

(7) Accepts the acetyl group from acetyllipoamide

(8) Provides a long, flexible arm that conveys intermediates to different enzyme components

(9) Oxidizes $FADH_2$

9. Which of the following enzymes have impaired activity in vitamin B_1 deficiency?

(a) Succinate dehydrogenase
(b) Pyruvate dehydrogenase
(c) Isocitrate dehydrogenase
(d) α-Ketoglutarate dehydrogenase
(e) Dihydrolipoyl transacetylase
(f) Transketolase

10. What are the potential advantages of a multienzyme complex with respect to the isolated enzyme components? Explain.

_____

_____

11. Which of the following statements concerning the enzymatic mechanism of citrate synthase is *correct*?

(a) Citrate synthase uses an NAD^+ cofactor.
(b) Acetyl CoA binds to citrate synthase before oxaloacetate.
(c) The histidine residues at the active site of citrate synthase participate in the hydrolysis of acetyl CoA.
(d) After citryl CoA is formed, additional structural changes occur in the enzyme.
(e) Each of the citrate synthase subunits binds one of the substrates and brings the substrates into close proximity to each other.

12. Citrate synthase binds acetyl CoA, condenses it with oxaloacetate to form citryl CoA, and then hydrolyzes the thioester bond of this intermediate. Why doesn't citrate synthase hydrolyze acetyl CoA?

_____

_____

13. If the methyl carbon atom of pyruvate is labeled with ^{14}C, which of the carbon atoms of oxaloacetate would be labeled after one turn of the citric acid cycle? (See the lettering scheme for oxaloacetate in Figure 16-1.)

(a) None. The label will be lost in CO_2.
(b) α
(c) β

(d) γ
(e) δ

14. Draw the structures of succinate, malate, and malonate, and indicate which of the molecules are chiral or prochiral.

Succinate	Malate	Malonate
_____	_____	_____

15. It has been shown that there are two kinds of NAD$^+$ (or NADP$^+$) dehydrogenases: A-stereospecific and B-stereospecific. Which of the following statements about these dehydrogenases is *correct*?

(a) The A-stereospecific dehydrogenases react with the R configuration and the B-stereospecific dehydrogenases react with the S configuration of the nicotinamide ring.

(b) The A- and B-stereospecific dehydrogenases have distinct active site pockets: one for the binding of NAD$^+$ and another for the binding of NADP$^+$.

(c) The A-stereospecific dehydrogenases transfer two hydrogen atoms, and the B-stereospecific dehydrogenases transfer a hydride ion.

(d) Both types of enzymes bind the same cofactors and carry out the same kinds of reactions, but they bind the cofactors with a different spatial orientation for the nicotinamide ring relative to the ribose ring.

The Citric Acid Cycle as a Source of Biosynthetic Precursors and the Glyoxylate Cycle

16. Match the intermediates of the citric acid cycle in the left column with their biosynthetic products in mammals, listed in the right column.

(a) Isocitrate _____

(b) α-Ketoglutarate _____

(c) Succinyl CoA _____

(d) *cis*-Aconitate _____

(e) Oxaloacetate _____

(1) Aspartic acid
(2) Glutamic acid
(3) Cholesterol
(4) Porphyrins
(5) None

17. Anaplerotic reactions

(a) are necessary because the biosynthesis of certain amino acids requires citric acid cycle intermediates as precursors.

(b) can convert acetyl CoA to oxaloacetate in mammals.

(c) can convert pyruvate into oxaloacetate in mammals.

    (d) are not required in mammals because mammals have an active glyoxylate cycle.

    (e) include the pyruvate dehydrogenase reaction operating in reverse.

18. Pyruvate carboxylase

    (a) catalyzes the reversible decarboxylation of oxaloacetate.

    (b) requires thiamine pyrophosphate as a cofactor.

    (c) is allosterically activated by NADH.

    (d) requires ATP.

    (e) is found in the cytoplasm of eucaryotic cells.

19. Malate synthase, an enzyme of the glyoxylate cycle, catalyzes the condensation of glyoxylate with acetyl CoA. Which enzyme of the citric acid cycle carries out a similar reaction? Would you expect the binding of glyoxylate and acetyl CoA to malate synthase to be sequential? Why?

_____

_____

_____

20. All organisms require 3- and 4-carbon precursor molecules for biosynthesis, yet bacteria can grow on acetate whereas mammals cannot. Explain why this is so.

_____

_____

21. Starting with acetyl CoA, what is the yield of high-energy phosphate bonds (net ATP formed) via the glyoxylate cycle?

    (a) 3

    (b) 6

    (c) 9

    (d) 12

    (e) 15

22. If a ^{14}C label were introduced in the carbonyl carbon of oxaloacetate, where would the label appear after one round of the glyoxylate cycle?

    (a) In the methylene carbons of succinate

    (b) In the carboxylate carbons of succinate

    (c) In the carbonyl carbon of glyoxylate

    (d) In the carboxylate carbon of glyoxylate

Control of the Pyruvate Dehydrogenase Complex and the Citric Acid Cycle

23. Indicate which of the following control mechanisms regulate the activity of the pyruvate dehydrogenase complex.

    (a) Product inhibition by acetyl CoA and NADH.

    (b) Feedback activation by ATP

    (c) Feedback inhibition by citrate

(d) Inhibition by the phosphorylation of the pyruvate dehydrogenase component
(e) Activation by dephosphorylation of the dihydrolipoyl transacetylase component

24. First select the enzymes in the left column that regulate the citric acid cycle. Then match those enzymes with the appropriate control mechanisms in the right column.

(a) Citrate synthase _____

(b) Aconitase _____

(c) Isocitrate dehydrogenase _____

(d) α-Ketoglutarate dehydrogenase _____

(e) Succinyl CoA synthetase _____

(f) Succinate dehydrogenase _____

(g) Fumarase _____

(h) Malate dehydrogenase _____

(1) Feedback inhibited by succinyl CoA and NADH
(2) Allosterically activated by ADP
(3) Allosterically inhibited by NADH and ATP
(4) Regulated by the availability of acetyl CoA and oxaloacetate

25. Although the ATP/ADP ratio and the availability of substrates and cycle intermediates are very important factors affecting the rate of the citric acid cycle, the $NADH/NAD^+$ ratio is of paramount importance. Explain why.

_____

_____

History and Evolution

26. Which of the following statements provide evidence for the *cyclic* functioning of the citric acid cycle in tissues?

(a) All the cycle enzymes are found in mitochondria.
(b) The addition of malonate inhibits CO_2 evolution from any of the intermediates.
(c) The addition of catalytic amounts of any of the intermediates greatly stimulates CO_2 evolution from endogenous substrates.
(d) The addition of citrate to liver slices leads to formation of small amounts of malate, fumarate, and oxaloacetate.
(e) Radiolabeled acetate (1-^{14}C-acetate) is converted to radiolabeled CO_2 very rapidly.

27. Which enzyme of the citric acid cycle is inhibited by malonate?

(a) Citrate synthase
(b) Succinyl CoA synthase
(c) Succinate dehydrogenase
(d) Fumarase
(e) Malate dehydrogenase

PROBLEMS

1. A cell is deficient in pyruvate dehydrogenase phosphate phosphatase. How would such a deficiency affect cellular metabolism?

2. ATP is an important source of energy for muscle contraction. Pyruvate dehydrogenase phosphate phosphatase is activated by calcium ion, which increases greatly in concentration during exercise. Why is activation of the phosphatase consistent with the metabolic requirements of muscle during contraction?

3. The oxidation of a fatty acid with an even number of carbon atoms yields a number of molecules of acetyl CoA, whereas the oxidation of an odd-numbered fatty acid yields not only molecules of acetyl CoA but also propionyl CoA, which then gives rise to succinyl CoA. Why does only the oxidation of odd-numbered fatty acids lead to the *net* synthesis of oxaloacetate?

4. In addition to its role in the action of pyruvate dehydrogenase, thiamine pyrophosphate (TPP) also serves as a cofactor for other enzymes, such as pyruvate decarboxylase, which catalyzes the *nonoxidative* decarboxylation of pyruvate. Propose a mechanism for the reaction catalyzed by pyruvate decarboxylase. What product would you expect? Why, in contrast to pyruvate dehydrogenase, are lipoamide and FAD not needed as cofactors for pyruvate decarboxylase?

5. The flavins FAD and FMN (flavin mononucleotide) are both bright yellow compounds; in fact, the name "flavin" was taken from *flavus*, the Latin word for yellow. The corresponding reduced compounds, $FADH_2$ and $FMNH_2$, are nearly colorless. What portion of a flavin molecule accounts for these color changes? The enzyme glucose oxidase, found in fungi, catalyzes the conversion of free glucose to gluconic acid. Glucose oxidase utilizes two molecules of FAD as cofactors. How could you use the light-absorbing properties of flavins as a means of monitoring glucose oxidase activity?

6. Some microorganisms can grow using ethanol as their sole carbon source. Propose a pathway for the utilization of this two-carbon compound; the pathway should convert ethanol into one or more molecules that can be used for energy generation and as biosynthetic precursors.

7. In the early 1900s, Thunberg proposed a cyclic pathway for the oxidation of acetate. In his scheme, two molecules of acetate are condensed, with reduction, to form succinate, which in turn is oxidized to yield oxaloacetate. The decarboxylation of oxaloacetate to pyruvate followed by the oxidative decarboxylation of pyruvate to acetate complete the cycle. Assuming that electron carriers like NAD^+ and FAD would be part of the scheme, compare the energy liberated by the Thunberg scheme to that liberated by the now-established citric acid cycle. Which of the steps in Thunberg's scheme were not found in subsequent studies?

8. When an isolated rat heart is perfused with sodium fluoroacetate, the rate of glycolysis decreases and hexose monophosphates accumulate. In cardiac cells, fluoroacetate is condensed with oxaloacetate to give fluorocitrate. Under these conditions, cellular citrate concentrations increase while the levels of other citric acid cycle components de-

crease. What enzyme is inhibited by fluorocitrate? How can you account for the decrease in glycolysis and the buildup of hexose monophosphates?

9. Transient phosphorylation of succinyl CoA synthetase in vitro can be achieved by incubating the enzyme with labeled GTP or with labeled inorganic phosphate and succinyl CoA. One can also observe the reversible formation of succinyl phosphate during the reaction. Propose a pathway for the steps catalyzed by the enzyme. How is the reaction catalyzed by succinyl CoA synthetase similar to that carried out by glyceraldehyde 3-phosphate dehydrogenase?

10. Recent studies suggest that succinate dehydrogenase activity is affected by oxaloacetate. Would you expect the enzyme activity to be enhanced or inhibited by oxaloacetate?

11. In addition to the carboxylation of pyruvate, there are other anaplerotic reactions that help to maintain appropriate levels of oxaloacetate. For example, the respective amino groups of glutamate and aspartate can be removed to yield the corresponding α-keto acids. How can these α-keto acids be used to replenish oxaloacetate levels?

12. Lipoic acid and FAD serve as prosthetic groups in the enzyme isocitrate dehydrogenase. Describe their possible roles in the reaction catalyzed by the enzyme.

ANSWERS TO SELF-TEST

Introduction

1. b, e

2. d

Reactions of the Citric Acid Cycle

3. (a) A: α-ketoglutarate; B: oxaloacetate
   (b) See the structure of isocitrate in the margin.
   (c) Reaction 5
   (d) Citrate synthase
   (e) Step 1, pyruvate dehydrogenase; step 4, isocitrate dehydrogenase; step 5, α-ketoglutarate dehydrogenase; step 7, succinate dehydrogenase; and step 9, malate dehydrogenase
   (f) step 6; the enzyme is succinyl CoA synthetase; the products of the reaction are succinate, CoA, and GTP.
   (g) Step 1, dihydrolipoyl dehydrogenase component of the pyruvate dehydrogenase complex. Step 5, dihydrolipoyl dehydrogenase component of the α-ketoglutarate dehydrogenase complex. Step 7, succinate dehydrogenase.
   (h) Step 1, pyruvate dehydrogenase; Step 4, isocitrate dehydrogenase; and Step 5, α-ketoglutarate dehydrogenase.

4. a, c

$$COO^-$$
$$CH_2$$
$$H-C-COO^-$$
$$HO-C-H$$
$$COO^-$$

Isocitrate

5. Although this step is energetically unfavorable at standard conditions, in mitochondria the concentrations of malate and NAD^+ are relatively high and those of the products, oxaloacetate and NADH, are quite low, so the overall $\Delta G'$ for this reaction is negative.

6. b

7. c. From glucose to lactate 2 ATP are formed; from glucose to CO_2 and H_2O over 30 ATP are formed.

The Pyruvate Dehydrogenase Complex, Citrate Synthase, and Stereospecific Enzymatic Reactions

8. (a) 3, 7 (b) 2, 9 (c) 1, 5 (d) 2, 6 (e) 3, 4, 8

9. b, d, f

10. A multienzyme complex can carry out the coordinated catalysis of a complex reaction. The intermediates in the reaction remain bound to the complex and are passed from one enzyme component to the next, which increases the overall reaction rate and minimizes side reactions. In the case of isolated enzymes, the reaction intermediates would have to diffuse randomly between enzymes.

11. d

12. Citrate synthase binds acetyl CoA only after oxaloacetate has been bound and the enzyme structure has rearranged to create a binding site for acetyl CoA. After citryl CoA is formed, there are further structural changes that bring an aspartate residue and a water molecule into the vicinity of the thioester bond for the hydrolysis step. Thus, acetyl CoA is protected from hydrolysis.

13. c and d. Both of the middle carbons of oxaloacetate will be labeled because succinate is a symmetrical molecule.

14. See the structures in the margin. Succinate is a prochiral molecule; the prochiral centers are the methylene carbons. Malate is a chiral and prochiral molecule; the carbon atom bearing the OH group is a chiral center, and the methylene carbon is a prochiral center. Malonate is neither chiral nor prochiral.

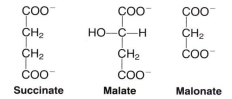

Succinate   Malate   Malonate

15. d

The Citric Acid Cycle as a Source of Biosynthetic Precursors and the Glyoxylate Cycle

16. (a) 5 (b) 2 (c) 4 (d) 5 (e) 1

17. a, c

18. a, d

19. The condensation of glyoxylate and acetyl CoA carried out by malate synthase in the glyoxylate cycle is similar to the condensation of oxaloacetate and acetyl CoA carried out by citrate synthase in the citric acid cycle. The binding of glyoxylate first, which induces structural changes in the enzyme that allow the subsequent binding of acetyl CoA, would be expected in order to prevent the premature hydrolysis of acetyl CoA. See Question 12.

20. Bacteria are capable, via the glyoxylate cycle, of synthesizing 4-carbon precursor molecules for biosynthesis (e.g., malate) from acetate or acetyl CoA. Mammals do not have an analogous mechanism; in

the citric acid cycle, the carbon atoms from acetyl CoA are released as CO_2, and there is no net synthesis of 4-carbon molecules.

21. a

22. a

Control of the Pyruvate Dehydrogenase Complex and the Citric Acid Cycle

23. a, d

24. a, c, d. (a) 4 (c) 2, 3 (d) 1
    The regulation of citrate synthase is explained differently in different texts. Although ATP is given as an allosteric inhibitor of citrate synthase by Stryer, other texts do not take this view. There is a general agreement, however, that citrate synthase is quite sensitive to the levels of available oxaloacetate and acetyl CoA.

25. The oxidized cofactors NAD^+ and FAD are absolutely required as electron acceptors in the various dehydrogenation reactions of the citric acid cycle. When these oxidized cofactors are not available, as when their reoxidation stops in the absence of O_2 or respiration, the citric acid cycle also stops.

History and Evolution

26. b, c

27. c

ANSWERS TO PROBLEMS

1. Pyruvate dehydrogenase phosphate phosphatase removes a phosphoryl group from pyruvate dehydrogenase, activating the enzyme complex and accelerating the rate of synthesis of acetyl CoA. Cells deficient in phosphatase activity cannot activate pyruvate dehydrogenase, so that the rate of entry of acetyl groups into the citric acid cycle will decrease, as will aerobic production of ATP. Under such conditions, stimulation of glycolytic activity and a subsequent increase in lactate production would be expected, as the cell responds to continued requirement for ATP synthesis.

2. As discussed in problem 1, the phosphatase activates pyruvate dehydrogenase, stimulating the rate of both glycolysis and the citric acid cycle. Calcium-mediated activation of pyruvate dehydrogenase therefore promotes increased production of ATP, which is then available for muscle contraction.

3. The entry of acetyl groups from acetyl CoA into the citric acid cycle does not contribute to the net synthesis of oxaloacetate because two carbons are lost as CO_2 in the pathway from citrate to oxaloacetate. Only the entry of compounds with three or more carbons, like succinate, can increase the relative number of carbon atoms in the pathway. Thus while odd-numbered fatty acids contribute to the net synthesis of oxaloacetate, those compounds with an even number of fatty acids do not.

4. The mechanism is similar to that shown on page 380 of Stryer, in which the C-2 carbanion of TPP attacks the α-keto group of pyruvate.

The subsequent decarboxylation of pyruvate is enhanced by the delocalization of electrons in the ring nitrogen of TPP. The initial product is hydroxyethyl-TPP, which is cleaved upon protonation to yield acetaldehyde and TPP. In contrast to the reaction catalyzed by pyruvate dehydrogenase, no net oxidation occurs, so lipoamide and FAD, which serve as electron acceptors, are not needed.

5. The conjugated π-system in the isoalloxazine ring of oxidized flavins like FAD and FMN accounts for their intense yellow color. The reduced molecules are partially saturated and their remaining double bonds are not conjugated, making them nearly colorless. To monitor glucose oxidase activity, one could use spectrophotometry to determine the wavelength of maximum absorption for FAD and then monitor the oxidation of the flavin in the enzyme by observing changes in absorption. As glucose is oxidized and FAD is reduced to $FADH_2$, one will observe a corresponding decrease in absorption.

6. The microorganism first converts ethanol to acetic acid, or acetate, by carrying out two successive oxidations, with acetaldehyde as an intermediate. Two molecules of a reduced electron carrier such as NADH will also be produced. Next, acetate is activated through the action of acetyl CoA synthetase to form acetyl CoA, and then the acetyl group is transferred to oxaloacetate to form citrate. After citrate is converted to isocitrate, two enzymes of the glycolytic pathway, isocitrate lyase and malate synthase, assist in the net formation of oxaloacetate from isocitrate and another molecule of acetyl CoA, as discussed in Stryer's text on page 389. Oxaloacetate can then be used for generation of energy as well as production of biosynthetic intermediates. Note that a small amount of oxaloacetate and other intermediates of the citric acid cycle must be present initially in order for acetate to enter the pathway.

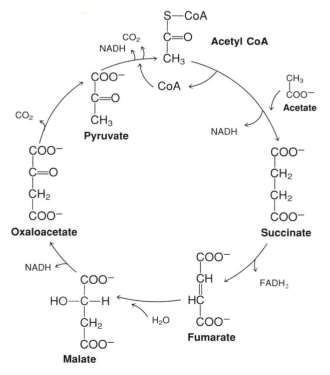

Figure 16-2
Thunberg's cycle.

7. As shown in Figure 16-2, there are at least four steps that generate reduced electron carriers. For each acetate group consumed, three NADH and one $FADH_2$ are generated, and their subsequent reoxidation in the electron transport chain provides energy for the generation of eleven molecules of ATP. The same number of reduced electron carriers is generated though the action of pyruvate dehydrogenase and the enzymes of the citric acid cycle, so that the energy liberated by both schemes is the same. Each of the reactions shown in Thunberg's scheme is known to occur, except for the condensation of two acetyl groups to form succinate.

8. The accumulation of citrate and the decrease in the levels of other citric acid cycle intermediates suggest that aconitase is inhibited by fluorocitrate. Excess citrate inhibits phosphofructokinase, causing a decrease in the rate of glycolysis and an accumulation of hexose monophosphates such as glucose 6-phosphate and fructose 1,6-bisphosphate.

9. The enzyme carries out three reversible steps leading to the formation of succinate and GTP. The first is the hydrolysis of succinyl CoA; the free energy of hydrolysis is used in the second step to form succinyl phosphate, which is noncovalently bound to the enzyme. Finally, the phosphate group is then transferred to an amino acid side chain in the enzyme. (Studies have shown that a specific histidine residue is phosphorylated.) The phosphate group is then transferred to GDP, yielding GTP. The overall reaction is another example of a substrate-level phosphorylation, in which the energy for the transfer of the phosphate is derived from a chemical reaction, rather than electron transfer. A similar reaction is catalyzed by glyceraldehyde 3-phosphate dehydrogenase.

10. Oxaloacetate is derived from succinate; it is a product of the reaction catalyzed by succinate dehydrogenase. When levels of oxaloacetate are high, one would expect the activity of the enzyme to be reduced. Low levels of oxaloacetate would call for an increase in succinate production.

11. Examination of the structures of the α-keto acid analogs of glutamate and aspartate shows that they are in fact both citric acid cycle intermediates, α-ketoglutarate and oxaloacetate. Glutamine, when it is deaminated, thus contributes directly to the insertion of additional molecules of oxaloacetate. α-Ketoglutarate, as a component of the citric acid cycle, is a precursor of oxaloacetate.

12. Lipoic acid contains a sulfhydryl group that could act as an acceptor for electrons from isocitrate. Those electrons could then be transferred to NAD^+ via FAD. The roles of the prosthetic groups would be similar to those they play in the reaction catalyzed by pyruvate dehydrogenase.

CHAPTER 17

Oxidative Phosphorylation

The reduced NADH and $FADH_2$ formed in glycolysis (Chapter 15) and in the citric acid cycle (Chapter 16) ultimately transfer their electrons to oxygen and release a large amount of useful energy in the process. The energy from this electron transfer is coupled via a transmembrane proton gradient to the synthesis of ATP. This process, known as oxidative phosphorylation, produces the bulk of the ATP in aerobic organisms. Stryer begins this chapter with an introduction to the concept of reduction potential as a measure of the free energy of chemical reactions (see Chapters 8 and 13) involving electron transfers. He then describes the components of the electron-transfer chain, their location in the mitochondria of eucaryotes, and the sequence of electron-transfer reactions that occur between the reduced cofactors and the final electron acceptor, O_2. Next, Stryer discusses the generation of a proton gradient as a consequence of the electron flow and the use of the resulting proton-motive force in the synthesis of ATP. Since ATP synthesis occurs inside mitochondria whereas a majority of the reactions that utilize ATP take place in the cytosol, the membrane shuttle systems for ATP-ADP and other cofactors and biomolecules are described. Finally, the control of oxidative phosphorylation, the mechanisms that uncouple electron transfer and oxidative phosphorylation, and the enzyme systems that degrade toxic, partially reduced products of oxygen are discussed.

When you have mastered this chapter, you should be able to complete the following objectives.

Introduction (Stryer pages 397–399)

1. Define *oxidative phosphorylation* and *respiration*.

2. Describe the *compartments* and *membranes of mitochondria* and locate the *respiratory assemblies* in them.

Redox Potentials and Free-Energy Changes (Stryer pages 399–401)

3. Relate the concepts of *redox potential* and *free-energy change*.

4. Describe the meaning and the measurement of the redox potential (E_0) for a *redox couple* relative to the *standard reference half-cell*.

5. Explain the meaning of E_0'.

6. Calculate $\Delta G^{\circ\prime}$ for *oxidation-reduction reactions* from the redox potentials for the individual redox couples.

The Respiratory Chain (Stryer pages 401–409)

7. List the components of the *respiratory chain* and the *electron-carrying groups*.

8. Describe the entry of electrons from *NADH* into *NADH-Q reductase* and their path through this *enzyme complex*.

9. Discuss the role of *ubiquinone (coenzyme Q or Q)* as a *mobile electron carrier* between NADH-Q reductase and *cytochrome reductase*.

10. Describe the entry of electrons into the respiratory chain from *flavoproteins* such as *succinate dehydrogenase, glycerol phosphate dehydrogenase,* and *fatty acyl CoA dehydrogenase*.

11. Describe the structures and properties of the *cytochromes* and distinguish among cytochromes b, c_1, c, a, and a_3.

12. List the components of the *cytochrome reductase complex*, and explain how *ubiquinol* transfers its electrons to cytochromes c_1 and b (b-566 and b-562) and ultimately to cytochrome c.

13. Explain how a two-electron carrier, ubiquinol, can interact with a one-electron carrier, the *Fe-S cluster*.

14. Describe *cytochrome oxidase* and its electron-carrying groups.

15. Outline the mechanism for the reduction of O_2 to H_2O on cytochrome oxidase. Describe the path of the electrons from the a-Cu_A cluster to the a_3-Cu_B cluster and note the changes in the oxidation states of Fe and O.

16. Describe the salient features of the three-dimensional structure of cytochrome c and relate them to its interaction with cytochrome reductase and cytochrome oxidase.

17. Discuss the conservation of the structure of cytochrome c in evolution.

Coupling of Oxidation and Phosphorylation by a Proton-Motive Force (Stryer pages 410–417)

18. Describe the *chemiosmotic model* of oxidative phosphorylation and give the evidence that supports it.

19. Correlate the sites of the electron-transport chain where *proton gradients* are generated with the *ATP yield* from various substrates, *P:O ratios,* estimates of $\Delta G^{\circ\prime}$ for oxidation-reduction reactions, and *specific inhibitors of electron flow.*

20. Discuss the possible origins of the protons that are translocated during electron transport.

21. Describe the mitochondrial location, the subunit structure, and the function of eucaryotic *ATP synthase.*

22. Outline the proposed mechanism of ATP synthesis by ATP synthase during proton flow. Relate *catalytic cooperativity* to the *binding-change mechanism.*

Mitochondrial Transport Systems (Stryer pages 417–421)

23. Explain the roles of the *glycerol-phosphate* and *malate-aspartate shuttles* in the oxidation of cytoplasmic NADH. Calculate the ATP yield in each case.

24. Describe the *ATP-ADP translocase* mechanism.

25. List the most important *mitochondrial transport systems* for ions and metabolites.

26. Calculate the net yield of ATP from the complete oxidation of glucose, taking into account the different shuttles for the cytoplasmic reducing equivalents of NADH.

Control of Oxidative Phosphorylation, Uncouplers, Scavengers of Toxic Derivatives of O_2, and Power Transmission by Proton Gradients (Stryer pages 421–423)

27. Describe *respiratory control.*

28. Explain the effect of *uncouplers* on oxidative phosphorylation. Note how regulated uncoupling can be used for *thermogenesis.*

29. Explain why oxygen is a potentially toxic substance. Summarize the reactions and the biological roles of *superoxide dismutase, catalase,* and the *peroxidases.*

30. List examples of *energy conversions* by proton gradients.

SELF-TEST

Introduction

1. Explain the difference between oxidative phosphorylation and substrate level phosphorylation as seen in glycolysis and the citric acid cycle.

2. Which of the following statements regarding mitochondria and their components are *correct*?

(a) Mitochondria are approximately 20 μm in diameter.
(b) The matrix compartment contains the enzymes of glycolysis.
(c) Mitochondria have two membrane systems: an inner membrane and an outer membrane.
(d) The inner membrane contains pores and is readily permeable to most small metabolites.
(e) The inner membrane has a large surface area because it is highly folded.

Redox Potentials and Free-Energy Changes

3. Which of the following statements about the redox potential for a reaction are *correct*?

(a) It is used to describe phosphate group transfers.
(b) It is unrelated to the free energy of the reaction.
(c) It can be used to predict when a given compound will reduce another, given a suitable catalyst.
(d) It can be used to predict whether a given oxidation will provide sufficient energy for the formation of ATP from ADP + P_i.
(e) It can be used to predict the rate of O_2 uptake at the expense of a given substrate.

4. The equation for the reduction of cytochrome *a* by cytochrome *c* is as follows:

$$\text{Cyt } a \,(+3) + \text{cyt } c \,(+2) \longrightarrow \text{cyt } a \,(+2) + \text{cyt } c \,(+3)$$

where

$$\text{Cyt } a \,(+3){:}\text{cyt } a \,(+2) \qquad\qquad E_0' = 0.27 \text{ V}$$
$$\text{Cyt } c \,(+3){:}\text{cyt } c \,(+2) \qquad\qquad E_0' = 0.22 \text{ V}$$

Under standard conditions—that is, 1 M concentrations of reactants and products—this reaction

(a) proceeds spontaneously.
(b) yields sufficient energy for ATP synthesis.
(c) does not alter the absorption spectra of the cytochromes.
(d) involves the transfer of two electrons.

5. Calculate $\Delta G^{\circ\prime}$ for the reaction:

$$\text{Succinate} + \text{FAD} \longrightarrow \text{fumarate} + \text{FADH}_2$$

$$\text{Fumarate}{:}\text{succinate} \qquad\qquad E_0' = 0.03 \text{ V}$$
$$\text{FAD}{:}\text{FADH}_2 \qquad\qquad E_0' = 0 \text{ V}$$

(a) -1.38 kcal/mol
(b) -0.69 kcal/mol

(c) 0.14 kcal/mol
(d) 0.69 kcal/mol
(e) 1.38 kcal/mol

6. The parameters $\Delta G^{\circ\prime}$ and $\Delta E_0^\prime$ can be used to predict the direction of chemical reactions at standard conditions. On the other hand, $\Delta G^\prime$ can be used for any concentrations of reactants and products to predict whether a chemical reaction will proceed. Using the expressions

$$\Delta G^\prime = \Delta G^{\circ\prime} + 2.3 \, RT \log_{10} \frac{[\text{Products}]}{[\text{Reactants}]}$$

$$\Delta G^{\circ\prime} = -nF \, \Delta E_0^\prime$$

derive an expression for $\Delta E^\prime$. Explain the significance of this redox potential.

7. Obtain $\Delta G^\prime$ and $\Delta E^\prime$ for the reaction given in Question 5 when the succinate concentration is 2×10^{-3} M, the fumarate concentration is 0.5×10^{-3} M, the FAD concentration is 2×10^{-3} M, the FADH$_2$ concentration is 0.2×10^{-3} M, and the temperature is 37°C. ($R = 1.98$ cal/mol $\cdot$ K)

The Respiratory Chain

8. Place the following respiratory chain components in their proper sequence. Also, indicate which are mobile carriers of electrons.

(a) Cytochrome c
(b) NADH-Q reductase
(c) Cytochrome oxidase
(d) Ubiquinone
(e) Cytochrome reductase

9. Match the enzyme complexes of the respiratory chain in the left column with the appropriate electron carrying groups from the right column.

(a) Cytochrome oxidase _____
(b) Cytochrome reductase _____
(c) NADH-Q reductase _____
(d) Succinate-Q reductase _____

(1) Heme c_1
(2) FAD
(3) Heme a_3
(4) Heme b-562
(5) Fe-S
(6) Cu_A and Cu_B
(7) FMN
(8) Heme a
(9) Heme b-566

10. Which of the following statements about the enzyme complexes of the electron-transport system are *correct*?

(a) They are located in the mitochondrial matrix.
(b) They cannot be isolated from one another in functional form.
(c) They have very similar visible spectra.
(d) They are integral membrane proteins located in the inner mitochondrial membrane.
(e) They interact with each other via mobile carriers of electrons.

11. Which of the following statements about ubiquinol are *correct*?

(a) It is the mobile electron carrier between cytochrome reductase and cytochrome oxidase.
(b) It is an integral membrane protein.
(c) Its oxidation involves the simultaneous transfer of two electrons to the Fe-S center of cytochrome reductase.
(d) It is oxidized to ubiquinone via a semiquinone intermediate.
(e) It is a lipid soluble molecule.

12. Which cytochrome has a protoporphyrin IX heme that is not covalently bound to protein?

(a) Cytochrome a
(b) Cytochrome a_3
(c) Cytochrome b
(d) Cytochrome c
(e) Cytochrome c_1

13. Explain the roles of cytochrome c_1 and the b cytochromes (b-562 and b-566) in the oxidation of ubiquinol to ubiquinone.

_____

_____

_____

_____

14. In the reduction of O_2 to H_2O by cytochrome oxidase, four electrons and four protons are used. How can this occur when a single electron is transferred by heme iron and by copper at a time?

_____

_____

_____

15. Which of the following statements about the structure and properties of cytochrome *c* is *incorrect*?

(a) It has negatively charged patches on its surface that permit interactions with cytochrome reductase and cytochrome oxidase.
(b) It contains very little α-helix or β-pleated-sheet secondary structure.
(c) It contains a heme with an Fe bonded to a histidine and a methionine residue.
(d) It has retained a highly conserved conformation throughout evolution.
(e) It is soluble in water.

Coupling of Oxidation and Phosphorylation by a Proton-Motive Force

16. Which of the following experimental observations provide evidence supporting the chemiosmotic model of oxidative phosphorylation?

(a) A closed membrane or vesicle compartment is required for oxidative phosphorylation.
(b) A system of bacteriorhodopsin and ATPase can produce ATP in synthetic vesicles when light causes proton pumping.
(c) A proton gradient is generated across the inner membrane of mitochondria during electron transport.
(d) Submitochondrial particles show a symmetric distribution of the respiratory chain components and ATP synthase across the inner mitochondrial membrane.
(e) ATP is synthesized when a proton gradient is imposed on mitochondria.

17. Which of the following statements about a mitochondrial preparation in which the reduced substrate is succinate are *correct*?

(a) The P:O ratio will be 2.
(b) There will be a net synthesis of three ATP per succinate oxidized to fumarate.
(c) The addition of CN^- will result in the synthesis of one ATP per succinate.
(d) Reduction of NADH-Q reductase will take place.
(e) Reduction of cytochrome reductase will occur.

18. Match each inhibitor in the left column with its *primary effect* from the right column.

(a) Azide _____

(b) Atractyloside _____

(c) Rotenone _____

(d) Dinitrophenol _____

(e) Carbon monoxide _____

(f) Oligomycin _____

(g) Antimycin A _____

(1) Inhibition of electron transport
(2) Uncoupling of electron transport and oxidative phosphorylation
(3) Inhibition of ADP-ATP translocation
(4) Inhibition of ATP synthase

19. Predict the oxidation-reduction states of NAD^+, NADH-Q reductase, ubiquinone, cytochrome c_1, cytochrome *c*, and cytochrome *a* in

liver mitochondria that are amply supplied with isocitrate as substrate, P_i, ADP, and oxygen but are inhibited by

(a) rotenone: _____

_____

(b) antimycin A: _____

_____

(c) cyanide: _____

_____

20. The reduction potential for methylene blue is similar to that of oxygen. Explain why massive doses of methylene blue are sometimes given for cyanide poisoning.

_____

_____

_____

21. Explain how one can predict the sites in the electron-transport chain where the coupling of oxidation to phosphorylation occurs on the basis of the redox potentials of the electron-transport chain components.

_____

_____

22. Which of the following statements about the mitochondrial ATP-synthesizing complex are *correct*?

(a) It contains more than 10 subunits.
(b) It is located in the intermembrane space of mitochondria.
(c) It contains an F_1 unit that constitutes the proton channel.
(d) It is sensitive to oligomycin inhibition.
(e) It translocates ATP through the mitochondrial membranes.

23. Match the major units of the ATP-synthesizing system in the left column with the appropriate components and functions from the right column.

(a) F_0 _____

(b) F_1 _____

(c) Stalk between F_0 and F_1 _____

(1) Regulates proton flow and ATP synthesis
(2) Contains the proton channel
(3) Contains the catalytic sites for ATP synthesis
(4) Contains F_1 inhibitor
(5) Contains α, β, γ, δ, and ϵ subunits
(6) Contains the protein that confers oligomycin sensitivity

24. Which of the following statements about the proposed mechanism for ATP synthesis by ATP synthase are *correct*?

(a) ATP synthase only forms ATP when protons flow through the complex.

(b) ATP synthase contains sites that change in their affinity for ATP as protons flow through the complex.

(c) ATP synthase binds ATP more tightly when protons flow through the complex.

(d) ATP synthase has two active sites per complex.

(e) ATP synthase has active sites that are functionally nonequivalent at a given time.

Mitochondrial Transport Systems

25. The inner mitochondrial membrane contains translocases—that is, specific transport proteins—for which pairs of substances?

(a) NAD^+ and NADH

(b) Glycerol 3-phosphate and dihydroxyacetone phosphate

(c) AMP and ADP

(d) Citrate and pyruvate

(e) Glutamate and aspartate

26. Explain why the rate of eversion of the binding site from the matrix to the cytosolic side is more rapid for ATP than for ADP when the ATP-ADP translocase functions in the presence of a proton gradient.

_____

_____

27. How many ATP are formed for each extramitochondrial NADH that is oxidized to NAD^+ by O_2 via the electron-transport chain. Assume that the glycerol-phosphate shuttle is operating.

(a) 1

(b) 2

(c) 3

(d) 4

28. How many ATP molecules are generated during the complete oxidative degradation of each of the following to CO_2 and H_2O? Assume the malate-aspartate shuttle is operating.

(a) Acetyl CoA _____

(b) Phosphoenolpyruvate _____

(c) glyceraldehyde 3-phosphate _____

Control of Oxidative Phosphorylation, Uncouplers, Scavengers of Toxic Derivatives of O_2, and Power Transmission by Proton Gradients

29. What is meant by the term _respiratory control_?

_____

_____

30. The rate of flow of electrons through the electron-transport chain is regulated by

(a) the ATP:ADP ratio.

(b) the concentration of acetyl CoA.

(c) the rate of oxidative phosphorylation.

(d) feedback inhibition by H_2O.

(e) the catalytic rate of cytochrome oxidase.

31. Uncouplers, such as dinitrophenol (DNP) or thermogenin, uncouple electron transport and phosphorylation by

    (a) inhibiting cytochrome reductase.
    (b) dissociating the F_0 and F_1 units of ATP synthase.
    (c) dissipating electron transport.
    (d) dissipating the proton gradient.
    (e) blocking the ATP-ADP translocase.

32. Which of the following are the products of the reaction of superoxide dismutase?

    (a) $O_2^- \cdot$
    (b) H_2O
    (c) H_2O_2
    (d) O_2
    (e) H_3O^+

PROBLEMS

1. Nitrite (NO_2^-) is toxic to many microorganisms. It is therefore often used as a preservative in processed foods. However, members of the genus *Nitrobacter* oxidize nitrite to nitrate (NO_3^-), using the energy released by the transfer of electrons to oxygen to drive ATP synthesis. Given the E_0' values below, calculate the maximum ATP yield per mole of nitrate oxidized.

$$NO_3^- + 2\,H^+ + 2\,e^- \longrightarrow NO_2^- + H_2O \qquad E_0' = +0.42 \text{ V}$$

$$\tfrac{1}{2}O_2 + 2\,H^+ + 2\,e^- \longrightarrow H_2O \qquad E_0' = +0.82 \text{ V}$$

2. A newly discovered compound called *coenzyme U* is isolated from mitochondria.

    (a) Several lines of evidence are presented in advancing the claim that coenzyme U is a previously unrecognized carrier in the electron-transport chain:
        (1) When added to a mitochondrial suspension, coenzyme U is readily taken up by mitochondria.
        (2) Removal of coenzyme U from mitochondria results in a decreased rate of oxygen consumption.
        (3) Alternate oxidation and reduction of coenzyme U when it is bound to the mitochondrial membrane can be easily demonstrated.
        (4) The rate of oxidation and reduction of coenzyme U in mitochondria is the same as the overall rate of electron transport.
        Which of the lines of evidence do you find the most convincing? Which are the least convincing? Why?
    (b) In addition to the evidence cited in (a), the following observations were recorded when coenzyme U was incubated with a suspension of submitochondrial particles:
        (1) The addition of NADH caused a rapid reduction of coenzyme U.
        (2) Reduced coenzyme U caused a rapid reduction of added cytochrome *c*.

(3) In the presence of antimycin A, the reduction of coenzyme U by added NADH took place as rapidly as in the absence of antimycin A. However, the reduction of cytochrome c by reduced coenzyme U was blocked in the presence of the inhibitor.

(4) The addition of succinate caused a rapid reduction of coenzyme U.

Assign a tentative position for coenzyme U in the electron-transport chain.

3. Coenzyme Q can be selectively removed from mitochondria using lipid solvents. If these mitochondria are then incubated in the presence of oxygen with an electron donor that is capable of reducing NAD^+, what will be the redox state of each of the carriers in the electron-transport chain?

4. The treatment of submitochondrial particles with urea removes F_1 subunits. When these treated particles are incubated in air with an oxidizable substrate and calcium ion, the concentration of calcium inside the particles increases.

(a) What do these observations tell you about the source of free energy required for accumulation of calcium ion?

(b) Would you expect the accumulation of calcium ion to be more sensitive to DNP or to oligomycin? Why?

5. In plants, the synthesis of succinate from acetyl CoA in the glyoxylate cycle generates NADH. Glyoxysomes do not have an electron-transport chain. Suggest a way in which NADH could be reoxidized in a plant cell so that the glyoxylate cycle can continue.

6. Analysis of the electron-transport pathway in a pathogenic gram-negative bacterium reveals the presence of five electron-transport molecules with the redox potentials listed in Table 17.1.

Table 17.1
Reduction potentials for pathogenic gram-negative bacterium

Oxidant	Reductant	Electrons transferred	$E_0'(V)$
NAD^+	NADH	2	−0.32
Flavoprotein b (oxidized)	Flavoprotein b (reduced)	2	−0.62
Cytochrome c (+3)	Cytochrome c (+2)	1	+0.22
Ferroprotein (oxidized)	Ferroprotein (reduced)	2	+0.85
Flavoprotein a (oxidized)	Flavoprotein a (reduced)	2	+0.77

(a) Predict the sequence of the carriers in the electron-transport chain.

(b) How many molecules of ATP can be generated under standard conditions when a pair of electrons is transported along the pathway?

(c) Why is it unlikely that oxygen is the terminal electron acceptor?

7. Calculate the minimum value of $\Delta E_0'$ that must be generated by a pair of electron carriers to provide sufficient energy for ATP synthesis. Assume that a pair of electrons is transferred.

8. The value of E_0' for the reduction of $NADP^+$ is -0.32 V.

(a) Calculate the equilibrium constant for the reaction catalyzed by NADPH dehydrogenase:

$$NADP^+ + NADH \longrightarrow NADPH + NAD^+$$

(b) What function could NADPH dehydrogenase serve in the cell?

9. Mitchell and Moyle carried out an elegant series of experiments using a sensitive pH meter to measure changes in hydrogen-ion concentration in an anaerobic suspension of rat mitochondria. Explain each of the following observations:

(a) The anaerobic suspension was incubated with β-hydroxybutyrate, an oxidizable substrate. When a small amount of oxygen was then injected into the suspension, an immediate decrease in pH was observed followed by a gradual return to the original hydrogen-ion concentration.

(b) When the experiment described in (a) was carried out in the presence of Triton X, a detergent that disrupts membranes, no decrease in pH was observed when oxygen was injected into the suspension.

(c) The anaerobic suspension was incubated in the absence of an oxidizable substrate. When a small amount of ATP was injected into the suspension, an immediate decrease in pH was observed followed by a gradual return to the original pH.

10. Why is it important for the value of E_0' for the $NAD^+ : NADH$ redox couple to be less negative than those for the redox couples of oxidizable compounds that are components of the glycolytic pathway and the citric acid cycle?

11. Arsenate, $AsO_4{}^{3-}$, is an uncoupling reagent for oxidative phosphorylation, but unlike DNP it does not transport protons across the inner mitochondrial membrane. How might arsenate function as an uncoupler?

12. A newly isolated soil bacterium grows without oxygen but requires ferric ion in the growth medium. Succinate suffices as a carbon source, but neither hexoses nor pyruvate can be utilized. The bacteria require riboflavin as a growth supplement. Neither niacin nor thiamin are required, and neither substance nor compounds derived from them can be found in the cells. The electron carriers found in the bacteria are cytochrome b, cytochrome c, FAD, and coenzyme Q.

(a) Propose a reasonable electron-transport chain that takes these observations, including the requirement for ferric ion, into account.

(b) Which of the other observations help to explain why the bacterium cannot utilize pyruvate or hexoses?

(c) Why is riboflavin required for the growth of the bacterium?

ANSWERS TO SELF-TEST

Introduction

1. Oxidative phosphorylation is the process in which ATP is formed as electrons are transferred from NADH or $FADH_2$ to O_2. The energy

coupling occurs through a proton gradient. In the substrate level phosphorylation, the syntheses of ATP or GTP are coupled to other chemical reactions that release sufficient energy to drive the formation of ATP or GTP from ADP and GDP, respectively.

2. c, e

Redox Potentials and Free Energy Changes

3. c, d

4. a. The $\Delta E'_0$ for the reaction is 0.05 volts. Calculating $\Delta G^{\circ\prime}$:

$$\Delta G^{\circ\prime} = -nF \, \Delta E'_0$$

where n, the number of electrons transferred, is 1, and F, the energy per volt and per mole, is 23.06 kcal/V · mol.

$$\Delta G^{\circ\prime} = -1 \times 23.06 \text{ kcal/V} \cdot \text{mol} \times 0.05 \text{ V}$$

$$= -1.15 \text{ kcal/mol}$$

Therefore, the reaction will proceed spontaneously, but it is not sufficiently exergonic to drive ATP synthesis, which requires -7.3 kcal/mol under standard conditions.

5. e. The $\Delta E'_0$ for this reaction is -0.03 volts. Calculating $\Delta G^{\circ\prime}$:

$$\Delta G^{\circ\prime} = -nF \, \Delta E'_0$$

$$= -2 \times 23.06 \text{ kcal/V} \cdot \text{mol} \times (-0.03 \text{ V})$$

$$= 1.38 \text{ kcal/mol}$$

6. The same proportionality constants that relate $\Delta G^{\circ\prime}$ and $\Delta E'_0$—that is, $\Delta G^{\circ\prime} = -nF \, \Delta E'_0$—can be used to relate $\Delta G'$ and $\Delta E'$. F is only the caloric equivalent of a volt, and n is the number of electrons participating in the redox reaction. Therefore,

$$\Delta G' = -nF \, \Delta E'$$

Substituting this in the expression for $\Delta G'$

$$\Delta G' = \Delta G^{\circ\prime} + 2.3 \, RT \log_{10} \frac{[\text{Products}]}{[\text{Reactants}]}$$

$$-nF \, \Delta E' = -nF \, \Delta E'_0 + 2.3 \, RT \log_{10} \frac{[\text{Products}]}{[\text{Reactants}]}$$

$$\Delta E' = \Delta E'_0 - \frac{2.3RT}{nF} \log_{10} \frac{[\text{Products}]}{[\text{Reactants}]}$$

$\Delta E'$ is a measure of the direction in which an oxidation-reduction reaction will proceed for any given concentrations of reactants and products.

7. Substituting into the equation for $\Delta G'$

$$\Delta G' = \Delta G^{\circ\prime} + 2.3 \, RT \log_{10} \frac{[\text{Products}]}{[\text{Reactants}]}$$

$$= 1.38 \text{ kcal/mol} + 2.3 \times 1.98 \text{ cal/mol} \cdot \text{K} \times$$
$$310 \text{ K} \log_{10} \frac{(0.5 \times 10^{-3})(0.2 \times 10^{-3})}{(2 \times 10^{-3})(2 \times 10^{-3})}$$

$$= 1.38 \text{ kcal/mol} - 2.26 \text{ kcal/mol}$$

$$= -0.88 \text{ kcal/mol}$$

Since $\Delta G' = -nF \, \Delta E'$

$$\Delta E' = \frac{-\Delta G'}{nF}$$

$$= -\frac{(-0.88 \text{ kcal/mol})}{(2)(23 \text{ kcal/mol} \cdot \text{V})}$$

$$= 0.019 \text{ V}$$

Alternatively, $\Delta E'$ could be calculated first using the expression derived in Question 6, and $\Delta G'$ could be subsequently determined from $\Delta G' = -nF \, \Delta E'$.

The Respiratory Chain

8. The proper sequence is b, d, e, a, and c. The mobile carriers are (a) and (d).

9. (a) 3, 6, 8 (b) 1, 4, 5, 9 (c) 5, 7 (d) 2, 5

10. d, e

11. d, e

12. c

13. See Figure 17-10 in Stryer. Ubiquinol transfers one electron to cytochrome c_1 via an Fe-S cluster on cytochrome reductase. The semiquinone derived from Q in this process donates an electron to cytochrome b-566, giving rise to ubiquinone. In turn, the electron from cytochrome b-566 is transferred to cytochrome b-562, which then reduces another semiquinone to ubiquinol. Thus, the b cytochromes act as a recycling device that allows ubiquinol, a two-electron carrier, to transfer its electrons, one at a time, to the Fe-S cluster of cytochrome reductase. Cytochrome c_1 accepts the electrons from the Fe-S cluster and transfers them to cytochrome c.

14. See Figure 17-15 in Stryer. Molecular O_2 is bound between the Fe^{2+} and Cu^+ ions of the a_3-Cu_B center of cytochrome oxidase. The oxygen remains bound while four electrons and four protons are sequentially added to its various intermediates, resulting in the net release of 2 H_2O. The a-Cu_A center supplies the electrons for this process. Although four electrons are used in the reduction of O_2 to 2 H_2O, the individual steps of the reaction cycle involve single electrons.

15. a

Coupling of Oxidation and Phosphorylation by a Proton-Motive Force

16. a, b, c, e

17. a, e

18. (a) 1 (b) 3 (c) 1 (d) 2 (e) 1 (f) 4 (g) 1

19. Upstream from the inhibitor, reduced respiratory-chain components will accumulate; downstream, oxidized components will be present. The point of inhibition is the crossover point.

   (a) Rotenone inhibits the step at NADH-Q reductase; therefore, NADH and NADH-Q reductase will be more reduced, and ubiquinone, cytochrome c_1, cytochrome c, and cytochrome a will be more oxidized.

(b) Antimycin A blocks electron flow between cytochromes b and c_1; therefore, NADH, NADH-Q reductase, and ubiquinol will be more reduced, and cytochrome c_1, cytochrome c, and cytochrome a will be more oxidized.

(c) Cyanide inhibits the transfer of electrons from cytochrome oxidase to O_2, so all the components of the respiratory chain will be more reduced.

20. Cyanide blocks the transfer of electrons from cytochrome oxidase to O_2. Therefore, all the respiratory-chain components become reduced and electron transport ceases; consequently, phosphorylation stops. An artificial electron acceptor with a redox potential similar to that of O_2, such as methylene blue, can reoxidize various components of the respiratory chain, reestablish a proton gradient, and thereby restore ATP synthesis.

21. The free-energy change ($\Delta G^{\circ\prime}$) for each electron-transfer step of the respiratory chain can be calculated from the redox-potential change ($\Delta E_0^{\prime}$) for that step, using the equation $\Delta G^{\circ\prime} = -nF\,\Delta E_0^{\prime}$. Since $\Delta G^{\circ\prime}$ for the synthesis of ATP from ADP is +7.3 kcal/mol, $\Delta G^{\circ\prime}$ for an electron-transfer reaction in which the coupling of oxidation to phosphorylation could occur must be more negative than -7.3 kcal/mol.

22. a, d

23. (a) 2 (b) 3, 5 (c) 1, 4, 6

24. b, e

Mitochondrial Transport Systems

25. b and e. Answer (a) is incorrect because only the electrons of NADH are transported, not the entire molecule. Answers (c) and (d) are incorrect because these pairs of compounds are not transported by the same translocase.

26. The proton gradient and the membrane potential make the cytosolic side of the inner mitochondrial membrane more positive than the matrix side; therefore, ATP, which has one more negative charge than ADP, is more attracted to the cytosolic side than is ADP.

27. b. The oxidation of NADH by the electron transport chain leads to the synthesis of three ATP. However, when the reducing equivalents of an extramitochondrial NADH enter the mitochondrial matrix via the glycerol-phosphate shuttle, they give rise to $FADH_2$, which only yields two ATP.

28. (a) 12 ATP; citric acid cycle (3 NADH, 1 $FADH_2$, 1 GTP)
    (b) 16 ATP; citric acid cycle (12 ATP), pyruvate dehydrogenase reaction (1 NADH, intramitochondrial), pyruvate kinase reaction (1 ATP)
    (c) 20 ATP; citric acid cycle (12 ATP), pyruvate dehydrogenase reaction (1 NADH), pyruvate kinase reaction (1 ATP), phosphoglycerate kinase reaction (1 ATP), glyceraldehyde 3-phosphate dehydrogenase reaction (1 NADH, extramitochondrial, which yields 3 ATP by the malate-aspartate shuttle)

Control of Oxidative Phosphorylation, Uncouplers, Scavengers of Toxic Derivatives of O_2, and Power Transmission by Proton Gradients

29. The regulation of the rate of oxidative phosphorylation by the availability of ADP is referred to as *respiratory control*.

30. a

31. d

32. c, d

ANSWERS TO PROBLEMS

1. The two reactions that generate nitrate are

$$NO_2^- + H_2O \longrightarrow NO_3^- + 2\,H^+ + 2\,e^- \qquad\qquad E_0' = -0.42\ V$$

$$\tfrac{1}{2}O_2 + 2\,H^+ + 2\,e^- \longrightarrow H_2O \qquad\qquad E_0' = +0.82\ V$$

The net reaction is

$$NO_2^- + \tfrac{1}{2}O_2 \longrightarrow NO_3^- \qquad\qquad E_0' = +0.40\ V$$

The Nernst equation is used to calculate the free energy liberated by the oxidation of nitrite under standard conditions:

$$\Delta G^{\circ\prime} = -nF\,\Delta E_0'$$

$$= -2(23.06\ kcal/V \cdot mol)(0.40\ V)$$

$$= -18.45\ kcal/mol$$

Therefore, each mole of nitrite oxidized yields 2.52 moles of ATP under standard conditions ($\Delta G^{\circ\prime} = -7.3$ kcal/mol).

2. (a) If coenzyme U is a component of the electron-transport chain, it should undergo successive reduction and oxidation in the mitochondrion, and its rate of electron transfer should be close to the overall rate. Observations (3) and (4) are therefore the most convincing. In addition to electron carriers, other compounds can be taken up by mitochondria, and some of them, such as pyruvate, can affect the rate of oxygen consumption because they are substrates that donate electrons to carriers. Therefore, observations (1) and (2) are less convincing.

(b) The first two observations show that coenzyme U lies along the electron-transport chain between NADH, which can reduce it, and cytochrome c, which is reduced by it. The fact that antimycin A blocks cytochrome c reduction by coenzyme U suggests that the carrier lies before site 2 (cytochrome reductase). Succinate, which can transfer electrons to Q, can also transfer electrons to coenzyme U, so the position of coenzyme U in the chain is similar to that of Q.

3. The removal of ubiquinone from the electron-transport chain means that no electrons can be transferred beyond Q in the pathway. You would therefore expect all carriers preceding Q to be more reduced and those beyond Q to be more oxidized. The removal of Q gives results similar to those observed by Chance in his crossover experiments (see Stryer, p. 413).

4. (a) The treated submitochondrial particles can carry out electron transport and can establish a proton gradient, but because the

F_1 subunits have been removed, they cannot synthesize ATP. The source of free energy for calcium accumulation must therefore be the proton-motive force generated by electron transport.

(b) DNP disrupts the proton-motive force by carrying protons across the mitochondrial membrane, thereby dissipating the free energy required for calcium accumulation. Oligomycin inhibits ATP synthase in mitochondria; because the treated particles do not depend upon ATP as a source of energy for calcium accumulation, oligomycin will have little effect.

5. Plant cells could use shuttle mechanisms to carry electrons from the glyoxysome to the mitochondrion. For example, oxaloacetate could be used as an acceptor of electrons from NADH, yielding malate, which could then be transferred to mitochondria, where the oxaloacetate would be regenerated. The transamination reaction discussed by Stryer on page 418 of the text could be used to return the four-carbon unit to the glyoxysome. The actual mechanism for regeneration of NAD^+ in the glyoxysome is unknown.

6. (a) The carrier with the most negative reduction potential has the weakest affinity for electrons and so transfers them most easily to an acceptor. The carrier with the most positive reduction potential will be the strongest oxidizing substance and will have the greatest affinity for electrons. A carrier should be able to pass electrons to any carrier having a more positive reduction potential. Thus, the probable order of the carriers in the chain is flavoprotein b, NADH, cytochrome c, flavoprotein a, and ferroprotein.

(b) The maximum amount of free energy is released by the transfer of two electrons from flavoprotein b, which has the most negative reduction potential, to the ferroprotein, which has the most positive reduction potential. The value of $\Delta E_0'$ is found by subtracting the reduction potential of flavoprotein b from that of the ferroprotein:

$$\Delta E_0' = +0.85 \text{ V} - (-0.62 \text{ V}) = 1.47 \text{ V}$$

The total amount of free energy released by the transfer of electron is

$$\Delta G^{\circ\prime} = -nF \, \Delta E_0'$$
$$= -(2)(23.06 \text{ kcal/V} \cdot \text{mol})(1.47 \text{ V})$$
$$= -67.80 \text{ kcal/mol}$$

Because $+7.3$ kcal/mol is required to drive ATP synthesis under standard conditions, the number of molecules of ATP synthesized per pair of electrons is $34.98/7.3 = 4.79$.

(c) It is unlikely that oxygen is the terminal electron acceptor because the reduction potential for the ferroprotein is slightly more positive than that of oxygen, so under standard conditions, the ferroprotein could not transfer electrons to oxygen.

7. The minimum amount of free energy that is needed to drive ATP synthesis under standard conditions is -7.3 kcal/mol. The value of $\Delta E_0'$ needed to generate this amount of free energy can be determined using the equation

$$\Delta G^{\circ\prime} = -nF\,\Delta E_0'$$

$$\Delta E_0' = \frac{-\Delta G^{\circ\prime}}{nF}$$

If a pair of electrons is transferred, then

$$\Delta E_0' = \frac{-(-7.3\ \text{kcal/mol})}{2(23.06\ \text{kcal/V}\cdot\text{mol})}$$

$$= +0.159\ \text{V}$$

8. (a) The transfer of a pair of electrons from NADH to $NADP^+$ occurs with no release of free energy:

$$NAD^+ + H^+ + 2\,e^- \longrightarrow NADH \qquad\qquad E_0' = -0.32\ \text{V}$$

$$NADP^+ + H^+ + 2\,e^- \longrightarrow NADPH \qquad\qquad E_0' = -0.32\ \text{V}$$

For the overall reaction, $\Delta E_0' = 0.00$ volt. Because

$$\Delta G^{\circ\prime} = -nF\,\Delta E_0'$$

then

$$\Delta G^{\circ\prime} = 0 = 2.303\ RT\log_{10} K_{eq}'$$

$$\log_{10} K_{eq}' = 0$$

$$K_{eq}' = 1$$

This means that the concentrations of the products equals the concentrations of the reactants; the reaction is at equilibrium.

(b) In the cell, NADPH dehydrogenase serves to replenish NADPH when the reduced cofactor is needed for biosynthetic reactions. On the other hand, metabolites such as isocitrate and glucose 6-phosphate are substrates for $NADP^+$-linked dehydrogenases. NAD^+ can accept reducing equivalents generated through the action of these enzymes.

9. (a) The injection of oxygen allowed electrons to be transferred from the oxidizable substrate to oxygen via the electron-transport chain in the inner mitochondrial membrane. Protons were pumped into the medium surrounding the organelles, resulting in a drop in pH. As the protons moved back into the mitochondria through ATP synthase or other routes, the hydrogen-ion concentration returned to its original level.

(b) The detergent disrupted the integrity of the closed membrane compartment in the mitochondria, making a proton gradient impossible to establish.

(c) ATP is a substrate for ATP synthase, which can also catalyze the hydrolysis of ATP (see Stryer's text, p. 414). During ATP hydrolysis, protons were pumped back through the proton channel of the synthase complex from the inner mitochondrial membrane into the medium, resulting in an elevation of the hydrogen-ion concentration.

10. NADH is a primary source of electrons for the respiratory chain. Oxidizable substrates must have a more negative reduction potential to donate electrons to NAD^+. A more negative redox potential for the NAD^+:NADH couple would make it unsuitable as an electron acceptor.

11. Arsenate chemically resembles inorganic phosphate; therefore, it

can enter into many of the same biochemical reactions as P_i (see p. 368 of Stryer's text). Arsenate can replace phosphate during oxidative phosphorylation, presumably forming an arsenate anhydride with ADP. Such compounds are unstable and are rapidly hydrolyzed, effectively causing the uncoupling of electron transport and oxidative phosphorylation.

12. (a) Because neither niacin nor NAD^+ are found in these cells, the transfer of electrons must procede from succinate to FAD and then to those carriers that have successively more positive redox potentials. In order, these are coenzyme Q, cytochrome b, cytochrome c, and ferric ion, which serves as the terminal electron acceptor in the absence of oxygen.

   (b) In order to utilize hexoses in the glycolytic pathway, NAD^+ is required as a cofactor for glyceraldehyde 3-phosphate dehydrogenase. Because the bacterium has no NAD^+, the glycolytic pathway is not operating. Similarly, the fact that thiamin is neither required nor found in the cells means that pyruvate dehydrogenase cannot be used to convert pyruvate to acetyl CoA.

   (c) Riboflavin is a precursor of FAD, which is required as an acceptor of electrons from succinate.

CHAPTER 18

Pentose Phosphate Pathway and Gluconeogenesis

The preceding chapters (Chapters 15, 16, and 17) introduced you to the metabolic pathways that produce the bulk of ATP in aerobic organisms from glucose and other fuel molecules. In this chapter, Stryer describes two additional metabolic pathways that involve glucose: the pentose phosphate pathway and gluconeogenesis. The role of the pentose phosphate pathway is to produce reduced nicotinamide adenine dinucleotide phosphate (NADPH), which is the currency of reducing power utilized for most reductive biosyntheses. In addition, this pathway generates ribose 5-phosphate and produces other three-, four-, five-, six-, and seven-carbon sugars from glucose. The pentose phosphate pathway is linked to glycolysis (Chapter 15) by the common intermediates glucose 6-phosphate, fructose 6-phosphate, and glyceraldehyde 3-phosphate. Gluconeogenesis is a biosynthetic pathway that generates glucose from noncarbohydrate precursors. All of the reversible steps of glycolysis also occur in gluconeogenesis; only the essentially irreversible enzymatic steps of glycolysis are bypassed by different reactions in gluconeogenesis. A few of these reactions take place in the mitochondrial matrix and involve oxaloacetate and malate, which are also intermediates of the citric acid cycle (Chapter 16).

When you have mastered this chapter, you should be able to complete the following objectives.

The Pentose Phosphate Pathway (Stryer pages 427–433)

1. Distinguish the structures of *NADH, NADPH,* and *ATP,* and contrast their functions in metabolism.

2. List the major physiological functions of the *pentose phosphate pathway,* which is also known as the *pentose shunt,* the *hexose monophosphate pathway,* and the *phosphogluconate oxidative pathway.*

3. Describe the reactions of the *oxidative branch* of the pentose phosphate pathway and the regulation of *glucose 6-phosphate dehydrogenase* by $NADP^+$ levels.

4. Explain how the pentose phosphate pathway and the glycolytic pathway are linked through reactions catalyzed by *transaldolase* and *transketolase.*

5. Outline the sugar interconversions of the *nonoxidative branch* of the pentose phosphate pathway.

6. Describe the different product stoichiometries obtained from the pentose phosphate pathway under conditions in which (1) more *ribose 5-phosphate* than NADPH is needed; (2) there is a balanced requirement for both; and (3) more NADPH than ribose 5-phosphate is needed, and the carbon atoms give rise to either CO_2 or pyruvate.

7. Explain the use of *radiolabeled glucose* in the elucidation of the relative activities of glycolysis and the pentose phosphate pathway in various tissues.

Enzymatic Mechanisms of Transketolase and Transaldolase, Glucose 6-Phosphate Dehydrogenase, and the Formation of Reduced Glutathione
(Stryer pages 434–438)

8. Compare the role of *thiamine pyrophosphate (TPP)* in transketolase with its role in *pyruvate dehydrogenase* and *α-ketoglutarate dehydrogenase.* Outline the enzymatic mechanisms of transketolase and transaldolase.

9. Discuss the effects of *glucose 6-phosphate dehydrogenase deficiency* on red cells in drug induced hemolytic anemia, and relate them to the biological roles of *glutathione.*

10. Describe the reduction of glutathione by *glutathione reductase.*

Gluconeogenesis (Stryer pages 438–442)

11. Describe the physiological significance of *gluconeogenesis.* List the primary precursors of gluconeogenesis.

12. List the irreversible reactions of glycolysis and the enzymatic steps of gluconeogenesis that bypass them.

13. Name the major organs that carry out gluconeogenesis. Locate the various enzymes of gluconeogenesis in cell compartments.

14. Describe the enzymatic steps in the conversion of *pyruvate* to *phosphoenolpyruvate*. Name the enzymes, intermediates, and cofactors involved in these reactions.

15. Explain the role of *biotin* as a carrier for *activated CO_2* in the *pyruvate carboxylase* reaction. Describe the control of pyruvate carboxylase by *acetyl CoA* and its role in maintaining the level of citric acid cycle intermediates.

16. Calculate the number of high-energy phosphate bonds consumed during gluconeogenesis and compare it to the number formed during glycolysis.

Regulation of Gluconeogenesis and the Cori Cycle (Stryer pages 442–445)

17. Describe the coordinated control of glycolysis and gluconeogenesis at the level of the *phosphofructokinase* and *fructose 1,6-bisphosphatase* reactions and the *pyruvate kinase* and *pyruvate carboxylase* reactions. Contrast the role of *fructose 2,6-bisphosphate* in the regulation of glycolysis and gluconeogenesis.

18. Explain how *substrate cycles* may *amplify metabolic signals* or *produce heat.*

19. Outline the *Cori cycle* and explain its biological significance.

20. Contrast the properties and roles of the *H* and *M isozymes* of *lactate dehydrogenase.*

SELF-TEST

The Pentose Phosphate Pathway

1. Which of the following properties are common to NADH and NADPH?

   (a) They can be used interchangeably as cofactors for dehydrogenases.
   (b) Both have the same net charge at pH 7.0.
   (c) Both contain two ribose moieties, a nicotinamide ring, an adenine base, and a phosphate anhydride bond.
   (d) They have equivalent oxidation-reduction potentials.
   (e) Both are readily translocated through mitochondrial membranes.

2. Which of the following compounds is not a product of the pentose phosphate pathway?

   (a) NADPH
   (b) Glycerate 3-phosphate
   (c) CO_2
   (d) Ribulose 5-phosphate
   (e) Sedoheptulose 7-phosphate

3. Figure 18-1 shows the first four reactions of the pentose phosphate pathway. Use it to answer the following questions:

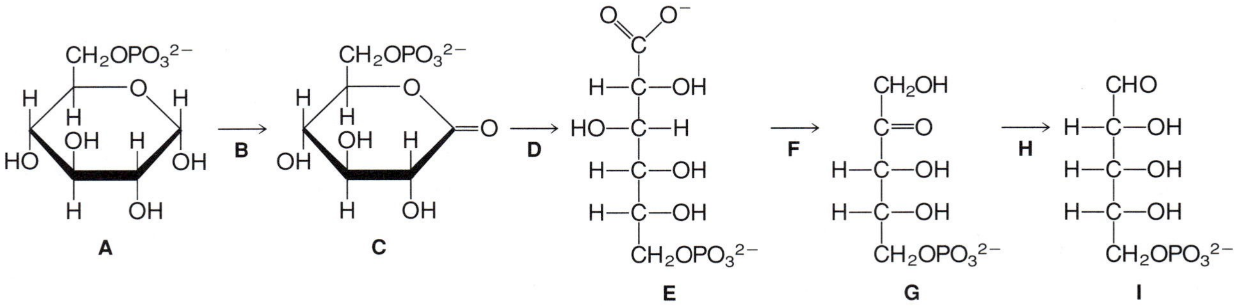

Figure 18-1
Reactions of the pentose phosphate pathway.

(a) Which reactions produce NADPH? _____

(b) Which reaction produces CO_2? _____

(c) Which compound is ribose 5-phosphate? _____

(d) Which compound is 6-phosphoglucono-δ-lactone?

_____

(e) Which compound is 6-phosphogluconate? _____

(f) Which reaction is catalyzed by phosphopentose isomerase?

_____

(g) Which enzyme is deficient in drug-induced hemolytic anemia?

_____

(h) Which compound can be a group acceptor in the transketolase

reaction? _____

4. Which of the following statements about glucose 6-phosphate dehydrogenase are *correct*?

(a) It catalyzes the committed step in the pentose phosphate pathway.
(b) It is regulated by the availability of NAD^+.
(c) One of its products is 6-phosphogluconate.
(d) It contains thiamine pyrophosphate as a cofactor.
(e) It is important in the metabolism of glutathione in erythrocytes.

5. The nonoxidative branch of the pentose phosphate pathway *does not* include which of the following reactions?

(a) Ribulose 5-P $\rightleftharpoons$ ribose 5-P
(b) Xylulose 5-P + ribose 5-P $\rightleftharpoons$

sedoheptulose 7-P + glyceraldehyde 3-P

(c) Ribulose 5-P + glyceraldehyde 3-P $\rightleftharpoons$ sedoheptulose 7-P

(d) Sedoheptulose 7-P + glyceraldehyde 3-P $\rightleftharpoons$
$\qquad$ fructose 6-P + erythrose 4-P

(e) Ribulose 5-P $\rightleftharpoons$ xylulose 5-P

6. Even if glucose 6-phosphate dehydrogenase is deficient, the synthesis of ribose 5-phosphate from glucose 6-phosphate can proceed normally. Explain how this is possible.

_____

_____

_____

7. Which of the following conversions take place in a metabolic situation that requires much more NADPH than ribose 5-phosphate, as well as complete oxidation of glucose 6-phosphate to CO_2? The arrows represent one or more enzymatic steps.

(a) Glucose 6-phosphate $\longrightarrow$ $\longrightarrow$ ribulose 5-phosphate

(b) Fructose 6-phosphate + glyceraldehyde 3-phosphate $\longrightarrow$ $\longrightarrow$
$\qquad$ ribose 5-phosphate

(c) Ribose 5-phosphate $\longrightarrow$ $\longrightarrow$
$\qquad$ fructose 6-phosphate + glyceraldehyde 3-phosphate

(d) Glyceraldehyde 3-phosphate $\longrightarrow$ $\longrightarrow$ pyruvate

(e) Fructose 6-phosphate $\longrightarrow$ glucose 6-phosphate

8. If glucose is labeled on the C-2 carbon with ^{14}C, will the glyceraldehyde 3-phosphates resulting from conversions occurring in the glycolytic and the pentose phosphate pathways be labeled in the same position? Explain your answer.

_____

_____

_____

_____

9. Liver synthesizes fatty acids and lipids for export to other tissues. Would you expect the pentose phosphate pathway to have a low or a high activity in this organ? Explain your answer.

_____

_____

_____

Enzymatic Mechanisms of Transketolase and Transaldolase, Glucose 6-Phosphate Dehydrogenase, and the Formation of Reduced Glutathione

10. Transaldolase and transketolase have which of the following similarities?

(a) Both require thiamine pyrophosphate.

(b) Both form a Schiff base with the substrate.

(c) Both use an aldose as a group donor.

(d) Both use a ketose as a group donor.

(e) Both form a covalent addition compound with the donor substrate.

11. Match the cofactors in the left column with the appropriate enzymes from the right column.

(a) Thiamine
    pyrophosphate _____

(b) NADP$^+$ _____

(c) Biotin _____

(d) GTP _____

(1) Glucose 6-P dehydrogenase
(2) 6-Phosphogluconate dehydrogenase
(3) Pyruvate dehydrogenase
(4) Pyruvate carboxylase
(5) Transketolase
(6) Phosphoenolpyruvate
    carboxykinase

12. Which of the following statements regarding reduced glutathione is *incorrect*?

(a) It contains one γ-carboxyglutamate, one cysteine, and one glycine residue.
(b) It keeps the cysteine residues of proteins in their reduced states.
(c) It is regenerated from oxidized glutathione by glutathione reductase.
(d) It reacts with hydrogen peroxide and organic peroxides.
(e) It is decreased relative to oxidized glutathione in glucose 6-phosphate dehydrogenase deficiency.

13. Suggest reasons why glucose 6-phosphate dehydrogenase deficiency may be manifested in red blood cells but not in adipocytes, which also require NADPH for their metabolism.

_____

_____

_____

Gluconeogenesis

14. Which of the following statements about gluconeogenesis are *correct*?

(a) It occurs actively in the muscle during periods of exercise.
(b) It occurs actively in the liver during periods of exercise or fasting.
(c) It occurs actively in adipose tissue during feeding.
(d) It occurs actively in the kidney during periods of fasting.
(e) It occurs actively in the brain during periods of fasting.

15. Glucose can be synthesized from which of the following noncarbohydrate precursors?

(a) Adenine
(b) Alanine
(c) Lactate
(d) Palmitic acid
(e) Glycerol

16. The following is the sequence of reactions of gluconeogenesis from pyruvate to phosphoenolpyruvate:

Pyruvate $\rightleftharpoons$ oxaloacetate $\rightleftharpoons$ malate $\rightleftharpoons$ oxaloacetate $\rightleftharpoons$ phosphoenolpyruvate
        A   **B**   **C**   **D**

Match the capital letters indicating the reactions of the gluconeogenic pathway with the following statements:

(a) Occurs in the mitochondria _____

(b) Occurs in the cytosol _____

(c) Produces CO_2 _____

(d) Consumes CO_2 _____

(e) Requires ATP _____

(f) Requires GTP _____

(g) Is regulated by acetyl CoA _____

(h) Requires a biotin cofactor _____

(i) Is also a reaction of the citric acid cycle _____

(j) Is an anaplerotic reaction _____

17. Explain why a CO_2 is added to pyruvate in the pyruvate carboxylase reaction only to be subsequently removed by the phosphoenolpyruvate carboxykinase reaction. Identify the high-energy intermediate in the carboxylation reaction.

_____

_____

_____

18. Glycogen is the storage form of glucose in the liver and muscles. In terms of absolute amounts, muscles contain more total glycogen than does liver, yet muscle glycogen is not a direct precursor of blood glucose. Explain this fact.

_____

_____

19. How many "high-energy" bonds are required to convert oxaloacetate to glucose?

(a) 2
(b) 3
(c) 4
(d) 5
(e) 6

Regulation of Gluconeogenesis and the Cori Cycle

20. Which of the following statements *correctly* describe what happens when acetyl CoA is abundant?

(a) Pyruvate carboxylase is activated.
(b) Phosphoenolpyruvate carboxykinase is activated.
(c) Phosphofructokinase is activated.
(d) If ATP levels are high, oxaloacetate is diverted to gluconeogenesis.
(e) If ATP levels are low, oxaloacetate is diverted to gluconeogenesis.

21. In the coordinated control of phosphofructokinase (PFK) and fructose 1,6-bisphosphatase (F-1,6-bisPase),

    (a) citrate inhibits PFK and stimulates F-1,6-bisPase.
    (b) fructose 2,6-bisphosphate inhibits PFK and stimulates F-1,6-bisPase.
    (c) acetyl CoA inhibits PFK and stimulates F-1,6-bisPase.
    (d) AMP inhibits PFK and stimulates F-1,6-bisPase.
    (e) NADPH inhibits PFK and stimulates F-1,6-bisPase.

22. Which of the following statements about the Cori cycle and its physiological consequences are *correct*?

    (a) It involves the synthesis of glucose in muscle.
    (b) It involves the release of lactate by muscle.
    (c) It involves lactate synthesis in the liver.
    (d) It involves ATP synthesis in muscle.
    (e) It involves the release of glucose by the liver.

23. Explain how a substrate cycle can generate heat.

_____

_____

_____

24. In muscle lactate dehydrogenase produces lactate from pyruvate, whereas in the heart it preferentially synthesizes pyruvate from lactate. Explain how this is possible.

_____

_____

_____

PROBLEMS

1. The conversion of glucose 6-phosphate to ribose 5-phosphate via the enzymes of the pentose phosphate pathway and glycolysis can be summarized as follows:

5 Glucose 6-phosphate + ATP $\longrightarrow$ 6 ribose 5-phosphate + ADP + H$^+$

Which enzyme uses the molecule of ATP shown in the equation?

2. Aminotransferases are enzymes that catalyze the removal of amino groups from amino acids to yield α-keto acids. How could the action of such enzymes contribute to gluconeogenesis? Consider the utilization of alanine, aspartate, and glutamate in your answer.

3. Citrate inhibits phosphofructokinase and stimulates fructose 1,6-bisphosphatase. Why are these effects consistent with cellular strategies for regulating glucose synthesis and degradation?

4. Dietary biotin is primarily linked to the ϵ-amino groups of lysine residues in proteins. Biotinidase, an enzyme found in most mammalian

tissues, hydrolyses the ∈-amino acid link, releasing free biotin, which can then be used in the formation of biotin-requiring enzymes. Deficiencies in biotinidase, although rare, are nevertheless well known. What effect would a biotinidase deficiency have on gluconeogenesis?

5. Liver and other organ tissues contain relatively large quantities of nucleic acids. During digestion, nucleases hydrolyze RNA and DNA, and among the products is ribose 5-phosphate.

(a) How can this molecule be used as a metabolic fuel?

(b) Another product formed by the degradation of nucleic acids is 2-deoxyribose 5-phosphate. Can this molecule be converted to glycolytic intermediates through the action of the pentose phosphate pathway? Why?

6. Bacteria can carry out gluconeogenesis using components of the citric acid cycle or amino acids, but unlike mammals, they can generate glucose from acetate via the glyoxylate pathway. Because bacteria have no organelles, the enzymes of the citric acid cycle and the glyoxylate pathway are not compartmentalized.

(a) Why is regulation of the glyoxylate pathway necessary in bacteria?

(b) Studies show that bacterial isocitrate lyase is allosterically inhibited by high concentrations of phosphoenolpyruvate. Would you expect to see the inhibition of isocitrate lyase when bacteria are utilizing glucose as a sole carbon source? Why?

(c) Would you expect the glyoxalate pathway to be more active than the citric acid cycle when bacteria are growing on acetate? Why?

(d) Would you expect to find glucose 6-phosphatase in bacteria?

7. You have glucose that is radioactively labeled with ^{14}C at C-1, and you have an extract that contains the enzymes that catalyze the reactions of the glycolytic and the pentose phosphate pathways, along with all the intermediates of the pathways.

(a) If the enzymes of the *oxidative* branch of the pentose phosphate pathway are *not* active in your extract, is it possible to obtain labeled sedoheptulose-7-phosphate using glucose labeled with ^{14}C at C-1? Explain.

(b) Suppose that in a second experiment *all* the enzymes of both the oxidative branch and nonoxidative branch of the pentose phosphate pathway are active. Will the labeling pattern of sedoheptulose 7-phosphate be different? Explain.

(c) Can sedoheptulose 7-phosphate form a heterocyclic ring?

8. Why is the pentose phosphate pathway more active in cells that are dividing than in cells that are not?

9. Liver cells can carry out gluconeogenesis in the absence of oxygen provided that ATP levels are sufficient. What cofactor must be continuously generated to ensure the continued generation of glucose from pyruvate?

10. A bacterium isolated from a soil culture can utilize ribose as a sole source of carbon when grown anaerobically. Experiments show that in the anaerobic pathways leading to ATP production, three molecules of ribose are converted to five molecules of CO_2 and five molecules of ethanol. These organisms also use ribose for the production of

NADPH. The assimilation of ribose begins with its conversion to ribose 5-phosphate, with ATP serving as a phosphoryl donor.

(a) Explain how ribose can be converted to CO_2 and ethanol under anaerobic conditions. Write the overall reaction, showing how much ATP can be produced per pentose utilized.

(b) Write an equation for the generation of NADPH using ribose as a sole source of carbon.

11. Mature erythrocytes, which lack mitochondria, metabolize glucose at a high rate. In response to the increased availability of glucose, erythrocytes generate lactate and also evolve carbon dioxide.

(a) Why is generation of lactate necessary to ensure the continued utilization of glucose?

(b) In erythrocytes, what pathway is likely to be used for the generation of carbon dioxide from glucose? Can glucose be completely oxidized to CO_2 in erythrocytes? Explain.

12. In liver, V_{max} for fructose bisphosphatase is three to four times higher than V_{max} for phosphofructokinase, whereas in muscle it is only about 10 percent of that of phosphofructokinase. Explain this difference.

ANSWERS TO SELF-TEST

The Pentose Phosphate Pathway

1. c, d

2. b

3. (a) B, F (b) F (c) I (d) C (e) E (f) H (g) B (h) I

4. a, e

5. c

6. Ribose 5-phosphate can be synthesized from fructose 6-phosphate and glyceraldehyde 3-phosphate, both of which are glycolytic products of glucose 6-phosphate. These reactions are carried out by transketolase and transaldolase in a reversal of the nonoxidative branch of the pentose phosphate pathway and do not involve glucose 6-phosphate dehydrogenase.

7. a, c, and e. Glucose 6-phosphate is converted to ribulose 5-phosphate, producing CO_2 and NADPH in the process. Then ribulose 5-phosphate, via ribose 5-phosphate, is transformed into fructose 6-phosphate and glyceraldehyde 3-phosphate. These two glycolytic intermediates are converted back to glucose 6-phosphate, and the cycle is repeated until the equivalent of six carbon atoms from glucose 6-phosphate are converted to CO_2.

8. No, the glyceraldehyde 3-phosphates resulting from the different pathways will be labeled on different carbons. For conversion via the glycolytic pathway, the ^{14}C label will appear on the C-2 of glyceraldehyde 3-phosphate. Conversion via the pentose phosphate pathway will

result in ribose 5-phosphate and xylulose 5-phosphate that are labeled on their C-1 carbons. These two sugars will give rise to sedoheptulose 7-phosphate and fructose 6-phosphate, which will be labeled on both their C-1 and C-3 carbons. Upon the conversion of the fructose 6-phosphate to glyceraldehyde 3-phosphate the labels will be on C-1 and C-3.

9. The activity of the pentose phosphate pathway in the liver is high. The biosynthesis of fatty acids and lipids requires reducing equivalents in the form of NADPH. In all organs that carry out reductive biosyntheses, the pentose phosphate pathway supplies a large proportion of the required NADPH. In Chapter 20 of Stryer (p. 488), you will learn that a reaction catalyzed by malate enzyme produces part of the NADPH required for fatty acid synthesis.

Enzymatic Mechanisms of Transketolase and Transaldolase, Glucose 6-Phosphate Dehydrogenase, and the Formation of Reduced Glutathione

10. d, e

11. (a) 3, 5 (b) 1, 2 (c) 4 (d) 6. (For a review of the cofactors of pyruvate dehydrogenase, see Stryer, pp. 379–383.)

12. a. Answer (a) is incorrect because the glutamate residue in glutathione is not γ-carboxyglutamate; rather, the glutamate in gluthathione forms a peptide bond with the adjacent cysteine residue *via* its γ-carboxyl group. Recall the structure and role of γ-carboxyglutamate residues in clotting factors (Stryer, p. 252).

13. The glucose 6-phosphate dehydrogenase in erythrocytes and that in adipocytes are specified by distinct genes; they have the same function but different structures—that is, they are isozymes. Furthermore, NADPH synthesis by the pentose phosphate pathway may not be as critical in the cells of other tissues as it is in erythrocytes because other tissues have other sources of NADPH (see the answer to Question 9).

Gluconeogenesis

14. b, d

15. b, c, e

16. (a) A, B (b) C, D (c) D (d) A (e) A (f) D (g) A (h) A (i) B (j) A. For (i), answer C is incorrect because this reaction occurs in the cytosol and not in the mitochondria, where the enzymes of the citric acid cycle are located.

17. The carboxylation reaction produces an activated carboxyl group in the form of a high-energy *carboxybiotin intermediate*. The cleavage of this bond and release of CO_2 in the phosphoenolpyruvate carboxykinase reaction or the transfer of the CO_2 to acceptors in other reactions in which biotin participates allows endergonic reactions to proceed. Thus, the formation of phosphoenolpyruvate from oxaloacetate is driven by the release of CO_2 ($\Delta G^{\circ\prime} = -4.7$ kcal/mol) and the hydrolysis of GTP ($\Delta G^{\circ\prime} = -7.3$ kcal/mol).

18. Muscle lacks glucose 6-phosphatase; therefore, glucose 6-phosphate cannot be dephosphorylated in muscle for export into the blood. Muscle stores glycogen for its own use as a source of readily available fuel during periods of activity.

19. c. The two steps in gluconeogenesis that consume GTP or ATP are

$$\text{Oxaloacetate} + \text{GTP} \longrightarrow \text{phosphoenolpyruvate} + \text{GDP} + \text{CO}_2$$

$$\text{3-Phosphoglycerate} + \text{ATP} \longrightarrow \text{1,3-bisphosphoglycerate} + \text{ADP}$$

Since two oxaloacetate molecules are required to synthesize one glucose molecule, a total of four "high-energy" bonds are required.

Regulation of Gluconeogenesis and the Cori Cycle

20. a, d

21. a

22. b, d, e

23. A substrate cycle, such as the one established by phosphofructokinase and fructose 1,6-bisphosphatase, causes the net hydrolysis of ATP. The energy from the hydrolysis of ATP into ADP and P_i is released as heat.

24. Muscle and heart have distinct lactate dehydrogenase isozymes. Heart lactate dehydrogenase contains mostly H-type subunits. This enzyme has higher affinity for substrates and is inhibited by high concentrations of pyruvate; that is, it is designed to form pyruvate from lactate. In contrast, muscle lactate dehydrogenase, which consists of M-type subunits, is more effective for forming lactate from pyruvate.

ANSWERS TO PROBLEMS

1. Phosphofructokinase uses ATP to convert fructose 6-phosphate to fructose 1,6-bisphosphate, which is then cleaved by aldolase to yield dihydroxyacetone phosphate (DHAP) and glyceraldehyde 3-phosphate. The conversion of DHAP to a second molecule of glyceraldehyde 3-phosphate provides the molecules that are needed for the synthesis of ribose 5-phosphate.

2. Examination of the structures of the α-keto acid analogues of alanine, asparatate, and glutamate shows that each can be used for gluconeogenesis. Specific transaminases convert alanine to pyruvate, aspartate to oxaloacetate, and glutamate to α-ketoglutarate. These amino acids, along with others whose carbon skeletons can be used for the synthesis of glucose, are termed glycogenic amino acids.

3. When citrate levels are high, the citric acid cycle has ample fuel molecules that can be oxidized for the generation of ATP. Under such conditions, gluconeogenesis is stimulated and the rate of glycolysis is decreased. Citrate helps achieve these goals by inhibiting phosphofructokinase, which catalyzes a rate-determining step in glycolysis, and by increasing the activity of fructose 1,6-bisphosphatase, a key enzyme in gluconeogenesis.

4. Biotin is required as a cofactor for pyruvate carboxylase, a key enzyme in gluconeogenesis. The lack of biotin will impair the response required of biotin-requiring enzymes when glucose levels are low. Infants with biotinidase deficiency often have high serum levels of pyru-

vate and lactate. A number of other enzymes that catalyze carboxylations, including acetyl CoA carboxylase, require biotin, so a biotinidase deficiency also causes elevations of other organic acids that are metabolites of the carboxylation reactions. All these symptoms can be alleviated by injections of free biotin.

5. (a) The most direct route for the oxidative degradation of ribose 5-phosphate is its conversion to glycolytic intermediates by the nonoxidative enzymes of the pentose phosphate pathway, as described by Stryer on pages 429 and 430 of his text. The overall reaction is

3 Ribose 5-phosphate $\longrightarrow$
   2 fructose 6-phosphate + glyceraldehyde 3-phosphate

(b) The formation of glycolytic intermediates from 2-deoxyribose 5-phosphate is not possible because, unlike ribose 5-phosphate, 2-deoxyribose 5-phosphate lacks a hydroxyl group at C-2. It is therefore not a substrate for phosphopentose isomerase, whose action is required to convert ketopentose phosphates to substrates that can be utilized by other enzymes of the pentose phosphate pathway. Most deoxyribose phosphate molecules are used in salvage pathways to form deoxynucleotides (Stryer, pp. 618 and 619).

6. (a) When ATP levels are low in the bacterial cell, the flux of carbon atoms through the citric acid cycle must increase to generate the reduced cofactors needed for the generation of ATP. Because the glyoxylate cycle bypasses the steps that generate NADH and $FADH_2$, it must be suppressed when ATP synthesis is needed. On the other hand, high levels of ATP and an abundance of components of the citric acid cycle cause the glyoxylate cycle to accelerate, leading to the synthesis of succinate and other biosynthetic intermediates.

(b) Glucose is converted to biosynthetic intermediates via the glycolytic and pentose phosphate pathways as well as the citric acid cycle. When ample glucose is available, isocitrate lyase activity can be inhibited because the glyoxylate pathway is not needed to generate such intermediates. High concentrations of PEP generated by glycolysis inhibit isocitrate lyase activity. You would expect the glyoxylate cycle to be more active.

(c) When acetate is the sole carbon source, acetyl CoA is formed. The net synthesis of succinate and other intermediates of the citric acid cycle cannot be accomplished from acetyl CoA. Instead, succinate is generated by the glyoxylate cycle, which is activated because low levels of PEP do not inhibit isocitrate lyase.

(d) No. Glucose 6-phosphatase activity is found in cells that export glucose to other cells. Bacteria usually utilize glucose as a fuel and as a source of carbon atoms and therefore have no need for an enzyme that allows glucose to leave the bacterial cell.

7. (a) Yes. The most direct route would be the conversion of glucose to fructose 6-phosphate, followed by the condensation of fructose 6-phosphate with erythrose 4-phosphate to form sedoheptulose 7-phosphate and glyceraldehyde 3-phosphate. The labeled carbon of glucose becomes the C-1 of fructose 6-phosphate and the C-1 of sedoheptulose 7-phosphate.

(b) The labeling pattern will be the same, although the amount of labeled carbon incorporated into the heptose will be reduced. In the oxidative branch of the pentose phosphate pathway, the labeled glucose is converted to glucose 6-phosphate with the ^{14}C label on C-1. Glucose 6-phosphate then undergoes successive oxidations and decarboxylation to form ribulose 5-phosphate. The label is lost when the C-1 carbon is removed during decarboxylation.

(c) Sedoheptulose 7-phosphate is a ketose and can form a heterocyclic ring through a hemiketal linkage. The most likely link would be between the keto group at C-2 and the hydroxyl group at C-6.

8. Cells have a high rate of nucleic acid biosynthesis when they grow and divide. Among the precursors needed is ribose 5-phosphate, which is synthesized through the action of the enzymes of the glycolytic and the pentose phosphate pathways. Biosynthetic reactions requiring NADPH occur at a high rate in growing and dividing cells. For these reasons, the enzymes of the pentose phosphate pathway will be extremely active in dividing cells.

9. The other cofactor that must be continuously generated is NADH, which is required by glyceraldehyde 3-phosphate dehydrogenase in the glycolytic pathway.

10. (a) To generate ATP, ethanol, and CO_2, ribose must first be converted to ribose 5-phosphate, with ATP serving as a phosphate donor. Then, in the nonoxidative branch of the pentose phosphate pathway, three molecules of ribose 5-phosphate are converted to two molecules of fructose 6-phosphate and one molecule of glyceraldehyde 3-phosphate. Two molecules of ATP are required for the production of fructose 1,6-bisphosphate from fructose 6-phosphate. The formation of a total of five molecules of glyceraldehyde 3-phosphate is achieved through the action of aldolase and triose phosphate isomerase. These five molecules are converted to five molecules of pyruvate, yielding ten ATP molecules and 5 NADH molecules. To keep the anaerobic cell in redox balance, the pyruvate molecules are converted to five molecules of ethanol, with the production of five CO_2 molecules and five NAD^+. The overall reaction is

$$3 \text{ Ribose} + 5 \text{ ADP} + 5 \text{ P}_i \longrightarrow 5 \text{ ethanol} + 5 \text{ CO}_2 + 5 \text{ ATP}$$

(b) Ribose 5-phosphate molecules must first be converted to glucose 6-phosphate for the oxidative enzymes of the pentose pathway to generate NADPH. The stoichiometry of the reactions is

$$6 \text{ Ribose 5-phosphate} \longrightarrow$$
$$4 \text{ fructose 6-phosphate} + 2 \text{ glyceraldehyde 3-phosphate}$$

$$4 \text{ Fructose 6-phosphate} \longrightarrow 4 \text{ glucose 6-phosphate}$$

$$2 \text{ Glyceraldehyde 3-phosphate} \longrightarrow \text{glucose 6-phosphate} + \text{P}_i$$

$$5 \text{ Glucose 6-phosphate} + 10 \text{ NADP}^+ + 5 \text{ H}_2\text{O} \longrightarrow$$
$$5 \text{ ribose 5-phosphate} + 10 \text{ NADPH} + 10 \text{ H}^+ + 5 \text{ CO}_2$$

The net reaction is

$$\text{Ribose 5-phosphate} + 10 \text{ NADP}^+ + 5 \text{ H}_2\text{O} \longrightarrow$$
$$10 \text{ NADPH} + 10 \text{ H}^+ + 5 \text{ CO}_2 + \text{P}_i$$

11. (a) Because erythrocytes lack mitochondria, they cannot use the citric acid cycle to regenerate the NAD^+ needed to sustain glycolysis. Instead, they regenerate NAD^+ by reducing pyruvate through the action of lactate dehydrogenase; NAD^+ is then reduced in the reaction catalyzed by glyceraldehyde 3-phosphate dehydrogenase during glycolysis. Failure to oxidize the NADH generated in the glycolytic pathway will cause a reduction in the rate of glucose breakdown.

(b) In erythrocytes, the pentose phosphate pathway is the only route available to yield CO_2 from glucose. Glucose can be completely oxidized by first entering the oxidative branch of the pathway, generating NADPH and ribose 5-phosphate. Transaldolase and transketolase then convert the pentose phosphates to fructose 6-phosphate and glyceraldehyde 3-phosphate. Part of the gluconeogenic pathway is used to convert both of the products to glucose 6-phosphate. The net reaction is

$$Glucose\ 6\text{-}P + 12\ NADP^+ + 7\ H_2O \longrightarrow$$
$$6\ CO_2 + 12\ NADPH + 12\ H^+ + P_i$$

12. In contrast to muscle tissue, which oxidizes glucose to yield energy, liver tissue generates glucose primarily for export to other tissues. Thus, one would expect the rate of gluconeogenesis in the liver to be greater than the rate of glycolysis. Therefore, the relative catalytic capacity (as measured by V_{max}) of fructose bisphosphatase, a key enzyme in gluconeogenesis, should be expected to exceed that of phosphofructokinase, which is a regulatory enzyme of the glycolytic pathway.

CHAPTER 19

Glycogen Metabolism

The topic of carbohydrate metabolism developed in Chapters 15 and 18 is concluded in this chapter with a discussion of the metabolism of glycogen, the storage form of glucose. Glycogen is important in the metabolism of higher animals because its glucose residues can be easily mobilized by the liver to maintain blood glucose levels and by muscle to satisfy the energy needs of muscle contraction during bursts of activity. Stryer first reviews the structure and the physiological roles of glycogen. You were already introduced to the structure of glycogen, a polymer of glucose, in the chapter on carbohydrates (Chapter 14). Next, Stryer presents the enzymatic reactions of glycogen degradation and synthesis. This is followed by a discussion of the control of these reactions by the phosphorylation and dephosphorylation of the key enzymes in response to hormonal signals. Stryer describes the structures and the control mechanisms of phosphorylase, phosphorylase kinase, and glycogen synthase. He concludes the chapter with discussions of the reaction cascade that controls the synthesis and degradation of glycogen, the role of phosphatases in this process, the effects of glucose on phosphorylase a, and the biochemical bases of several glycogen storage diseases.

When you have mastered this chapter, you should be able to complete the following objectives.

Introduction (Stryer pages 449–451)

1. Describe the structure of *glycogen* and its role in the *liver* and *muscle*.

2. Discuss the properties of *glycogen granules* and their location within cells.

Reactions of Glycogen Degradation and Synthesis (Stryer pages 451–457)

3. Explain the advantage of the *phosphorolytic cleavage* of glycogen over its *hydrolytic cleavage*.

4. Describe the roles of *pyridoxal 5'-phosphate* and the *carbonium ion intermediate* in the mechanism of action of *glycogen phosphorylase*.

5. Outline the steps in the degradation of glycogen, and relate them to the action of phosphorylase, *transferase*, and *α-1,6-glucosidase* (which is also known as the *debranching enzyme*).

6. Compare the reaction mechanisms of *phosphoglucomutase* and *phosphoglyceromutase*.

7. Explain the importance of *glucose 6-phosphatase* in the release of glucose by the liver. Note the absence of this enzyme in the brain and muscle.

8. Contrast the synthesis of glycogen with its degradation. Explain the role of *UDP-glucose* in the synthesis of glycogen.

9. Outline the steps in the synthesis of glycogen, name the pertinent enzymes, and note the requirement for a *primer*.

10. Discuss the efficiency of glycogen as a storage form of glucose.

Control of Glycogen Metabolism, and Structural Features of the Enzymes of Glycogen Metabolism (Stryer pages 458–462)

11. Explain the reasons for the coordinated control of glycogen synthesis and degradation. Compare the effects of *insulin, glucagon,* and *epinephrine* on glycogen metabolism in the liver and muscle.

12. Describe the structure of *cyclic AMP*, its synthesis by *adenylate cyclase*, and its effect on glycogen metabolism.

13. Describe the phosphorylation of phosphorylase by *phosphorylase kinase*.

14. Explain the relationships between *phosphorylase a* and *phosphorylase b*, the *T (tense)* and *R (relaxed)* forms of each, and the allosteric effectors that mediate their interconversions in skeletal muscle.

15. Describe the overall structure of glycogen phosphorylase. Note the variety of binding sites, their functional roles, and the critical location of the phosphorylation and AMP binding sites next to the subunit interface.

16. Discuss the major structural features of phosphorylase kinase. Explain the effects of *calmodulin* and Ca^{2+} on glycogen metabolism in muscle.

17. Distinguish between the active and inactive forms of glycogen synthase.

Reaction Cascade in the Phosphorylation and Dephosphorylation of Glycogen Synthase and Phosphorylase (Stryer pages 462–464)

18. List the sequence of events from the binding of hormones by their receptors to the phosphorylation of glycogen synthase and phosphorylase.

19. Explain the roles of *phosphatase 1* and *inhibitor 1* in the control of the activities of glycogen phosphorylase and synthase.

20. Describe how cyclic AMP is degraded.

Regulation of Blood Glucose Levels, and Glycogen Storage Diseases (Stryer pages 464–466)

21. Describe the events that lead to the inactivation of phosphorylase and the activation of glycogen synthase by glucose in the liver. Note the role of phosphorylase *a* as the glucose sensor in liver cells.

22. Give examples of glycogen storage diseases, and relate the biochemical defects with the clinical observations.

SELF-TEST

Introduction

1. Answer the following questions about the glycogen fragment in Figure 19-1:

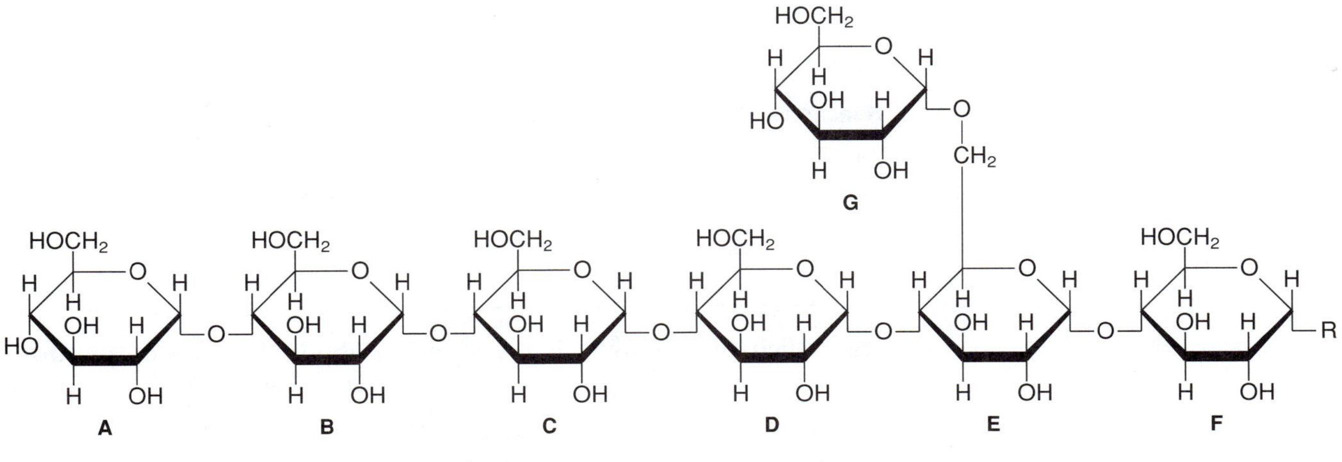

Figure 19-1
Fragment of glycogen. (R represents the rest of the glycogen molecule.)

(a) Which residues are at nonreducing ends? _____

(b) An α-1,6-glycosidic linkage occurs between which residues?

_____

(c) An α-1,4-glycosidic linkage occurs between which residues?

_____

(d) Is this a substrate for phosphorylase a? _____

Explain: _____

_____

(e) Is this a substrate for the debranching enzyme? _____

Explain: _____

_____

(f) Is this a substrate for the branching enzyme? _____

Explain: _____

_____

2. Which of the following statements about glycogen storage are *incorrect*?

(a) Glycogen is stored in muscles and liver.
(b) Glycogen is stored in brain and adipose tissue.
(c) Glycogen reserves can exceed the equivalent of 10,000 kcal in the average human.
(d) Glycogen essentially fills cells that specialize in glycogen storage.
(e) Glycogen storage occurs in the form of dense granules in the cytoplasm of cells.

3. How do the multiple enzymes bound to glycogen granules differ from multienzyme complexes, such as pyruvate dehydrogenase?

_____

_____

Reactions of Glycogen Degradation and Synthesis

4. Explain why the phosphorolytic cleavage of glycogen is more energetically advantageous than its hydrolytic cleavage.

_____

_____

5. Which of the following statements about the role of pyridoxal phosphate in the mechanism of action of phosphorylase are *correct*?

(a) It interacts with orthophosphate.
(b) It acts as a general acid-base catalyst.
(c) It orients the glycogen substrate in the active site.

(d) It donates a proton to the O-4 of the departing glycogen chain.
(e) It binds water at the active site.

6. Match the enzymes that degrade glycogen in the left column with the appropriate properties from the right column.

(a) Phosphorylase _____

(b) α-1,6-Glucosidase _____

(c) Transferase _____

(1) is part of a single polypeptide chain with two activities.
(2) cleaves α-1,4-glucosidic bonds.
(3) releases glucose.
(4) releases glucose 1-phosphate.
(5) moves three sugar residues from one chain to another.
(6) requires ATP.

7. The phosphoglucomutase reaction is similar to the phosphoglyceromutase reaction of the glycolytic pathway. Which of the following properties are common to both enzymes?

(a) Both have a phosphoenzyme intermediate.
(b) Both use a glucose 1,6-bisphosphate intermediate.
(c) Both contain pyridoxal phosphate, which donates its phosphate group to the substrate.
(d) Both transfer the phosphate group from one position to another on the same molecule.

8. The activity of which of the following enzymes is *not required* for the release of large amounts of glucose from liver glycogen?

(a) Glucose 6-phosphatase
(b) Fructose 1,6-bisphosphatase
(c) α-1,6-Glucosidase
(d) Phosphoglucomutase
(e) Glycogen phosphorylase

9. Which of the following are common features of both glycogen synthesis and glycogen breakdown?

(a) Both require UDP-glucose.
(b) Both involve glucose 1-phosphate.
(c) Both are driven in part by the hydrolysis of pyrophosphate.
(d) Both occur on cytoplasmic glycogen granules.
(e) Both use the same enzyme for branching and debranching.

10. Explain why the existence of distinct biosynthetic and degradative pathways for glycogen is important for the metabolism of liver and muscle cells.

_____

_____

11. Is it *true* or *false* that branching in the structure of glycogen increases the rates of its synthesis and degradation? _____

Explain: _____

12. Answer the following questions about the enzymatic degradation of amylose, a linear α-1,4 polymer of glucose that is a storage form of glucose in plants.

(a) Would phosphorylase act on amylose? Explain.

_____

_____

(b) What would be the rate of glucose release from amylose by phosphorylase relative to that for a glycogen molecule having an equivalent number of glucose monomers? Explain.

_____

_____

(c) How could the rate of degradation of amylose be increased?

_____

_____

13. Starting from a glucose residue in glycogen, how many net ATP molecules will be formed in the glycolysis of the residue to pyruvate?

(a) 1
(b) 2
(c) 3
(d) 4
(e) 5

Control of Glycogen Metabolism, and Structural Features of the Enzymes of Glycogen Metabolism

14. Which of the following statements about the hormonal regulation of glycogen synthesis and degradation are *correct*?

(a) Insulin increases the capacity of the liver to synthesize glycogen.
(b) Insulin is secreted in response to low levels of blood glucose.
(c) Glucagon and epinephrine have opposing effects on glycogen metabolism.
(d) Glucagon stimulates the breakdown of glycogen, particularly in the liver.
(e) The effects of all three of the regulating hormones are mediated by cyclic AMP.

15. Which of the following statements about cyclic AMP are *correct*?

(a) It has a different net charge than does AMP at pH 7.0.
(b) It has a phosphodiester bond between the C-5 and C-2 carbons of the ribose moiety.
(c) Its synthesis is stimulated by the hydrolysis of PP_i.
(d) Its hydrolysis is highly exergonic.
(e) It is inactivated by a specific phosphodiesterase that yields P_i and adenosine.

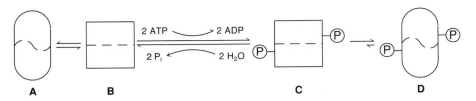

Figure 19-2
Conformational states of phosphorylase.

16. Consider the diagram of the different conformational states of glycogen phosphorylase in Figure 19-2. Then answer the following questions:

    (a) Which are the active forms of phosphorylase? _____

    (b) Which form requires high levels of AMP to become activated?

      _____

    (c) Which conversion is inhibited by ATP and glucose 6-phosphate? _____

    (d) What enzyme catalyzes the conversion of C to B? _____

    (e) What compound can convert D to C? _____

17. Indicate which of the following substances have binding sites on phosphorylase. For those that do, give their major roles or effects.

    (a) Calmodulin _____

    (b) Glycogen _____

    (c) Pyridoxal phosphate _____

    (d) Ca^{2+} _____

    (e) AMP _____

    (f) P_i _____

    (g) ATP _____

    (h) Glucose _____

18. Explain the role of calmodulin in the control of phosphorylase kinase in the liver.

_____

_____

19. Which of the following statements about glycogen synthase are *correct*?

    (a) It is activated when it is dephosphorylated.
    (b) It is activated when it is phosphorylated.
    (c) It is activated when it is phosphorylated and in the presence of high levels of glucose 6-phosphate.
    (d) It is activated when it is phosphorylated and in the presence of high levels of AMP.

Reaction Cascade in the Phosphorylation and Dephosphorylation of Glycogen Synthase and Phosphorylase

20. Place the following steps of the reaction cascade of glycogen metabolism in the proper sequence.

    (a) Phosphorylation of protein kinase
    (b) Formation of cyclic AMP by adenylate cyclase
    (c) Phosphorylation of phosphorylase *b*
    (d) Hormone binding to target cell receptors
    (e) Phosphorylation of glycogen synthase *a* and phosphorylase kinase

21. Explain the effect of insulin on the activity of protein phosphatase 1 and the subsequent effects on glycogen metabolism.

_____

_____

_____

22. Why are enzymatic cascades, such as those that control glycogen metabolism and the clotting of blood, of particular importance in metabolism?

_____

_____

Regulation of Blood Glucose Levels, and Glycogen Storage Diseases

23. Which of the following are effects of glucose on the metabolism of glycogen in the liver.

    (a) The binding of glucose to phosphorylase *a* converts this enzyme to the inactive T form.
    (b) The T form of phosphorylase *a* becomes susceptible to the action of phosphatase.
    (c) The R form of phosphorylase *b* becomes susceptible to the action of phosphorylase kinase.
    (d) When phosphorylase *a* is converted to phosphorylase *b*, the bound phosphatase is released.
    (e) The free phosphatase dephosphorylates and activates glycogen synthase.

24. Explain how a defect in phosphofructokinase in muscle can lead to increased amounts of glycogen having a normal structure. Patients with this defect are normal except for having a limited ability to perform strenuous exercise.

_____

_____

25. For the defect in Question 24, explain why there is not a massive accumulation of glycogen?

_____

_____

1. A patient can perform nonstrenuous tasks but becomes fatigued upon physical exertion. Assays from a muscle biopsy reveal that glycogen levels are slightly elevated relative to normal. Using crude extracts from muscle, the activity of glycogen phosphorylase at various levels of calcium ion for the patient and a normal individual were determined. The results of those assays are shown in Figure 19-3. Briefly explain the clinical and biochemical findings for the patient.

2. A certain strain of mutant mice is characterized by limited ability to engage in prolonged exercise. After a high carbohydrate meal, one of these mice can exercise on a treadmill for only about 30 percent of the time a normal mouse can. At exhaustion, blood glucose levels in the mutant mouse are quite low, and they increase only marginally after rest. When liver glycogen in fed mutant mice is examined before exercise, the polymers have chains that are highly branched, with average branch lengths of about 10 glucose residues in either α-1,4 or α-1,6 linkage. Glycogen from exhausted normal mice has the same type of structure. Glycogen from exhausted mutant mice is still highly branched, but the polymer has an unusually large number of single glucose residues with α-1,6 linkages. Practically all the chains with α-1,4 linkages are still about 10 residues in length, although some have 13 to 15 residues. Explain the metabolic and molecular observations concerning the mutant mice.

3. Your colleague discovers a fungal enzyme that can liberate glucose residues from cellulose. The enzyme is similar to glycogen phosphorylase in that it utilizes inorganic phosphate for the phosphorolytic cleavage of glucose residues from the nonreducing ends of cellulose. Why would you suspect that other types of cellulases may be important in the rapid degradation of cellulose?

4. Consider a patient with the following clinical findings: fasting blood glucose level is 25 mg per 100 ml (normal values are from 80 to 100 mg per 100 ml); feeding the patient glucose results in a rapid elevation of blood glucose level, followed by a normal return to fasting levels; feeding the patient galactose or fructose results in the elevation of blood glucose to normal levels; the administration of glucagon fails to generate hyperglycemia; biochemical examination of liver glycogen reveals a normal glycogen structure.

   (a) Which of the enzyme deficiencies described on page 465 of Stryer's text could account for these clinical findings?
   (b) What additional experiments would you conduct to provide a specific diagnosis for the patient?

5. Vigorously contracting muscle often becomes anaerobic when the demand for oxygen exceeds the amount supplied though the circulation. Under such conditions, lactate may accumulate in muscle. There is some evidence that under anaerobic conditions a certain percentage of lactate can be converted to glycogen in muscle. One line of evidence involves the demonstration of activity for malic enzyme, which can use CO_2 to convert pyruvate to malate, using NADPH as an electron donor.

   (a) Why is lactate produced in muscle when the supply oxygen is insufficient?

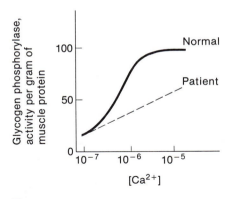

Figure 19-3
Response of glycogen phosphorylase to calcium ion in a patient and in a normal individual.

(b) In muscle, pyruvate carboxylase activity is very low. How could malic enzyme activity facilitate the synthesis of glycogen from lactate?

(c) Why would you expect the conversion of lactate to glycogen to occur only after vigorous muscle contraction ceases?

(d) Is there any energetic advantage to converting lactate to glycogen in muscle rather than sending the lactate to the liver, where it can be resynthesized to glucose and then returned to muscle for glycogen synthesis?

6. Cyclic nucleotide phosphatases are inhibited by caffeine. What effect would this have on glycogen metabolism when epinephrine levels are dropping in the blood?

7. During the degradation of branched chains of glycogen, a transferase shifts a chain of three glycosyl residues from one branch to another, exposing a single remaining glycosyl residue to α-1,6-glucosidase activity. Free glucose is released, and the now unbranched chain can be further degraded by glycogen phosphorylase.

(a) Estimate the free-energy change of the transfer of glycosyl residues from one branch to another.

(b) About 10 percent of the glycosyl residues of normal glycogen are released as glucose, whereas the remainder are released as glucose 1-phosphate. Give two reasons why it is desirable for cells to convert most of the glycosyl residues in glycogen to glucose 1-phosphate.

(c) Patients who lack liver glycosyl transferase have been studied. Why would you expect liver extracts from such individuals also to lack α-1,6-glucosidase activity?

8. You are studying a patient with McArdle's disease, which is described on page 466 of Stryer's text. Explain what you would expect to find when you carry out each of the following analyses:

(a) Fasting level of blood glucose

(b) Structure and amount of liver glycogen

(c) Structure and amount of muscle glycogen

(d) Change in blood glucose levels upon feeding the patient fructose

(e) Change in blood lactate levels after vigorous exercise

9. An investigator has a sample of purified muscle phosphorylase b that he knows is relatively inactive.

(a) Suggest two in vitro methods that could be employed to generate active phosphorylase from the inactive phosphorylase b.

(b) After the phosphorylase is activated, the investigator incubates the enzyme with a sample of unbranched glycogen in a buffered solution. He finds that no glycosyl residues are cleaved. What else is needed for the cleavage of glycosyl residues by active phosphorylase?

10. Arsenate can substitute for many reactions in which phosphate is the normal substrate. However, arsenate esters are far less stable than phosphate esters, and they decompose to arsenate and an alcohol:

$$R\text{—}OAsO_3{}^{2-} + H_2O \longrightarrow R\text{—}OH + AsO_4{}^{2-}$$

(a) In which of the steps of glycogen metabolism could arsenate be used as a substrate?

(b) What are the energetic consequences of utilizing arsenate as a substrate in glycogen degradation?

11. As described on page 464 of Stryer's text, the ratio of glycogen phosphorylase to protein phosphatase I is approximately ten to one. Suppose that in some liver cells the overproduction of the phosphatase results in a ratio of one of one. How will such a ratio affect the cell's response to an infusion of glucose?

ANSWERS TO SELF-TEST

Introduction

1. (a) A and G
   (b) G and E
   (c) All of the bonds are α-1,4-glycosidic linkages except for the one between residues G and E.
   (d) No. The two branches are too short for phosphorylase cleavage. Phosphorylase stops cleaving four residues away from a branch point.
   (e) Yes. Residue G can be hydrolyzed by α-1,6-glucosidase (the debranching enzyme).
   (f) No. The branching enzyme transfers a block of around seven residues from a nonreducing end of a chain at least eleven residues long. Furthermore, the new α-1,6-glycosidic linkage must be at least four residues away from a preexisting branch point at a more internal site. The fragment of glycogen in Figure 19-1 does not fulfill these requirements.

2. b, c, d

3. Unlike the enzymes in such multienzyme complexes as pyruvate dehydrogenase, the enzymes in glycogen granules, which catalyze the synthesis and degradation of glycogen and regulate some of the reactions of these processes, are not present in defined stoichiometric ratios. Rather, they bind to some of the many available sites on glycogen granules.

Reactions of Glycogen Degradation and Synthesis

4. The phosphorolytic cleavage of glycogen produces glucose 1-phosphate, which can enter into the glycolytic pathway after conversion to glucose 6-phosphate. These reactions do not require ATP. On the other hand, the hydrolysis of glycogen would produce glucose, which would have to be converted to glucose 6-phosphate, requiring the expenditure of an ATP. Therefore, the phosphorolytic cleavage of glycogen is more energetically advantageous.

5. a, b

6. (a) 2, 4 (b) 1, 3 (c) 1, 2, 5. None of these enzymes requires ATP.

7. a

8. b

9. b, d

10. The separate pathways for the synthesis and degradation of glycogen allow the synthesis of glycogen to proceed despite a high ratio of orthophosphate to glucose 1-phosphate, which energetically favors the degradation of glycogen. In addition, the separate pathways allow the coordinated reciprocal control of glycogen synthesis and degradation by hormonal and metabolic signals.

11. True. Since degradation and synthesis occur at the nonreducing ends of glycogen, the branched structure allows simultaneous reactions to occur at many nonreducing ends, thereby increasing the rates of degradation or biosynthesis.

12. (a) Yes, phosphorylase would act on amylose by removing one glucose residue at a time from the nonreducing end.
    (b) The rate of degradation of amylose would be much slower than that of glycogen because amylose would have only a single nonreducing end available for reaction, whereas glycogen has many ends where reaction can proceed simultaneously.
    (c) The rate of degradation of amylose could be increased through the addition of an endosaccharidase, an enzyme that cleaves glycosidic bonds at internal sites. This would accelerate degradation by providing new nonreducing ends that are susceptible to the action of phosphorylase.

13. c. A glucose molecule that is degraded in the glycolytic pathway yields 2 ATP; however, the formation of glucose 1-P from glycogen does not consume the ATP that is required for the formation of glucose 6-P from glucose. Thus, the net yield of ATP for a glucose residue derived from glycogen is 3 ATP.

Control of Glycogen Metabolism, and Structural Features of the Enzymes of Glycogen Metabolism

14. a, d

15. a, c, d

16. (a) A and D
    (b) B
    (c) B to A
    (d) Protein phosphatase 1
    (e) Glucose
    The phosphorylated form of glycogen phosphorylase is phosphorylase *a*, which is mostly present in the active conformation designated as D in Figure 19-2. In the presence of high levels of glucose, phosphorylase *a* adopts a strained, inactive conformation, designated as C in the figure. The dephosphorylated form of the enzyme is called phosphorylase *b*. Phosphorylase *b* is mostly present in an inactive conformation, labeled B in the figure. When AMP binds to the inactive phosphorylase *b*, the enzyme changes to an active conformation, designated as A in the figure. The effects of AMP can be reversed by ATP or glucose 6-phosphate.

17. (b) Glycogen as the substrate binds to the active site; there is also a glycogen-particle binding site that keeps the enzyme attached to the glycogen granule.
    (c) Pyridoxal phosphate is the prosthetic group that positions orthophosphate for phosphorolysis and acts as a general acid-base catalyst.
    (e) AMP binds to an allosteric site and activates phosphorylase *b*.

(f) P_i binds to the pyridoxal phosphate at the active site and cleaves the α-1,4-glycosidic bond. Another P_i is covalently bound to serine 14 by phosphorylase kinase. This phosphorylation converts phosphorylase b into active phosphorylase a.

(g) ATP binds to the same site as AMP and blocks its effects.

(h) Glucose binds to the catalytic site differently from glycogen; it inhibits phosphorylase a in the liver by changing the conformation of the enzyme to the inactive T form.

Answers (a) and (d) are incorrect because calmodulin and Ca^{2+} bind to phosphorylase kinase rather than to phosphorylase.

18. Phosphorylase kinase can be activated by the binding of Ca^{2+} to calmodulin, which is a subunit of this enzyme. Upon binding Ca^{2+}, calmodulin undergoes conformational changes that activate the phosphorylase kinase, which in turn activates glycogen phosphorylase. These effects lead to glycogen degradation in active muscle.

19. a, c

Reaction Cascade in the Phosphorylation and Dephosphorylation of Glycogen Synthase and Phosphorylase

20. d, b, a, e, c

21. Insulin decreases the level of phosphorylated inhibitor 1, which binds to protein phosphatase 1 and prevents its action. Thus, insulin activates the phosphatase, which dephosphorylates phosphorylase, protein kinase, and glycogen synthase. These changes result in a decrease in glycogen degradation and the stimulation of glycogen synthesis.

22. Enzymatic cascades lead from a small signal, due to one or a few molecular interactions, to a large subsequent enzymatic response. Thus small biological signals can be greatly amplified in a short time, and their effects can be regulated at various levels.

Regulation of Blood Glucose Levels, and Glycogen Storage Diseases

23. a, b, d, e

24. Since a defect in phosphofructokinase does not impair the ability of muscle to synthesize and degrade glycogen normally, the structure of glycogen will be normal. However, the utilization of glucose 6-phosphate in the glycolytic pathway is impaired; therefore, some net accumulation of glycogen will occur. The inability to perform strenuous exercise is due to the impaired glycolytic pathway in muscle and the diminished production of ATP.

25. Because of the impaired use of glucose 6-phosphate in glycolysis, the accumulation of this key intermediate will lead to the storage of extra glycogen. However, glycogen storage will not become excessive because glucose 6-phosphate will inhibit hexokinase and hence the entry of glucose into muscle.

ANSWERS TO PROBLEMS

1. Calcium ion normally activates muscle phosphorylase kinase, which in turn phosphorylates muscle phosphorylase. In the patient, glycogen

phosphorylase activity is less responsive to Ca^{2+} than it is in the normal individual. It is likely that Ca^{2+} cannot activate phosphorylase kinase in the patient, perhaps because the δ subunit of the enzyme is altered in some way. As a result, there are too few molecules of enzymatically active glycogen phosphorylase to provide the rate of glycogen breakdown that is needed to sustain vigorous muscle contraction. Elevated levels of muscle glycogen should be expected when glycogen phosphorylase activity is lower than normal.

2. The longer chains of glucose residues in α-1,4 linkage and the unusually high number of single glucose residues in α-1,6 linkage suggest that while transferase activity is present, α-1,6-glucosidase activity is deficient in the mutant strain. For such chains, far fewer ends with glucose residues are available as substrates for glycogen phosphorylase. (Recall that glycogen phosphorylase cannot cleave α-1,6 linkages.) This limited ability to mobilize glucose residues means that less energy is available for prolonged exercise.

3. Cellulose is an unbranched polymer of glucose residues with β-1,4 linkages. Therefore, each chain has only one nonreducing end that is available for phosphorolysis by the fungal enzyme. Compared to the rate of breakdown of molecules of glycogen, whose branched chains provide more sites for the action of glycogen phosphorylase, the generation of glucose phosphate molecules from cellulose via the fungal enzyme alone could be quite slow. Therefore, you should expect to find other cellulases that generate additional nonreducing ends in cellulose chains.

4. (a) A low fasting blood glucose level indicates a failure either to mobilize glycogen or to release glucose from the liver. However, the elevations in blood glucose levels after feeding the patient glucose, galactose, or fructose indicate that the liver can release glucose derived from the diet or formed from other monosaccharides. The lack of response to glucagon indicates that the enzymatic cascade for glycogen breakdown is defective. Therefore, you would suspect a deficiency of liver glycogen phosphorylase or phosphorylase kinase. Of the diseases described in Table 19-1 of Stryer's text, both type VI and type VIII could account for the findings, which include normal glycogen structure.

   (b) The direct assay of the activities of glycogen phosphorylase and phosphorylase kinase would enable you to make a specific diagnosis. For these purposes, a liver biopsy would be necessary.

5. (a) Muscle cells produce lactate from pyruvate under anaerobic conditions in order to generate NAD^+, which is required to sustain the activity of glyceraldehyde 3-phosphate dehydrogenase in the glycolytic pathway.

   (b) The low activity of muscle pyruvate carboxylase means that other pathways for the synthesis of oxaloacetate must be available. The formation of malate, which is then converted to oxaloacetate (see p. 441 of Stryer's text), enables the muscle cell to carry out the synthesis of glucose 6-phosphate via gluconeogenesis. Glycogen can then be synthesized through the conversion of glucose 6-phosphate to glucose 1-phosphate, the formation of UDP-glucose, and the transfer of the glucose residue to a glycogen primer chain.

(c) Energy for vigorous muscle contraction under anaerobic conditions is derived primarily from the conversion of glycogen and glucose to lactate. The simultaneous conversion of lactate to glycogen would simply result in the unnecessary hydrolysis of ATP.

(d) In liver, the conversion of two lactate molecules to a glucose residue in glycogen through gluconeogenesis requires 7 high-energy phosphate bonds; six are required for the formation of glucose 6-phosphate from 2 molecules of lactate, and one is needed for the synthesis of UDP-glucose from glucose 1-phosphate. The conversion of lactate to glycogen in muscle requires two fewer high-energy bonds because the formation of oxaloacetate through the action of malic enzyme does not require ATP. Recall that pyruvate carboxylase requires ATP for the synthesis of oxaloacetate from pyruvate.

6. When epinephrine levels in the blood decrease, the synthesis of cyclic AMP decreases. Existing cyclic AMP is degraded by cyclic nucleotide phosphatases. The inhibition of these enzymes by caffeine prolongs the degradation of glycogen because the remaining cyclic AMP continues to activate protein kinase, which in turn activates phosphorylase kinase.

7. (a) Because the bonds broken and formed during the transferase action are both α-1,4-glycosidic bonds, the free-energy change is likely to be close to zero.

(b) The generation of glucose 1-phosphate rather than glucose means that one less ATP equivalent is required for the conversion of a glucose residue to two molecules of lactate. In addition, the phosphorylation of glucose ensures that the molecule cannot diffuse across the cell membrane before it is utilized in the glycolytic pathway.

(c) The glucosidase and the transferase activities are both found on the same 160-kd polypeptide chain. A significant alteration in the structure of the domain for glucosidase could impair the functioning of the transferase domain.

8. (a) In McArdle's disease, muscle phosphorylase is deficient, but liver phosphorylase is normal. Therefore, you would expect glucose and glycogen metabolism in the liver to be normal and the control of blood glucose by the liver also to be normal.

(b) Normal glycogen metabolism in the liver means that both the amount of liver glycogen and its structure would be the same as in unaffected individuals.

(c) Defective muscle glycogen phosphorylase means that glycogen breakdown is impaired. Moderately increased concentrations of muscle glycogen could be expected, although the structure of the glycogen should be similar to that in normal individuals.

(d) Fructose can be converted to glucose 6-phosphate in the liver, which can then export glucose to the blood. Because defective muscle phosphorylase has no effect on fructose metabolism, you would expect similar elevations in blood glucose after the ingestion of fructose in normal and affected individuals.

(e) During vigorous exercise, blood lactate levels normally rise as muscle tissue exports the lactate generated through glycogen breakdown. The defect in muscle phosphorylase limits the extent to which glycogen is degraded in the muscle. This in turn reduces the amount of lactate exported during exercise, so the rise in blood lactate levels would not be as great in the affected individual.

9. (a) The investigator can activate the phosphorylase by adding AMP to the sample or by using active phosphorylase kinase and ATP to phosphorylate the enzyme.

(b) Inorganic phosphate is also required for the conversion of glycosyl residues in glycogen to glucose 1-phosphate molecules.

10. (a) Arsenate can substitute for inorganic phosphate in the glycogen phosphorylase reaction, generating glucose arsenate esters.

(b) When P_i is used as a substrate for glycogen phosphorylase, glucose 1-phosphate is generated. The glucose 1-arsenate esters that are generated when arsenate is used as a substrate spontaneously degrade to yield glucose and arsenate. The conversion of glucose to pyruvate requires one more ATP equivalent than does the conversion of glucose 1-phosphate to pyruvate.

11. The normal ten to one ratio means that glycogen synthase molecules are activated only after most of the phosphorylase a molecules are converted to the inactive b form, which ensures that the simultaneous degradation and synthesis of glycogen does not occur. A phosphorylase to phosphatase ratio of one to one means that, as soon as a few phosphorylase molecules are inactivated, phosphatase molecules that are no longer bound to phosphorylase begin to convert glycogen synthase molecules to the active form. Glycogen degradation and synthesis then occur simultaneously, resulting in the wasteful hydrolysis of ATP.

CHAPTER 20

Fatty Acid Metabolism

In his consideration of the generation and storage of metabolic energy, Stryer has thus far focused on the carbohydrates (Chapters 15, 18, and 19). In Chapter 20, he turns to the fatty acids as metabolic fuels. After describing the nomenclature of fatty acids, he explains why they are the most concentrated energy stores. The pathway of the oxidation of fatty acids, which liberates the energy of fatty acids and makes it available to the cell, is then presented. The oxidation of unsaturated fatty acids is described and the formation and role of the ketone bodies as interorgan acetyl transport molecules are discussed. Stryer then describes how both saturated and unsaturated fatty acids are synthesized. The energetics of the oxidation and synthesis of fatty acids are given, and an outline of the control of these processes is provided. A review of Chapter 12 will remind you of the structural role of lipids in membranes and of the effect of the fatty acids in determining membrane fluidity.

LEARNING OBJECTIVES

When you have mastered this chapter, you should be able to complete the following objectives.

Introduction (Stryer pages 469–471)

1. List the three major physiological functions of the *fatty acids.*

2. Derive the structure of a *saturated* or *unsaturated fatty acid* from its systematic name. Specify the α, β, and ω carbon atoms and designate the position of a double bond in a fatty acid given either its Δ or ω number.

3. Recognize the structures of *palmitate, stearate, palmitoleate, linoleate,* and *linolenate.*

Fatty Acid Oxidation (Stryer pages 471–480)

4. Explain why *triacylglycerols* are highly concentrated forms of *stored metabolic energy.*

5. Describe the *lipolysis* of triacylglycerols by *lipases.* Explain the role of *cyclicAMP* in the regulation of lipase in *adipose cells.* Account for the conversion of glycerol to *glycerol 3-phosphate* and *dihydroxyacetone phosphate.*

6. Recount the experiments of Franz Knoop that showed fatty acid catabolism occurs by β-*oxidation.*

7. Describe the reaction that links *coenzyme A (CoA)* to a fatty acid.

8. Explain the involvement of *carnitine* in the transport of fatty acids from the cytoplasm into mitochondria.

9. List the four reactions of the β-oxidation cycle and identify their substrates and products. Explain the function of NAD^+ and FAD in this pathway.

10. Indicate the two reactions, in addition to those of the β-oxidation pathway, used to oxidize naturally occurring unsaturated fatty acids.

11. Describe the complete oxidation of an *odd-numbered fatty acid.*

12. Calculate the *energy yield,* in terms of ATP molecules, for the β-oxidation of a given fatty acid.

Ketone Bodies (Stryer pages 478–480)

13. Explain the consequences of limiting *oxaloacetic acid* concentrations on the oxidation of fatty acids. Name and identify the structures of the *ketone bodies.*

14. Describe the synthesis and normal catabolism of the ketone bodies. Explain why only the liver exports *acetoacetate* and *3-hydroxybutyrate.*

15. Describe the effect of high levels of *acetoacetate* on fat metabolism in adipose tissue.

16. Provide the biochemical basis for the inability of animals to convert fatty acids into glucose.

17. Contrast fatty acid oxidation and *fatty acid synthesis.*

18. Name the substrates and products of the committed step in fatty acid synthesis and describe its regulatory mechanism.

19. Indicate the common component of *acyl carrier protein (ACP)* and CoA, give its functions, and indicate the overall functions of ACP and CoA in fatty acid metabolism.

20. Describe the four reactions of the elongation cycle of fatty acid synthesis. Explain how *malonyl CoA* provides the driving force for the condensation of acetyl units with the growing acyl chain.

21. Calculate the energy cost of the synthesis of a given fatty acid and explain the origins of the *NADPH* used in the reductive steps.

22. Contrast the enzymatic machinery for fatty acid biosynthesis in bacteria with that in eucaryotes. Outline the movements of the elongating acyl chain on the mammalian *fatty acid synthesase* dimer during fatty acid biosynthesis.

23. Outline the *citrate-malate* and *malate-pyruvate* pathways and show how they transport acetyl groups and reducing equivalents across the mitochondrial inner membrane.

24. Describe the elongations and *desaturations* that can occur on preformed fatty acids. Explain why linoleate and linolenate are essential in the diet.

Control of Fatty Acid Metabolism (Stryer pages 490–491)

25. Integrate the short- and long-term mechanisms of control of fatty acid oxidation and biosynthesis. Recount the effects of malonyl CoA, acetyl CoA, palmitoyl CoA, citrate, cyclicAMP, and ATP on the metabolism of fatty acids.

SELF-TEST

Introduction

1. Which of the following are major physiological functions of free fatty acids?

   (a) They stabilize the structure of membranes.
   (b) They serve as precursors of phospholipids and glycolipids.
   (c) They serve as fuel molecules.
   (d) They are precursors and constituents of triacylglycerols.

2. For the following four naturally occurring fatty acids, give the systematic name, the common name, and the abbreviations for each. Use the Δ convention for the name and abbreviations of the unsaturated compounds. Also indicate the position of the double bond closest to the methyl end of the chain using the ω convention.

(a) $CH_3(CH_2)_{14}CO_2H$

_____

(b) 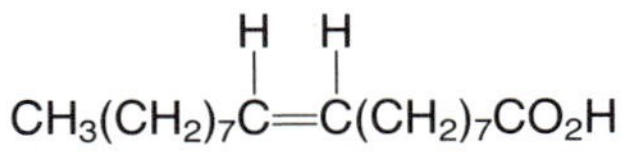 $CH_3(CH_2)_7\overset{H}{\underset{}{C}}=\overset{H}{\underset{}{C}}(CH_2)_7CO_2H$

_____

(c) $CH_3(CH_2)_4\overset{H}{\underset{}{C}}=\overset{H}{\underset{}{C}}CH_2\overset{H}{\underset{}{C}}=\overset{H}{\underset{}{C}}(CH_2)_7CO_2H$

_____

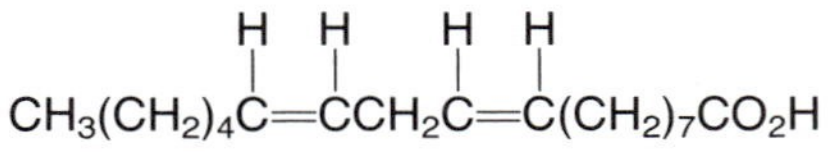

(d) 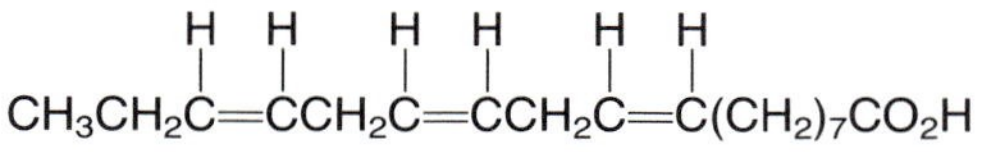 $CH_3CH_2\overset{H}{\underset{}{C}}=\overset{H}{\underset{}{C}}CH_2\overset{H}{\underset{}{C}}=\overset{H}{\underset{}{C}}CH_2\overset{H}{\underset{}{C}}=\overset{H}{\underset{}{C}}(CH_2)_7CO_2H$

_____

Fatty Acid Oxidation

3. What two properties make triacylglycerols more efficient than glycogen for the storage of metabolic energy?

_____

_____

4. Which of the following statements about the triacylglycerols stored in adipose tissue are *correct*?

(a) They are hydrolyzed to form fatty acids and dihydroxyacetone.
(b) They are hydrolyzed by a lipase that is activated by covalent modification.
(c) They release fatty acids that can be oxidized to CO_2 and H_2O to provide energy to the cell.
(d) They can yield a precursor of glucose.
(e) They are mobilized by epinephrine or glucagon.

5. Predict the residual acids resulting from the oxidation of phenylpentanoic and phenylhexanoic acids in a dog. What is the biochemical meaning of this finding, which was made by Knoop?

_____

_____

_____

6. (a) Draw a thioester bond and explain its role in the β-oxidation of fatty acids.

_____

_____

(b) What is the $\Delta G^{\circ\prime}$ value for the hydrolysis of acetyl coenzyme A? What is the significance of this value with respect to fatty acid metabolism?

_____

_____

(c) Describe the mechanism for the formation of acyl CoA.

_____

_____

_____

(d) How is pyrophosphatase involved in the activation of fatty acids for β-oxidation?

_____

_____

7. Place the following incomplete list of reactions or relevant locations during the β-oxidation of fatty acids in the proper order.

(a) Reaction with carnitine
(b) Fatty acid in cytosol
(c) Activation of fatty acid by joining to CoA
(d) Hydration
(e) NAD⁺-linked oxidation
(f) Thiolysis
(g) Acyl CoA in mitochondrion
(h) FAD-linked oxidation
(i) Electron transport and oxidative phosphorylation

8. Explain the involvement of carnitine in the β-oxidation of fatty acids.

_____

_____

_____

9. Calculate the yield in ATP molecules of the complete oxidation of hexanoic acid (C6:0).

_____

_____

10. In the process of hydrogenating certain oils to saturate the double bonds in their fatty acids, some of the *cis* double bonds are converted to the *trans* conformation. Predict what would happen if a monoenoic fatty acid with a *trans*-Δ^{10} bond were produced, injested, and degraded by the β-oxidation pathway. If another of the injested fatty acids contained a *cis*-Δ^{11} double bond, what would be the outcome of these processes? What effect would the presence of the double bond have on the yield of ATP obtained by the β-oxidation of these fatty acids?

_____

_____

_____

Ketone Bodies

11. Which of the following statements about acetoacetate and 3-hydroxybutyrate are *correct*?

(a) They are normal fuels for heart muscle and the renal cortex.
(b) They are not formed during fasting because fatty acids are rapidly converted to acetyl CoA.
(c) Both can give rise to acetone.
(d) Each contains four carbon atoms and requires three acetyl CoA molecules for its synthesis.

Fatty Acid Synthesis

12. Match the reactant or characteristic in the right column with the appropriate pathway in the left column.

(a) Fatty acid
    oxidation _____

(b) Fatty acid
    synthesis _____

(1) Acyl CoA
(2) Occurs in the cytosol
(3) Uses NAD$^+$
(4) D-3-Hydroxyacyl derivative involved
(5) Pantetheine involved
(6) Malonyl CoA
(7) Single polypeptide with multiple activities involved
(8) Uses FAD

13. Explain the requirement for bicarbonate in fatty acid biosynthesis.

_____

_____

_____

14. Which of the following statements about citrate are *correct?*

(a) It transports reducing power from the mitochondria into the cytosol.
(b) It inhibits gluconeogenesis.
(c) It activates the first enzyme of fatty acid biosynthesis.
(d) It transports acetyl groups from the mitochondria into the cytosol.
(e) It supplies the CO_2 required for formation of malonyl CoA.

15. The fatty acid synthase of mammals is a dimer consisting of identical subunits, each of which contains all the activities necessary to synthesize fatty acids from malonyl CoA and acetyl CoA. Why is a single subunit unable to carry out the reactions?

_____

_____

16. Possible advantages of multifunctional polypeptide chains, that is, polypeptide chains having more than one active site, include which of the following?

(a) Enhanced stability beyond that expected for a noncovalent complex of the same activities on separate polypeptide chains.
(b) Fixed stoichiometric relationships among the different enzymatic activities because of their coordinate synthesis.
(c) Enhanced specificity and decreased side reactions because the product of each active site is in the immediate vicinity of the active site carrying out the next reaction in the sequence.

(d) Enhanced versatility because the product of any one active site could be used by any other active site in its immediate vicinity to generate a variety of products.

(e) Accelerated overall reaction rate because of the proximity of the active sites.

17. The major product of the fatty acid synthase complex in mammals is

(a) oleate.
(b) stearate.
(c) stearoyl CoA.
(d) linoleate.
(e) palmitate.
(f) palmitoyl CoA.

18. Which of the following statements about desaturases in humans are *correct?*

(a) They cannot introduce double bonds into a fatty acid that already contains a double bond.
(b) They cannot introduce double bonds between the Δ^9 position and the ω end of the chain.
(c) They convert the essential fatty acid linoleate into arachidonate.
(d) They use an isozyme of the FAD-linked dehydrogenase of the β-oxidation cycle to form double bonds.

19. Calculate the ATP and NADPH requirements for the synthesis of lauric acid (C12:0) from acetyl CoA.

_____

_____

Control of Fatty Acid Metabolism

20. Match the appropriate compound or process in the right column with its ability to control the process in the left column.

(a) β-Oxidation of fatty acids _____

(b) Fatty acid biosynthesis _____

(1) Citrate
(2) Palmitoyl CoA
(3) Acetyl CoA
(4) NADH
(5) Malonyl CoA
(6) Hormones stimulating the formation of cyclic AMP in adipose tissue
(7) Adaptive control through the synthesis and degradation of enzymes

PROBLEMS

1. Many plants have enzyme systems that catalyze the formation of a *cis* double bond in oleic acid at one or more positions between C-9 and

the terminal methyl group. The fact that these enzyme systems exist in plants is of great significance to animals. Why?

2. (a) Suppose that the normal mechanism for the oxidation of fatty acids in mammals were through oxidation at the α-carbon rather than the β-carbon. What results would Knoop have obtained in his experiments?

(b) Stumpf and his colleagues have described an α-oxidation system in plant leaves and seeds. Molecular oxygen is used in the α-oxidative decarboxylation of a free fatty acid, which yields a fatty aldehyde that is one carbon shorter than the original fatty acid. The fatty aldehyde is in turn oxidized to the corresponding fatty acid, with NAD^+ serving as an electron acceptor. These steps are repeated, resulting in the complete oxidation of the fatty acid. Suppose that the NADH generated through the α-oxidation of palmitate is reoxidized in the mitochondrial electron-transport chain. Compare the yield of ATP generated by the α-oxidation of palmitate with that generated by β-oxidation of the same fatty acid. Assume that the products of the final round of oxidation are carbon dioxide and acetic acid.

(c) If fatty acid oxidation occurs via the α-oxidation route, will odd-numbered fatty acids be glucogenic? Why?

3. Although most components of the diet contain fatty acids with unbranched chains, some plant tissues contain fatty acids with methyl groups at odd-numbered carbons in the acyl chain. These fatty acids cannot be broken down through β-oxidation.

(a) Which step in β-oxidation is likely to be blocked when branched chain fatty acids are substrates?

(b) Some tissues, including brain tissue, can carry out the limited α-oxidation of a fatty acid with one or more methyl groups at odd-numbered carbons. Using the pathway discussed in Problem 2, show how one round of α-oxidation enables a cell to bypass the block to β-oxidation. Use as your substrate a molecule of palmitate with a methyl branch at C-3.

4. The microbial oxidation of long-chain alkanes, which are found in crude oil, is the subject of intense study because of concern about oil spills. In many bacteria, alkane oxidation occurs within the outer membrane. A monooxygenase enzyme uses molecular oxygen and an oxidizable substrate, such as NADH, to convert an alkane to a primary alcohol. Studies show that three additional reactions are required for the primary alcohol to undergo β-oxidation. Propose a pathway for the conversion of a long-chain primary alcohol to a substrate that can undergo β-oxidation. Include cofactors and electron acceptors that may be required.

5. Malonyl CoA, labeled with ^{14}C in the methylene carbon, is used in excess as a substrate in an in vivo system for the synthesis of palmitoyl CoA, which is catalyzed by a yeast fatty acid synthase complex. Acetyl CoA and other substrates are also present in the system, but acetyl CoA carboxylase is not. Which carbons in palmitoyl CoA will be labeled?

6. A deficiency of carnitine acyltransferase I in human muscle causes cellular damage and recurrent muscle weakness, especially during fasting or exercise. A deficiency of the enzyme in the liver causes an enlarged and fatty liver, hypoglycemia, and a reduction in the levels of ketone bodies in blood. Explain the causes of these symptoms.

7. One intermediate in the conversion of propionyl CoA to succinyl CoA is methylmalonyl CoA, the structure of which is shown in the margin. This compound is an analog of malonyl CoA. In individuals who are unable to convert propionyl CoA to succinyl CoA, high levels of methylmalonyl CoA are observed. What effect could such levels of methylmalonyl CoA have on fatty acid metabolism?

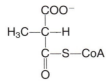

L-Methylmalonyl CoA

8. Animals cannot synthesize glucose from even-numbered fatty acids, which make up the bulk of the fatty acids in their diet.

   (a) How can odd-numbered fatty acids be used for the net synthesis of glucose in animals?
   (b) Triacylglycerols can be used as precursors of glucose. Give two reasons why this is possible.
   (c) Why are most of the fatty acids found in animal tissues composed of an even number of carbon atoms?
   (d) Some bacteria synthesize odd-numbered fatty acids. What CoA derivative is required, in addition to acetyl CoA and malonyl CoA, for the synthesis of an odd-numbered fatty acid?

9. (a) Describe how malonyl CoA effects the balance between the rates of synthesis and β-oxidation of fatty acids in a liver cell.
   (b) Show that failure to regulate these two processes reciprocally results in the wasteful hydrolysis of ATP.

10. Plant seeds contain triacylglycerols in organelles called spherosomes. During germination, lipases located in the spherosome membrane convert triacylglycerol to monoacylglycerols, free fatty acids, and glycerol. Both free fatty acids and monoacylglycerols enter the glyoxysome, whereas most of the glycerol is metabolized in the plant-cell cytosol. A membrane-bound lipase in the glyoxysome converts monoacylglycerols to free fatty acids and glycerol.

   (a) Describe two possible metabolic fates of glycerol in the cytosol.
   (b) What is the fate of fatty acids in the glyoxysome?
   (c) When a germinating plant begins to carry out photosynthesis, the number of glyoxysomes in the germinating plant decreases rapidly. Why?
   (d) Plant tissues with high numbers of mitochondria also have high concentrations of carnitine, but there is little correlation between numbers of glyoxysomes and carnitine concentrations in germinating tissue. What does this observation suggest about the role of carnitine in fatty acid metabolism in these two organelles?
   (e) Another difference between plant glyoxysomes and plant mitochondria is that glyoxysomes cannot oxidize acetyl CoA whereas mitochondria can. How is this observation related to the metabolism of fatty acids in these two organelles?

11. Individuals concerned about weight must not only pay attention to triacylglyceride intake but also to the consumption of starch, glucose, and other carbohydrates. Although carbohydrates can be converted to glycogen in liver, muscle, and other tissues, only about 5 percent of the energy stored in the body is present as glycogen. What happens to most carbohydrates that are consumed in caloric excess?

12. Wakil's pioneering studies on fatty acid synthesis included the crucial observation that bicarbonate is required for the synthesis of palmi-

toyl CoA. He was surprised to find that very low levels of bicarbonate could sustain palmitate synthesis; that is, there was no correlation between the amount of bicarbonate required and the amount of palmitate produced. Later he also found that ^{14}C-labeled bicarbonate is not incorporated into palmitate. Explain these observations.

13. Many of the enzymes of the β-oxidation pathway have relatively broad specificities for fatty acyl chain lengths. Why is this important for the economy of the cell?

14. Liver tissue carries out the synthesis of ketone bodies from fatty acids. Suppose a liver cell converts palmitic acid to acetoacetate and then exports it to the circulation. How many molecules of ATP per molecule of palmitate converted to acetoacetate are available to the liver cell?

15. In tissue culture, cells that are deficient in NADP$^+$-linked malate enzyme can be isolated. They exhibit a slightly lower rate of fatty acid synthesis when compared with normal cells. However, cell lacking citrate lyase are very difficult to isolate. Why?

16. An unusual sphingolipid contains a 22-carbon, polyunsaturated fatty acid called clupanodonic acid, or 7,10,13,16,19-docosapentaenoic acid. In mammals, both the mitochondrial and endoplasmic-reticular acyl chain-elongation and desaturation systems can synthesize clupanodonate from linolenate.

(a) What steps are required to synthesize clupanodonate from linolenate?
(b) Why are mammals unable to synthesize clupanodonate from linoleate?

ANSWERS TO SELF-TEST

Introduction

1. b, c, d

2. (a) Hexadecanoic acid; palmitic acid; C16:0
   (b) *cis*-Δ^9-Octadecenoic acid; oleic acid; C18:1 *cis*-Δ^9; ω-9
   (c) *cis,cis*-Δ^9,Δ^{12}-Octadecadienoic acid; linoleic acid; C18:2 *cis,cis*-Δ^{9,12}; ω-6
   (d) *cis,cis,cis*- (or all *cis*-) Δ^{9,12,15}-Octadecatrienoic acid; linolenic acid; C18:3 *cis,cis,cis*-Δ^{9,12,15}; ω-3

Note that Question 2 shows the undissociated form of the fatty acids, whereas Table 20-1 in Stryer names their ionized or carboxylate forms.

Fatty Acid Oxidation

3. Triacylglycerols contain a high proportion by weight of fatty acids. Fatty acids are highly reduced and have a higher energy content (9 kcal/g) than glycogen (4 kcal/g), which is composed of carbohydrate residues containing numerous oxygen atoms. In addition, fats are anhydrous, whereas glycogen is hydrated, so on the basis of actual storage weight, triacylglycerols contain six times more calories per gram than glycogen does.

4. b, c, d, and e. The hydrolysis of triacylglycerols yields glycerol, not dihydroxyacetone, and glycerol can be converted into glyceraldehyde 3-phosphate, which can ultimately give rise to glucose. Several hormones affect the hormone sensitive lipase of adipose tissue *via* a cyclic-AMP–modulated phosphorylation that activates the enzyme.

5. The compound with the phenyl group bearing the odd-numbered fatty acid chain would give rise to benzoic acid. In contrast, phenylhexanoic acid would be degraded to phenylacetic acid. These results suggest that the fatty acids are being metabolized from the carboxyl terminus and more importantly, that the carbon atoms are being removed two at a time. Thus, oxidation of the β-carbon atom of the fatty acid was a likely step in the degradation.

6. (a) A thioester bond is shown in the margin. The thioester linkage joins the fatty acid to CoA, which acts as a tag or handle by which the enzymes of the β-oxidation path can recognize, bind, and act upon the saturated alkane chains of the fatty acids. Furthermore, since the thioester linkage is a "high-energy" bond, it can transfer the acyl group to carnitine—a reaction that is necessary to deliver the acyl group from the cytosol to the mitochondrial matrix for oxidation.

$$\underset{\text{Thioester bond}}{R-\overset{\overset{\displaystyle O}{\|}}{C}-SR'}$$

(b) $\Delta G^{\circ\prime} = -7.5$ kcal/mol (see p. 322 in Stryer). The relatively large and negative value for the free energy of hydrolysis for acetyl CoA indicates that energy must be supplied to synthesize it and that, conversely, it is an "activated" donor of acetyl groups.

(c) The carboxyl group of the fatty acid is first activated by reaction with ATP to form an acyl adenylate, which contains a mixed anhydride linkage between the carboxylate and the 5′-phosphate of AMP, with the release of PP_i. In a second step, also catalyzed by acyl CoA synthetase, the acyl group is transferred to the sulfhydryl group of CoA to form the thioester bond and release AMP.

(d) The hydrolysis of PP_i couples the cleavage of a second high-energy bond to the formation of the thioester bond to make its formation exergonic.

7. b, c, a, g, h, d, e, f, i

8. Acyl CoA is formed in the cytosol and the enzymes of the β-oxidation pathway are in the matrix of the mitochondrion. The mitochondrial inner membrane is impermeable to CoA and its acyl derivatives. However, a translocase protein can shuttle carnitine and its acyl derivatives across the inner mitochondrial membrane. The acyl group is transferred to carnitine on the cytosol side of the inner membrane and back to CoA on the matrix side. Thus, carnitine, acts as a transmembrane carrier of acyl groups.

9. After activation of hexanoic acid to hexanoyl CoA, two rounds of β-oxidation are required to produce 3 acetyl CoA molecules. The two cycles also produce 2 $FADH_2$ and 2 NADH molecules. Each acetyl CoA yields 12 ATP upon complete oxidation, each $FADH_2$ produces 2 ATP, and each NADH makes 3 ATP, a total of 46 ATP. Activation uses 2 ATP molecules because AMP is one of the products of the reaction, so the net yield is 44 ATP.

10. Four rounds of β-oxidation of a fatty acid with a *trans*-Δ^{10} double bond would yield a *trans*-Δ^2-enoyl CoA derivative. This compound is the natural intermediate formed by an acyl CoA dehydrogenase. It

would be hydrated by enoyl CoA hydratase to form the L-3-hydroxyacyl CoA derivative. For the fatty acid with a *cis*-Δ^{11} double bond, four rounds of β-oxidation would produce a *cis*-Δ^3 double bond, which would not serve as a substrate for enoyl CoA hydratase. An isomerase would convert this bond into the *trans*-Δ^2 configuration to allow subsequent metabolism. Since the double bond already exists in the fatty acids and does not arise from β-oxidations, one less $FADH_2$ would be formed. Consequently, two fewer ATP would be produced for each preexisting double bond.

Ketone Bodies

11. a, c, and d. Choice (b) is incorrect because the high levels of acetyl CoA arising from the oxidation of fatty acids coupled with the lower levels of oxaloacetate because of its conversion to glucose are the conditions that favor the formation of ketone bodies.

Fatty Acid Synthesis

12. (a) 1, 3, 5, and 8 (b) 1, 2, 4, 5, 6, and 7. Acyl CoA is involved in both the synthesis and the oxidation of fatty acids.

13. The irreversible and committed step of fatty acid biosynthesis is the formation of malonyl CoA from acetyl CoA and HCO_3^- by acetyl CoA carboxylase. HCO_3^- is fixed to form a dicarboxylic acid at the expense of an ATP cleavage. This facilitates the subsequent condensation reactions with activated acyl groups to form an acetoacyl ACP by releasing CO_2 to help drive the reaction.

14. c, d

15. The reactions of elongation require the interactions of domains from different subunits of the dimer in order to form active sites at the interfaces of the subunits.

16. a, b, c, and e. (For a discussion relevant to answer (e) see p. 382 in Stryer.)

17. e

18. b and c. For (c) additional elongations as well as desaturations are required.

19. For a C12:0–fatty acid, 6 acetyl CoA molecules are required. One serves as a primer forming the ω end of the chain and 5 undergo condensation reactions as their malonyl CoA derivatives. Each malonyl CoA requires 1 ATP and each cycle of elongation uses 2 NADPH molecules. Thus, 5 ATP and 10 NADPH are required.

Control of Fatty Acid Metabolism

20. (a) 3, 4, 5, and 6. Answer (6) is correct because, although indirect, it is a mechanism that supplies fatty acids for oxidation.
    (b) 1, 2, 6, 7

ANSWERS TO PROBLEMS

1. Because they lack an enzyme that can introduce double bonds beyond the C-9 position in a fatty acid, animals cannot synthesize linoleate

and linolenate. These unsaturated fatty acids are precursors for a number of other important fatty acids as well as the prostaglandins. Animals therefore rely on their diet as the source of linoleate and linolenate, which are synthesized only in plants.

2. (a) Because the α-oxidation of a fatty acid results in the removal of a single terminal carbon atom during each round of oxidation, Knoop would have found that the degradation of both phenylbutyrate and phenylpropionate yields benzoate.
(b) The net yield from the β-oxidation of palmitate is 129 molecules of ATP, as discussed on page 477 of Stryer's text. If 1 molecule of NADH is generated for 15 of the 16 carbons of palmitate, then the yield of ATP is 3×15, or 45. For α-oxidation, activation of the acetate molecule to acetyl CoA requires 2 ATP. Subsequent oxidation of acetyl CoA generates 12 ATP molecules. Thus, 55 molecules of ATP are generated by the α-oxidation of a molecule of palmitate.
(c) In β-oxidation of odd-numbered fatty acids, the products include propionyl CoA, which can be converted to succinyl CoA, a glucogenic substrate. However, α-oxidation of an odd-numbered fatty acid would yield carbon dioxide as well as a single molecule of acetate or acetyl CoA, neither of which is glucogenic.

3. (a) As shown in Figure 20-1, the oxidation of a fatty acid with a methyl group at C-3 procedes to the formation of the L-hydroxymethylacyl CoA derivative. Subsequent oxidation of the β-carbon to the ketoacyl derivative is blocked by the methyl group. Compare this pathway with the one shown in Figure 20-5 (p. 475) of Stryer's text.
(b) The oxidation of the α-carbon in the palmitate derivative followed by decarboxylation of the molecule yields a fatty acid that has a methyl group at an even-numbered carbon. Activation of the fatty acid to form an acyl CoA derivative followed by oxidation at the β-carbon results in the generation of propionyl CoA and a shortened acyl derivative, lauroyl CoA (see Figure 20-2).

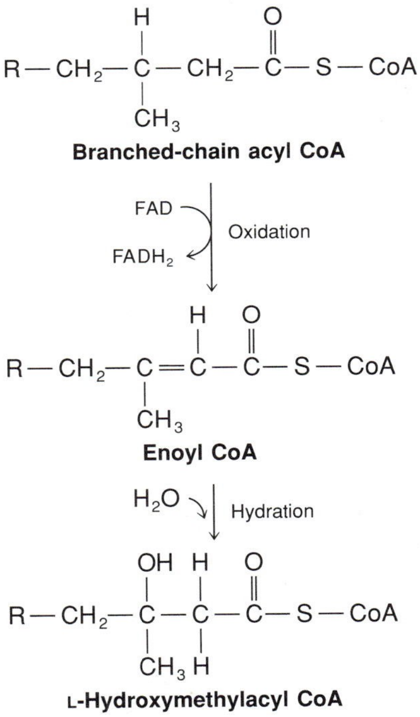

Figure 20-1
Formation of L-hydroxymethylacyl CoA through β-oxidation of branched-chain acyl CoA.

4. The normal route for β-oxidation in bacteria utilizes acyl CoA derivatives, which are formed from free fatty acids. To convert a primary alcohol to a free fatty acid, two oxidative steps are needed, each requiring an electron acceptor. In *Corynebacterium*, NAD^+-dependent dehydrogenases catalyze the sequential conversion of a primary alcohol to a fatty acid, with the corresponding aldehyde as an intermediate. Conversion of the free fatty acid to an acyl CoA derivative requires two equivalents of ATP (because ATP is converted to AMP and PP_i), as well as Coenzyme A. The reaction is catalyzed by acyl CoA synthase.

5. As shown in Figure 20-13 on page 484 of Stryer's text, acetyl-ACP and malonyl-ACP condense to form acetoacetyl-ACP. Carbons 4 and 3 of acetoacetyl-ACP are not labeled because they are derived from acetyl CoA. These two carbons will become carbons 15 and 16 of palmitate. Only C-2 of acetoacetyl-ACP will be labeled because it is derived from the methylene carbon of malonyl-ACP. When the second round of synthesis begins, butyryl ACP condenses with a second molecule of methylene-labeled malonyl-ACP, which contributes C-1 and C-2 of the newly formed six-carbon ACP derivative. In this compound, C-2 and C-4 will be labeled. Chain elongation continues until palmitoyl-ACP is formed.

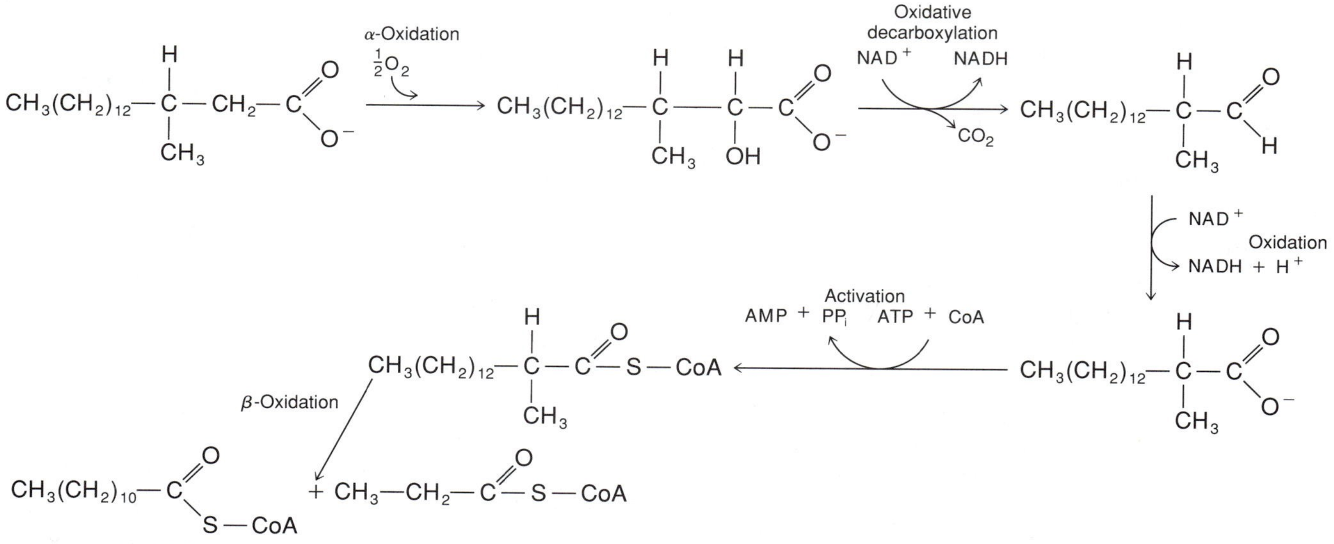

CH$_3$(CH$_2$)$_{12}$—C(H)(CH$_3$)—CH$_2$—C(=O)—O$^-$ →[α-Oxidation, $\frac{1}{2}$O$_2$] CH$_3$(CH$_2$)$_{12}$—C(H)(CH$_3$)—C(H)(OH)—C(=O)—O$^-$ →[Oxidative decarboxylation, NAD$^+$ → NADH, CO$_2$] CH$_3$(CH$_2$)$_{12}$—C(H)(CH$_3$)—C(=O)H

→[Oxidation, NAD$^+$ → NADH + H$^+$] CH$_3$(CH$_2$)$_{12}$—C(H)(CH$_3$)—C(=O)—O$^-$

→[Activation, AMP + PP$_i$ ← ATP + CoA] CH$_3$(CH$_2$)$_{12}$—C(H)(CH$_3$)—C(=O)—S—CoA

→[β-Oxidation]

CH$_3$(CH$_2$)$_{10}$—C(=O)—S—CoA   **Acyl CoA**

+ CH$_3$—CH$_2$—C(=O)—S—CoA   **Propionyl CoA**

Figure 20-2
The α-oxidation and oxidative decarboxylation of a branched-chain fatty acid allow generation of intermediates which can enter normal oxidative pathways.

Each even-numbered carbon atom, except for carbon 16 (at the ω-end), will be labeled.

6. Carnitine acyltransferase I facilitates the transfer of long-chain fatty acids into the mitochondrion by catalyzing the formation of fatty acyl carnitine molecules. The failure to form such molecules means that long-chain fatty acids are not available for cellular oxidation. In muscle, exercise or fasting increases dependence on fatty acids as a source of energy, so the inability to metabolize them interferes with cellular functions, causing cramps, weakness, and muscle damage. Liver cells also require formation of fatty acyl carnitine molecules to oxidize fats in mitochondria. If fatty acids cannot be utilized, they will remain in the cytosol, where high concentrations of them cause cell enlargement and interfere with other functions. Liver cells then must use glucose as a source of energy instead of exporting it to other cells. Because liver cells use acetyl CoA, which is derived primarily from fatty acid oxidation, as a precursor of ketone bodies, the failure to oxidize fatty acids will result in a reduction in the rate of ketone body synthesis. This in turn will exacerbate the symptoms of hypoglycemia because tissues, such as cardiac muscle, that normally use ketone bodies as a source of energy will rely more heavily on glucose as a source of energy.

7. Because malonyl CoA is a substrate for fatty acid synthase, competition from methylmalonyl CoA could cause a decrease in the rate of palmitoyl CoA synthesis in the cytosol, which could in turn lead to an increase in the amount of acetyl CoA carboxylase activity (recall that palmitoyl CoA inhibits acetyl CoA carboxylase). In addition, high levels of methylmalonyl CoA could interfere with transport of long-chain fatty acyl chains into mitochondria by inhibiting carnitine acyltransferase, as does malonyl CoA. Thus, both the synthesis and the oxidation of fatty acids could be inhibited by methylmalonyl CoA.

8. (a) The oxidation of an odd-numbered fatty acid yields a number of acetyl CoA molecules as well as one molecule of propionyl CoA, which can be converted to succinyl CoA, a component of

the citric acid cycle. Although two-carbon compounds like acetyl CoA cannot be used for the net synthesis of glucose, succinyl CoA can contribute net carbons to the citric acid cycle, enabling oxaloacetate, and ultimately glucose, to be formed through gluconeogenesis.

(b) Triacylglycerols are converted to glycerol and three free fatty acids through the action of lipases. Glycerol can be converted to glucose via dihydroxyacetone phosphate. Odd-numbered fatty acids found in triacylglycerols can also be used for net synthesis of glucose, whereas even-numbered fatty acids cannot.

(c) During fatty acid synthesis, most organisms use acetyl CoA as a source of the ω carbon and its adjacent carbon in the acyl chain. Two of the three carbons of malonyl CoA are incorporated during each cycle of acyl chain elongation. Thus the resulting fatty acid will contain an even number of carbon atoms.

(d) To produce an odd-numbered fatty acid, at least one odd-numbered CoA intermediate must be incorporated in its entirety during fatty acid synthesis. Propionyl CoA can be used by certain bacteria for the initial condensation step with malonyl CoA in fatty acid synthesis. The resulting five-carbon acyl intermediate is then extended in two-carbon units to yield an odd-numbered fatty acid.

9. (a) Malonyl CoA is a key substrate for the synthesis of fatty acids; when it is abundant, synthesis is stimulated. In addition, high levels of this intermediate inhibit carnitine acyltransferase I, thereby limiting the entry of fatty acyl chains into the mitochondrion, where they are oxidized. A decrease in the concentration of malonyl CoA leads to a decrease in the rate of fatty acid synthesis and an increase in the rate of fatty acid oxidation in the mitochondrion.

(b) The overall equation for the synthesis of palmitoyl CoA is

$$8 \text{ Acetyl CoA} + 7 \text{ ATP} + 14 \text{ NADPH} \longrightarrow$$
$$\text{palmitoyl CoA} + 14 \text{ NADP}^+ + 7 \text{ CoA} + 7 \text{ H}_2\text{O} + 7 \text{ ADP} + 7 \text{ P}_i$$

The overall equation for the oxidation of palmitoyl CoA is

$$\text{Palmitoyl CoA} + 7 \text{ FAD} + 7 \text{ NAD}^+ + 7 \text{ CoA} + 7 \text{ H}_2\text{O} \longrightarrow$$
$$8 \text{ acetyl CoA} + 7 \text{ FADH}_2 + 7 \text{ NADH} + 7 \text{ H}^+$$

Assuming that NADPH is equivalent in reducing power to NADH, that a molecule of FADH_2 yields 2 ATP during electron transport and oxidative phosphorylation, and that a molecule of NADH yields 3 ATP, then 49 ATP molecules are required to synthesize a molecule of palmitoyl CoA, whereas 35 ATP are generated by the conversion of palmitoyl CoA to 8 molecules of acetyl CoA. There is a net loss of 14 ATP molecules if the two processes occur simultaneously.

10. (a) Glycerol is converted to dihydroxyacetone phosphate, which in turn can serve as a source of glucose or can be converted to acetyl CoA.

(b) Fatty acids serve as a source of acetyl CoA, which is used in the glyoxylate cycle and gluconeogenesis.

(c) The primary function of glyoxysomes is to utilize fatty acids from triacylglycerols for the synthesis of glucose, which is used as a source of other molecules by the developing plant. Once leaf development enables the plant to generate glucose by photosynthesis, glyoxysomes are no longer needed.

(d) Carnitine functions in the transport of long-chain fatty acids from the cytosol to the interior of the mitochondrion. The observation suggests that although carnitine may be important in mitochondrial transport of fatty acyl chains, the compound is not involved in the movement of fatty acyl chains into the glyoxysome. It is also possible that glyoxysomes metabolize fatty acids with shorter acyl chains, for which transport facilitated by carnitine is not necessary.

(e) The fate of fatty acids is different in glyoxysomes and in mitochondria. Both organelles carry out β-oxidation of fatty acids to acetyl CoA; however, in glyoxysomes, acetyl CoA is a precursor of glucose, whereas mitochondria oxidize acetyl CoA to CO_2 and H_2O to generate ATP.

11. Carbohydrates consumed in caloric excess are converted to acetyl CoA, which in turn serves as a source of fatty acids. The concurrent synthesis of glycerol from carbohydrates such as glucose and fructose provides the second precursor needed for the synthesis of triacylglycerols, which are the primary storage form of energy in humans.

12. Bicarbonate is a source of carbon dioxide for the reaction catalyzed by acetyl CoA carboxylase, in which malonyl CoA is formed. Malonyl CoA is then used as a source of two carbon units for fatty acyl chain elongation, and the carbon atom derived originally from bicarbonate is released as CO_2. Carbon dioxide is then rapidly converted to bicarbonate, which is used again for the synthesis of another molecule of malonyl CoA. Thus, the carbon atom derived from bicarbonate can be used many times for the production of malonyl CoA, but it is never incorporated into the growing acyl chain, so it does not appear in palmitate.

13. If each enzyme could operate only on fatty acyl CoA derivatives of a particular chain length, then as many as eight sets of enzymes would be required to carry out the β-oxidation of palmitate. The fact that most enzymes of the β-oxidation pathway can use acyl CoA molecules of different chain lengths as substrates means that the cell needs to synthesize fewer types of proteins in order to carry out fatty acid oxidation.

14. In order to synthesize acetoacetate from palmitate, liver cells must carry out β-oxidation of the 16-carbon acyl chain, generating 8 molecules of acetyl CoA, which will in turn generate 4 molecules of acetoacetate. A total of 7 NADH and 7 $FADH_2$ molecules are generated per molecule of palmitate converted to acetyl CoA. The 14 reduced cofactors are equivalent to 35 ATP molecules. Because 2 molecules of ATP are needed to activate palmitate, the net yield of ATP per palmitate is 33.

15. Both malate enzyme and citrate lyase are part of the shuttle system that transports two-carbon units from the mitochondrion to the cytosol. Malate enzyme also generates reducing power in the form of NADPH, which is used for fatty acid synthesis; however, the phosphogluconate pathway also serves as a source of NADPH so that fatty acid synthesis can continue even if malate enzyme is deficient. Recall from page 441 of Stryer's text that malate can cross the mitochondrial membrane. Citrate lyase is more critical to fatty acid synthesis because it is required to

generate acetyl CoA from citrate in the cytosol. Without cytosolic acetyl CoA, fatty acid synthesis cannot take place, and the cells cannot grow and divide.

16. (a) To synthesize clupanodonate from linolenate, the acyl chain must be elongated from 18 to 22 carbons, and two new double bonds must be introduced into the chain. Although the details of the various mammalian desaturation systems are not completely understood, it appears that a double bond at C-6 can be introduced when a double bond at C-9 is available and a double bond at C-5 can be introduced when one at C-8 is available. Thus, the probable sequence of reactions includes the introduction of a double bond at C-6 of linolenate (yielding a 18:4 cis-Δ^6, Δ^9, Δ^{12}, Δ^{15} acyl chain) followed by chain elongation to a 20-carbon derivative. The introduction of a double bond at C-5 then gives an acyl chain denoted as 20:5 cis-Δ^5, Δ^8, Δ^{11}, Δ^{14}, Δ^{17}. The final reaction required to yield clupanodonate is chain elongation to the 22-carbon fatty acyl chain.

(b) Linoleate has cis double bonds at C-9 and at C-12. Elongation to a 22-carbon chain would yield an acyl chain with double bonds at C-13 and C-16. To form clupanodonate, a double bond at C-19 is needed, but mammals lack the enzymes required to introduce double bonds beyond C-9. Thus, linoleate cannot be used for the synthesis of clupanodonate in mammals.

Amino Acid Degradation and
the Urea Cycle

The organism derives useful energy from both stored and exogenous fuels. Stryer has described how carbohydrates (Chapter 15) and fats (Chapter 20) may be catabolized to form NADH and $FADH_2$ (Chapter 16). Electron transport and oxidative phosphorylation harvest some of the free energy of these compounds to form ATP (Chapter 17). In Chapter 21, Stryer explains how proteins fit into this scheme. Although proteins are not stored as fuels as carbohydrates and fats are, supplies of proteins in excess of those needed for biosynthesis are degraded for energy or are converted to fats or carbohydrates. Each amino acid contains at least one nitrogen atom. Stryer describes how the α-amino groups of most amino acids are transferred to α-ketoglutarate to form glutamate by transamination and how the α-amino group of glutamate is converted to ammonia by an oxidative deamination. In addition, he describes the formation of aspartate by transamination between oxaloacetate and glutamate. He then describes the urea cycle, in which ammonia, the α-amino group of aspartate, and CO_2 are condensed to form urea—a nontoxic excretory product in higher animals. He also explains that some other organisms excrete nitrogen as uric acid or ammonia.

Because there are 20 amino acids, the catabolic pathways of their carbon skeletons are varied. Stryer describes how each amino acid gives rise to one or more of seven primary products. Two of these, acetyl CoA and acetoacetyl CoA, can be converted to ketone bodies (Chapter 20), and the remaining five can be converted into glucose (Chapter 18). The two groups of products lead to the glycogenic-ketogenic classification of the amino acids. Throughout the chapter, Stryer explains the roles of several cofactors, including pyridoxal phosphate, tetrahydrobiopterin, and vitamin B_{12}, in the mechanisms of particular enzymes involved in amino-acid metabolism. The pathological consequences of defects or deficiencies in some of the enzymes involved in catabolism of amino acids and the synthesis of urea are also described.

LEARNING OBJECTIVES

After mastering this chapter, you should be able to complete the following objectives.

Introduction (Stryer page 495)

1. Describe the fates of exogenously supplied amino acids that are not used for protein synthesis.

Amino-Group Metabolism (Stryer pages 495–499)

2. Name the major organ of amino acid degradation in mammals.

3. Describe the reactions catalyzed by the *aminotransferases (transaminases)* and state the major function of these reactions. Explain how the α-amino groups are removed from serine and threonine.

4. Describe the reaction catalyzed by *glutamate dehydrogenase*, outline its regulation, and state its major function. Note that the participation of NAD^+ in the reaction links nitrogen metabolism and energy generation.

5. Recognize the structures of *pyridoxal phosphate (PLP)* and *pyridoxamine phosphate (PMP)*, indicate the reactive functional groups on each, and name the dietary precursor of PLP.

6. Describe the aminotransferase reaction mechanism and explain the involvement of a *Schiff base* and PLP. List the other kinds of reactions catalyzed by PLP and describe the common features of PLP catalysis.

Urea Synthesis (Stryer pages 500–502)

7. Define the terms *ureotelic, uricotelic,* and *ammonotelic.*

8. List the immediate precursors of the nitrogen and carbon atoms of *urea* and name the carrier molecule on which urea is assembled during the *urea cycle.*

9. Name the enzymes and the other components of the urea cycle, note their intracellular locations, and indicate the molecular connection between this cycle and the citric acid cycle. Account for the ATP requirement of the cycle.

Carbon-Atom Metabolism (Stryer pages 503–512)

10. Name the seven major metabolic products of the catabolism of the carbon skeletons of the amino acids.

11. Describe the basis for the *glycogenic-ketogenic* designation of the amino acids and classify each amino acid accordingly.

12. List the amino acids that give rise to pyruvate. Consider the role of NAD^+ in these conversions.

13. List the amino acids that give rise to *oxaloacetate.*

14. List the amino acids that give rise to *fumarate.*

15. List the amino acids that give rise to α-*ketoglutarate.* Describe the role of *tetrahydrofolate* in one of the conversions.

16. List the amino acids that give rise to *succinyl CoA*. Describe the roles of CoA, biotin, and *coenzyme B₁₂* in these conversions.

17. Describe the structure of coenzyme B₁₂ and outline its formation from B₁₂ᵣ. Indicate its general function in the three types of reactions it catalyzes.

18. Explain the cause of *pernicious anemia,* and outline the normal sequence of events in the absorption of *cobalamin* and its conversion to coenzyme B₁₂.

19. List the amino acids that give rise to *acetyl CoA* and to *acetoacetyl CoA.* Describe the roles of biotin, CoA, NAD^+, FAD, thiamine pyrophosphate, lipoate, and *tetrahydrobiopterin* in these conversions.

20. Distinguish between *monoxygenases* and *dioxygenases.* Explain the roles of O_2 and *glutathione* in the catabolism of aromatic amino acids.

Inborn Errors of Metabolism (Stryer pages 513–514)

21. Describe *alcaptonuria* and relate Garrod's hypothesis concerning this disorder.

22. Explain the biochemical bases of *phenylketonuria, methylmalonic aciduria,* and *maple syrup urine disease.* Describe some of the consequences of these diseases.

SELF-TEST

Introduction

1. Surplus dietary amino acids may be converted into

   (a) proteins.
   (b) fats.
   (c) ketone bodies.
   (d) glucose.
   (e) a variety of biomolecules for which they are precursors.

Amino-Group Metabolism

2. Which of the following compounds serves as an acceptor for the amino groups of many amino acids during catabolism?

   (a) Glutamine
   (b) Asparagine
   (c) α-Ketoglutarate
   (d) Oxalate

3. The removal of α-amino groups from amino acids for conversion to urea in animals may occur by

   (a) transamination.
   (b) reductive deamination.
   (c) oxidative deamination.
   (d) transamidation.

4. Which of the following amino acids have their α-amino groups removed by dehydratases?

   (a) Histidine
   (b) Tryptophan
   (c) Serine
   (d) Glutamine
   (e) Threonine

5. The products of an aminotransferase-catalyzed reaction between pyruvate and glutamate would be

   (a) aspartate and oxaloacetate.
   (b) aspartate and α-ketoglutarate.
   (c) alanine and oxaloacetate.
   (d) alanine and α-ketoglutarate.

6. Explain the role of pyridoxal phosphate in aminotransferase reactions. Be sure to describe the Schiff base and the ketimine that are involved in the mechanism.

_____

_____

_____

_____

_____

Urea Synthesis

7. Considering all forms of life, which of the following are major excretory forms of the α-amino groups of amino acids?

   (a) Urea
   (b) Uracil
   (c) Ammonia
   (d) Uric acid

8. How many moles of ATP are required to condense 2 moles of nitrogen and 1 mole of CO_2 into 1 mole of urea *via* the urea cycle? How many high-energy bonds are used in this process? Do both atoms of nitrogen enter the cycle as NH_4^+?

_____

_____

_____

9. Describe the role of ornithine in the urea cycle.

_____

_____

Carbon-Atom Metabolism

10. Match the catabolic products in the right column with the amino acids in the left column from which they can be derived.

(a) Alanine _____

(b) Aspartate _____

(c) Glutamine _____

(d) Phenylalanine _____

(e) Leucine _____

(f) Valine _____

(1) Succinyl CoA
(2) Acetoacetyl CoA
(3) α-Ketoglutarate
(4) Oxaloacetate
(5) Pyruvate
(6) Acetyl CoA
(7) Fumarate

11. Classify the following amino acids as glycogenic (G), ketogenic (K), or both (GK).

(a) Leucine _____   (e) Histidine _____

(b) Alanine _____   (f) Isoleucine _____

(c) Tyrosine _____   (g) Aspartate _____

(d) Serine _____   (h) Phenylalanine _____

12. What common feature is shared by the catabolism of fatty acids having an odd number of carbon atoms and the catabolism of the amino acids isoleucine, methionine, and valine?

_____

_____

13. Cobalt is a component of which of the following coenzymes?

(a) FAD
(b) NAD^+
(c) Dihydrobiopterin
(d) CoA
(e) Coenzyme B_{12}
(f) Pyridoxal phosphate

14. Catabolism of which of the following amino acids requires the direct involvement of O_2?

(a) Histidine
(b) Phenylalanine
(c) Tyrosine
(d) Isoleucine
(e) Glutamine

15. Describe the reactions of the oxygenases and distinguish between monoxygenases and dioxygenases.

_____

_____

_____

16. Match the functions from the right column with the appropriate coenzymes, listed in the left column, that are involved in amino-acid catabolism.

(a) Pyridoxal phosphate _____

(b) Coenzyme B_{12} _____

(c) Tetrahydrobiopterin _____

(d) NAD^+ _____

(e) Biotin _____

(1) Accepts electrons
(2) Provides free radicals
(3) Carries amino groups
(4) Carries CO_2

Inborn Errors of Metabolism

17. Match the enzyme or protein in the right column with the disorder in the left column that ordinarily produces a deficiency of the enzyme or protein.

(a) Methylmalonic
    aciduria _____

(b) Pernicious
    anemia _____

(c) Hyperammonemia _____

(d) Phenylketonuria _____

(e) Alcaptonuria _____

(f) Maple syrup
    urine disease _____

(1) Carbamoyl phosphate synthetase
(2) Homogentisate oxidase
(3) Intrinsic factor
(4) Phenylalanine hydroxylase
(5) Branched-chain α-keto acid dehydrogenase
(6) Transferase that converts B_{12s} into deoxyadenosyl-cobalamin

---

PROBLEMS

1. Birds require arginine in their diet. Would you expect to find the production of urea in these animals? Why?

2. Would you expect a defect in fumarase activity or a defect in alanine aminotransferase to have a greater effect on the rate of urea biosynthesis?

3. Pyridoxal phosphate or related metabolites are required growth factors for *Lactobacillus* species. For example, when amino acids such as alanine or glutamate are used as the sole source of nutrition, these bacilli do not grow nor do they generate metabolic energy unless pyridoxal phosphate or its metabolites are supplied. Explain these observations.

4. A male infant six weeks of age exhibited symptoms of pronounced hyperammonemia, which included vomiting, fever, irritability, and screaming episodes interspersed with periods of lethargy. The infant had low levels of blood urea, elevated serum transaminase, and generalized hyperaminoacidemia and aminoaciduria. The levels of citrulline, argininosuccinic acid, and arginine were relatively low in both blood and urine. Enzymatic assays of liver tissue established that the level of mitochondrial carbamoyl phosphate synthetase (CPS) was approximately 20% of normal; the enzyme was active only in the presence of relatively high concentrations of N-acetyl glutamate. All other urea-

cycle enzyme levels were relatively normal. The infant was treated with a supplement containing arginine, pyridoxine, and α-keto analogs of essential amino acids (those that cannot be synthesized in humans). As part of the therapy, dietary protein was also restricted.

(a) Why were blood levels of ammonia high in this infant?

(b) Explain why infants with CPS deficiency normally exhibit relatively low blood levels of citrulline, argininosuccinic acid, arginine, and urea.

(c) Why is the administration of supplemental arginine recommended for this infant?

(d) Why were α-keto analogs of essential amino acids and pyridoxine administered?

(e) Why was dietary protein restricted?

(f) Why would you expect to see elevated concentrations of glutamate, glutamine, and alanine in the blood and urine of this infant?

5. Why is glutamate dehydrogenase a logical point for the control of ammonia production in cells?

6. During the process of glomerular filtration in the kidney, amino acids, as well as other metabolites, enter the lumen of the kidney tubule. Normally, a large proportion of these amino acids are reabsorbed into the blood through the action of membrane-bound carrier systems that are specific for different classes of amino acids. Cystinuria is a disorder whose symptoms include urinary excretion with unusually high concentrations of cystine as well as excess amounts of ornithine, lysine, and arginine. Patients with this disorder often have urinary tract stones, which are caused by the limited solubility of cystine. A related disorder found in other individuals is characterized by the appearance of ornithine, lysine, and arginine in the urine although the levels of urinary cystine are normal.

(a) What is the most likely source of cystine in cells?

(b) What common structural feature of the four amino acids—cystine, ornithine, lysine, and arginine—is recognized by the carrier in the kidney tubule membrane?

(c) How many carrier systems may exist for these molecules?

(d) Other aminoacidurias are due to a deficiency in one or more of the enzymes in the catabolic pathway for an amino acid. This leads to higher concentrations of the amino acid in the blood and a corresponding increase in the concentrations in the glomerular filtrate. In this case, the capacity of the reabsorption system is surpassed, causing some amino acid to be lost in the urine. How could you distinguish between a defect in amino-acid metabolism and a defect in a renal transport system?

7. Why is the catabolism of isoleucine said to be both glucogenic and ketogenic?

8. Brain cells take up tryptophan, which is then converted to 5-hydroxytryptophan by tryptophan hydroxylase, an enzyme whose activity is similar to that of phenylalanine hydroxylase. Aromatic amino acid decarboxylase then catalyzes the formation of the potent neurotransmitter 5-hydroxytryptamine, also called serotonin. In the blood, tryptophan is bound to serum albumin, with an affinity such that about 10% of the tryptophan is freely diffusable. The rate of tryptophan uptake by brain cells depends upon the concentration of free tryptophan.

In these cells, tryptophan concentration is normally well below that of the K_M for tryptophan hydroxylase. Aspirin and other drugs displace tryptophan from albumin, thereby increasing the concentration of free tryptophan.

    (a) What cofactor is required for the activity of tryptophan hydroxylase?

    (b) Dietary deficiencies in pyridoxine and related metabolites can induce a number of symptoms, including those that appear to be related to derangements in serotonin metabolism. What enzyme could be affected by a deficiency of vitamin B_6?

    (c) What effect does aspirin have on tryptophan metabolism in brain cells?

9. Sodium borohydride reduces imines to stable secondary amines, as shown in the following general reaction:

$$\begin{matrix} R_1 \\ \diagdown \\ C=N-R_3 \\ \diagup \\ R_2 \end{matrix} \xrightarrow{\text{NaBH}_4} \begin{matrix} R_1 \\ \diagdown \\ CH-NH-R_3 \\ \diagup \\ R_2 \end{matrix}$$

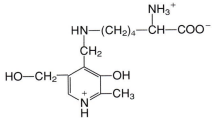

A purified sample of alanine aminotransferase is treated with sodium borohydride and is then subjected to partial acid hydrolysis. Among the amino acids isolated is the fluorescent derivative shown in the margin. Identify the derivative, and describe the role of its parent compound in the catalytic action of the aminotransferase.

10. In many microorganisms, glutamate dehydrogenase (GDH) participates in the catabolism of glutamate by generating ammonia and α-ketoglutarate, which undergoes oxidation in the citric acid cycle. However, when *E. coli* is grown with glutamate as the sole source of carbon, the synthesis of GDH protein is strongly repressed. Under these conditions, aspartase, an enzyme that catalyzes the removal of ammonia from aspartate to form fumarate, is required in order for the cell to grow in glutamate. Propose a cyclic pathway for the catabolism of glutamate that includes aspartate.

    When *E. coli* is grown in glucose and ammonia, GDH synthesis is accelerated and the enzyme is active. Under these conditions, what role does GDH play in bacterial metabolism?

11. The formation of the aldimine linkage between an amino acid substrate and the pyridoxal phosphate in alanine aminotransferase can be portrayed as a condensation. However, when the enzyme reacts with its substrate in ^3H_2O, less than 1% of tritium is incorporated into the product. What does this observation tell you about the substrate-enzyme interaction?

12. Certain forms of maple syrup urine disease respond to the administration of thiamine at 50 to 100 times the normal recommended level. Why?

13. After an overnight fast, muscle tissue proteolysis generates free amino acids, many of which pass into the blood. Among the amino acids that are found in the blood are alanine, glutamate, and glutamine, all of which are rapidly taken up by the liver. What happens to these amino acids when they enter hepatic cells?

14. Propionyl CoA and methylmalonyl CoA both inhibit *N*-acetyl glutamate synthase activity in slices of liver tissue. What clinical symptom

would you expect to see in patients suffering from methylmalonic aciduria as a result of this inhibition?

15. Dialysis of purified glutamate aminotransferase can be used to remove pyridoxal phosphate from the enzyme, but the dissociation of the cofactor from the enzyme is very slow. Why would the addition of glutamate to the enzyme solution increase the rate of dissociation of the cofactor from the enzyme?

ANSWERS TO SELF-TEST

Introduction

1. All are correct. The amino acids that are needed for protein synthesis and as precursors for other biomolecules are used directly for these purposes. The carbon skeletons of any excess can be converted into acetyl CoA or glucose, depending upon the particular amino acid, and thus into the products that are derivable from these two basic molecules.

Amino-Group Metabolism

2. c. The transamination of several different amino acids with α-ketoglutarate forms glutamate and α-keto acids that can subsequently be catabolized.

3. a, c

4. c and e. Serine and threonine are deaminated by dehydratases that take advantage of the β-hydroxyl of these amino acids to carry out a dehydration followed by a rehydration to release NH_4^+.

5. d. A five-carbon amino acid will yield a five-carbon α-keto acid as a result of a transamination; in this case, glutamate yields α-ketoglutarate. The other partner in the reaction, pyruvate, will yield alanine as a product.

6. Pyridoxal phosphate (PLP) acts as an amino carrier in transamination reactions. It is covalently bound to the ϵ-amino group of a lysine residue in the enzyme by a Schiff-base or aldimine bond, that is, by a carbon–nitrogen double bond between the ϵ-amino group of the lysine and a carbon of PLP (see p. 497 of Stryer). The enzyme catalyzes a displacement of the ϵ-amino group of the lysine of the enzyme and forms an analogous aldimine bond with the α-amino group of an amino acid. Through a deprotonation and a reprotonation, the aldimine is isomerized to a ketimine in which the carbon–nitrogen bond is between the α-amino nitrogen and the α-carbon of the amino acid (see Figure 21-2 in Stryer). The addition of H_2O to the ketimine releases the carbon skeleton of the amino acid as an α-keto acid and leaves the coenzyme in the enzyme-bound pyridoxamine form, which now contains the α-amino group of the amino acid. After dissociation of the α-keto acid, a different α-keto acid binds to the enzyme to form a new ketimine and the overall process reverses, resulting in the transfer of the enzyme-bound amino group to the keto acid to form a new amino acid and regenerate the pyridoxal phosphate.

Urea Synthesis

7. a, c, and d. Uracil is a pyrimidine component of RNA and is not a nitrogen excretory product. Uric acid is a purine derivative excreted by uricotelic organisms.

8. Three moles of ATP are directly involved in the synthesis of urea. Two are converted to ADP and P_i by carbamoyl phosphate synthetase and one is converted to AMP and PP_i by argininosuccinate synthetase. Two ATP would be required to convert AMP back into ATP, so a total of four high-energy bonds are used. Only one molecule of nitrogen enters as NH_4^+; the other enters in the α-amino group of aspartate, which can be formed by a transamination between oxaloacetate and glutamate.

9. Ornithine serves as the carrier on which the urea molecule is constructed. The δ-amino group of ornithine has a carbamoyl group added to it by ornithine transcarbamoylase to form citrulline. Subsequent steps of the cycle add aspartate to bring in the second nitrogen atom and also regenerate ornithine when arginase cleaves arginine to form urea.

Carbon-Atom Metabolism

10. (a) 5 (b) 4, 7; aspartate can be converted to fumarate *via* the urea cycle. (c) 3 (d) 2, 7 (e) 2, 6 (f) 1

11. (a) K (b) G (c) GK (d) G (e) G (f) GK (g) G (h) GK

12. Both give rise to methylmalonyl CoA, which can, in turn, be converted into succinyl CoA and ultimately into glucose. (See Stryer, p. 478 and p. 506.)

13. e

14. b and c. Monoxygenases and dioxygenases are involved in the conversion of phenylalanine to tyrosine and the subsequent opening of the aromatic ring during catabolism.

15. Oxygenases use O_2 as a substrate and incorporate one or both of the atoms into the oxidized product. Monoxygenases, also called mixed-function oxygenases, incorporate one atom of the oxygen into the product; the other appears in water. Dioxygenases incorporate both atoms of O_2 into the product. Monoxygenases are often involved in hydroxylation, and dioxygenases are often involved in the opening of aromatic rings.

16. (a) 3 (b) 2 (c) 1 (d) 1 (e) 4

Inborn Errors of Metabolism

17. (a) 6 (b) 3 (c) 1 (d) 4 (e) 2 (f) 5

ANSWERS TO PROBLEMS

1. In the urea cycle, arginine is cleaved to yield urea and ornithine. The fact that birds require arginine in their diet indicates that they are unable to synthesize it for utilization in protein synthesis. As a result,

they are also unable to synthesize urea to dispose of ammonia; instead, they synthesize uric acid. Birds do have carbamoyl phosphate synthetase activity; however, it is located in the cytosol and it catalyzes the formation of carbamoyl phosphate, which is then utilized for pyrimidine synthesis.

2. Fumarase activity has an effect on the urea cycle because it is needed, along with malate dehydrogenase, for the regeneration of oxaloacetate, which in turn undergoes transamination to form aspartate. The amino group of aspartate contains one of the two nitrogen atoms that are used to synthesize urea. Alanine aminotransferase is one of a number of aminotransferases that can transfer amino groups from amino acids to α-ketoglutarate to generate glutamate. Subsequent deamination of glutamate provides ammonia for the urea cycle. If all the other aminotransferases in the cell are active, then alanine aminotransferase would not be particularly essential. Thus, a defect in fumarase activity would have the greater effect on the rate of urea biosynthesis.

3. In order to utilize amino acids as sources of oxidative energy or to generate glucose through gluconeogenesis, the bacilli must carry out transamination reactions to dispose of ammonia (or to use it for the biosynthesis of other nitrogen-containing compounds) as well as to generate α-keto acids that can be used in the citric acid cycle or other pathways. Pyridoxal phosphate is a required cofactor for the aminotransferase enzymes, and in bacteria in which it cannot be synthesized, it must be derived from the growth medium. Otherwise, amino acids cannot be metabolized. In this case, where an amino acid is the only source of carbon and of nitrogen, the bacilli will not be able to introduce the amino acid into any catabolic pathway. Pyridoxal phosphate also functions as a cofactor for a large number of other enzymes, including decarboxylases, racemases, aldolases, and deaminases. Therefore, deficiencies in pyridoxal phosphate would also adversely affect a large number of other pathways.

4. (a) Carbamoyl phosphate synthetase utilizes ammonia and bicarbonate to synthesize carbamoyl phosphate, which enters the urea cycle. A deficiency in the activity of this enzyme leads to an accumulation of ammonia in blood and in urine.

(b) Because carbamoyl phosphate condenses with ornithine to form citrulline, it is in effect a precursor of all the components of the urea cycle. When its synthesis is depressed, the rate of synthesis of other urea-cycle components will be decreased.

(c) Arginine is needed as a component of most proteins; it must therefore be supplied to avoid arginine deficiency, which would result in a decrease in the rate of protein synthesis. In the urea cycle, arginine serves as a precursor of ornithine, which in turn serves as an acceptor of the relatively small number of carbamoyl groups that enter the urea cycle. Thus, supplemental arginine could accelerate the rate of urea synthesis in an attempt to drive the reaction catalyzed by the partially active CPS enzyme toward the net formation of carbamoyl phosphate. Finally, recent findings suggest that arginine is a feed-forward activator for the synthesis of *N*-acetyl glutamate, which activates CPS. Recall that the deficient enzyme appeared to be heavily dependent on levels of *N*-acetyl glutamate. Supplying arginine indirectly activates CPS, thereby increasing the rate of ammonia utilization.

(d) The administration of α-keto analogs of essential amino acids serves two functions. First, these substrates undergo transamination with glutamate as the amino donor. The corresponding increase in the synthesis of glutamate, through the action of glutamate dehydrogenase, utilizes more ammonia, removing at least some of it from the blood. Second, the transamination of the α-keto acids generates essential amino acids, which are used for protein synthesis. This is especially important in a situation in which dietary protein intake is restricted (see the following).

(e) When dietary protein is hydrolyzed to its component amino acids, subsequent catabolic steps include transamination and deamination, which produce free ammonia. The less protein consumed, the lower the production of ammonia from the breakdown of amino acids. A reduction in the blood ammonia level is the primary goal of the therapy for this infant.

(f) Increased levels of ammonia in the cells will drive the reaction catalyzed by glutamate dehydrogenase toward the net formation of glutamate and will also stimulate the synthesis of glutamine, which is in effect a carrier of two molecules of ammonia. With the high concentrations of glutamate, the equilibria for most transamination reactions would be shifted toward the net formation of other amino acids, such as alanine, a transaminated form of pyruvate. Concentrations of these and other amino acids would therefore be increased in both the blood and the urine.

5. The deamination of amino acids occurs through the action of transaminases as well as through the action of glutamate dehydrogenase (GDH). GDH is the only enzyme that catalyzes the oxidative deamination of an L-amino acid. It deaminates glutamate, whose precursor, α-ketoglutarate, is the ultimate acceptor of amino groups from almost all of the amino acids. In addition, while the reactions catalyzed by the transaminases are freely reversible, the GDH reaction is far from equilibrium; it is therefore a logical activity to control because large changes in its velocity can be achieved with small changes in the concentrations of allosteric effectors like ATP or NADH. Thus, GDH is the enzyme of choice for the control of ammonia synthesis.

6. (a) Cystine is a dibasic amino acid, composed of two cysteine molecules joined by a disulfide linkage. Disulfide linkages exist in many proteins, and when they are hydrolyzed by proteases to yield free amino acids, cystine is often one of the products.

(b) Like cystine, the other three amino acids—arginine, ornithine, and lysine—have two basic groups. The carrier systems probably recognize and bind these groups for transport.

(c) From the disorders described, it is likely that there are two transport systems. One carries all four dibasic acids; when it is defective, reabsorption of all four species fails, allowing all four to spill over into the urine. Another system transports ornithine, arginine, and lysine but not cystine; its failure to function accounts for the appearance of these three amino acids but not cystine in the urine.

(d) A defect in the catabolic pathway for a particular amino acid causes elevation in the concentration of that single amino acid in blood and in urine as well, unless other amino acids carried by the same kidney transport system are lost to the urine. A defect in renal reabsorption means that all those amino acids that share the affected carrier system will be lost to the urine. Their concentration in blood will be lower than normal, while their concentration in urine will be higher.

7. The catabolic pathway for isoleucine leads to the formation of acetyl CoA and propionyl CoA. Acetyl CoA can be utilized for the net synthesis of fatty acids or ketone bodies, but it cannot be used for the net synthesis of glucose; thus, it is said to be ketogenic. In contrast to acetyl CoA, propionyl CoA is converted to succinyl CoA, which can be utilized through part of the citric acid cycle and the gluconeogenic pathway to give the net formation of glucose. The distinction between the two types of substrates is somewhat arbitrary, however. For example, succinyl CoA can also be converted via pyruvate and acetyl CoA to citrate, which, when transported to the cytosol, serves as the source of carbons for the synthesis of fatty acids. Thus, glucogenic substrates can, under certain conditions, be ketogenic; however, ketogenic substrates can not be glucogenic, unless a cell has a functional glyoxylate pathway.

8. (a) Tetrahydrobiopterin is utilized as a reductant by many hydroxylase enzymes, including phenylalanine hydroxylase and tryptophan hydroxylase.

(b) Pyridoxal phosphate, which is derived from dietary pyridoxine, is a cofactor for a number of enzymatic reactions that occur at the α-carbon of an amino acid. In this case, the cofactor participates in the decarboxylation of 5-hydroxytryptophan. A deficiency of vitamin B_6 would lead to a reduction in the rate of synthesis of serotonin.

(c) The higher the concentration of free tryptophan, the greater the rate of uptake of the amino acid by the brain cells. Because the normal concentration of tryptophan in these cells is below that of the K_M for tryptophan hydroxylase, an influx of more tryptophan into the cells provides more substrate for the enzyme. You would therefore expect an increase in the level of 5-hydroxytryptophan production.

9. The fluorescent derivative is ϵ-pyridoxyllysine, in which lysine is covalently linked to a reduced pyridoxyl group. At the active site of the native enzyme, the parent compound is pyridoxyl phosphate in Schiff-base linkage with the ϵ-amino group of a specific lysine residue. When an amino acid enters the active site, the ϵ-amino group of lysine is displaced and a new Schiff base is formed between PLP and the α-amino group of the amino acid. The formation of the new Schiff base initiates the transamination reaction.

10. A possible catabolic pathway for glutamate that includes aspartate is as follows:

Glutamate undergoes transamination with oxaloacetate to generate α-ketoglutarate and aspartate. Oxidation of α-ketoglutarate is carried out in the citric acid cycle, whereas aspartate is cleaved to yield ammonia and fumarate. Fumarate is converted to malate, which is then oxidized to oxaloacetate in the citric acid cycle so that α-ketoglutarate can be regenerated to serve as an acceptor of the amino group from glutamate. It should be noted that glutamate will also be used as a source of amino groups by the aminotransferases and other enzymes involved in biosynthesis.

In cells grown in glucose and ammonia, GDH catalyzes the assimilation of ammonia by incorporating it into oxaloacetate to yield glutamate, which serves as a source of amino groups for other biosynthetic reactions.

11. The fact that very little hydrogen exchange occurs when alanine binds to the enzyme suggests that a direct transaldiminization occurs. Pyridoxal phosphate is linked by a Schiff base to a specific lysine residue at the active site. Alanine displaces the cofactor, with a direct exchange of the Schiff-base linkage to the α-amino group of alanine. It appears that there is no intermediate formation of free aldehyde during the binding of the substrate. The small amount of tritium that is incorporated into the product may be due to the ionization of lysine.

12. Maple syrup urine disease is characterized by the appearance of large quantities of leucine, valine, and isoleucine and their corresponding α-keto acids in the blood and urine. This is due to a deficiency of branched-chain α-keto acid dehydrogenase, the enzyme that catalyzes the oxidative decarboxylation of the three aliphatic amino acids. The reaction is analogous to those catalyzed by pyruvate dehydrogenase and α-ketoglutarate dehydrogenase. These enzymes utilize thiamine pyrophosphate as a cofactor. A response to the administration of thiamine indicates that the deficient enzyme does not bind thiamine as efficiently as a normal enzyme does. Higher concentrations of the cofactor are therefore required for optimal enzyme activity.

13. In the liver, alanine, glutamate, and glutamine are utilized as sources of carbon for gluconeogenesis. Glutamine is deaminated to yield glutamate and ammonia. Glutamate undergoes oxidative deamination to form ammonia and α-ketoglutarate, a substrate for gluconeogenesis. Pyruvate is generated from the transamination of alanine, and carboxylation of the α-keto acid yields oxaloacetate, another source of carbon for gluconeogenesis. The ammonia generated by the conversion of the amino acids to their corresponding α-keto acids is used for the synthesis of urea. Glucose synthesized by liver can be returned through the blood to the muscle, where it serves as a source of energy. Thus, muscle uses the amino acids as a means of contributing to the generation of glucose in liver as well as a means of transporting ammonia to the liver, where the synthesis of urea can be carried out.

14. The activity of mitochondrial carbamoyl phosphate synthetase (CPS) depends upon the availability of N-acetyl glutamate, which is generated from glutamate and acetyl CoA. A reduction in the availability of the activating molecule will lead to a decrease in the activity of CPS, which utilizes ammonia and bicarbonate for the synthesis of carbamoyl phosphate. This in turn leads to an increase in the level of ammonia in blood and urine. Over two-thirds of patients with methylmalonic aciduria are hyperammonemic.

15. In the native enzyme, the dissociation of the cofactor from the enzyme during dialysis is very slow because PLP is covalently bound to a lysine residue. The addition of glutamate, a substrate for the transami-

nation reaction, leads to the formation of a Schiff base between gluta-mate and PLP, which means that the cofactor is no longer covalently attached to the enzyme molecule. Although PLP is bound to the en-zyme by noncovalent forces, such as those shown in Figure 21-3 of the text, these forces are not as strong as a covalent bond, so during dialysis, the rate of dissociation of the cofactor from the enzyme is increased.

Photosynthesis

Stryer ends Part III of *Biochemistry,* "Generation and Storage of Metabolic Energy," with a description of photosynthesis. To this point, he has dealt with the mechanisms by which organisms obtain energy from their environment by oxidizing fuels to generate ATP and reducing power. In this chapter, he describes how light energy is transduced into the same forms of chemical energy and how CO_2 is converted into carbohydrate by photosynthetic organisms.

The chloroplasts of green plants are specialized organelles that absorb light and, by a complicated series of reactions, form hexoses from H_2O and CO_2 with the release of O_2. Light energy is used to form both reducing potential as NADPH and a transmembrane proton gradient that synthesizes ATP. The ATP and NADPH are used, in part, to synthesize the carbohydrate. Stryer starts with a description of the chloroplast and the historical observations that established the basic characteristics of photosynthesis. He then describes the structures and functions of the chlorophylls, the photoreceptors. The roles of two macromolecular complexes, photosystems I and II, in generating light-induced separations of opposite electrical charges whose potential energy can be used to produce NADPH, oxidize H_2O to O_2, and generate a proton gradient to drive ATP synthesis are presented. Stryer explains the structural and functional interrelationships of the components and the regulation of the photosynthetic apparatus. He also describes the photoreceptors in cyanobacteria and red algae, which absorb light at wavelengths different from those absorbed by green plants and thus allow these organisms to occupy distinct ecological niches.

The final part of the chapter details the Calvin cycle, which fixes CO_2 and leads to the formation of hexoses. Reactions also found in glycolysis, gluconeogenesis, and the pentose phosphate pathway are involved in these processes. The key roles and reactions of ribulose 1,5-bisphosphate carboxylase are given. Stryer discusses the energetics of photosynthesis and describes the specialized biosynthetic pathway, the C4 pathway, that has evolved in many tropical plants to enhance their photosynthetic efficiency.

A review of the basic concepts of metabolism in Chapter 13; mitochondrial structure, redox potentials, the proton-motive force, and free-energy changes in Chapter 17; and the pentose phosphate pathway in Chapter 18 will help you to understand this chapter.

LEARNING OBJECTIVES

When you have mastered this chapter, you should be able to complete the following objectives.

The Chloroplast, Chlorophylls, and Basic Reactions of Photosynthesis
(Stryer pages 517–521)

1. Write the basic equation of photosynthesis.

2. Provide an overview of the roles of *light, chlorophylls a* and *b, reaction centers, photosystems I* and *II*, and the *Calvin cycle* in photosynthesis.

3. Describe the structure of the *chloroplast*. Locate the *outer, inner,* and *thylakoid membranes;* the *thylakoid space;* the *granum;* and the *stroma.* Associate these structures with the functions they perform.

4. Describe the properties of the thylakoid membrane.

5. Relate the findings of Priestley, Ingenhousz, Senebier, de Saussure, and Mayer to our present understanding of photosynthesis.

6. List the structural components of chlorophylls *a* and *b*, and explain why chlorophylls are effective *photoreceptors*.

7. Explain the significance of the observation that the rate of photosynthesis reaches a limiting value as the intensity of the incident light increases.

8. Distinguish between the *light* and *dark reactions* of photosynthesis.

9. Define the *photosynthetic unit* and describe the function of the *reaction centers*. Explain how the components of the photosynthetic unit interact to funnel light to the reaction centers.

The Splitting of H_2O, the Evolution of O_2, and the Separation of Charge
(Stryer pages 521–527)

10. Name the molecular source of O_2 evolved in photosynthesis, and write the general formula for the *splitting of the hydrogen donor* proposed by Van Niel.

11. Write the equation of the *Hill reaction*, and relate its significance to our understanding of photosynthesis.

12. Describe the *red drop* and the *enhancement phenomenon*, and recount the proposal of Emerson to explain them.

13. List the *oxidants* and *reductants* produced by the light reactions of photosystems I and II.

14. Describe the roles of P680, *plastoquinone*, and *pheophytin* in the *separation of charge* brought about by photosystem II. Explain the function of the *manganese center* in the H_2O-splitting reaction.

15. Describe the composition and function of the *cytochrome* bf *complex*, and outline the roles of *plastocyanin, Cu^{2+},* and *Fe-S centers* in the formation of a *transmembrane proton gradient.*

16. Indicate the reactions of photosystem I, including the roles of P700, A_0, *ferredoxin, FAD, NADPH,* and plastocyanin(Cu^+) in these processes.

17. Write the net reaction carried out by the combined actions of photosystem II, the cytochrome *bf* complex, and photosystem I.

Photophosphorylation and ATP Synthesis (Stryer pages 527–533)

18. Explain how photosystem I can synthesize *ATP* without forming *NADPH* or O_2.

19. Contrast the formation of ATP by *photophosphorylation* and by *oxidative phosphorylation*. Describe the structure of the *ATP synthase* of chloroplasts.

20. Rationalize the differences in the *photosynthetic assemblies* in the *stacked* and *unstacked* regions of the thylakoid membranes.

21. Explain the molecular and ecological functions of *phycobilisomes* in *cyanobacteria* and *red algae*.

The Formation of Hexoses by CO_2 Fixation (Stryer pages 533–540)

22. Describe *Calvin's experiments* with *Chorella* using $^{14}CO_2$.

23. Describe formation of *3-phosphoglycerate* by *ribulose 1,5-bisphosphate carboxylase (Rubisco)*.

24. Detail the formation of *phosphoglycolate* by the *oxygenase reaction* of Rubisco, and outline its subsequent metabolism. Define *photorespiration*.

25. Outline the conversion of 3-phosphoglycerate into fructose 6-phosphate and the *regeneration of ribulose 1,5-bisphosphate*.

26. Write a balanced equation for the *Calvin cycle*, and account for the ATP and NADPH expended to form a hexose molecule.

27. Estimate the efficiency of photosynthesis.

28. Describe the role of *thioredoxin* in coordinating the light and dark reactions of photosynthesis.

29. Describe the C_4 *pathway* and its adaptive value to tropical plants. Explain how CO_2 transport suppresses the oxygenase reaction of Rubisco.

SELF-TEST

The Chloroplast, Chlorophylls, and Basic Reactions of Photosynthesis

1. Write the basic reaction for photosynthesis in green plants:

_____

2. Assign each function or product from the right column to the appropriate structure or pathway in the left column.

(a) Chlorophyll _____   (1) O_2 generation
                                  (2) ATP synthesis
(b) Light-harvesting   (3) Light collection
    complex _____   (4) NADPH synthesis
                                  (5) Separation of charge
(c) Photosystem I _____

(d) Photosystem II _____

(e) Calvin cycle and gluconeogenic pathway _____

(6) 3-Phosphoglycerate formation
(7) Light absorption
(8) Hexose formation
(9) Transmembrane proton gradient

3. Thylakoid membranes contain which of the following?

(a) Light-harvesting complexes
(b) Reaction centers
(c) ATP synthase
(d) Electron-transport chains
(e) Galactolipids
(f) Sulfolipids
(g) Phospholipids

4. For each observation about photosynthesis in the left column, indicate the investigator from the right column who made it.

(a) CO_2 is consumed. _____

(b) Plants revitalize air in which candles have been burned. _____

(c) H_2O is a reactant. _____

(d) Sunlight is needed to revitalize air in which candles have been burned. _____

(e) Light energy is converted into chemical energy. _____

(1) Priestley
(2) Ingenhousz
(3) Senebier
(4) De Saussure
(5) Mayer

5. Which of the following are constituents of chlorophylls?

(a) Substituted tetrapyrrole
(b) Plastoquinone
(c) Mg^{2+}
(d) Fe^{2+}
(e) Phytol
(f) Iron porphyrin

6. Why do chlorophylls absorb visible light efficiently?

_____

_____

7. Which of the following statements about the photosynthetic unit are *true*?

(a) It is a chlorophyll molecule.
(b) It collects light energy through the absorption of light by chlorophyll molecules.
(c) It contains a reaction center with a specialized chlorophyll that contributes to the transduction of light energy into chemical energy.
(d) It contains chlorophyll molecules that transfer energy from one to another by direct electromagnetic interactions.
(e) It is the product of Planck's constant, h, and the frequency of the incident light, ν.

The Splitting of H_2O, the Evolution of O_2, and the Separation of Charge

8. Write the equation with which Van Niel indicated that H_2O is split by light energy in green plants to provide hydrogen atoms for CO_2

reduction. What did studies with $H_2^{18}O$ reveal about this phenomenon?

_____

_____

_____

9. Which of the following statements about the Hill reaction are *true*?

   (a) It confirmed that H_2O is the source of the O_2 evolved during photosynthesis.
   (b) It demonstrated the central role of the chloroplast in photosynthesis.
   (c) It demonstrated that electron acceptors other than CO_2 can be used in the generation of O_2.
   (d) It revealed the role of Fe^{2+} in photosynthesis.
   (e) It connected light absorption to an endergonic oxidation-reduction reaction.

10. Describe the observations that suggested that two interacting light reactions occur during photosynthesis and that they use light of different wavelengths. What are the molecular bases of the two light reactions?

_____

_____

_____

_____

11. Which of the following statements about photosystem II are *correct*?

   (a) It is a multimolecular transmembrane assembly containing several polypeptides, more than 200 chlorophyll *a* and *b* molecules, antenna chlorophylls, a special chlorophyll (P680), pheophytin, and plastoquinones.
   (b) It transfers electrons to photosystem I via the cytochrome *bf* complex.
   (c) It uses light energy to create a separation of charge whose potential energy can be used to oxidize H_2O and to produce a reductant, plastoquinol.
   (d) It uses an Fe^{2+}-Cu^+ center as a charge accumulator to form O_2 without generating potentially harmful hydroxyl radicals, superoxide anions, or H_2O_2.

12. The interaction of photosystems I and II in photosynthesis by green plants can be represented as follows:

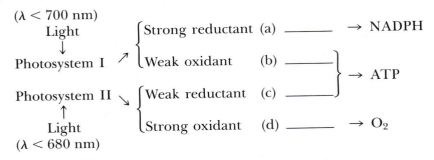

From among the following, identify the oxidants and reductants in the scheme:

(1) Ferredoxin$_{reduced}$ (4) Ferredoxin$_{oxidized}$
(2) Plastoquinol   (5) Plastoquinone
(3) Chlorophyll P680$^+$ (6) Plastocyanin(Cu^{2+})

13. Explain how plastocyanin and plastoquinol are involved in ATP synthesis.

_____

_____

_____

14. Write the net equation of the reaction catalyzed by photosystem I, and describe how NADPH is formed. What is the role of FAD in this process?

_____

_____

_____

Photophosphorylation and ATP Synthesis

15. Which of the following statements about cyclic photophosphorylation are *correct*?

    (a) It doesn't involve NADPH formation.
    (b) It uses electrons supplied by photosystem II.
    (c) It is activated when NADP$^+$ is limiting.
    (d) It does not generate O$_2$.
    (e) It leads to ATP production via the cytochrome *bf* complex.
    (f) It involves a substrate-level phosphorylation.

16. Describe the experiment by which Jagendorf showed that chloroplasts could synthesize ATP in the dark when an artificial pH gradient was created across the thylakoid membrane. What have subsequent studies shown concerning the stoichiometry of protons transported per ATP synthesized?

_____

_____

_____

_____

17. Which of the following statements about the thylakoid membrane are *correct*?

    (a) It contains photosystem I and ATP synthase in the unstacked regions.
    (b) It contains the cytochrome *bf* complex in the unstacked regions only.
    (c) It contains photosystem II mostly in the stacked regions.
    (d) It facilitates communication between photosystems I and II by

the circulation of plastoquinones and plastocyanins in the thylakoid space.

(e) It allows direct interaction between P680* and P700* reaction centers through its differentiation into stacked and unstacked regions.

(f) It can respond to environmental signals by varying its content, the degree of stacking, and the proportions of different photosynthetic assemblies.

18. Cyanobacteria and red algae are photosynthetic algae that live more than a meter underwater, where little blue or red light reaches them. Explain how they can carry out photosynthesis.

_____

_____

_____

The Formation of Hexoses by CO_2 Fixation

19. The observation that the incubation of photosynthetic algae with $^{14}CO_2$ in the light for a very brief time (5 seconds) led to the formation of ^{14}C-labeled 3-phosphoglycerate suggested that the $^{14}CO_2$ was condensing with some two-carbon acceptor. That acceptor was

(a) acetate.
(b) acetyl CoA.
(c) acetyl phosphate.
(d) acetaldehyde.
(e) glycol phosphate.
(f) none of the above.

20. Which of the following statements about ribulose 1,5-bisphosphate carboxylase (Rubisco) are *correct*?

(a) It is present at low concentrations in the chloroplast.
(b) It is activated by the addition of CO_2 to the ϵ-amino group of a specific lysine to form a carbamate that then binds a divalent metal cation.
(c) It catalyzes, as one part of its reaction sequence, an extremely exergonic reaction, the cleavage of a six-carbon diol derivative of arabinitol to form two three-carbon compounds.
(d) It catalyzes a reaction between ribulose 1,5-bisphosphate and O_2 that decreases the efficiency of photosynthesis.
(e) It catalyzes the carboxylase reaction more efficiently and the oxygenase reaction less efficiently as the temperature increases.

21. The Rubisco-catalyzed reaction of O_2 with ribulose 1,5-bisphosphate forms which of the following?

(a) 3-Phosphoglycerate
(b) 2-Phosphoacetate
(c) Phosphoglycolate
(d) Glycolate
(e) Glyoxylate

22. Describe photorespiration and explain why it decreases the efficiency of photosynthesis.

_____

_____

_____

23. Outline the synthesis of fructose 6-phosphate from 3-phosphoglycerate.

_____

_____

_____

24. Which of the following statements about the Calvin cycle are *true?*

(a) It regenerates the ribulose 1,5-bisphosphate consumed by the Rubisco reaction.
(b) It forms glyceraldehyde 3-phosphate, which can be converted to fructose 6-phosphate.
(c) It requires ATP and NADPH.
(d) It is exergonic because light energy absorbed by the chlorophylls is transferred to Rubisco.
(e) It is comprised of enzymes, several of which can be activated through reduction of disulfide bridges by reduced thioredoxin.
(f) It is controlled, in part, by the rate of the Rubisco reaction.
(g) Its rate decreases as the level of illumination increases because both the pH and the level of Mg^{2+} of the stroma decrease.

25. How many moles of ATP and NADPH are required to convert 6 moles of CO_2 to fructose 6-phosphate?

_____

_____

26. It is said that the C_4 pathway increases the efficiency of photosynthesis. What is the justification for this statement when more than 1.6 times as much ATP is required to convert 6 moles of CO_2 to a hexose when this pathway is used in contrast with the pathway used by plants lacking the C_4 apparatus? Account for the extra ATP molecules used in the C_4 pathway.

_____

_____

_____

_____

PROBLEMS

1. Suppose that you were designing spectrophotometric assays for chlorophyll *a* and chlorophyll *b*. What wavelengths would you use for the detection of each? (See Figure 22-5, p. 520 of Stryer.) Explain your answer very briefly.

2. Green light has a wavelength of approximately 520 nm. Explain why solutions of chlorophyll appear to be green. (See Figure 22-5, p. 520 of Stryer.)

3. If you were going to extract chlorophylls *a* and *b* from mashed

spinach leaves, would you prefer to use acetone or water as a solvent? Explain your answer briefly.

4. The spectrophotometric absorbance, A, of a species is given by the Beer-Lambert law:

$$A = E \times l \times [c]$$

where $[c]$ is the molar concentration of the absorbing species, l is the length of the light path in centimeters, and E is the absorbance of a 1 M solution of the species in a 1-cm cell. Suppose that a mixture of chlorophylls a and b in a 1-cm cell gives an absorbance of 0.2345 at 430 nm and an absorbance of 0.161 at 455 nm. Calculate the molar concentration of each chlorophyll in the mixture. Use the values in the margin for E. (Hint: The absorbance of a mixture is equal to the sum of the absorbance of its constituents.)

	E_{430}	E_{455}
Chlorophyll a	1.21×10^5	0.037×10^5
Chlorophyll b	0.53×10^5	1.55×10^5

5. Predict what product(s) will be rapidly labeled with ^{18}O when the following additions are made to actively photosynthesizing systems.

   (a) $H_2{}^{18}O$ is added to green plants.
   (b) $C^{18}O_2$ is added to green plants.
   (c) $C^{18}O_2$ is added to green sulfur bacteria.

6. Would you expect oxygen to be evolved when $NADP^+$ is added to an illuminated suspension of isolated chloroplasts? Explain briefly.

7. Would your answer to Problem 6 change if the chloroplasts were illuminated with extremely monochromatic light of 700 nm? Explain the basis for your answer.

8. $NADP^+$ and A_0 are two components in the electron-transport chain associated with photosystem I (see Stryer, Figure 22-13, p. 525). A_0^- carries a single electron whereas NADPH carries two electrons. Write the overall reaction that occurs, and calculate the $\Delta E_0'$ and $\Delta G^{\circ\prime}$ for the reduction of $NADP^+$ by A_0^- using the thermodynamic data presented by Stryer on pages 400, 401, and 526.

9. Calculate the maximum free-energy change, $\Delta G^{\circ\prime}$, that occurs as a pair of electrons is transferred from photosystem II to photosystem I. Estimate the E_0' values from Figure 22-13 on page 525 of Stryer. Then compare your answer to the free-energy change that occurs in mitochondria as a pair of electrons is transferred from $NADH + H^+$ to oxygen. (See Stryer, p. 401.)

10. Explain the defect(s) in the hypothetical scheme for the light reactions of photosynthesis depicted in Figure 22-1.

11. The following overall equation may be written for photosynthesis in green plants:

$$6 CO_2 + 6 H_2O + n(Nh\nu) = glucose + 6 O_2 \qquad \Delta G^{\circ\prime} = +686 \text{ kcal mol}^{-1}$$

where n is the number of moles of photons required, N is Avogadro's number, and $Nh\nu$ is the energy one mole of photons.

   (a) Give the minimum value for n, and explain your answer briefly.
   (b) The energy of a photon is given by the relationship

$$E = h\nu = h(c/\lambda)$$

   where h is Planck's constant, ν is the frequency, λ is the wavelength, and c is the velocity of light. Calculate the thermodynamic efficiency of photosynthesis when light of 550 nm is used

Figure 22-1
A hypothetical scheme for
photosynthesis.

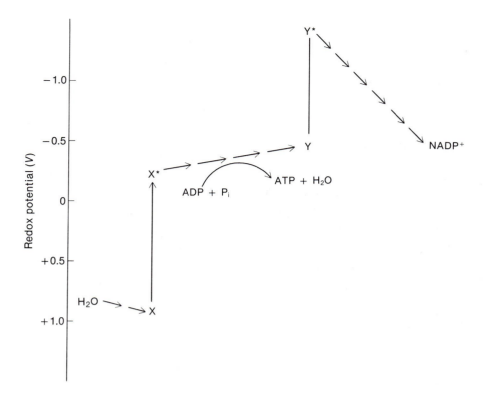

to drive the synthesis of one mole of glucose. Then repeat the
calculation when light of 720 nm is used. Which has the greater
thermodynamic efficiency? Use the physical constants given in
Appendix A on page 1044 of Stryer.

(c) In practice, using extremely monochromatic light of 720 nm to
drive photosynthesis would not be a good idea? Why not?

(d) In view of the fact that neither chlorophyll *a* nor chlorophyll *b*
absorbs light at 550 nm, explain how any photosynthesis may
occur when light of 550 nm is used. (See Stryer, Figure 22-5,
p. 520.)

12. Explain why ATP synthesis requires a larger pH gradient across
the thylakoid membrane of a chloroplast than across the inner membrane of a mitochondrion.

13. When a short pulse of $^{14}CO_2$ is added to systems that are actively
carrying out photosynthesis, in which position(s) of glucose will the
label first appear? Explain.

14. If $^{18}O_2$ were added to C_3 plants on a bright, sunny day, would you
expect glycine subsequently isolated from the leaves to be labeled? Explain.

15. (a) Write a balanced equation for the conversion of oxaloacetate to
malate in the mesophyll cell of a C_4 plant.

(b) What serves as the likely source of reducing power?

16. Give the source of the 12 NADPH used in the biosynthesis of glucose in a bundle-sheath cell of a C_4 plant.

The Chloroplast, Chlorophylls, and Basic Reactions of Photosynthesis

1. The basic reaction for photosynthesis in green plants is

$$H_2O + CO_2 \xrightarrow{\text{Light}} (CH_2O) + O_2$$

where (CH_2O) represents carbohydrate.

2. (a) 3, 5, 7 (b) 3, 7 (c) 2, 3, 4, 5, 7, 9 (d) 1, 2, 3, 5, 7, 9 (e) 6, 8. Chlorophylls are involved in light absorption, light collection in the antenae, and reaction center chemistry. Photosystems I and II cooperate to generate a transmembrane proton-motive force that can synthesize ATP. The enzymes of the Calvin cycle and a gluconeogenic pathway combine to form hexoses.

3. a, b, c, d, e, f, g

4. (a) 3 (b) 1 (c) 4 (d) 2 (e) 5

5. a, c, e

6. The polyene structure (alternating single and double bonds) of chlorophylls causes them to have strong absorption bands in the visible region of the spectrum. Their peak molar absorption coefficients are higher than 10^5 cm^{-1} M^{-1}.

7. b, c, d

The Splitting of H$_2$O, the Evolution of O$_2$, and the Separation of Charge

8. Van Niel's equation for photosynthesis in green plants is

$$CO_2 + 2\,H_2O \xrightarrow{\text{Light}} (CH_2O) + O_2 + H_2O$$

Photosynthesis carried out in ^{18}O-labeled H_2O generated $^{18}O_2$, showing that water was oxidized to oxygen by the process.

9. a, b, c, e

10. When the rate of photosynthesis as a function of the wavelength of the incident light was examined, the rate declined precipitously for wavelengths longer than 680 nm even though chlorophyll continues to absorb light in the range from 680 to 700 nm. If there were only a single species of photoreceptor, the quantum yield of the process should remain constant over its entire absorption band. The fact that it did not suggested that more than one species of photoreceptor was involved in accumulating light energy. A second observation was that light at a shorter wavelength, such as 600 nm, could augment light in the 680- to 700-nm range such that the rate of photosynthesis in the presence of both was greater than the sum of the rates obtained with light of each wavelength separately. This red drop and enhancement phenomenon indicated that there were two separate but interacting light reactions in photosynthesis.

Photosystems I and II have different molecular compositions (e.g., different chlorophylls) and thus absorb light maximally at different wavelengths. They catalyze the two different light reactions, but interact with one another through the cytochrome *bf* complex.

11. a, b, and c. Answer (d) is incorrect because a cluster of four manganese ions serves as a charge accumulator by interactions with the strong oxidant P680$^+$ and H$_2$O to form O$_2$.

12. (a) 1 (b) 6 (c) 2 (d) 3

13. Two electrons from the weak reductant plastoquinol (QH$_2$) are transferred to two molecules of the weak oxidant plastocyanin (PC) in a reaction catalyzed by the transmembrane cytochrome *bf* complex; in the process, two protons are pumped across the thylakoid membrane to acidify the thylakoid space with respect to the stroma. The transmembrane proton gradient is used to synthesize ATP.

14. The net reaction catalyzed by photosystem I is

$$PC(Cu^+) + ferredoxin_{oxidized} \longrightarrow PC(Cu^{2+}) + ferredoxin_{reduced}$$

where PC is plastocyanin. Reduced ferredoxin is a powerful reductant. Two reduced ferredoxins reduce NADP$^+$ with the uptake of a proton to form NADPH and two oxidized ferredoxins in a reaction catalyzed by ferredoxin-NADP$^+$ reductase. FAD is a prosthetic group on the enzyme that serves as an adaptor to collect two electrons from two reduced ferredoxin molecules for their subsequent transfer to a single NADP$^+$ molecule.

Photophosphorylation and ATP Synthesis

15. a, c, d, and e. Answer (b) is incorrect because photosystem I provides the electrons for photophosphorylation.

16. In Jagendorf's experiment, chloroplasts were equilibrated with a buffer at pH 4 to acidify their thylakoid spaces. The suspension was then rapidly brought to pH 8 and ADP and P$_i$ were added. The pH of the stroma suddenly increased to 8 whereas that of the thylakoid space remained at 4, resulting in a pH gradient across the thylakoid membrane. Jagendorf observed that ATP was synthesized as the pH gradient dissipated and that the synthesis occurred in the dark. Later studies of the stoichiometry of ATP synthesis revealed that approximately three protons traverse the membrane for each ATP synthesized.

17. a, c, d, and f. Answer (b) is incorrect because the cytochrome *bf* complex is uniformly distributed throughout the thylakoid membrane. Answer (e) is incorrect because the differentiation into stacked and unstacked regions probably prevents direct interaction between the excited reaction center chlorophylls P680* and P700*.

18. Cyanobacteria and red algae contain protein assemblies called phycobilisomes that absorb some of the green and yellow light that reaches them to perform photosynthesis.

The Formation of Hexoses by CO$_2$ Fixation

19. f. The $^{14}CO_2$ actually condenses with the five-carbon compound ribulose 1,5-bisphosphate to form a six-carbon compound that is rapidly hydrolyzed to form two molecules of 3-phosphoglycerate, one of which is ^{14}C-labeled.

20. b, c, and d. Answer (e) is incorrect because the rate of the oxygenase reaction increases relative to that of the carboxylase reaction as the

temperature increases; the altered ratio of the two reaction rates decreases the efficiency of photosynthesis as the temperature increases.

21. a, c

22. The oxygenase reaction of Rubisco and the salvage reactions that convert two resulting phosphoglycolate molecules into serine are called photorespiration because CO_2 is released and O_2 is consumed in the process. Unlike genuine respiration, no ATP or NADPH is produced by photorespiration. Ordinarily, no CO_2 is released during photosynthesis, and all the fixed CO_2 can be used to form hexoses. During photorespiration, no CO_2 is fixed and the products into which ribulose 1,5-bisphosphate is converted by the oxygenase reaction of Rubisco cannot be completely recycled into carbohydrate because of the loss of CO_2 in the phosphoglycolate salvage reactions.

23. Phosphoglycerate kinase converts 3-phosphoglycerate, the initial product of photosynthesis, to the glycolytic intermediate 1,3-bisphosphoglycerate, which is then converted to glyceraldehyde 3-phosphate (G3P) by an NADPH-dependent G3P dehydrogenase in the chloroplast. Triosephosphate isomerase converts G3P to dihydroxyacetone phosphate, which aldolase can condense with another G3P to form fructose 1,6-bisphosphate. The phosphate ester at C-1 is hydrolyzed to give fructose 6-phosphate. The result of this pathway, which is functionally equivalent to the gluconeogenic pathway, is the conversion of the CO_2 fixed by photosynthesis into a hexose. (See Figure 22-31, p. 536 in Stryer.)

24. a, b, c, e, f

25. Eighteen moles of ATP and 12 moles of NADPH are required to fix 6 moles of CO_2. Two moles of ATP are used by phosphoglycerate kinase to form 2 moles of 1,3-bisphosphoglycerate, and 1 mole of ATP is used by ribulose 5-phosphate kinase to form 1 mole of ribulose 1,5-bisphosphate per mole of CO_2 fixed. Two moles of NADPH are used by G3P dehydrogenase to form 2 moles of G3P per mole of CO_2 incorporated. Therefore, 3 moles of ATP and 2 moles of NADPH are used for each mole of CO_2 fixed. (See Figure 22-32, p. 537 in Stryer.)

26. Plants lacking the C_4 pathway cannot compensate for the relative increase in the rate of the oxygenase reaction of Rubisco with respect to the rate of the carboxylase reaction that occurs as the temperature rises. Plants with the C_4 pathway increase the concentration of CO_2 in the bundle-sheath cell, where the Calvin cycle occurs, thereby increasing the ability of CO_2 to compete with O_2 as a substrate for Rubisco. As a result, more CO_2 is fixed and less ribulose 1,5-bisphosphate is degraded into phosphoglycolate, which cannot be efficiently converted into carbohydrate. Thus, the Calvin cycle functions more efficiently in these specialized plants under conditions of high illumination and at higher temperatures than it would otherwise.

The concentration of CO_2 is increased by an expenditure of ATP. The collection of 1 CO_2 molecule and its transport on C_4 compounds from the mesophyll cell into the bundle-sheath cell is brought about by the conversion of 1 ATP to AMP and PP_i in a reaction in which pyruvate is phosphorylated to PEP. The PP_i is hydrolyzed, and 2 ATP are required to resynthesize ATP from AMP. Thus, an *extra* 2 ATP/CO_2 × 6 CO_2/hexose = 12 ATP/hexose are used by the C_4 pathway.

ANSWERS TO PROBLEMS

1. You would use 430 nm for chlorophyll a and 455 nm for chlorophyll b. These are the wavelengths of maximum absorbance, so they would provide the most sensitive spectrophotometric assays.

2. Chlorophyll appears to be green because it has no significant absorption in the green region of the spectrum and therefore transmits green light.

3. Acetone is the preferred solvent. Because of the hydrophobic porphyrin ring and the very hydrophobic phytol tail of the chlorophylls, they are soluble in organic solvents like acetone but are insoluble in water.

4. The total absorbance at each wavelength is the sum of the absorbances due to chlorophylls a and b separately. At 430 nm,

$$0.2345 = (1.21 \times 10^5)[\text{chl } a] + (0.53 \times 10^5)[\text{chl } b]$$

At 455 nm,

$$0.161 = (0.037 \times 10^5)[\text{chl } a] + (1.55 \times 10^5)[\text{chl } b]$$

Solving these two equations for the two unknowns yields

$$[\text{chl } a] = 1.5 \times 10^{-6} \text{ M and } [\text{chl } b] = 1.0 \times 10^{-6} \text{ M}.$$

5. (a) $^{18}O_2$ will be evolved. The oxygen produced by green plants comes from water split by photosystem II.
   (b) Carbohydrate and water will both be labeled with ^{18}O. (See the Van Niel equation on p. 521 of Stryer.) The oxygen liberated derives from water. Therefore, of the two oxygen atoms of CO_2, one is incorporated into carbohydrate and the other into water.
   (c) Again, both carbohydrate and water will be labeled with ^{18}O.

6. Oxygen would be evolved. In this reaction, $NADP^+$ would be the counterpart of the artificial electron acceptors used by Hill (see Stryer, pp. 521–522).

7. Yes. Little oxygen would be evolved when 700-nm light is used. This is the red drop (see Stryer, p. 522). Oxygen is evolved by photosystem II, which contains P680 and is therefore not maximally excited by 700-nm light.

8. A_0^- is the stronger reductant in the system because it is associated with the half-cell having the more negative standard reducing potential. A_0^- will therefore reduce $NADP^+$ to NADPH under standard conditions. Following the convention for redox problems presented by Stryer on page 401, we first write the partial reaction for the reduction involving the weaker reductant (the half-cell with the more positive standard reducing potential):

$$NADP^+ + H^+ + 2\,e^- \longrightarrow NADPH \qquad E_0' = -0.32 \text{ V} \quad (1)$$

Next, we write, *again as a reduction,* the partial reaction involving the stronger reductant (the half-cell with the more negative standard reducing potential):

$$A_0 + e^- \longrightarrow A_0^- \qquad E_0' = -1.1 \text{ V} \quad (2)$$

To get the overall reaction that occurs, we must equalize the number of electrons by multiplying equation 2 by 2. (We do not, however, multiply the half-cell potential by 2.)

$$2\,A_0 + 2\,e^- \longrightarrow 2\,A_0^- \qquad\qquad E_0' = -1.1\ \text{V} \quad (3)$$

Then we *subtract* equation 3 from equation 1. This yields

$$NADP^+ + H^+ + 2\,A_0^- \longrightarrow NADPH + 2\,A_0 \qquad \Delta E_0' = +0.78\ \text{V}$$

In calculating $\Delta E_0'$ values, do not make the mistake of multiplying half-cell reduction potentials by factors used to equalize the number of electrons. Remember that $\Delta E_0'$ is a *potential difference* and hence, at least for our purposes, is independent of the *amount* of electron flow. For example, in a house with an adequate electrical power supply, the potential difference measured at the fusebox is approximately 117 V regardless of whether the house is in total darkness or all the lights are turned on.

To get the free-energy change for the overall reaction, we start with the relationship given on page 401 of Stryer:

$$\Delta G^{\circ\prime} = -nF\,\Delta E_0'$$

Substitution yields

$$\Delta G^{\circ\prime} = -2 \times 23.06 \times 0.78 = -36.0\ \text{kcal/mol}$$

9. In this process, electrons are transferred down an electron-transport chain from P680* to P700. The E_0' value for P680* is approximately -0.8 V, and that for P700 is approximately 0.4; $\Delta E_0'$ is therefore $+1.2$ V. The free-energy change is calculated from the relationship given on page 401 of Stryer:

$$\Delta G^{\circ\prime} = -nF\,\Delta E_0'$$
$$= -2 \times 23.06 \times 1.2$$
$$= -55.3\ \text{kcal/mol}$$

In the mitochondrial electron-transport chain, the free-energy change as a pair of electrons is transferred from $NADH + H^+$ to oxygen is -52.6 kcal/mol (see p. 401 of Stryer).

10. In the scheme in Figure 22-1, electrons are shown flowing "uphill" from X* to Y as ATP is being formed. This is a thermodynamic impossibility. For electrons to flow spontaneously from X* to Y, the redox potential of X* must be more negative than that of Y. For ATP to be formed as electron transfer occurs, the free-energy change must be of sufficient magnitude to allow for ATP biosynthesis. In order to make electrons flow from X* to Y as depicted in the hypothetical scheme, ATP would be consumed, not generated.

11. (a) The minimum value of $n = 48$. In each turn of the Calvin cycle NADPH are required and 1 CO_2 is fixed (see Figure 22-32, p. 537 of Stryer). Therefore, for 6 CO_2 to be reduced to the oxidation level of carbohydrate, 12 NADPH are required. Each NADPH generated has received two electrons from photosystem I, requiring the absorption of two photons. The electrons donated by photosystem I are replaced by photosystem II, requiring the absorption of another two photons for a total of four photons for each NADPH.

(b) The energy of a mole of 550-nm photons is given by

$$E = Nh(c/\lambda)$$

where N is Avogadro's number. Substitution yields

$$E = (6.022 \times 10^{23} \text{ mol}^{-1})(1.584 \times 10^{-34} \text{ cal s})\frac{(2.998 \times 10^{17} \text{ nm s}^{-1})}{550 \text{ nm}}$$

$$= 52 \text{ kcal}$$

Since 48 moles of photons are required for the synthesis of one mole of glucose, the total energy input is $48 \times 52 = 2496$ kcal, and the efficiency is $(686/2496) \times 100\% = 27\%$.

When light of 720 nm is used, the total energy input is $39.5 \times 48 = 1896$ kcal, and the efficiency is $(686/1896) \times 100\% = 36\%$. Therefore, light of longer wavelength gives higher thermodynamic efficiency.

(c) Even though photosynthesis is more efficient thermodynamically with 720-nm light, the rate of photosynthesis under these conditions would be negligible due to the red drop (see Stryer, p. 522).

(d) Accessory pigments absorb light of 550 nm and transfer the light energy to the reaction centers. Note that the quantum yield of photosynthesis is significant at 550 nm. (See Stryer, p. 522, Figure 22-9. Also see the discussion of light antennae in cyanobacteria and red algae on p. 531.)

12. The synthesis of ATP in both the chloroplast and the mitochondrion is driven by the proton-motive force across the membrane. The proton-motive force, Δp is given by the equation

$$\Delta p = E_{\text{m}} - \Delta\text{pH}$$

where E_{m} is the membrane potential, and ΔpH is the pH gradient across the membrane. In mitochondria, a membrane potential of 0.14 V is established during electron transport. In chloroplasts, the light-induced potential is close to 0. Therefore, there must be a greater pH gradient in the chloroplast to give the same free-energy yield. (See Stryer, p. 411 and p. 528.)

13. The label will first appear in positions 3 and 4 of glucose. Phosphoglycerate labeled in the carboxyl carbon will be formed, and this carbon becomes carbon atoms 3 and 4 of glucose.

14. Yes. C_3 plants photorespire, and oxygen is incorporated into the carboxyl group of phosphoglycolate as shown in Figure 22-28 on page 535 of Stryer. Since the carboxyl group of phosphoglycolate will become the carboxyl group of glycine, the glycine will be labeled. (See Figure 22-29 on page 536 of Stryer and the accompanying text.)

15. (a) The equation for the conversion of oxaloacetate to malate is

$$\text{Oxaloacetate} + \text{NADPH} + \text{H}^+ \longrightarrow \text{malate} + \text{NADP}^+$$

(b) The NADPH is furnished by the light reactions of photosynthesis.

16. Six of the 12 NADPH are provided by the oxidative decarboxylation of malate:

$$6 \text{ Malate} + 6 \text{ NADP}^+ \longrightarrow 6 \text{ pyruvate} + 6 \text{ NADPH} + 6 \text{ H}^+ + 6 \text{ CO}_2$$

The other 6 NADPH come from the light reactions of photosynthesis. Note, however, that the malate is originally produced in the mesophyll cell by the reduction of oxaloacetate by $\text{NADPH} + \text{H}^+$ coming from photosystem I. (See the answer to Problem 15.) Thus, all of the NADPH required for the reduction of CO_2 ultimately arises from the light reactions.

CHAPTER 23

Biosynthesis of Membrane Lipids and Steroid Hormones

This chapter describes the biosynthesis of membrane lipids, steroid hormones, and other important lipid molecules, such as bile salts, vitamin D, and polymers of isoprene units. As background material for this chapter, you should review the earlier chapters on cell membranes (Chapter 12) and fatty acid metabolism (Chapter 20), paying particular attention to the structure and properties of lipids and the central role of acetyl CoA in the metabolism of lipids. Stryer begins Chapter 23 with a discussion of the formation of triacylglycerols, phosphoglycerides, and sphingolipids from the simple precursors glycerol 3-phosphate, fatty acyl CoAs, certain polar alcohols (e.g., choline), serine, and sugars. He then describes the synthesis of cholesterol from acetyl CoA via the important intermediate isopentenyl pyrophosphate. Cholesterol is the precursor of bile salts as well as steroid hormones. The cholesterol and triacylglycerols synthesized in the liver and intestines are transported to peripheral tissues by lipoproteins; therefore, the classification of lipoproteins, their properties, and the mechanisms by which they deliver lipids to cells are discussed next. Finally, Stryer describes the synthesis of steroid hormones and vitamin D from cholesterol and introduces a variety of isoprenoid lipids that are derived from isopentenyl pyrophosphate.

LEARNING OBJECTIVES

When you have mastered this chapter, you should be able to complete the following objectives.

Synthesis of Triacylglycerols and Phosphoglycerides (Stryer pages 547–552)

1. Describe the roles of *glycerol 3-phosphate, lysophosphatidate,* and *phosphatidate* in the synthesis of *triacylglycerols* and *phosphoglycerides.*

2. Describe the formation of triacylglycerols from phosphatidate.

3. Contrast the biosynthesis of phosphoglycerides in bacteria and mammals. Note the significance of *CDP-diacylglycerol* and *CDP-choline* or *CDP-ethanolamine,* the activated precursors in these biosyntheses.

4. Compare the biosynthesis of *phosphatidyl serine, phosphatidyl ethanolamine, phosphatidyl choline,* and *phosphatidyl inositol* in bacteria and mammals.

5. State the physiological role of phosphatidyl inositol and its degradation products, *inositol 1,4,5-trisphosphate* and *diacylglycerol.*

6. Compare the structures and biosynthetic pathways of the *glyceryl ether phospholipids,* including *platelet-activating factor* and *plasmalogens,* with those of the *glyceryl ester phospholipids.*

7. List the *phospholipases* together with their specificities for the ester bonds of phosphoglycerides. Describe the biological roles of phospholipases.

Synthesis of Sphingolipids (Stryer pages 552–554)

8. Summarize the steps in the biosynthesis of *sphingosine* from *palmitoyl CoA* and *serine.*

9. Outline the synthesis of *sphingomyelin, cerebrosides,* and *gangliosides* from sphingosine. Note the use of activated sugars and acidic sugars.

10. Discuss the general degradation pathway of gangliosides and the biochemical basis of *Tay-Sachs disease.*

Synthesis of Cholesterol (Stryer pages 554–559)

11. Describe the physiological roles of *cholesterol.*

12. List the major stages in cholesterol biosynthesis and give the key intermediates.

13. Compare the synthetic paths leading from acetyl CoA to *mevalonate* and to the *ketone bodies.* Note the role of *3-hydroxy-3-methylglutaryl CoA reductase (HMG CoA reductase)* as the major regulatory enzyme in cholesterol biosynthesis.

14. Describe the conversion of mevalonate into *isopentenyl pyrophosphate.*

15. Outline the condensation reactions leading from isopentenyl pyrophosphate to *squalene.* Describe the mechanisms of these condensation reactions.

16. Discuss the *cyclization* of squalene and the formation of cholesterol from *lanosterol.*

17. Describe the physiological roles and the general structures of the *bile salts*.

Lipoproteins (Stryer pages 560–564)

18. Describe the regulation of liver HMG CoA reductase by dietary cholesterol.

19. List the various classes of *lipoproteins* together with their lipid and protein components. Describe their lipid-transport functions.

20. Summarize the steps in the delivery of cholesterol to cells via the *low-density-lipoprotein (LDL) receptor*. Discuss the regulation of cellular functions by the *LDL pathway*.

21. Describe the proposed domain structure of the LDL receptor derived from the primary sequence of this protein.

22. Discuss the biochemical defects of the LDL receptor that result in *familial hypercholesterolemia*.

Synthesis of Steroid Hormones (Stryer pages 565–569)

23. Give the numbering scheme for the carbon atoms of cholesterol and its derivatives, and distinguish between the α and β-*oriented* groups and *cis* or *trans ring fusions*.

24. List the five major classes of *steroid hormones*, their physiological functions, and their sites of synthesis. Outline their biosynthetic relationships.

25. Describe the *hydroxylation reactions* involving *cytochrome* P_{450}. Indicate the role of these *monooxygenase reactions* in steroid biosynthesis, the *detoxification* of *xenobiotic compounds*, and the generation of *carcinogens*.

26. Describe the synthesis of *pregnenolone* from cholesterol, the conversion of pregnenolone into *progesterone*, and the subsequent reactions leading to *cortisol* and *aldosterone*.

27. Outline the synthesis of *androgens* and *estrogens* from progesterone.

28. Explain the biochemical basis for both the virilization and the Na^+ loss observed in *21-hydroxylase deficiency*.

Vitamin D and Derivatives of Isopentenyl Pyrophosphate (Stryer pages 569–571)

29. Discuss the synthesis and the physiological role of *vitamin D*.

30. Give examples of biomolecules that contain *isoprene units*.

SELF-TEST

Synthesis of Triacylglycerols and Phosphoglycerides

1. Which of the following reactions are sources of glycerol 3-phosphate that is used in lipid synthesis?

(a) Reduction of dihydroxyacetone phosphate
(b) Oxidation of glyceraldehyde 3-phosphate
(c) Phosphorylation of glycerol
(d) Dephosphorylation of 1,3-bisphosphoglycerate
(e) Reductive phosphorylation of pyruvate

2. Match the lipids in the left column with the major synthetic precursors or intermediates listed in the right column.

(a) Triacylglycerol _____

(b) Phosphatidyl ethanolamine (bacteria) _____

(c) Phosphatidyl ethanolamine (mammals) _____

(1) Phosphatidate
(2) Diacylglycerol
(3) Acyl CoA
(4) Glycerol 3-phosphate
(5) CDP-diacylglycerol
(6) CDP-ethanolamine

3. Explain the role of the CDP derivatives in the synthesis of phosphoglycerides.

_____

_____

4. Calculate the number of "high energy" phosphate bonds that are expended in the formation of phosphatidyl choline from diacylglycerol and choline in mammals.

_____

5. Which of the following is a common reaction for the formation of phosphatidyl ethanolamine in bacteria?

(a) Decarboxylation of phosphatidyl serine
(b) Reaction of CDP-ethanolamine with a diacylglycerol
(c) Demethylation of phosphatidyl choline
(d) Reaction of ethanolamine with CDP-diacylglycerol
(e) Reaction of CDP-ethanolamine with CDP-diacylglycerol

6. Which of the following is a lipid with a signal-transducing activity?

(a) Phosphatidyl choline
(b) Phosphatidyl serine
(c) Plasminogen activator
(d) Phosphatidyl inositol 4,5-bisphosphate
(e) Phospholipase A_2

7. For the lipid classes listed in the left column, select the characteristic structural components or properties from the right column.

(a) Glyceryl ester phospholipids _____

(b) Plasmalogens _____

(c) Platelet activating factor _____

(1) Two long hydrocarbon chains
(2) Acetyl group
(3) Phosphate group
(4) Ether linkage
(5) $\alpha-\beta$ Double bond
(6) Glycerol group
(7) Long fatty acyl chain
(8) Relatively high solubility in water

8. Which of the following phospholipases *would not* cleave the phospholipid shown below?

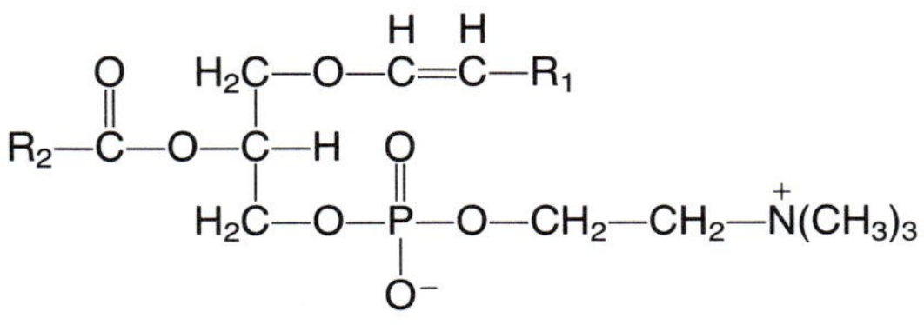

Figure 23-1
A phospholipid.

(a) Phospholipase A_1
(b) Phospholipase A_2
(c) Phospholipase C
(d) Phospholipase D

Synthesis of Sphingolipids

9. Which of the following is *not* a precursor or intermediate in the synthesis of sphingomyelin?

(a) Palmitoyl CoA
(b) Lysophosphatidate
(c) CDP-choline
(d) Acyl CoA
(e) Serine

10. Match the lipids in the left column with the appropriate activated precursors from the right column.

(a) Sphingomyelin _____

(b) Ganglioside _____

(c) Phosphatidyl serine _____

(1) Acyl CoA
(2) CDP-choline
(3) CDP-diacylglycerol
(4) CMP-*N*-acetylneuraminate
(5) UDP-sugar

11. In which compartment of the cell does ganglioside G_{M2} accumulate in Tay-Sachs patients? What is the biochemical defect?

_____

_____

Synthesis of Cholesterol

12. From the following compounds, identify the intermediates in the synthesis of cholesterol and list them in their proper sequence.

(a) Geranyl pyrophosphate
(b) Squalene
(c) Isopentenyl pyrophosphate
(d) Mevalonate
(e) Cholyl CoA
(f) Farnesyl pyrophosphate
(g) Lanosterol

13. Which of the following are common features of the syntheses of mevalonate and ketone bodies?

(a) Both involve 3-hydroxy-3-methylglutaryl CoA (HMG CoA).
(b) Both require NADPH.
(c) Both require the HMG CoA cleavage enzyme.
(d) Both occur in the mitochondria.
(e) Both occur in liver cells.

14. Select the appropriate characteristics from the right column for the three stages in the synthesis of cholesterol in the left column.

(a) Mevalonate to isopentenyl pyrophosphate _____

(b) Isopentenyl pyrophosphate to squalene _____

(c) Squalene to cholesterol _____

(1) Releases PP_i
(2) Requires NADPH
(3) Requires O_2
(4) Releases CO_2
(5) Requires ATP

15. Yeast cells growing aerobically are able to synthesize sterols and incorporate them into membranes. However, under anaerobic conditions yeast cells do not survive unless they are provided with an exogenous source of sterols. Explain the metabolic basis of this nutritional requirement.

_____

_____

16. The physiological roles of bile salts include which of the following?

(a) They aid in the digestion of lipids.
(b) They aid in the digestion of proteins.
(c) They facilitate the absorption of sugars.
(d) They facilitate the absorption of lipids.
(e) They provide a means for excreting cholesterol.

17. Which of the following are common features in the structures of cholesterol and glycocholate?

(a) Both have three hydroxyl groups.
(b) Both contain four fused rings.
(c) Both have a hydrocarbon side chain.
(d) Both contain a carboxylate group.
(e) Both contain double bonds.

18. Explain the structural characteristics of bile salts that make them effective biological detergents.

_____

_____

Lipoproteins

19. Match the appropriate components or properties in the right column with the lipoproteins in the left column.

(a) Chylomicron _____

(b) VLDL _____

(1) contains apoprotein B-100.
(2) contains apoprotein B-48.
(3) contains apoprotein A-1.

(c) LDL _____

(d) HDL _____

(4) transports endogenous cholesterol esters.

(5) transports dietary triacylglycerols.

(6) transports endogenous triacylglycerols.

(7) is degraded by lipoprotein lipase.

(8) is taken up by cells via receptor-mediated mechanisms.

(9) is a precursor of LDL.

(10) may remove cholesterol from cells.

20. Which of the following events occur in the LDL pathway in fibroblasts? Place them in their proper sequential order.

(a) Breakdown of LDL in lysosomes

(b) Endocytosis of LDL along with LDL receptors

(c) Regulation of the synthesis of HMG CoA reductase and LDL receptors

(d) Degradation of LDL receptors in lysosomes

(e) Binding of LDL to LDL receptors

21. Exons in the gene for the LDL receptor give rise to structurally diverse domains. What is the likely function of the cysteine-rich amino-terminal domain, which contains a cluster of negatively charged side chains?

(a) Carbohydrate binding

(b) Membrane attachment

(c) Lipoprotein binding

(d) Growth-factor binding

(e) Clathrin binding

22. Explain how LDL regulates the cholesterol content in fibroblasts.

_____

_____

23. Assume that LDL is produced normally in a patient but that the apoprotein B-100 domain that recognizes the receptor is functionally defective, which prevents the binding of LDL to its receptor. What outcome would this defect have on cholesterol metabolism in peripheral cells?

_____

_____

Synthesis of Steroid Hormones

24. For the sterol structure in the margin, complete the following:

(a) Name this sterol: _____

(b) It is synthesized from _____ via three hydroxylation reactions.

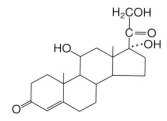

Figure 23-2
A sterol.

(c) It has how many fewer carbon atoms than cholesterol?

_____

(d) Its concentration will be diminished if there is a deficiency of 21-hydroxylase—true or false? _____

Explain why. _____

_____

25. Hydroxylation reactions involving cytochrome P_{450} have which of the following characteristics?

(a) They require a proton gradient.
(b) They involve electron transport from NADPH to O_2.
(c) They activate O_2 by binding it to adrenodoxin.
(d) They transfer one oxygen atom from O_2 to the substrate and form water from the other.
(e) They occur in adrenal mitochondria and liver microsomes.

26. Explain how foreign aromatic compounds are detoxified and excreted by mammals.

_____

_____

_____

27. Match the steroid hormones in the left column with the characteristics in the right column that distinguish them from one another.

(a) Aldosterone _____

(b) Estrogen _____

(c) Testosterone _____

(1) has 18 carbon atoms.
(2) has 19 carbon atoms.
(3) has 21 carbon atoms.
(4) contains an aromatic ring.
(5) contains an aldehyde group at C-18.

Vitamin D and Derivatives of Isopentenyl Pyrophosphate

28. Which of the following statements about active vitamin D is _incorrect_?

(a) It has the same fused ring system as cholesterol.
(b) It requires hydroxylation reactions for its synthesis from cholecalciferol.
(c) It is important in the control of calcium and phosphorus metabolism.
(d) It can be synthesized from cholesterol in the presence of UV light.
(e) It can be derived from the diet.

29. Which of the following lipids does not contain isoprene units?

(a) Coenzyme Q
(b) Carotene
(c) Vitamin K
(d) Arachidonate
(e) Phytol side chain of chlorophyll

1. Why is de novo cholesterol synthesis dependent on the activity of citrate lyase?

2. An infant has an enlarged liver and spleen, cataracts and anemia and exhibits general retardation of development. Mevalonate is found in the urine. Investigation reveals a deficiency of mevalonate kinase, which catalyzes the formation of 5-phosphomevalonate from mevalonate.

   (a) Why is urinary excretion of mevalonate consistent with a deficiency of mevalonate kinase?
   (b) How would a deficiency of mevalonate kinase affect cholesterol synthesis in this infant?
   (c) What level of activity, relative to normal, would you expect to find for HMG CoA reductase in cells isolated from the infant? Briefly explain your answer.

3. Normally, most of the bile acids that are secreted into the intestine undergo reabsorption and are returned to the liver. Cholestyramine is a positively charged resin that binds bile acids in the intestinal lumen and prevents their reabsorption.

   (a) To examine the effects of cholestyramine on LDL metabolism, two fractions of LDL were prepared: one was covalently labeled on tyrosine residues with ^{125}I; the other was labeled with ^{131}I and treated with cyclohexanedione, which interferes with binding to the LDL receptor. When rabbits were given cholestyramine, hepatic uptake of ^{125}I-labeled LDL was enhanced relative to normal, whereas the uptake of ^{131}I-labeled LDL was unchanged relative to that in rabbits that had not been given cholestyramine. Briefly explain the relationship between the action of cholestyramine and LDL uptake in the liver.
   (b) The administration of cholestyramine usually results in a 15 to 20 percent reduction in levels of circulating LDL, whereas the administration of a combination of cholestyramine and mevinolin can often yield a 30 to 40 percent reduction. Why?

4. The presence of apoprotein E in lipoproteins enables them to be taken up by hepatic cells. Provide a brief explanation for each of the following observations made of an individual with a deficiency in apoprotein E synthesis.

   (a) Elevated levels of plasma triacylglycerols and cholesterol, coupled with the presence of chylomicron remnants and IDL. These latter particles persist in the bloodstream much longer than in normal individuals.
   (b) Abnormally low levels of LDL in the blood.
   (c) Abnormally high levels of LDL receptors in liver cells.
   (d) A marked reduction in levels of circulating chylomicron remnants and IDL when the diet is low in cholesterol and fat.

5. Pregnant women often have increased rates of triacylglycerol breakdown and, as a result, have elevated levels of ketone bodies in their blood. Why do they also often exhibit an increase in plasma lipoprotein levels?

6. Hopanoids, previously described in Problem 6 of Chapter 12, are pentacyclic molecules that are found in bacteria and in some plants. These organisms utilize a pathway similar to that of cholesterol synthesis to make hopanoids. The hopane biosynthetic pathway includes the formation of squalene.

   (a) How many molecules of mevalonic acid are required for the synthesis of hopane?

   (b) Squalene can undergo concerted cyclization to form hopane in a reaction that is catalyzed by a unique type of squalene cyclase. The reaction is initiated by a proton and does not require oxygen. Compare this step with the formation of lanosterol from squalene. Why could it be argued that the synthesis of hopanoids preceded the synthesis of sterols in evolution?

7. Your colleague has discovered a compound that is a very powerful inhibitor of HMG CoA reductase, and she has evidence that the drug will completely block the synthesis of mevalonate in liver. Why is this compound unlikely to be useful as a drug?

8. The liver is the site of the synthesis of plasma phospholipids and lipoproteins. Rats maintained on a diet deficient in choline often develop fat deposits in liver tissue. How could choline deficiency be related to this aberration in lipid metabolism?

9. Among the sugar residues found in a blood group ganglioside is fucose. Experiments utilizing isolated Golgi membranes and ribonucleoside triphosphates show that fucose can be incorporated into the ganglioside only when GTP is available. What is the role of GTP in fucose incorporation?

10. Suppose a cell is deficient in phosphatidate phosphatase, which catalyzes the formation of diacylglycerol from phosphatidate. What effects on lipid metabolism would you expect?

11. (a) Would you expect virilization among patients who have desmolase deficiency?

    (b) Why is enlargement of the adrenal glands common among such individuals?

12. In the adult form of Gaucher's disease, glucosylcerebrosides accumulate in liver, spleen, and bone-marrow cells. Although the common galactosylceramides and their derivatives are found in the tissues of affected individuals, accumulations of galactosylcerebrosides or their metabolites are not found, nor do ceramides accumulate. What enzyme activity is probably deficient in patients with Gaucher's disease?

13. Cells of the adrenal cortex have very high concentrations of LDL receptors. Why?

14. At low concentrations of phospholipid substrates in water, the reaction catalyzed by a phospholipase occurs at a rather low rate. The reaction rate accelerates when the concentrations of the phospholipid substrates increase to the point that micelles are formed. How is this property of phospholipases related to their activity in the cell?

15. Glucagon has been shown to reduce the activity of HMG CoA reductase. Why is this observation consistent with the overall effect of glucagon on cellular metabolism?

Synthesis of Triacylglycerols and Phosphoglycerides

1. a, c

2. (a) 1, 2, 3, 4 (b) 1, 3, 4, 5 (c) 1, 2, 3, 4, 6

3. CDP-diacylglycerol and the CDP-alcohols are activated intermediates that allow the formation of phosphate ester bonds in phosphoglycerides, a process that is otherwise highly exergonic. ATP supplies the energy to form these compounds. Recall that UDP-sugars are used in a similar manner in the synthesis of carbohydrates (see Stryer, p. 455).

4. Summing the individual reactions:

Choline + ATP $\longrightarrow$ phosphorylcholine + ADP

Phosphorylcholine + CTP $\longrightarrow$ CDP-choline + PP$_i$

CDP-choline + diacylglycerol $\longrightarrow$ CMP + phosphatidyl choline

Choline + ATP + CTP + diacylglycerol $\longrightarrow$
ADP + PP$_i$ + CMP + phosphatidyl choline

Two "high-energy" bonds (from ATP and CTP) are consumed in these reactions. In addition, pyrophosphate is hydrolyzed by pyrophosphatase, driving the net reaction further to the right. In fact, three high-energy bonds would have to be formed to regenerate ATP and CTP from ADP and CMP.

5. a

6. d

7. (a) 1, 3, 6, 7 (b) 1, 3, 4, 5, 6, 7 (c) 2, 3, 4, 6, 8

8. a. Phospholipases are specific for *ester* bonds; therefore, phospholipase A$_1$ will not cleave the *ether* bond on the C-1 carbon of the given plasmalogen.

Synthesis of Sphingolipids

9. b

10. (a) 1, 2 (b) 1, 4, 5 (c) 1, 3

11. The degradative enzymes for gangliosides are located in lysosomes; therefore, ganglioside G$_{M2}$ will accumulate in the lysosomes of Tay-Sachs patients. The enzyme that removes the terminal sugar, GalNAc, from the ganglioside is deficient in these individuals.

Synthesis of Cholesterol

12. All of the compounds given are intermediates in the biosynthesis of cholesterol except for (e) cholyl CoA, which is a catabolic derivative of cholesterol and a precursor of bile salts. The proper sequence is d, c, a, f, b, and g.

13. a, e

14. (a) 4, 5 (b) 1, 2 (c) 2, 3

15. A key intermediate in the biosynthesis of cholesterol and related sterols is squalene, an open-chain isoprenoid hydrocarbon. It is converted to squalene 2,3-epoxide, which in turn is converted to lanosterol. The conversion of squalene to the 2,3-epoxide is catalyzed by a monooxygenase, and molecular oxygen is a required component for this reaction. Yeast cells under anaerobic conditions cannot synthesize sterols because they lack oxygen.

16. a, d, e

17. b, e

18. Bile salts are effective detergents because they contain both polar and nonpolar regions. They have several hydroxyl groups, all on one side of the ring system, and a polar side chain that allow interactions with water. The ring system itself is nonpolar and can interact with lipids or other nonpolar substances. Bile salts are planar amphipathic molecules, in contrast with such detergents as sodium dodecyl sulfate (see Stryer, p. 292), which are linear.

Lipoproteins

19. (a) 2, 3, 5, 7 (b) 1, 6, 7, 9 (c) 1, 4, 8 (d) 3, 4, 10

20. All the events except (d) occur in the LDL pathway. LDL receptors are usually recycled to the cell surface after endocytosis. The proper sequence is e, b, a, and c.

21. c

22. Cholesterol released during the degradation of LDL suppresses the formation of HMG CoA reductase, the enzyme that controls the endogenous synthesis of cholesterol. In addition, the synthesis of new LDL receptors is suppressed, decreasing the uptake of exogenous cholesterol by the cell.

23. A defect in apoprotein B-100 that prevents the binding of LDL to the cell-surface receptor would result in the stimulation of the synthesis of endogenous cholesterol and LDL receptors and a decrease in the synthesis of cholesterol esters via the ACAT reaction. Indeed, the cellular and physiological consequences of such a mutation may be similar to those seen in familiar hypercholesterolemia

Synthesis of Steroid Hormones

24. (a) Cortisol
    (b) Progesterone
    (c) Six
    (d) True. A deficiency of 21-hydroxylase will result in an inability to introduce an hydroxyl group at C-21 of progesterone, which will prevent the synthesis of cortisol and mineralocorticoids from progesterone.

25. b, d, e

26. In mammals, foreign aromatic molecules are hydroxylated by the cytochrome-P_{450}–dependent monooxygenases that are present in the endoplasmic reticulum of the liver cells. The hydroxylated derivatives are more water soluble and have functional groups for the attachment of very polar substances, such as glucuronate, that allow them to be excreted in urine.

27. (a) 3, 5 (b) 1, 4 (c) 2

Vitamin D and Derivatives of Isopentenyl Pyrophosphate.

28. a

29. d

ANSWERS TO PROBLEMS

1. Citrate lyase catalyzes the formation of acetyl CoA in the cytosol (see pp. 487–488 of Stryer's text). Acetyl CoA is utilized for the synthesis of HMG CoA, which gives rise to mevalonate for the synthesis of cholesterol in the cytosol.

2. (a) A deficiency of mevalonate kinase activity means that mevalonate cannot be utilized as a precursor of 5-phosphomevalonate. If there are no other pathways where mevalonate can be used, its concentration in the liver will increase until it spills over into the blood and then into the urine. Furthermore, because the activity of HMG CoA reductase is increased in this infant, see answer (c), the rate of mevalonate synthesis will be stimulated.

   (b) You would expect the rate of cholesterol synthesis to be depressed because the pathway is blocked at the step in which 5-phosphomevalonate is formed.

   (c) You would expect to find a higher than normal level of HMG CoA reductase activity. A depressed rate of cholesterol synthesis lowers the amount of cholesterol in the cell. HMG CoA reductase activity increases because inhibition of its synthesis and its activity by cholesterol is reduced.

3. (a) The experiments show that rabbits given cholestyramine have higher rates of removal of LDL from the blood and that hepatic uptake of LDL depends on the ability of the lipoprotein to bind to the LDL receptor. One explanation for this is that cholestyramine interferes with the return of bile acids to the liver, stimulating the synthesis of more bile acids from cholesterol. An increased demand for cholesterol stimulates the synthesis of LDL receptors, which take up more cholesterol-containing LDL particles from the blood.

   (b) Although cholestyramine stimulates the hepatic uptake of cholesterol-containing LDL, it has no direct effect on de novo cholesterol synthesis in the liver. Mevinolin inhibits HMG CoA reductase, thereby depressing the rate of cholesterol biosynthesis. The subsequent requirement for cholesterol leads to a further increase in the number of LDL receptors, which in turn can take up more LDL from the circulation.

4. (a) Both chylomicron remnants and IDL normally contain apoprotein E. A deficiency of the apoprotein means that hepatic uptake of chylomicron remnants and IDL is impaired, so these particles persist in the circulation. Because both of these types of particles contain triacylglycerols and cholesterol, circulating levels of these compounds are also elevated.

(b) Both chylomicron remnants and IDL particles serve as precursors of LDL in the liver. When the uptake of the LDL precursors by hepatic tissue is impaired by an apoprotein E deficiency, the rate of synthesis and export of LDL particles is reduced.

(c) As discussed in answer (b), LDL synthesis in the liver is impaired. Additional LDL receptors are synthesized because their synthesis is no longer repressed by LDL-derived cholesterol.

(d) A diet low in cholesterol and fat will reduce the rate of formation of chylomicrons, which are precursors of chylomicron remnants and IDL.

5. Increased levels of ketone bodies, such as acetoacetate, imply that the levels of acetyl CoA and HMG CoA, both precursors of cholesterol, are also elevated. Cholesterol synthesis is stimulated by an increase in the availability of these substrates. The subsequent decreased demand for dietary cholesterol results in an elevation in cholesterol-containing lipoproteins.

6. (a) Mevalonic acid, a 6-carbon compound, is a precursor of isopentenyl pyrophosphate (IPP), which contains 5-carbon atoms. IPP serves as the basic unit for the formation of squalene, a 30-carbon compound. Six molecules of IPP are needed for the synthesis of a molecule of squalene, which is in turn the precursor of hopane. Thus, six molecules of mevalonic acid are required.

(b) Aerobic processes, such as the synthesis of sterols, probably evolved later than anaerobic processes and only after free oxygen became available. Thus, the synthesis of hopane from squalene, an anaerobic process, probably preceded the synthesis of sterols, such as lanosterol, from squalene.

7. The synthesis of mevalonate is required not only for the synthesis of cholesterol but also for the synthesis of a number of other important compounds derived from isopentenyl pyrophosphate, including ubiquinone, an important component of the electron-transport chain. Therefore, the complete blockage of mevalonate synthesis, even if adequate cholesterol is available in the diet, is not advisable.

8. Choline, which is synthesized only to a limited extent in mammals, is a constituent of phosphatidyl choline, an important component of membranes and lipoproteins. A deficiency in choline could interfere with the synthesis and export of lipoproteins like VLDL, which is a carrier of triacylglycerols to peripheral tissues. Failure to export fats such as triacylglycerols leads to their accumulation in the liver.

9. Nucleotide sugars, such as UDP-glucose, serve as donors during the incorporation of sugar residues into gangliosides. In this case, it appears that the donor of fucose residues is GDP-fucose, which is synthesized from fucose and GTP.

10. You would expect to see reduced rates of synthesis of triacylglycerols, which use diacylglycerols as acceptors of activated acyl groups. In addition, phosphatidyl choline synthesis is dependent on the availability of diacylglycerols as acceptors of choline phosphate from CDP-choline.

11. (a) You would not expect desmolase deficiency to lead to virilization. An increase in androgen production, which causes virilization in both males and females, is due to elevated levels of 17α-hydroxyprogesterone. The pathway from cholesterol to 17α-

hydroxyprogesterone includes a step catalyzed by desmolase, which removes a C_6 unit from a derivative of cholesterol to form pregnenolone. A deficiency of desmolase would therefore decrease the rate of androgen synthesis.

(b) Pregnenolone is a precursor of the glucocorticoids, which exert feedback control on the activity of the adrenal cortex. Desmolase deficiency leads to diminished production of glucocorticoids. Failure of the normal feedback mechanism leads to increased ACTH production and to enlargement of the adrenal glands.

12. Because glucosylcerebrosides accumulate but galactosylcerebrosides do not, you would suspect that the defect involves ganglioside breakdown rather than ganglioside synthesis. The defect involves the step that removes glucose from the cerebroside to yield free ceramide, or N-acyl sphingosine. The enzyme that carries out this step is a glycosyl hydrolase; it is also called B-glucosidase.

13. Cells of the adrenal cortex utilize cholesterol for the synthesis of a number of steroid hormones, including cortisol. Although these cells can themselves synthesize cholesterol, it is often also necessary for additional cholesterol to be obtained from plasma lipoproteins. A high concentration of LDL receptors enables cortical cells to take up LDL, which contains cholesterol, rapidly.

14. In the cell, a phospholipase would most often encounter substrates that are part of an aggregate, such as those phospholipids found in membranes. Thus, the enzyme should be expected to function at a higher rate with aggregates or assemblies of lipid molecules.

15. The presence of glucagon is a signal that carbohydrate and triacylglycerol catabolism is needed to generate energy in the cell. Under such conditions, one would expect biosynthetic reactions to be suppressed. The inhibition of HMG CoA reductase by glucagon suppresses cholesterol biosynthesis.

CHAPTER 24

Biosynthesis of Amino Acids and Heme

In this chapter, Stryer explains the biosynthetic origins of the amino acids. He deals first with how atmospheric nitrogen in the form of N_2 is converted to NH_4^+ by the process of nitrogen fixation. He then explains how NH_4^+ is incorporated into the covalent structures of the amino acids glutamate and glutamine. Next, a major portion of the chapter discusses how the individual carbon-atom skeletons of most of the amino acids are constructed from a few common intermediates and how nitrogen from glutamate and glutamine is incorporated into their structures, primarily in α-amino groups. He explains that the lack of some biosynthetic pathways in humans has led to the dietary requirement for nine amino acids. The roles of tetrahydrofolate and S-adenosylmethionine as carriers of single carbon atoms during metabolism is also explained. Stryer includes a general discussion of the control of metabolic pathways and uses examples from amino acid metabolism to exemplify the relevant principles.

Stryer emphasizes the important role of the amino acids as precursors of many biomolecules by listing several such derivatives. He explains in detail the synthesis of glutathione and describes its roles in the transport of amino acids, as a sulfhydryl buffer, and in the detoxification of peroxides. He concludes the chapter with a discussion of the formation and degradation of another important nitrogen containing molecule—heme. A review of the discussions of amino acid structure in Chapter 2, heme structure in Chapter 7, enzyme control in Chapter 10, and amino acid catabolism in Chapter 21 of Stryer will facilitate your understanding of this chapter.

When you have mastered this chapter, you should be able to complete the following objectives.

The Conversion of N₂ into Organic Nitrogen (Stryer pages 575–578)

1. Define *nitrogen fixation* and name the groups of organisms that can carry out this conversion. Name the molecular form of nitrogen that is used by higher organisms to synthesize biomolecules.

2. Describe the *nitrogenase complex* and explain the roles of its *reductase* and *nitrogenase* components.

3. Explain the energy requirement for nitrogen fixation and write an equation giving the *stoichiometry* of the overall reaction.

4. Outline the key roles of *glutamate* and *glutamine* in the assimilation of NH_4^+ into amino acids and describe the reactions of *glutamate dehydrogenase, glutamine synthetase,* and *glutamate synthase.* Recognize the functions of *ATP* and *NADPH* in these processes.

Nonaromatic Amino Acid Biosyntheses and One-Carbon Atom Metabolism (Stryer pages 578–584)

5. Identify the *essential* amino acids for humans and explain why they are essential.

6. Classify the amino acids into six *biosynthetic families* and identify their *seven precursors.*

7. Describe the single step biosyntheses of alanine, aspartate, asparagine, and tyrosine.

8. Outline the syntheses of glutamate, glutamine, proline, and arginine from *α-ketoglutarate.*

9. Outline the syntheses of serine, glycine, and cysteine from *3-phosphoglycerate.*

10. Explain the roles of *pyridoxal phosphate, tetrahydrofolate,* and *S-adenosylmethionine* in amino acid biosyntheses.

11. Identify the structure of tetrahydrofolate and indicate the reactive part of the molecule. Describe the sources of this cofactor in humans.

12. Draw the structures of the single-carbon groups that can be carried on tetrahydrofolate and provide examples of reactions that generate and use them.

13. Draw the structure of *S-*adenosylmethionine and describe its synthesis. Indicate the reactive part of the molecule and describe the basis of its *high methyl-group transfer potential.*

14. Outline the *activated methyl cycle* and describe the roles of *methylcobalamin* and *ATP* in the cycle.

Biosynthesis of Aromatic Amino Acids and Histidine (Stryer pages 584–588)

15. Outline the biosyntheses of phenylalanine, tyrosine, and trypto-

phan in *E. coli.* Describe the roles of *phosphoenolpyruvate, erythrose 4-phosphate,* and *phosphoribosylpyrophosphate* in these reactions.

16. Outline the synthesis of histidine from phosphoribosylpyrophosphate, ATP, and glutamine.

Control of Amino Acid Biosynthesis (Stryer pages 588–591)

17. Define the *committed step* of a metabolic pathway and recognize that it is often the target of *feedback regulation.* Recount the salient features of control by *sequential feedback, enzyme multiplicity, concerted feedback,* and *cumulative feedback,* and suggest physiological situations in which such mechanisms would be useful to a cell.

18. Illustrate the cumulative feedback control of *glutamine synthetase* from *E. coli.* Explain the mechanisms and functions of the *reversible covalent modifications* and describe the advantage of employing an *enzymatic cascade* in regulating this reaction.

Amino Acids as Versatile Precursors of Biomolecules (Stryer pages 591–597)

19. Provide examples of important biomolecules that are derived from amino acids.

20. Draw the structure of *glutathione* and describe its synthesis. Indicate the functions of glutathione and describe the involvement of *selenium* in the glutathione peroxidase reaction.

21. Describe the *γ-glutamyl cycle* and explain how it can facilitate the transport of amino acids from one cell to another.

22. Name the two molecular precursors of the *porphyrins* in mammals and outline the synthesis of *heme.* Identify the key regulatory points and mechanisms in heme biosynthesis.

23. Distinguish between *congenital erythropoietic porphyria* and *acute intermittent porphyria* in terms of their underlying molecular pathologies.

24. Outline the catabolism of heme to *bilirubin* and describe the involvement of *UDP-glucuronate* in converting bilirubin to an excretory form in humans. Explain why *biliverdin* is reduced to bilirubin in mammals.

SELF-TEST

The Conversion of N_2 into Organic Nitrogen

1. Define nitrogen fixation and explain why it is crucial to the maintenance of all life.

_____

_____

_____

2. Place the following components, reactants, and products of the nitrogenase complex reaction in their correct sequence during the e^- transfers of nitrogen fixation.

(a) Oxidized ferredoxin
(b) Reductase component
(c) Nitrogenase component
(d) NH_4^+

(e) N_2
(f) Reduced ferredoxin
(g) Electron source

3. Write the net equation for nitrogen fixation and describe the sources of the electrons and ATP.

4. Match the enzyme with the reaction it catalyzes.

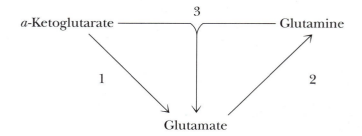

(a) Glutamine synthetase _____

(b) Glutamate dehydrogenase _____

(c) Glutamate synthase _____

5. Which of the reactions shown in Question 4 require the following?

(a) NH_4^+ _____

(b) ATP _____

(c) NADH _____

(d) NADPH _____

6. All organisms can incorporate NH_4^+ into glutamate and glutamine using glutamate dehydrogenase and glutamine synthetase. Why do procaryotes have an additional enzyme, glutamate synthase, to perform this function?

Nonaromatic Amino Acid Biosynthesis and One-Carbon Atom Metabolism

7. Propose a simple pathway to generate ^{14}C-labeled aspartate from pyruvate that has all of its carbon atoms radiolabeled.

8. Which of the following amino acids are essential dietary components for an adult human?

(a) Alanine
(b) Aspartate
(c) Histidine
(d) Tryptophan
(e) Leucine
(f) Phenylalanine
(g) Glutamine
(h) Asparagine
(i) Glutamate
(j) Threonine
(k) Methionine

9. Which of the following amino acids are derived from pyruvate?

(a) Phenylalanine
(b) Alanine
(c) Tyrosine
(d) Histidine
(e) Valine
(f) Leucine
(g) Cysteine
(h) Glycine

10. Which of the following amino acids are derived from α-ketoglutarate?

(a) Glutamate
(b) Proline
(c) Cysteine
(d) Aspartate
(e) Glutamine
(f) Arginine
(g) Ornithine
(h) Serine

11. Which of the following compounds provide the carbon skeletons of the six biosynthetic families of amino acids?

(a) Pyruvate
(b) Oxaloacetate
(c) α-Ketoglutarate
(d) Succinate
(e) 2-Deoxyribose
(f) 3-Phosphoglycerate
(g) Ribose 5-phosphate
(h) Glucose 6-phosphate
(i) Phosphoenolpyruvate
(j) Erythrose 4-phosphate
(k) α-Ketobutyrate

12. Three coenzymes are involved in carrying activated one-carbon units. Match the activated group in the right column with the appropriate coenzyme in the left column.

(a) Tetrahydrofolate _____

(b) *S*-Adenosylmethionine _____

(c) Biotin _____

(1) $-CH_3$
(2) $-CH_2-$
(3) $-CHO$
(4) $-CHNH$
(5) $-CH=$
(6) $-CO_2^-$

13. The major source of one-carbon units for the formation of the tetrahydrofolate derivative N^5,N^{10}-methylenetetrahydrofolate is the conversion of

(a) methionine to homocysteine.
(b) deoxyuridine 5′-phosphate to deoxythymidine 5′-phosphate.
(c) 3-phosphoglycerate to serine.
(d) serine to glycine.

14. Why does *S*-adenosylmethionine have a higher methyl-group transfer potential than N^5-methyltetrahydrofolate?

15. How many high-energy bonds are expended during the synthesis of S-adenosylmethionine from ATP and methionine?

_____

16. The conversion of homocysteine into methionine involves which of the following cofactors?

(a) N^5-Methyltetrahydrofolate
(b) N^5,N^{10}-Methylenetetrahydrofolate
(c) Methylcobalamin
(d) Pyridoxal phosphate

17. Fill the empty boxes in the pathways outlined in Figure 24-1 using the following compounds:

(1) S-Adenosylhomocysteine
(2) Serine
(3) N^5,N^{10}-Methylenetetrahydrofolate
(4) N^5-Methyltetrahydrofolate
(5) Cysteine
(6) S-Adenosylmethionine
(7) ATP

Figure 24-1
Some reactions of one-carbon metabolism.

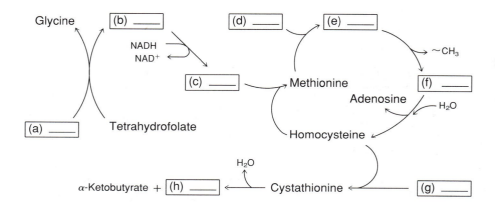

Biosynthesis of Aromatic Amino Acids and Histidine

18. Which of the following are aromatic amino acids?

(a) Leucine
(b) Phenylalanine
(c) Tryptophan
(d) Tyrosine
(e) Threonine

19. What is the ultimate source of most of the essential amino acids required by humans?

_____

20. Which of the following are intermediates in the pathway for the biosynthesis of both phenylalanine and tryptophan?

(a) Anthranilate
(b) Chorismate
(c) Shikimate
(d) Prephenate

21. Which of the following provide atoms to form histidine?

  (a) Erythrose 4-phosphate
  (b) Phosphoribosylpyrophosphate
  (c) Glutamine
  (d) Phosphoenolpyruvate
  (e) ATP

Control of Amino Acid Biosynthesis

22. In the following biosynthetic pathway

$$A \rightarrow B \rightarrow C \rightarrow D \rightarrow E \rightarrow F \rightarrow G$$

which is likely to be the committed step?

_____

23. In the biosynthetic pathway in Question 22, which compound is likely to inhibit the committed step?

_____

24. Since glutamine is an important source of nitrogen in biosynthetic reactions, the enzyme that synthesizes it is carefully regulated. Which of the following compounds act as inhibitors of glutamine synthetase in *E. coli*?

  (a) Tryptophan   (e) AMP
  (b) Histidine   (f) CTP
  (c) Carbamoyl phosphate (g) Alanine
  (d) Glucosamine 6-phosphate (h) Glycine

25. How does covalent modification contribute to the regulation of glutamine synthetase in *E. coli*?

_____

_____

_____

Amino Acids as Versatile Precursors of Biomolecules

26. Glutathione is composed of which of the following amino acids?

  (a) Glutamine
  (b) Glutamate
  (c) Methionine
  (d) Cysteine
  (e) Glycine

27. Glutathione

  (a) cycles between oxidized and reduced forms in the cell.
  (b) is involved in the detoxification of H_2O_2 and organic peroxides.
  (c) donates amide groups from its γ-glutamyl residue during biosynthetic reactions.
  (d) is involved in intercellular amino acid transport.

28. Which of the following are intermediates or precursors in the synthesis of heme?

(a) δ-Aminolevulinic acid
(b) Bilirubin
(c) Prophobilinogen
(d) Biliverdin
(e) Glycine
(f) Succinyl CoA
(g) Fe^{2+}

29. Which reaction of heme metabolism releases carbon monoxide?

_____

_____

PROBLEMS

1. The glyA⁻ mutation in Chinese hamster ovary cells in tissue culture makes these cells partially dependent on glycine. The mutation affects the mitochondrial form of serine transhydroxymethylase, which catalyzes the conversion of serine to glycine, with tetrahydrofolate serving as an acceptor of the hydroxymethyl group.

(a) Would you expect heme synthesis to be adversely affected in glyA⁻ mutants? Why?
(b) These cells appear to grow and divide much faster when choline is added to the medium. Why?

2. Hereditary coproporphyria (HCP) is a form of hereditary hepatic porphyria that is caused by a partial deficiency of coproporphyrinogen oxidase, which converts coproporphyrinogen III to protoporphyrin IX. The symptoms include abdominal pain, vomiting, and neuropsychiatric disorders.

(a) Why are patients with a total deficiency of coproporphyrinogen oxidase unlikely to be found?
(b) Which of the intermediates in heme biosynthesis would you expect to accumulate in patients with this disorder? What products would have lower than normal concentrations? Explain your answer not only in terms of the enzyme deficiency but also in terms of the regulation of heme synthesis.
(c) Why would you expect patients with HCP to be extremely sensitive to light?
(d) Why should patients with HCP not be given barbiturates or steroid drugs?

3. In early nutritional studies, cysteine was thought to be an essential amino acid. In 1937, Abraham White and E. F. Beach showed that cysteine could be removed from protein hydrolysates with cuprous mercaptide. Rats fed on such treated hydrolysates could grow, provided that sufficient methionine was supplied in the diet.

(a) What did this result reveal about cysteine metabolism?
(b) What would you expect the result to be if the rats were fed homocysteine along with the treated hydrolysates?

4. The Crigler-Najjar syndrome is characterized by the lack of glucuronyl transferase, which catalyzes the attachment of glucuronate

units to the two propionate side chains of bilirubin. Infants with this disorder have high concentrations of bilirubin and its breakdown products in their skin, and their serum concentrations of bilirubin are twenty to fifty times above normal, but no bilirubin is found in the bile of these infants. Can you explain these observations?

5. Deficiencies in pyridoxine or in pyridoxal phosphate (PLP) are relatively rare because these compounds are widespread in the normal diet. Individuals that have one of these deficiencies frequently exhibit symptoms related to pellagra, a disease often ascribed to an insufficiency of niacin or nicotinamide. In addition, there is often an accumulation of the compounds kynurenine and hydroxykynurenine in individuals with PLP deficiency. (The structure of these two compounds is shown in the margin.) Describe the relationship among pyridoxal phosphate, kynurenine, and pellagra.

6. Blood from rabbits that have been made anemic in order to produce a high percentage of reticulocytes can be used to study the biosynthesis of heme in vitro. Isotopically labeled 3-phosphoglycerate with ^{14}C at C-1, C-2, or C-3 can be prepared; the three possible structures are shown in the margin. Assuming that reticulocytes can synthesize the appropriate precursors for heme synthesis from glycolytic intermediates, which labeled form of 3-phosphoglycerate will yield a ^{14}C-labeled, nitrogen-containing compound that will be incorporated into heme with the least dilution of the isotope?

7. The essential amino acids are those that cannot be synthesized de novo in humans. Given an abundance of other amino acids in the diet, the α-keto acid analogs that correspond to the essential amino acids can substitute for these compounds in the diet.

    (a) What do these observations tell you about the steps in the synthesis of essential amino acids that may be missing in humans?

    (b) If ^{15}N-labeled alanine is supplied in the diet, many other amino acids in the body will contain at least a small amount of the label within 48 hours. What enzymes are primarily responsible for this observation?

8. In muscle, glutamine synthetase is very active, catalyzing the formation of glutamine from glutamate and ammonia at the expense of a molecule of ATP. In the liver, the rate of formation of glutamine is very low, but a high level of glutaminase activity, which generates ammonia and glutamate, is observed. Can you explain the difference in the levels of enzyme activity in these two organs?

9. Consider three forms of bacterial glutamine synthetase: GS, the deadenylylated form; GS-(AMP)$_1$, a form with one AMP unit per 12 subunits; and GS-(AMP)$_{12}$, the fully adenylated form.

    (a) Which of these forms is most sensitive to feedback inhibition by several of the final products of glutamine metabolism, such as tryptophan or histidine? Why is it important that the activity of the most sensitive form not be *completely* inhibited by tryptophan?

    (b) Which form has the lowest K_M for ammonia?

    (c) Why is it important that adenylyl transferase not carry out adenylylation and deadenylylation of glutamine synthetase at the same time?

    (d) Glutamine synthetase in mammals is not subject to the same type of complex regulation that is seen in bacteria. Why?

Tryptophan

Kynurenine

3-Hydroxykynurenine

1-^{14}C-3-Phosphoglycerate

2-^{14}C-3-Phosphoglycerate

3-^{14}C-3-Phosphoglycerate

10. For every acetyl CoA introduced into the citric acid cycle, two molecules of CO_2 are released during a complete turn of the cycle. Yet, ^{14}C-labelled acetyl CoA can be used in an in vitro system for the synthesis of radioactive heme. Why?

11. Which of the carbon atoms of glucose must be labeled with ^{14}C in order to synthesize ^{14}C-labeled N^5,N^{10}-methylenetetrahydrofolate that can in turn be used to generate ^{14}C-methyl-S-adenosylmethionine?

12. In addition to catalyzing the reduction of diatomic nitrogen to ammonia, nitrogenase carries out a number of other reactions, including the conversion of acetylene to ethylene, the conversion of cyanide to methane and ammonia, and the conversion of azide to diatomic nitrogen and ammonia.

  (a) Write balanced equations for these reactions, showing how many electrons are involved in each.
  (b) One of the requirements for nitrogenase activity is ATP. In bacterial organisms like *Rhizobium,* ATP is generated by electron transport and oxidative phosphorylation, both of which are processes that require oxygen. However, nitrogenase is extremely sensitive to oxygen. How could the presence of leghemoglobin in the bacterial membrane of *Rhizobium* resolve this paradox?

13. In many bacteria, all the enzymes of the histidine pathway are coordinately synthesized when the culture medium is deficient in histidine. Conversely, the synthesis of these enzymes is coordinately repressed when histidine is added to the medium. The genes for all ten enzymes are closely linked to each other on the bacterial chromosome. In many cases, a stop-codon mutation in the gene of one of these enzymes not only results in a defective enzyme but the mutation also reduces the rate of synthesis of those enzymes coded for by genes further along the chromosome.

  (a) What are the advantages of the coordinate synthesis and repression of the enzymes of the histidine pathway in bacteria?
  (b) Briefly explain the observations concerning stop-codon mutations.

14. Most of the proteins synthesized in mammals contain all twenty common amino acids. As stated on page 578 of the text, more protein is degraded than is synthesized when even one essential amino acid is missing from the diet.

  (a) Under such conditions, how could an increase in the rate of protein degradation provide the missing amino acid?
  (b) How does an increase in the rate of protein degradation contribute to increased levels of nitrogen excretion?

15. A pathway for the synthesis of ornithine from glutamate is shown in Figure 24-2.

  (a) Why can this pathway also be considered to be part of the de novo pathway for the synthesis of arginine?
  (b) Inspect the pathway for proline biosynthesis given at the bottom of page 579 of the text, and then explain why the N-acetylation of glutamate is required for the synthesis of ornithine.

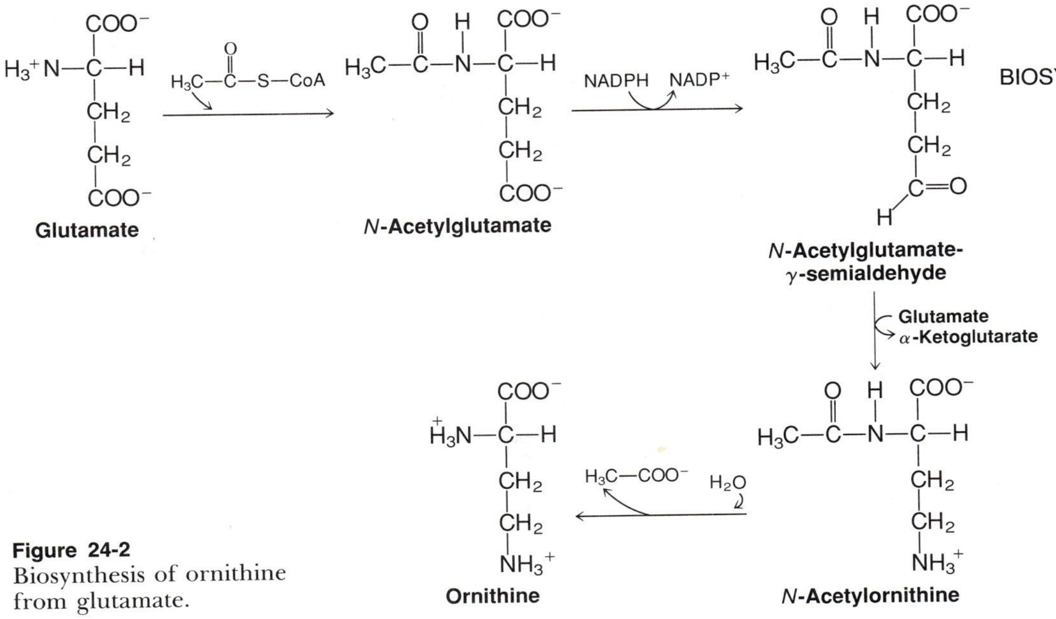

Figure 24-2
Biosynthesis of ornithine
from glutamate.

ANSWERS TO SELF-TEST

The Conversion of N_2 into Organic Nitrogen

1. Nitrogen fixation is the process by which nitrogen present in the atmosphere as N_2 is enzymatically converted to NH_4^+ by some bacteria and blue-green algae. This process is crucial to all other organisms because they can use only NH_4^+, and not N_2, as the source of nitrogen for biosynthesis.

2. g, a, f, b, c, e, d

3. The net equation for nitrogen fixation is

$$N_2 + 6\ e^- + 12\ ATP + 12\ H_2O \longrightarrow$$
$$2\ NH_4^+ + 12\ ADP + 12\ P_i + 4\ H^+$$

The 6 electrons needed to reduce each N_2 are supplied by oxidative processes in nonphotosynthetic nitrogen-fixing organisms and by light energy from the sun in photosynthetic nitrogen-fixing organisms. The ATP requirement is met by the usual oxidative or photosynthetic mechanisms of the cells.

4. (a) 2 (b) 1 (c) 3

5. (a) 1,2 (b) 2 (c) None (d) 1,3. Glutamate dehydrogenase uses NADPH when catalyzing reductive aminations and NAD^+ when carrying out oxidative deaminations.

6. Glutamate synthase catalyzes the reductive amination of α-ketoglutarate in a reaction with glutamine to form two glutamates. Glutamate can also be made from NH_4^+ and α-ketoglutarate using glutamate dehydrogenase. However, this route requires high concentrations of NH_4^+ because of the high K_M of the enzyme for NH_4^+. Procaryotes

can use glutamine synthetase, which has a low K_M for NH_4^+, to form glutamine when NH_4^+ concentrations are low. Thus, by using an additional enzyme, they can form glutamate from glutamine and α-ketoglutarate when NH_4^+ is scarce.

Nonaromatic Amino Acid Biosynthesis and One-Carbon Atom Metabolism

7. Pyruvate is converted to acetyl CoA by pyruvate dehydrogenase. Acetyl CoA condenses with oxaloacetate and passes through the TCA cycle to form labeled oxaloacetate, which undergoes a transamination reaction with glutamate to form labeled aspartate. Recall that the label in the acetyl portion of acetyl CoA is not eliminated as $^{14}CO_2$ in a single pass through the cycle. Alternatively, pyruvate carboxylase (Stryer, p. 439) forms oxaloacetate directly from pyruvate. Again, transamination with glutamate leads to labeled aspartate, but in this case, with a different pattern of labeled atoms.

8. c, d, e, f, j, k

9. b, e, f

10. a, b, e, f, and g. Recall that ornithine is a precursor of arginine in the urea cycle and is derived from α-ketoglutarate.

11. a, b, c, f, g, i, j

12. (a) 1, 2, 3, 4, 5 (b) 1 (c) 6

13. d. Furthermore, since 3-phosphoglycerate can give rise to serine, you can see how carbohydrates can provide activated one-carbon units *via* a glucose → 3-phosphoglycerate → serine → glycine pathway.

14. The positive charge on the sulfur atom of *S*-adenosylmethionine activates the methyl sulfonium bond and makes methyl-group transfer from *S*-adenosylmethionine energetically more favorable than from N^5-methyltetrahydrofolate.

15. Three high-energy bonds are expended. The adenosyl group of ATP is condensed with methionine to form a carbon to sulfur bond with the release of P_i and PP_i, which is hydrolyzed to 2 P_i.

16. a and c. Homocysteine transmethylase uses a vitamin B_{12} derived cofactor.

17. (a) 2 (b) 3 (c) 4 (d) 7 (e) 6 (f) 1 (g) 2 (h) 5

Biosynthesis of Aromatic Amino Acids and Histidine

18. b, c, d

19. The essential amino acids in the diet of humans are ultimately derived primarily from plants.

20. b, c

21. b, c, e

Control of Amino Acid Biosynthesis

22. A → B. Control of the first step conserves the first compound, A, in the sequence and also saves metabolic energy by preventing subsequent reactions in the pathway.

23. G. The end-product of a biosynthetic pathway often controls the committed step.

24. All the choices are correct. When all eight compounds are bound to the enzyme, it is almost completely inactive. The control of this enzyme is an excellent example of cumulative feedback inhibition.

25. Glutamine synthetase can be covalently modified by the attachment of an AMP to each of its twelve subunits. The more adenylylated the enzyme becomes, the more susceptible it is to feedback inhibition by the compounds listed in Question 24. Thus, covalent modification modulates the sensitivity of the enzyme to its effectors. An added level of control exists in this system; adenylyltransferase, the enzyme that adenylylates glutamine synthetase, is itself covalently modified. See pages 590–591 of Stryer for a description of how a cascade of modifications such as this can amplify the signals embodied in the changing levels of the effectors.

Amino Acids on Versatile Precursors of Biomolecules

26. b, d, e

27. a, b, d

28. a, c, e, f, g

29. Heme oxygenase uses O_2 and NADPH to open the tetrapyrrole heme structure and, in the process, releases one of the methene-bridge carbon atoms as CO. This is the first step in a pathway that leads to the formation of a diglucuronide derivative of bilirubin that is excreted in the bile.

ANSWERS TO PROBLEMS

1. (a) Glycine is an obligatory precursor of heme; in the reaction catalyzed by δ-aminolevulinate dehydrase, glycine condenses with succinyl CoA to form δ-aminolevulinate. The reduction in the concentration of glycine in the cell caused by the glyA⁻ mutation will cause a decrease in the rate of heme synthesis.

   (b) Choline cannot be synthesized as a free molecule in animal cells; however, it can be synthesized as part of a phosphatide by the successive methylation of ethanolamine, which is in phosphate linkage with diacylglycerol (see p. 549 of Stryer's text). The methyl donor for the formation of phosphatidylcholine is *S*-adenosylmethionine, which is converted to *S*-adenosylhomocysteine. Regeneration of *S*-adenosylmethionine depends on the availability of N^5-methyltetrahydrofolate, which in turn depends on the generation of one or another of the methylated forms of tetrahydrofolate. Serine transhydroxymethylase activity, which generates N^5,N^{10}-methylene tetrahydrofolate as well as glycine, is therefore required for the continued entry of methyl groups into tetrahydrofolate. A decrease in the rate of synthesis of methylated folate will in turn depress the rate of synthesis of compounds that require one or more methyl transfers. These compounds include choline, which must be added as a supplement to insure that sufficient levels of the molecule are available for optimal membrane synthesis.

2. (a) The heme molecule is required for a large number of important cellular functions, including the transport of oxygen in hemoglobin and the transport of electrons in the cytochromes. The total absence of any of the enzymes in the heme biosynthetic pathway, including coproporphyrinogen oxidase, would result in a failure to synthesize heme, a condition that is fatal.

(b) Because heme synthesis is blocked at a later step in the pathway, intermediates synthesized in the steps preceding the block, such as uroporphyrinogen III and coproporphyrinogen III, will accumulate, while those synthesized in the steps past the block will be present in low concentrations. Among these scarce substances will be heme itself, since the small quantity produced is rapidly incorporated into hemoglobin and other substances that require heme. Because the cellular concentration of heme is relatively low, both the synthesis and activity of δ-aminolevulinate synthetase are no longer under control. Heme synthesis is depressed, and the synthesis of δ-aminolevulinate and other early intermediates in the pathway will be stimulated.

(c) As noted in the answer to 2b, at least two types of porphyrin molecules accumulate as a result of the block in the pathway. These intermediates, which accumulate in the skin and teeth, strongly absorb light, resulting in extreme photosensitivity.

(d) Barbiturates and steroid drugs undergo hydroxylation in liver cells in a process mediated by cytochrome P-450, a heme-containing protein (see p. 566 of Stryer's text). Induction of cytochrome P-450 synthesis in response to an increase in drug concentrations will deplete the already low number of heme molecules available. A decrease in heme concentration will stimulate the synthesis and activity of δ-aminolevulinate synthetase, resulting in the synthesis of additional precursors of heme. This will in turn exacerbate the symptoms of HCP.

3. (a) The studies led White and Beach to conclude that methionine, which is another sulfur-containing amino acid, is a biosynthetic precursor of cysteine. Later work elucidated the roles of methionine in the active methyl cycle and in the synthesis of cysteine and confirmed their conclusion.

(b) In the active methyl cycle, S-adenosyl homocysteine is cleaved to yield adenosine and homocysteine. Although homocysteine can be converted to methionine, it can also condense with serine to form cystathionine, which is then cleaved to yield cysteine and α-ketobutyrate. Feeding rats homocysteine along with a treated hydrolysate will make supplementation with methionine unnecessary.

4. Bilirubin is an intermediate in the breakdown of heme. When it enters the circulation, it is immediately complexed with serum albumin, which serves as a carrier for relatively insoluble compounds like bilirubin and free fatty acids. In the liver, glucuronyl transferase makes bilirubin more soluble through the attachment of two glucuronate groups; bilirubin diglucuronide is then secreted into bile and eliminated. Unconjugated bilirubin cannot be secreted into bile. As the concentration

of bilirubin in the circulation increases, the capacity of serum albumin to absorb the molecule is surpassed, and the excess diffuses into the tissues. Such a condition is known as jaundice. In extreme cases, uncon-jugated bilirubin can enter nerve tissues, where it disrupts normal neu-rological function.

5. Examination of the structures of kynurenine and hydroxy-kynurenine shows they are both related to tryptophan. As mentioned on page 592 of the text, the nicotinamide ring of NAD^+ is synthesized from tryptophan. The fact that a deficiency in PLP results in an insuffi-ciency of this compound suggests that one or more of the steps in the pathway from tryptophan to nicotinamide involve PLP or one of its metabolites. The enzyme kynureninase is a PLP-dependent enzyme that catalyzes the cleavage of the side chain of 3-hydroxykynurenine to generate alanine and 3-hydroxyanthranilate:

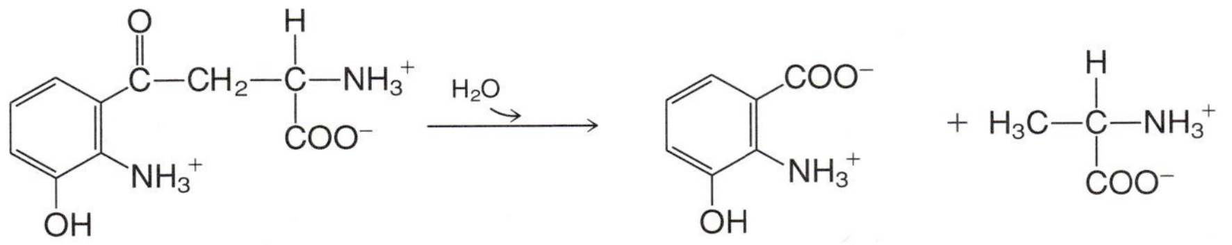

3-Hydroxykyurenine   **3-Hydroxyanthranilate**   **Alanine**

The latter compound is a necessary intermediate in NAD^+ synthesis. Failure to synthesize 3-hydroxyanthranilate leads to an insufficiency of nicotinamide, which in turn can lead to pellagra.

6. The most appropriate nitrogen-containing compound for the syn-thesis of heme is glycine, which condenses with succinyl CoA to form δ-aminolevulinate. Glycine is synthesized from 3-phosphoglycerate through a pathway that involves serine, as described on page 580 of the text. When serine is converted to glycine, the β-carbon of serine, which corresponds to C-3 in 3-phosphoglycerate, is transferred to tetrahydro-folate. The carboxyl group of glycine, which is derived from C-1 of 3-phosphoglycerate, is released as CO_2 when the amino acid condenses with succinyl CoA. The fact that only the α-carbon of glycine is incorpo-rated in δ-aminolevulinate means that you should choose $2\text{-}^{14}C\text{-}3\text{-}$phosphoglycerate as a means of labeling heme.

7. (a) The fact that α-keto acid analogs can substitute for essential amino acids means that the carbon skeletons of the essential amino acids are not synthesized in humans. Many studies have shown that one or more of the enzymes needed for the synthe-sis of these structures are missing.

   (b) The enzymes that are primarily responsible for the distribu-tion of the ^{15}N label among the other amino acids are the aminotransaminases, which catalyze the interconversions of amino acids and their corresponding α-keto acids. The redistri-bution of the label begins with the transamination of alanine, with α-ketoglutarate serving as the amino acceptor to yield py-ruvate and glutamate. Glutamate then serves as an amino donor for other α-keto acids. In order for an essential amino acid to be labeled, you must postulate the transamination of that amino acid to yield the corresponding α-keto acid analog, followed by the donation of a labeled amino group from gluta-mate or another donor of amino groups.

8. Ammonia, which is generated as part of the process of amino acid catabolism in muscle, is toxic and must be removed from the cells. This could be done through the synthesis of urea, but that process occurs only in the liver. In muscle cells, therefore, glutamine synthetase catalyzes the formation of glutamine, which is an efficient and nontoxic carrier of ammonia. This accounts for the high activity of that enzyme in muscle. The glutamine is transported by the blood to the liver, where glutaminase and aspartate aminotransferase work together to generate aspartate and two molecules of ammonia from glutamine, hence, the high activity of glutaminase in the liver. Aspartate and ammonia are both used by the liver for the synthesis of urea, a nontoxic and disposable form of ammonia.

9. (a) The fully adenylylated form of glutamine synthetase is the most sensitive to molecules like tryptophan and histidine. Because glutamine is utilized for the synthesis of a variety of compounds, complete inhibition of the enzyme by only one of those products, such as tryptophan, would inappropriately inhibit the production of all the others. Thus, the enzyme is cumulatively inhibited by at least eight different nitrogen-containing compounds.

(b) The deadenylylated form, which is not subject to cumulative feedback inhibition, is generated in response to increases in the cellular concentrations of α-ketoglutarate (the precursor of glutamate) and ATP. These molecules signal the need for glutamine synthesis, even when other nitrogen-containing compounds are present. Under these conditions, the deadenylylated form of the enzyme binds ammonia even when ammonia concentrations are relatively low; that is, it has a relatively low K_M for ammonia.

(c) The simultaneous adenylylation and deadenylylation of glutamine synthetase would result in a loss of feedback control of the enzyme because the adenylylated form is subject to cumulative inhibition whereas the deadenylylated form is not. In addition, it would lead to the wasteful hydrolysis of ATP since every round of adenylylation and deadenylylation generates AMP and inorganic pyrophosphate from ATP.

(d) Mammals acquire many nitrogen-containing compounds, such as tryptophan and histidine, in their diet rather than through de novo biosynthesis, so glutamine synthetase does not play as prominent a role in the nitrogen metabolism of mammals as it does in that of bacteria. Complex regulation of the enzyme is therefore not needed in mammals.

10. In the citric acid cycle, the two radioactive carbons of acetyl CoA combine with oxaloacetate to form citrate. As the cycle continues, two carbons are lost as carbon dioxide, but these atoms are not the two radioactive atoms that are derived directly from acetyl CoA (see page 378 of the text); the two radioactive carbons are lost in later turns of the cycle, when they appear in a different position in citrate. Thus, if radioactive succinyl CoA is generated directly from radioactive citrate, it could be incorporated immediately into δ-aminolevulinic acid, the precursor of heme. Inspection of the pathway for heme synthesis, which is outlined on pages 594–595 of the text, shows that many of the carbons in heme can indeed be labeled using radioactive acetyl CoA.

11. The methyl group of S-adenosylmethionine is derived from the methylene group of N^5,N^{10}-methylenetetrahydrofolate, which is in turn

derived from the hydroxymethyl side chain of serine. Serine is synthe-sized from glucose through the glycolytic pathway. The series of steps include the generation of 3-phosphoglyceric acid (3-PGA) from glu-cose, followed by oxidation, transamination, and phosphorolysis to yield serine (see p. 580 of the text). In order to generate serine with radioactive carbon in the hydroxymethyl side chain, $3\text{-}^{14}C\text{-}3\text{-PGA}$ is needed. Such a molecule can be generated by labeling the C-1 or C-6 of glucose, as shown in Figure 24-3.

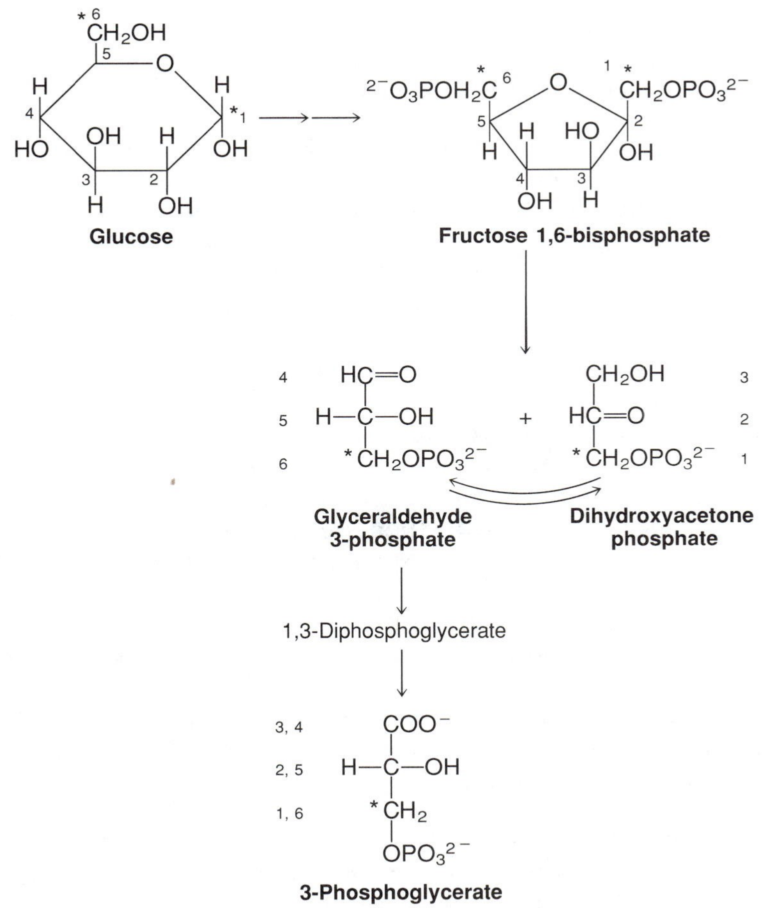

Figure 24-3
Synthesis of $3\text{-}^{14}C\text{-}3$-phosphoglycerate from glucose labeled at C-1 or C-6. Asterisks denote labeled carbon atoms, and numbers beside the three-carbon compounds correspond to the numbered carbons in glucose and in fructose 1,6-bisphosphate.

12. (a) For each of the reactions described, a different number of elec-trons is needed. The reduction of acetylene requires two elec-trons, whereas the conversion of cyanide to methane and am-monium ion requires six electrons along with eight hydrogen ions. The generation of diatomic nitrogen and ammonium ion requires two electrons and four hydrogen atoms.

$$HC{\equiv}CH + 2\,H^+ + 2\,e^- \longrightarrow H_2C{=}CH_2$$
Acetylene   **Ethylene**

$$CN^- + 8\,H^+ + 6\,e^- \longrightarrow CH_4 + NH_4^+$$
Cyanide   **Methane**   **Ammonia**

$$N_3^- + 4\,H^+ + 2\,e^- \longrightarrow N_2 + NH_4^+$$
Azide   **Nitrogen**   **Ammonia**

The role of these reactions in nitrogen-fixing organisms is not understood. It is worth noting that an agricultural field test for nitrogen fixation has been developed that takes advantage of the ability of nitrogenase to reduce acetylene.

(b) The presence of leghemoglobin in the bacterial membrane of organisms like *Rhizobium* functions in two ways. First, it binds oxygen with high affinity, so it thereby simply prevents oxygen from entering the cell and denaturing the nitrogenase molecule. Second, leghemoglobin supplies oxygen as an electron acceptor for reduced electron carriers in the bacterial membrane, allowing the ATP required for nitrogen fixation to be generated by oxidative phosphorylation.

13. (a) The ten enzymes of the histidine pathway catalyze the formation of intermediates that are needed only for the synthesis of histidine. The coordinate synthesis of all the histidine pathway enzymes insures that roughly equivalent numbers of enzyme molecules will be produced so that all the enzymes will be available as a group when the biosynthesis of histidine is required. On the other hand, if histidine is available, none of the enzymes are needed, so coordinate repression shuts off the synthesis of all the enzymes simultaneously.

(b) The effects of stop-codon mutations, coupled with the fact that all ten genes are closely linked, suggest that the genes for the enzymes of the histidine pathway are transcribed as a single message, which may be then used for protein synthesis without further processing. For such a message, the ribosomes begin translation at one end of the molecule and carry out the synthesis of a number of proteins as they proceed toward the other end. When the ribosomes encounter a stop-codon mutation, they terminate translation prematurely, and they cannot reinitiate translation further downstream. This could explain why enzymes having genes downstream from a stop-codon mutation are not synthesized.

14. (a) Many experiments have shown that under normal conditions cells continually synthesize and degrade proteins. Although both essential and nonessential amino acids are continually recycled during these processes, reutilization is not completely efficient; thus, additional amino acids are needed. In mammals, there are no reservoirs of free amino acids; the only sources of essential amino acids are dietary proteins or the proteins of the body tissues. If an essential amino acid is not available from the diet, cells appear to accelerate the hydrolysis of their own proteins, in order to generate the missing essential amino acid. How the rate of cellular proteolysis is accelerated in response to a deficiency of an essential amino acid is not understood.

(b) An increased rate of protein degradation generates a higher concentration of free amino acids. During the oxidation of those amino acids not used for synthesis of other proteins, ammonia will be produced. An elevation in ammonia concentration in the body stimulates the formation of urea, causing the level of nitrogen excretion to increase.

15. (a) Ornithine is a precursor of arginine, as part of the pathway for the synthesis of urea. Thus the pathway for the synthesis of ornithine from glutamine, along with part of the urea cycle

pathway, can together be considered as a de novo pathway for the synthesis of arginine. Arginine can in turn be used for the synthesis of urea, or it can serve instead as one of the amino acids used for polypeptide synthesis.

(b) In the pathway for proline biosynthesis, glutamic-γ-semialdehyde cyclizes with the loss of water to form Δ'-pyrroline-5-carboxylate. However, in the pathway for the formation of ornithine shown in Figure 24-2, an N-acetylated derivative of the semialdehyde molecule is formed. The N-acetyl group blocks the condensation of the amino group with the aldehyde group, thereby preventing the formation of the pyrroline ring. This allows the pathway to proceed toward the synthesis of ornithine.

CHAPTER 25

Biosynthesis of Nucleotides

In this chapter, Stryer completes his treatment of the biosyntheses of the major classes of macromolecular precursors by describing the synthesis of the purine and pyrimidine nucleotides. Besides being the precursors of RNA and DNA, these compounds serve a number of other important roles that are reviewed in the opening paragraph of the chapter. You should review the nomenclature of the nucleotides, which is given on page 72 of Stryer.

Stryer begins with the synthesis of the purine nucleotide, inosine 5'-monophosphate (IMP). He lists its precursors and describes the synthesis of the activated form of ribose 5-phosphate, 5-phosphoribosyl-1-pyrophosphate (PRPP). He then describes the steps in the construction of the purine ring on ribose 5-phosphate and the conversion of IMP, the initial product, to AMP or GMP. Next, Stryer describes the synthesis of the free pyrimidine orotate and its attachment to ribose 5-phosphate through reaction with PRPP to form the pyrimidine nucleotide. The reactions leading to the uracil and cytosine nucleotide derivatives are then given, followed by the reactions that interconvert the mono-, di-, and triphosphate derivatives of the nucleotides. The regulation of purine and pyrimidine synthesis is also covered.

Two reactions that are required to form the precursors of DNA are described in detail: ribonucleotide reductase converts ribonucleotides to deoxyribonucleotides, and thymidylate synthase methylates dUMP to form dTMP. Stryer presents the mechanisms and cofactors of these enzymes and explains how two important anti-cancer drugs inhibit the synthesis of dTMP and thus the growth of cancer cells. Nucleotides also serve important roles as constituents of NAD^+, $NADP^+$, FAD, and coenzyme A (CoA), so the syntheses of these cofactors are described. Stryer concludes the chapter with an explanation of how the purines and pyrimidines are catabolized and a discussion of two pathological conditions that arise from defects in the catabolic pathway of the purines.

When you have mastered this chapter, you should be able to complete the following objectives.

Introduction (Stryer pages 601–602)

1. List the major biochemical roles of the *nucleotides*.

2. Distinguish among the *purine* and *pyrimidine nucleosides* and *nucleotides*.

Biosynthesis of the Purine Nucleotides (Stryer pages 602–607)

3. List the precursors that provide the carbon and nitrogen atoms of the purine ring. Note the numbering of the ring atoms.

4. Describe the substrates and products of the *ribose phosphate pyrophosphokinase (PRPP synthetase)* reaction and explain the importance of *5-phosphoribosyl-1-pyrophosphate (PRPP)* in purine as well as amino acid metabolism.

5. Outline the synthesis of the purine base *hypoxanthine* as it occurs on ribose 5-phosphate to form *inosine 5'-monophosphate (IMP)*, noting the sources of the atoms and the cofactors involved. Note the steps that require ATP.

6. Describe the synthesis of *adenylate (AMP)* and *guanylate (GMP)* from IMP. List the cofactors and intermediates of the reactions.

7. Explain the mechanism for replacing a carbonyl oxygen with an amino group, and list the potential sources of the nitrogen atom in these reactions.

8. Describe the synthesis of purine nucleotides by the *salvage reactions*.

9. Outline the regulation of the biosynthesis of the purine nucleotides and name the committed step in the pathway.

Biosynthesis of the Pyrimidine Nucleotides (Stryer pages 607–609)

10. List the precursors that provide the carbon and nitrogen atoms of the pyrimidine ring. Note the numbering of the ring atoms.

11. Explain the role of *carbamoyl phosphate* in pyrimidine biosynthesis, and compare the two *carbamoyl phosphate synthetase* reactions that occur in eucaryotic cells.

12. Describe the *aspartate transcarbamoylase* reaction and outline the remaining reactions that form *orotate*. Outline the conversion of orotate to *uridylate (UMP)*.

13. Contrast the enzymes of pyrimidine biosynthesis in *E. coli* with those of higher organisms and list the potential advantages of *multifunctional enzymes*.

14. Explain how nucleoside *mono-* and *diphosphate kinases* can interconvert the nucleoside mono-, di-, and triphosphates.

15. Describe the reaction that converts UTP to *CTP* and note the different nitrogen atom sources used in bacteria and mammals.

16. Outline the regulation of pyrimidine biosynthesis and name the committed step. Note the steps that require ATP.

Formation of the Deoxyribonucleotides (Stryer pages 610–616)

17. Describe the *ribonucleotide reductase* reaction and outline its mechanism and regulation. Compare the reaction in *E. coli* and mammals with that in *Lactobacillus leichmanni* and note the *radical* mechanism used in all cases.

18. Describe the *thymidylate synthase* reaction. Account for the source of the methyl group and describe the change in the oxidation state of the transferred carbon atom that occurs during the reaction.

19. Relate the reactions of ribonucleotide reductase and thymidylate synthase to the hypothesis that RNA was the primordial informational molecule.

20. Explain the role of *dihydrofolate reductase* in the synthesis of deoxythymidylate.

21. Account for the ability of *fluorouracil* to act as a *suicide inhibitor*. Describe the inhibitory mechanism of *methotrexate* and *aminopterin*. Explain how these three compounds interfere with the growth of cancer cells. List the mechanisms by which a cell could become resistant to methotrexate.

Biosynthesis of NAD⁺, FAD, and CoA (Stryer pages 617–618)

22. Describe the synthesis of *NAD⁺* and account for the sources of the *nicotinamide* portion of the molecule.

23. Outline the syntheses of *FAD* and *CoA* and describe the sources of the *riboflavin* and *pantothenate* precursors.

Degradation of the Purines and Pyrimidines (Stryer pages 618–623)

24. Describe the reactions of the *nucleotidases* and *nucleoside phosphorylases.*

25. Outline the conversions of AMP and *guanine* to *uric acid.* Describe the role of *xanthine oxidase* in these processes.

26. Recognize that *allantoin, allantoate, urea,* and NH_4^+ are excretion products of the degradation of purines in organisms other than primates.

27. Name the excretory form of the nitrogen atoms of *thymine.*

28. Describe the major clinical findings in patients with *gout*, and explain the rationale for the use of *allopurinol* to alleviate the symptoms of the disease. Relate the functioning of *hypoxanthine-guanine phosphoribosyl transferase (HGPRT)* to the rate of de novo purine biosynthesis in these patients.

29. Give the biochemical lesion that leads to the *Lesch-Nyhan syndrome*, and describe the symptoms of the disease.

Introduction

1. Describe the physiological roles of the nucleotides.

_____

_____

_____

2. Cytosine is a

   (a) purine base.
   (b) pyrimidine base.
   (c) purine nucleoside.
   (d) pyrimidine nucleoside.

3. Guanosine is a

   (a) purine base.
   (b) pyrimidine base.
   (c) purine nucleoside.
   (d) pyrimidine nucleoside.

4. Which of the following are nucleotides?

   (a) Deoxyadenosine
   (b) Cytidine
   (c) Deoxyguanylate
   (d) Uridylate

Biosynthesis of the Purine Nucleotides

5. Which of the following compounds directly provide atoms to form the purine ring?

   (a) Aspartate
   (b) Carbamoyl phosphate
   (c) Glutamine
   (d) Glycine
   (e) CO_2
   (f) N^5,N^{10}-Methylenetetrahydrofolate
   (g) N^{10}-Formyltetrahydrofolate
   (h) NH_4^+

6. Which of the following statements about 5-phosphoribosyl-1-pyrophosphate (PRPP) are _true_?

   (a) It is an activated form of ribose 5-phosphate.
   (b) It is formed from ribose 1-phosphate and ATP.
   (c) It has a pyrophosphate group attached to the C-1 atom of ribose in the α configuration.
   (d) It is formed in a reaction in which PP_i is released.

7. The first product of purine nucleotide biosynthesis that contains a complete purine ring (hypoxanthine) is

   (a) AMP.
   (b) GMP.
   (c) IMP.
   (d) xanthylate (XMP).

8. The conversion of IMP to AMP requires which of the following?

   (a) ATP
   (b) GTP
   (c) Aspartate
   (d) Glutamine
   (e) NAD^+

9. The conversion of IMP to GMP requires which of the following?

   (a) ATP
   (b) GTP
   (c) Aspartate
   (d) Glutamine
   (e) NAD^+

10. Describe the general mechanism used by cells to replace a carbonyl group with an amino group.

   _____

   _____

   _____

11. Which of the following reactants and products are involved in the salvage reactions of purine biosynthesis?

   (a) IMP → AMP
   (b) IMP → GMP
   (c) Adenine → AMP
   (d) Guanine → GMP

12. During a purine salvage reaction, what is the source of the energy required to form the C—N glycosidic bond between the base and ribose?

   _____

13. Show which of the nucleotides in the right-hand column regulate each of the conversions in the left column.

   (a) Ribose 5-phosphate → PRPP _____   (1) AMP
                                                (2) GMP
   (b) PRPP → phosphoribosylamine _____   (3) IMP

   (c) Phosphoribosylamine → IMP _____

   (d) IMP → adenylosuccinate _____

   (e) IMP → xanthylate (XMP) _____

Biosynthesis of the Pyrimidine Nucleotides

14. Which of the following statements about the carbamoyl phosphate synthetase of mammals, which is used for pyrimidine biosynthesis, are *true*?

   (a) It is located in the mitochondria.
   (b) It is located in the cytosol.
   (c) It uses NH_4^+ as a nitrogen source.
   (d) It uses glutamine as a nitrogen source.
   (e) It requires *N*-acetylglutamate as a positive effector.

15. What is the committed step in pyrimidine biosynthesis?

_____

_____

16. Would orotidylate 5'-monophosphate (OMP) or UMP move more rapidly toward the anode during electrophoresis at pH 8? Explain.

_____

_____

17. How is orotate, a free pyrimidine, converted into a nucleotide?

_____

_____

18. Why might covalently linked enzymes, such as those of the pyrimidine biosynthetic pathway of mammals, be advantageous to an organism?

_____

_____

_____

19. Which of the following enzymes are involved in converting the nucleoside 5'-monophosphate (NMP) products of the purine or pyrimidine biosynthetic pathways into their 5'-triphosphate (NTP) derivatives?

   (a) Purine nucleotidase
   (b) Nucleoside diphosphate kinase
   (c) Nucleoside monophosphate kinases
   (d) Nucleoside phosphorylase

20. How is the exocyclic amino group on the N-4 position of cytosine formed?

_____

_____

_____

_____

21. Which of the following compounds are involved in the control of the pyrimidine biosynthetic pathway?

   (a) UTP
   (b) UMP
   (c) CTP
   (d) CMP

Formation of the Deoxyribonucleotides

22. Which of the following statements about ribonucleotide reductase are _true_?

   (a) It converts ribonucleoside diphosphates into 2'-deoxyribonucleoside diphosphates in humans.

(b) It catalyzes the homolytic cleavage of a bond.

(c) It accepts electrons directly from $FADH_2$.

(d) It receives electrons directly from either thioredoxin or glutaredoxin.

(e) It contains two kinds of allosteric regulatory sites; one for control of overall activity and another for control of substrate specificity.

23. Select from the following those compounds that are precursors of 2'-deoxythmidine-5-triphosphate (dTTP) in mammals and place them in their correct biosynthetic order.

(a) OMP   (f) dUDP

(b) UMP   (g) dUTP

(c) UDP   (h) dTMP

(d) UTP   (i) dTDP

(e) dUMP (j) dTTP

_____

24. Define *suicide inhibitor* and give an example from pyrimidine biosynthesis.

_____

_____

_____

25. Methotrexate, a folate antagonist, interferes with nucleic acid biosynthesis in bacteria. Would you expect it to inhibit purine or pyrimidine biosynthesis? Explain.

_____

_____

_____

Biosynthesis of NAD⁺, FAD, and CoA

26. Which of the following can serve as precursors of NAD^+ in humans?

(a) Riboflavin

(b) Pantothenate

(c) Tyrosine

(d) Tryptophan

(e) Niacin

Degradation of the Purines and Pyrimidines

27. Which of the following compounds would give rise to urate if they were catabolized completely in humans?

(a) ADP-glucose   (e) CoA

(b) GDP-mannose   (f) FAD

(c) CDP-choline   (g) UMP

(d) UDP-galactose

28. Which of the following statements about patients with gout are *true*?

(a) They have elevated levels of serum urate.
(b) They have elevated levels of serum PRPP.
(c) They sometimes have a partial deficiency of hypoxanthine-guanine phosphoribosyl transferase.
(d) They sometimes have abnormally high levels of PRPP synthetase.
(e) They can be treated with allopurinol to decrease de novo purine biosynthesis.

29. Explain why allopurinol is not effective in decreasing de novo purine biosynthesis in patients with Lesch-Nyhan syndrome when it does so in many patients with gout?

PROBLEMS

1. In the early 1950s, Rose and Schweigert carried out experiments that involved the injection of radioactive nucleosides into animals. They found that when cytidine with tritium-labeled cytosine and ^{14}C-labeled ribose was injected, the deoxyribonucleotide later appeared with the same ratio of specific activities in DNA isolated from the tissues of the injected animals. Why was this an important clue to biochemists studying the biosynthesis of deoxyribonucleotides?

2. Mammalian lymphocytes that lack adenosine deaminase (see Problem 12, p. 626 of Stryer's text) neither grow nor divide. The level of dATP in these cells is 100 times higher than that in normal lymphocytes, and the synthesis of DNA in the cells is impaired.

(a) How is adenosine converted to dATP? Assume that the first step is catalyzed by a specific nucleoside kinase.
(b) How does the elevation in dATP concentration in the abnormal lymphocytes affect the synthesis of DNA?

3. Clinicians who use F-dUMP and methotrexate *together* in cancer treatment find that the combined effects on cancer cells are not synergistic. Suggest why the administration of methotrexate could interfere with the action of F-dUMP.

4. Elevated levels of ammonia in the blood can be caused by a deficiency of mitochondrial carbamoyl phosphate synthetase or a deficiency of any of the urea cycle enzymes. These two types of disorders can be distinguished by the presence of orotic acid or related metabolites in the urine.

(a) Why is it possible to determine the basis of hyperammonemia in this way?
(b) Why would a deficiency of cytoplasmic carbamoyl phosphate synthetase not cause hyperammonemia? What problems would such an enzyme deficiency cause? How would you treat a pa-

tient who has a deficiency in cytoplasmic carbamoyl phosphate synthetase?

5. The degradation of thymine yields β-aminoisobutyrate, as shown on page 620 of Stryer's text. What cofactors are needed for the conversion of β-aminoisobutyrate to succinyl CoA?

6. In order to study the enzymes of the salvage reactions for nucleotide synthesis in *E. coli,* you plan to grow a large batch of cells in a liquid medium. You have a choice between a simple glucose-salts medium and a medium that includes a hydrolyzed extract prepared from yeast cells. Which of these media should you choose? Why?

7. You wish to prepare ^{14}C-labeled purines by growing bacteria in a medium containing a suitably labeled precursor. The only precursors available are amino acids that are all uniformly labeled to the same specific activity per carbon atom. Which of the amino acids would you use to obtain purine rings that are labeled to the highest specific activity?

8. A deficiency of glucose 6-phosphatase in the liver is responsible for the clinical disorder known as von Gierke's syndrome, as is discussed on page 465 of Stryer's text. Among the clinical findings in victims of this disease is hyperuricemia, an elevation of urate levels in the blood. Several explanations for this particular finding have been offered.

   (a) Lactate interferes with the secretion of urate by the kidney tubules. Why are lactate concentrations often elevated in patients with von Gierke's syndrome, and how could interference with the tubular secretion of urate cause hyperuricemia?
   (b) The incorporation of 1-^{14}C-labeled glycine into urate is higher in patients with von Gierke's disease than in normal individuals. This indicates that the rate of synthesis of purines is elevated in these patients. Can you suggest why?

9. One of the drugs used for cancer chemotherapy is 6-mercaptopurine, which must be converted to a nucleotide before it can have any effect on the rate of purine synthesis.

   (a) How can 6-mercaptopurine be converted to a nucleotide, and what would you call the product?
   (b) Suggest two ways in which the nucleotide derived from 6-mercaptopurine can inhibit de novo purine synthesis.

10. Hydroxyurea, a potent chelator of ferric ions, has been shown to interfere with DNA synthesis, and it is used as an antitumor agent. What is the target enzyme for hydroxyurea?

11. Infants with hereditary orotic aciduria type I are deficient in the enzymes orotate phosphoribosyltransferase and orotidylate decarboxylase. Among the clinical findings in these individuals is developmental retardation, anemia, and orotic acid in the urine. Treatment includes the oral administration of a source of pyrimidine nucleotides.

   (a) Would you choose cytosine, uridine, or UMP as a source of pyrimidine nucleotides for affected infants? Why?
   (b) Red blood cells from patients with orotic aciduria type I exhibit increased activities of aspartate transcarbamoylase and dihydroorotase. These activities return to a more normal level when a source of pyrimidine nucleotides is provided. Can you explain these findings?

6-Mercaptopurine

Hydroxyurea

12. Nucleoside phosphorylases catalyze the interconversion of bases and nucleosides through following reactions:

$$\text{Ribose 1-phosphate} + \text{base} \rightleftharpoons \text{ribonucleoside} + P_i$$

or

$$\text{Deoxyribose 1-phosphate} + \text{base} \rightleftharpoons \text{deoxyribonucleoside} + P_i$$

The equilibrium constant for each of these reactions is close to 1.

(a) The pathway for the incorporation of radioactive thymine into bacterial DNA includes a step catalyzed by nucleoside phosphorylase. It has often been observed that the incorporation of thymine into DNA is enhanced when deoxyadenosine or deoxyguanosine is added to the medium. Can you explain this observation? Why might deoxyguanosine be preferable to deoxyadenosine?

(b) In cells that cannot carry out de novo synthesis of IMP, inosine can be utilized to produce IMP but only through an indirect salvage route because of the absence of inosine kinase. Suggest an alternative pathway for the formation of IMP from inosine. Among the enzymes you will need are nucleoside phosphorylase and phosphoribomutase, which isomerizes ribose 1-phosphate to ribose 5-phosphate.

13. In contrast to the normal bacterial and mammalian enzymes, the ribonucleotide reductase coded for by bacteriophage T4 is *activated* by dATP. Why?

14. (a) Write out the steps for the formation of glycinamide ribonucleotide from glycine and 5-phosphoribosyl-1-amine, and show how ATP participates.

(b) Suppose that you incubate glycinamide ribonucleotide synthase with glycinamide ribonucleotide, ADP, and inorganic phosphate labeled with ^{18}O. Will you recover any ^{18}O-labeled glycine? Why?

15. Many multivitamin preparations contain nicotinamide. Most mammalian cells contain cytosolic enzymes that convert nicotinamide directly to NAD^+. What other substrates are required for the formation of NAD^+ from nicotinamide? How could PRPP and ATP be used as sources of those substrates?

ANSWERS TO SELF-TEST

Introduction

1. The nucleotides (1) are the activated precursors of DNA and RNA; (2) are the source of derivatives that are activated intermediates in many biosyntheses; (3) include ATP, the universal currency of energy in biological systems, and GTP, which powers many movements of macro-molecules; (4) include the adenine nucleotides, which are components of the major coenzymes NAD^+, FAD, and CoA; and (5) serve as metabolic regulators.

2. b

3. c. Nucleosides consist of a purine or pyrimidine base attached to a ribose or 2′-deoxyribose.

4. c and d. Nucleotides are nucleosides that contain one or more phosphate-group substituents on their ribose or deoxyribose.

Biosynthesis of the Purine Nucleotides

5. a, c, d, e, f, g

6. a, c

7. c

8. b, c

9. a, d, e

10. The carbonyl oxygen is converted into a form that can be readily displaced by an amino group through a reaction with a high-energy phosphate (ATP or GTP) to form a mono- or diphosphate ester. The phosphate or pyrophosphate group can then be displaced by the nucleophilic attack of the nitrogen atom from NH_3, the side-chain amide group of glutamine, or the α-amino group of aspartate. The resulting adduct is an amino group or can be converted into one.

11. c, d

12. The activated form of ribose 5-phosphate (R5P), PRPP, reacts with the purine base to form the nucleotide and release PP_i. The displacement and subsequent hydrolysis of PP_i drives the formation of the N-glycosyl bond. ATP ultimately provides the energy through its reaction with R5P to form PRPP.

13. (a) 1, 2, 3 (b) 1, 2, 3 (c) None (d) 1 (e) 2

Biosynthesis of the Pyrimidine Nucleotides

14. b and d. The mitochondrial carbamoyl phosphate synthetase used for urea synthesis is activated by N-acetylglutamate and uses NH_4^+ as the nitrogen source.

15. The formation of N-carbamoylaspartate by the aspartate transcarbamoylase (ATCase) reaction is the committed step in pyrimidine biosynthesis. See pages 234–239 of Stryer for a detailed description of the regulation.

16. Since OMP contains a carboxyl substituent on the pyrimidine ring that is lacking in UMP, it has an extra negative charge at pH 8, so it will move toward the anode more rapidly.

17. Orotate condenses with PRPP in a reaction catalyzed by orotate phosphoribosyl transferase to form the nucleotide orotidylate (OMP). Orotidylate decarboxylase converts OMP to the more abundant nucleotide UMP.

18. The clustering of two or more enzymes (or active sites) on a single polypeptide chain ensures that their synthesis is coordinated and helps assure that they will assemble into a coherent complex. Also, the proximity of the active sites means that side reactions are minimized as substrates are channeled from one of the active sites to another. Finally, a multifunctional complex with covalently linked active sites is likely to be more stable than a complex formed by noncovalent interactions.

19. b and c. Several different specific nucleoside monophosphate kinases phosphorylate dNMPs and NMPs, using ATP as the phosphoryl donor. A single enzyme, nucleoside diphosphate kinase, uses the phosphorylation potential of ATP to convert the dNDPs and NDPs to dNTPs and NTPs. The ubiquitous adenylate nucleotides are interconverted by adenylate kinase (myokinase).

20. CTP is formed by amination of UTP. The carbonyl oxygen at C-4 of UTP is replaced with an amino group *via* the formation of an enol phosphate ester intermediate (see the answer to Question 10). In *E. coli*, NH_4^+ serves as the source of the nitrogen atom that displaces the phosphate group, whereas the amide group of glutamine serves this purpose in mammals.

21. b and c. ATCase is regulated through feedback inhibition by CTP, the end-product of the pyrimidine pathway. See the answer to Question 15. Carbamoyl phosphate synthetase, which forms a substrate for ATCase, is inhibited by UMP.

Formation of the Deoxyribonucleotides

22. a, b, d, and e. NADPH provides electrons *via* thioredoxin or glutaredoxin. See page 613 of Stryer.

23. a, b, c, f, g, e, h, i, and j. In mammals, NDPs are converted to dNDPs by ribonucleotide reductase. Thus, UDP is converted to dUDP, which is converted to dUTP by nucleoside diphosphate kinase. A specific pyrophosphatase hydrolyzes dUTP to dUMP, which is then converted to dTMP by thymidylate synthase. The dTMP is converted to dTTP. A *priori,* you might have expected the dUDP product of the ribonucleotide reductase reaction to be converted directly to dUMP, but, in fact, cells contain a dUTP pyrophosphatase to prevent dUTP from serving as a DNA precursor, and it is this enzyme that functions in the dTTP biosynthetic pathway.

24. A suicide inhibitor is a substrate that is converted by an enzyme into a substance that is capable of reacting with and inactivating the enzyme. In pyrimidine biosynthesis, thymidylate synthase converts fluorouracil into a derivative that becomes covalently attached to the enzyme and thereby inactivates it.

25. Methotrexate and aminopterin, a similar compound, are analogs of dihydrofolate (DHF) and inhibitors of dihydrofolate reductase, an enzyme that converts DHF to tetrahydrofolate (THF). The thymidylate synthase reaction converts N^5, N^{10}-methylenetetrahydrofolate to DHF in the process of methylating dUMP to form dTMP. In the presence of one of the inhibitors, this reaction functions as a sink that reduces the THF level of the cell by converting THF to DHF. Since THF derivatives are substrates in two reactions of purine metabolism and one of pyrimidine metabolism, both pathways are affected by the inhibitor.

Biosynthesis of NAD⁺, FAD, and CoA

26. d and e. Niacin becomes a dietary requirement if the supply of the essential amino acid tryptophan is inadequate.

Degradation of the Purines and Pyrimidines

27. a, b, e, and f. Each of these compounds contains a heterocyclic purine base.

28. a, c, d, and e. Patients with gout have elevated serum urate levels that arise from increased intracellular concentrations of PRPP.

29. Allopurinol is converted to a ribonucleotide by hypoxanthine-guanine phosphoribosyl transferase (HGPRT). Allopurinol ribonucleotide inhibits the conversion of PRPP to phosphoribosylamine and thus inhibits de novo purine biosynthesis. Since patients with Lesch-Nyhan syndrome are deficient in HGPRT, they cannot form the inhibitor ribonucleotide, so de novo purine biosynthesis is not diminished.

ANSWERS TO PROBLEMS

1. The experiments showed that ribonucleosides are converted *intact* to deoxyribonucleotides. Had cleavage occurred to produce the free sugar and base, the ribose and cytosine would have been incorporated into pools of different sizes. The subsequently synthesized deoxyribonucleotides would then contain deoxyribose and cytosine with a different ratio of specific activities. These results also suggested that free ribose is not converted to deoxyribose before deoxyribonucleotides are synthesized. As a result of these experiments, biochemists began to search for enzymatic activities that convert intact ribonucleotides to deoxyribonucleotides. Their work led to the discovery of ribonucleotide reductase, which reduces ribonucleoside diphosphates to deoxynucleoside diphosphates.

2. (a) Adenosine is phosphorylated to AMP, with ATP serving as the phosphoryl donor, in a reaction carried out by a specific nucleoside kinase. The conversion of AMP to ADP through the action of a specific nucleoside monophosphate kinase is accomplished as ATP is again utilized as a phosphate donor. Ribonucleotide reductase catalyzes the reduction of ADP to dADP, which is then converted to dATP by nucleotide diphosphokinase.

   (b) High concentrations of dATP displace ATP from the overall activity site on ribonucleotide reductase, which lowers the rate of synthesis of all four deoxyribonucleoside diphosphates. This in turn leads to a depletion of deoxyribonucleoside triphosphates, which are the substrates for DNA synthesis.

3. Methotrexate blocks the regeneration of tetrahydrofolate from dihydrofolate, which is produced during the synthesis of thymidylate. The failure to regenerate tetrahydrofolate means that those biochemical reactions in the cell that depend on one-carbon metabolism cannot be carried out. One of the products of tetrahydrofolate is methylenetetrahydrofolate, which is used as a substrate by thymidylate synthetase and is required for inhibition of the enzyme by F-dUMP. A deficiency of methylenetetrahydrofolate means that F-dUMP cannot irreversibly inactivate thymidylate synthetase. Conversely, F-dUMP prevents the formation of dihydrofolate, thereby abolishing the adverse effects caused by the depletion of tetrahydrofolate in the cell.

4. (a) The presence of orotic acid, a precursor of pyrimidines, in the urine suggests that carbamoyl phosphate synthesized in mitochondria is not utilized there. Instead, carbamoyl phosphate

enters the cytosol, where it stimulates an increase in the rate of synthesis of precursors of pyrimidines, including orotic acid. An excess of carbamoyl phosphate arises in mitochondria whenever any of the urea cycle enzymes are deficient. Such a condition will lead to hyperammonemia, as well as to the accumulation of carbamoyl phosphate. Although a deficiency in mitochondrial carbamoyl phosphate synthetase leads to hyperammonemia, it cannot lead to an accumulation of mitochondrial carbamoyl phosphate and therefore does not stimulate pyrimidine synthesis in the cytosol.

(b) Cytoplasmic carbamoyl phosphate synthetase is involved primarily in the pathway for pyrimidine synthesis, not for the assimilation of ammonia. Recall that, in the cytosol, the substrate for the formation of carbamoyl phosphate is glutamine, not ammonia. A deficiency of carbamoyl phosphate synthesis in the cytosol would cause a depletion of pyrimidines. Such a deficiency is treated by administration of uracil or uridine, which are precursors of UMP and CMP.

5. The transamination of β-aminoisobutyrate to form methylmalonate semialdehyde requires pyridoxal phosphate as a cofactor. This reaction is similar to the conversion of ornithine to glutamate γ-semialdehyde. Then NAD^+ serves as an electron acceptor for the oxidation of methylmalonate semialdehyde to methylmalonate. The conversion of methylmalonate to methylmalonyl CoA requires coenzyme A. The final reaction, in which methylmalonyl CoA is converted to succinyl CoA, is catalyzed by methylmalonyl CoA mutase, an enzyme that contains a derivative of vitamin B_{12} as its coenzyme.

6. You should choose the medium that includes the extract prepared from yeast cells because it contains nucleotides, nucleosides, and purine and pyrimidine bases. All these compounds can be utilized by the enzymes of the bacterial salvage pathways to synthesize the nucleotides needed for growth and division, thereby making the de novo synthesis of purines and pyrimidines less necessary. Bacteria growing in a simple glucose-salts medium are required to carry out de novo synthesis of the nucleotides; in these bacteria the activities of the enzymes of the salvage reactions would be lower because, in rapidly growing cells, the concentrations of the breakdown products of nucleotide metabolism would be lower.

7. Examination of the pathway for purine synthesis shows that only glycine is incorporated intact into the purine ring at the C-4 and C-5 positions. Therefore, glycine is a good choice as the radiolabeled precursor. Serine can also be considered because it is a precursor of glycine and the ultimate donor of C-1 groups to tetrahydrofolate, and activated tetrahydrofolate derivatives participate in two reactions in the formation of purines. Whether serine is a better choice than glycine depends upon the relative amounts of the two unlabeled amino acids in the cell.

8. (a) A deficiency of glucose 6-phosphatase in the liver leads to increased concentrations of glucose 6-phosphate in that organ. This causes an increase in the rate of glycolysis, one result of which is the elevation of both lactate and pyruvate concentrations. Because lactate interferes with the tubular secretion of urate into the urine, increased levels of urate are found in the blood.

(b) Among the alternative pathways that the excess glucose 6-phosphate in the liver can enter is the pentose phosphate pathway, and among the products of that pathway is ribose 5-phosphate. One explanation for the increased rate of purine synthesis is that an elevation in the level of ribose 5-phosphate results in an increase in the synthesis of PRPP. Since PRPP is a precursor of the purines, their synthesis is also stimulated. This in turn leads to higher concentrations of urate, which is a breakdown product of the purine nucleotides.

9. (a) The structure of 6-mercaptopurine is very similar to that of hypoxanthine, which suggests that the drug can undergo condensation with PRPP to form 6-mercaptopurineriboside-5′-monophosphate. The reaction is catalyzed by hypoxanthine-guanine phosphoribosyltransferase.

   (b) Because it is an analog of IMP, 6-purineriboside-5′-monophosphate inhibits the normal pathways for the conversion of IMP to AMP and to GMP. In addition, high concentrations of the analog could inhibit the formation of PRPP and of 5′-phosphoribosylamine, the first two steps of purine synthesis. These two steps are subject to feedback inhibition by IMP.

10. Hydroxyurea inhibits ribonucleotide reductase. By sequestering ferric ions, hydroxyurea destabilizes the organic free radical in the B2 subunit of the enzyme. The inhibition of enzyme activity leads to a depletion of deoxyribonucleoside diphosphates, which are normally converted to deoxynucleoside triphosphates, the substrates for DNA synthesis.

11. (a) You should choose uridine, an uncharged molecule, that can pass across plasma membranes into the cytoplasm, where it can be converted to UMP through the action of a nucleoside kinase. The synthesis of CMP is dependent upon the availability of UMP. While cytosine can be used to generate CMP in the cell, it cannot be used to generate UMP. The administration of UMP is not recommended because its phosphate group is charged at neutral pH, making its transport across the plasma membrane difficult.

   (b) The observations suggest that pyrimidine nucleotides control either enzyme synthesis or catalytic activities of preexisting enzymes. Recall that carbamoyl phosphate synthetase, aspartate transcarbamoylase, and dihydroorotase are covalently joined on a single polypeptide chain, which in turn suggests that the activities are coded for by a single gene. Regulation of their synthesis by UTP offers a convenient way to regulate pyrimidine synthesis, as does allosteric control by UTP or another product of the pyrimidine pathway. Although it is known that UTP inhibits carbamoyl phosphate synthetase activity, more work needs to be done to establish the molecular basis for regulation of the pyrimidine biosynthetic pathway.

12. (a) Deoxyribonucleosides such as deoxyadenosine can be converted to the free base and deoxyribose 1-phosphate by nucleoside phosphorylase. Increased levels of deoxyribose 1-phosphate are then available for the formation of deoxythymidine from thymine in the reverse reaction catalyzed by nucleoside phosphorylase. Deoxyguanosine might be preferable to deoxy-

adenosine because the converison of elevated levels of deoxya-denosine to dAMP and then to dATP could lead to the inactivation of ribonucleotide reductase, which is sensitive to the concentration of dATP.

(b) Inosine is cleaved to produce hypoxanthine and ribose 1-phosphate through the action of nucleoside phosphorylase; note that inorganic phosphate is required for this reaction. Hypoxanthine-guanine phosphoribosyl transferase converts free hypoxanthine to IMP by condensation with PRPP. PRPP can be derived from ribose 1-phosphate in two steps: (1) the conversion of ribose 1-phosphate to ribose 5-phosphate, which is catalyzed by phosphoribomutase, and (2) the formation of PRPP from ribose 5-phosphate and ATP, which is catalyzed by PRPP synthetase.

13. In most organisms, the inhibition of ribonucleotide reductase is desirable when an adequate number of deoxyribonucleotides have been synthesized. In bacteria and in mammals the enzyme is inhibited by high concentrations of dATP. However, the goals of a viral infection include the rapid synthesis of large quantities of deoxyribonucleotides so that many copies of the viral genome can be synthesized. Evidently the need for the unrestricted synthesis of nucleotides can account for the fact that dATP stimulates, rather than inhibits, the production of deoxyribonucleoside diphosphates.

14. (a) The pathway for the synthesis of glycinamide ribonucleotide is shown in Figure 25-1. It includes the formation of glycinyl phosphate, an acyl phosphate. The amino group of phosphoribosylamine then displaces the phosphate group of glycinyl phosphate as an amide linkage is formed.

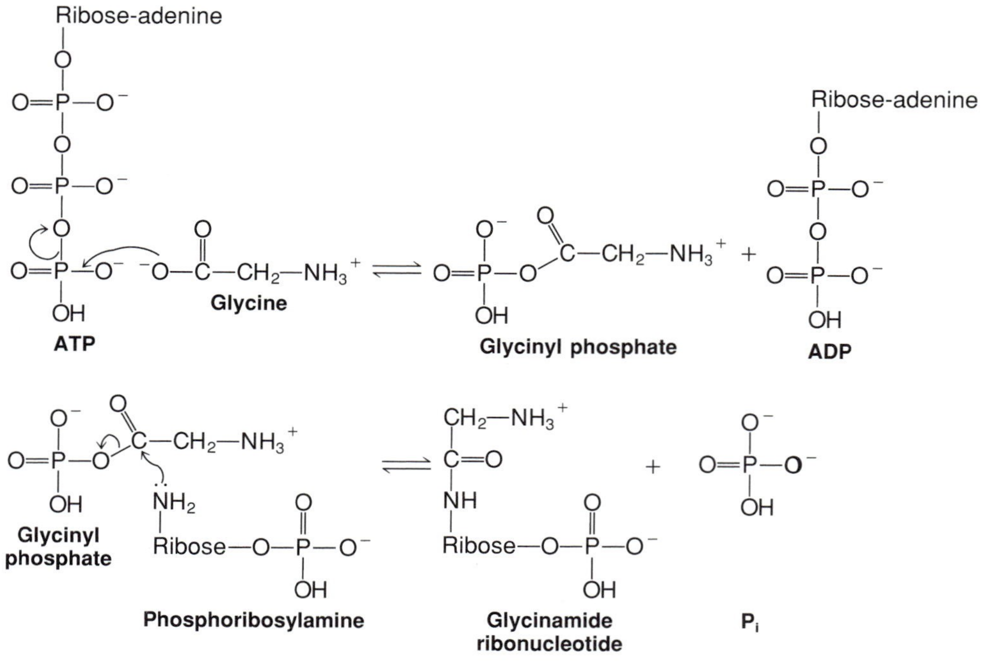

Figure 25-1
Formation of glycinamide ribonucleotide from glycine and 5-phosphoribosyl-1-amine. Glycinyl phosphate is an intermediate in the reaction.

(b) The reverse of the reaction that produces glycinamide ribonucleotide would yield ^{18}O-labeled glycinyl phosphate, which
could then react with ADP to form glycine and ATP. One of the
oxygens of the carboxylate group in glycine is thus derived
from ^{18}O-labeled inorganic phosphate. Experiments of this
type were performed to show that glycinyl phosphate is an intermediate in the pathway to glycinamide ribonucleotide.

15. Both ribose phosphate and an AMP moiety must be added to nicotinamide in order to convert it to NAD^+. The ribosephosphate is derived from PRPP when a phosphoribosyltranferase catalyzes the formation of nicotinamide ribonucleotide or nicotinamide mononucleotide.
The final step utilizes ATP, which serves as an adenyl donor for the
formation of the dinucleotide NAD^+.

CHAPTER 26

Integration of Metabolism

This chapter, which concludes the two major sections of the text devoted to metabolism, provides an integrated view of mammalian metabolism and a review of the principal themes of metabolism. Stryer starts with a recapitulation of the roles of ATP, NADPH, and the building-block molecules derived from fuels in biosynthesis and cellular processes; he also stresses the regulatory mechanisms that control metabolism. (Chapters 10 and 13 contain important background material for these sections.) Stryer then reviews the major metabolic pathways, from glycolysis through fatty acid metabolism (Chapters 15 through 20), and explains the roles of glucose 6-phosphate, pyruvate, and acetyl CoA as key intermediates at junctions between the various metabolic pathways. The metabolic characteristics of the major organs are presented next, followed by a description of the major hormonal effects on their metabolism. The special role of the liver in buffering blood glucose levels is also considered. Finally, the adaptations of metabolism to abnormal conditions are illustrated through two classical examples, prolonged starvation and diabetes.

LEARNING OBJECTIVES

When you have mastered this chapter, you should be able to complete the following objectives.

Strategy and Regulation of Metabolism (Stryer pages 627–630)

1. Review the sources of *ATP, NADPH,* and the *building-block molecules* and their roles in biosyntheses.

2. List the general mechanisms for the *regulation of metabolism*. Give examples in each of the major metabolic pathways.

3. Locate the enzymes of the major metabolic pathways in the *compartments* of a eucaryotic cell.

Major Metabolic Pathways and Key Intermediates (Stryer pages 630–634)

4. Describe the roles of *glycolysis,* name its products, and describe the regeneration of NAD^+ under aerobic and anaerobic conditions.

5. Outline the regulation of *phosphofructokinase* in the liver and in muscle.

6. Discuss the roles of the *citric acid cycle,* the *electron-transport chain,* and *oxidative phosphorylation* in the oxidative degradation of fuels. Explain the control of these pathways by the availability of *ADP*.

7. Describe the roles and the regulation of the *pentose phosphate pathway.*

8. Discuss the physiological role of *gluconeogenesis* and the reciprocal regulation of gluconeogenesis and glycolysis.

9. Outline the *synthesis and degradation of glycogen* and their coordinated control by *phosphorylation* and *dephosphorylation* and allosteric effectors.

10. Summarize the metabolism of *fatty acids*. Describe the regulation of *fatty acid β-oxidation* and *fatty acid synthesis*. Explain how compartmentation is involved in these processes.

11. Describe the fates of *glucose 6-phosphate* in cells. Outline the pathways that give rise to glucose 6-phosphate, and discuss the release of glucose into the blood.

12. Describe the sources and the key metabolic products of *pyruvate*.

13. Discuss the sources and fates of *acetyl CoA*.

Metabolic Profiles of the Major Organs, Hormonal Regulators, and the Control of Blood Glucose Levels (Stryer pages 634–639)

14. Summarize the fuel requirements of the *brain*.

15. Explain the use of fuels by resting and active *muscle*.

16. Describe the synthesis and turnover of *triacylglycerols* by *adipose tissue*. Indicate the role of glucose in this tissue.

17. Discuss the role of the *liver* in providing glucose to other tissues and in regulating lipid metabolism. List the fuels used by the liver for its own needs.

18. Explain and contrast the metabolic effects of *insulin, glucagon,* and *epinephrine* or *norepinephrine.*

19. Describe how *blood glucose levels* are controlled by the liver in response to glucagon and insulin.

Metabolic Adaptations in Starvation and Diabetes (Stryer pages 639–642)

20. List the approximate *fuel reserves* of a 70-kg man in *kilocalories.*

21. Describe the metabolic changes that occur after one and three days of *starvation.*

22. Discuss the metabolic adaptations that occur after prolonged starvation; note especially the shift in brain fuels and the decreased rate of protein degradation.

23. Explain the role of large *fat deposits* in migratory birds.

24. Describe the metabolic derangements in *diabetes mellitus* resulting from relative insulin insufficiency and glucagon excess.

25. Describe the formation of hemoglobin A_{Ic} and its use in monitoring blood glucose in diabetic patients.

SELF-TEST

Strategy and Regulation of Metabolism

1. Biosynthetic pathways that require NADPH include which of the following?

(a) Gluconeogenesis
(b) Fatty acid biosynthesis
(c) Ketone body formation
(d) Cholesterol biosynthesis
(e) Tyrosine biosynthesis

2. Match the biomolecules in the left column with their precursors from the right column.

(a) Phosphoglycerides _____   (1) *S*-Adenosylmethionine
                                (2) Acetyl CoA
(b) Porphyrins _____   (3) Ribose 5-phosphate
                                (4) Dihydroxyacetone
(c) Cholesterol _____   phosphate
                                (5) Tetrahydrofolate
(d) Triacylglycerols _____   (6) Succinyl CoA

(e) Nucleic acids _____

3. Consider the following examples of metabolic regulation:

(1) Fatty acid oxidation in mitochondria is diminished when fatty acid biosynthesis in the cytosol is active due to the inhibition of carnitine acyl transferase I by malonyl CoA.

(2) The synthesis of HMG CoA reductase in various cells is inhibited by low-density lipoproteins.

(3) Glucose 6-phosphatase is present in the liver and kidneys but not in muscle.

(4) Amidophosphoribosyl transferase, the enzyme that catalyzes the committed step in the biosynthesis of purine nucleotides, is inhibited by all purine nucleotides.

(5) The enzyme that catalyzes the synthesis and degradation of fructose 2,6-bisphosphate is phosphorylated and dephosphorylated in response to hormonal signals.

Indicate which of these examples apply to each of the following modes of metabolic regulation:

(a) Allosteric interactions _____

(b) Covalent modifications _____

(c) Enzyme levels _____

(d) Compartmentation _____

(e) Metabolic specialization of organs _____

4. In which compartment of a liver cell are ketone bodies synthesized?

(a) Plasma membrane
(b) Cytosol
(c) Lysosomes
(d) Mitochondria
(e) Endoplasmic reticulum

Major Metabolic Pathways and Key Intermediates

5. Match each metabolic pathway in the left column with its *major* role in metabolism from the right column.

(a) Glycolysis _____

(b) Gluconeogenesis _____

(c) Pentose phosphate pathway _____

(d) Glycogen synthesis _____

(e) Fatty acid degradation _____

(1) Control of glucose levels in blood
(2) Formation of NADH and $FADH_2$
(3) Storage of fuel
(4) Synthesis of NADPH and ribose 5-phosphate
(5) Production of ATP and building blocks of biomolecules

6. The control of phosphofructokinase in the liver and in muscle is different. Both epinephrine and glucagon initiate responses to low glucose levels, yet epinephrine stimulates glycolysis in muscle, whereas glucagon inhibits glycolysis in the liver. Explain this fact.

_____

_____

7. Explain how it is possible to have high $NADPH/NADP^+$ and low $NADH/NAD^+$ ratios in the cytosol of the same cells.

_____

_____

8. Regulation of fatty acid biosynthesis occurs at the enzymatic step catalyzed by

    (a) carnitine acyl transferase I.
    (b) acetyl CoA carboxylase.
    (c) pyruvate carboxylase.
    (d) citrate synthase.
    (e) citrate-malate translocase.

9. Match the three key metabolic intermediates in the left column with their major products from the right column. Indicate the most direct relationships, that is, those not separated by other key intermediates.

(a) Glucose 6-phosphate _____

(b) Pyruvate _____

(c) Acetyl CoA _____

    (1) Ketone bodies
    (2) Oxaloacetate
    (3) Pyruvate
    (4) Glycogen
    (5) CO_2
    (6) Lactate
    (7) Ribose 5-phosphate
    (8) Fatty acids
    (9) Alanine
   (10) Cholesterol
   (11) Acetyl CoA

Metabolic Profiles of the Major Organs, Hormonal Regulators, and the Control of Blood Glucose Levels

10. In adipose tissue, glucose 6-phosphate *is not* converted into which of the following?

    (a) Pyruvate
    (b) Glycogen
    (c) Glucose
    (d) Ribose 5-phosphate

11. In the liver, the major fates of pyruvate include the formation of which of the following?

    (a) Acetyl CoA
    (b) Lactate
    (c) Oxaloacetate
    (d) Alanine

12. Which of the following statements about the metabolism of the brain are *incorrect*?

    (a) It uses fatty acids as fuel in the fasting state.
    (b) It uses about 60% of the glucose consumed by the whole body in the resting state.
    (c) It lacks fuel reserves.
    (d) It can use acetoacetate and 3-hydroxybutyrate under starvation conditions.
    (e) It releases lactate during periods of intense activity.

13. Which of the following tissues converts pyruvate to lactate most effectively?

(a) Liver
(b) Muscle
(c) Adipose tissue
(d) Brain
(e) Kidney

14. Adipose cells constantly break down and resynthesize triacylglycerols, but synthesis cannot proceed without an external supply of glucose. Explain why.

_____

_____

15. Which of the following statements about the metabolism of adipose tissue are *correct*?

(a) It has an active pentose phosphate pathway.
(b) It contains a hormone-sensitive lipase that hydrolyzes triacylglycerols.
(c) It uses ketone bodies as its preferred fuel.
(d) It releases fatty acids to the blood as triacylglycerols that are packaged in VLDL.
(e) It is the most abundant source of stored fuel.

16. Select the statements from the right column that best describe the metabolism of each organ, tissue, or cell in the left column.

(a) Brain _____

(b) Muscle _____

(c) Adipose tissue _____

(d) Liver _____

(e) Red blood cell _____

(1) Releases glycerol and fatty acids into the blood during fasting periods.
(2) In a normal nutritional state, utilizes glucose as the exclusive fuel.
(3) Synthesizes ketone bodies when the supply of acetyl CoA is high.
(4) Can release lactate into the blood.
(5) Utilizes α-keto acids from amino acid degradation as an important fuel.
(6) Can store glycogen but cannot release glucose into the blood.
(7) Can synthesize fatty acids, triacylglycerols, and VLDL when fuels are abundant.

17. Which of the following statements about ketone bodies is *incorrect*?

(a) Thiolase is required for the utilization of ketone bodies as fuels.
(b) Ketone bodies are produced in the liver.
(c) The adipose tissue normally uses ketone bodies as fuels.
(d) Ketone bodies are the blood transport forms of acetyl units.
(e) During starvation, the brain derives approximately two-thirds of its fuel needs from ketone bodies.

18. When fuels are abundant, the liver does not degrade fatty acids; rather, it converts them into triacylglycerols for export as VLDL. Ex-

plain how β-oxidation of fatty acids and the formation of ketone bodies from fatty acids are prevented under these conditions.

_____

_____

19. Match each hormone in the left column with its properties, major target organ, and metabolic effects from the right column.

(a) Insulin _____

(b) Glucagon _____

(c) Epinephrine _____

(1) A catecholamine hormone
(2) A polypeptide hormone
(3) Secreted by the α cells of the pancreas
(4) Effects are mediated by cAMP
(5) The target organ is the liver
(6) The target organ is muscle
(7) Signals the fed state
(8) Secreted in response to low blood glucose levels
(9) Promotes the storage of fuels
(10) Promotes the breakdown of stored fuels

20. Use an "S" to indicate the following metabolic processes that are stimulated and an "I" to indicate those that are inhibited by the action of insulin.

(a) Gluconeogenesis in liver _____

(b) Entry of glucose into muscle and adipose cells _____

(c) Glycolysis in the liver _____

(d) Intracellular protein degradation _____

(e) Glycogen synthesis in liver and muscle _____

(f) Uptake of branched-chain amino acids by muscle _____

(g) Synthesis of triacylglycerols in adipose tissue _____

21. Explain the allosteric effects of glucose on glycogen metabolism.

_____

_____

22. The blood glucose level of a normal individual, measured after an overnight fast, is approximately 80 mg/100 ml. After a meal rich in carbohydrate, it rises to about 120 mg/100 ml and then declines to the fasting level. The approximate time course of these changes and the inflection points are shown in Figure 26-1. After examining the figure, complete the following sentences:

(a) The increase in the glucose level from A to B is due to

_____

(b) The decrease in the glucose level from B to C is due to

_____

_____

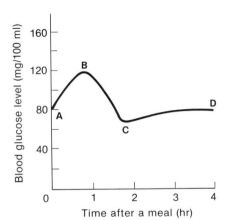

Figure 26-1
Blood glucose levels after a meal rich in carbohydrate.

(c) The leveling off of the glucose level from C to D is due to

_____

_____

(d) The slight overshoot that is sometimes observed at C can be explained by

_____

_____

Metabolic Adaptations in Starvation and in Diabetes

23. Estimate the ratio of glucose produced by gluconeogenesis starting with 1 g of protein to that produced starting with 1 g of triacylglycerol. Assume that the protein is made up entirely of Ala residues, which have a molecular mass of 72 daltons, that the molecular mass of the triacylglycerol is 890 daltons, and that the mass contributed by the glycerol moiety is 89 daltons. Also assume complete conversion.

24. Match the fuel storage forms in the left column with the most appropriate characteristics from the right column.

(a) Glycogen _____

(b) Triacylglycerols _____

(c) Protein _____

(1) Largest storage form of calories
(2) Most readily available fuel during muscular activity
(3) Major source of precursors for glucose synthesis during starvation
(4) Depleted most rapidly during starvation
(5) Not normally used as a storage form of fuel

25. Relative to the well-fed state, fuel utilization after three days of starvation shifts in which of the following ways?

(a) More glucose is consumed by the brain.
(b) Adipose tissue triacylglycerols are degraded to provide fatty acids to most tissues.
(c) The brain begins to use ketone bodies as fuels.
(d) Proteins are degraded in order to provide three-carbon precursors of glucose.
(e) Glycogen is stored as a reserve fuel.

26. Metabolic adaptations to prolonged starvation include which of the

following changes relative to the metabolic picture after three days of starvation?

(a) The rate of lipolysis in the adipose tissue increases.
(b) The glucose output by the liver decreases.
(c) The ketone body output by the liver decreases.
(d) The utilization of glucose by the brain decreases as the utilization of ketone bodies increases.
(e) The rate of degradation of muscle protein decreases.

27. Show the changes in blood glucose levels you would expect to see for an insulin-dependent diabetic patient after a meal rich in carbohydrate by plotting their time course on Figure 26-1. Explain your answer.

_____

_____

28. Which of the following occur in individuals with untreated diabetes?

(a) Fatty acids become the main fuel for most tissues.
(b) Glycolysis is stimulated and gluconeogenesis is inhibited in the liver.
(c) Ketone body formation is stimulated.
(d) Excess glucose is stored as glycogen.
(e) Triacylglycerol breakdown is stimulated.

29. Is it *true* or *false* that in diabetes the brain shifts to ketone bodies as its major fuel? _____

Explain. _____

_____

_____

30. Which of the following statements about hemoglobin A_{Ic} are *true*?

(a) It is a mutant hemoglobin that occurs in diabetic patients.
(b) It is absent in normal individuals.
(c) It involves the covalent modification of hemoglobin A by glucosyl transferase.
(d) It is useful in monitoring glucose levels in blood.
(e) It is an example of how reducing sugars at high concentrations may damage proteins.

PROBLEMS

1. Cardiac muscle exhibits a high demand for oxygen, and its functioning is severely impaired when coronary circulation is blocked.

(a) Considering the energy-generating substrates available to and used by the heart under normal circumstances, why is oxygen required by heart muscle?
(b) Suppose that in an intact animal you can measure the concentrations of biochemical metabolites in the arteries leading to the

cardiac muscle and in the veins carrying blood away from heart tissue. If the supply of oxygen to heart tissue is reduced, what differences in arterial and venous glucose concentrations will you observe? What metabolite will be elevated in heart muscle that has an insufficient supply of oxygen?

(c) Patients suffering from cardiac damage due to oxygen insufficiency exhibit an increase in the levels of certain enzymes in the plasma; among these are lactate dehydrogenase and several types of aminotransferase enzymes. During recovery from a heart attack, the plasma levels of these enzymes decrease. Explain these findings.

2. An infant suffering from a particular type of organic acidemia has frequent attacks of vomiting and lethargy, which are exacerbated by infections, fasting, and the consumption of protein or fat. During these episodes, the patient suffers from hypoglycemia, which can be alleviated by injections of β-hydroxybutyrate. In addition, concentrations of ketone bodies in the blood are extremely low. The patient also has elevated concentrations of a number of organic acids in both blood and urine. Among these acids are β-hydroxy-β-methylglutarate, β-methylglutaconate, and isovalerate.

(a) What enzyme deficiency could cause an elevation in the concentration of these organic acids?

(b) How could the same enzyme deficiency lead to a reduction in the concentration of ketone bodies?

(c) How does fasting exacerbate the symptoms of the disorder?

(d) The consumption of fat causes a noticeable increase in the concentration of β-hydroxy-β-methylglutarate. Why?

(e) How can the consumption of protein cause an increase in the levels of organic acids?

(f) How can the administration of β-hydroxybutyrate relieve hypoglycemia?

3. Patients who remain unconscious after a serious surgical operation are given 100 to 150 g of glucose daily through the intravenous administration of a 5% solution. This amount of glucose falls far short of the daily caloric needs of the patient. What is the benefit of the administration of glucose?

4. A biochemist in the Antarctic is cut off from his normal food supplies and is forced to subsist on a diet that consists almost entirely of animal fats. He decides to measure his own levels of urinary ketone bodies, beginning on the day he starts the high-fat diet. What changes in urinary ketone-body levels will he find?

5. In liver tissue, insulin stimulates the synthesis of glucokinase. What implications does this have for an individual who has an insulin deficiency?

6. Within a few days after a fast begins, nitrogen excretion accelerates to a relatively high level. After several weeks, the rate of nitrogen excretion falls to a lower level. The excretion of nitrogen then continues at a relatively constant rate until the body is depleted of triacylglycerol stores; then the rate of urea and ammonia excretion again rises to a very high level.

(a) What events trigger the initial surge of nitrogen excretion?

(b) Why does the nitrogen excretion rate decrease after several weeks of starvation?

(c) Explain the increase in nitrogen excretion that occurs when lipid stores are exhausted.

7. Some bacterial species can grow and divide with either alanine or leucine as a sole carbon source. Others exhibit a strong preference for alanine and grow poorly on leucine. Can you explain the difference between the two types of cells?

8. Suppose that genetic engineering techniques enable you to transfer the genes for the enzymes of the glyoxalate pathway to human tissues and to express those genes in mitochondria. Why might those wishing to lose weight rapidly be interested in such a system?

9. Among the difficulties caused by prolonged fasting are metabolic disorders caused by vitamin deficiencies. What vitamins are needed during starvation to insure that cells can continue to carry out the metabolic adaptations discussed on pages 639 and 640 of Stryer's text?

10. Describe the general fate of each of the following compounds in the mitochondria and in the cytosol of a liver cell.

   (a) Palmitoyl CoA
   (b) Acetyl CoA
   (c) Carbamoyl phosphate
   (d) NADH
   (e) Glutamate
   (f) Malate

11. Young men who are championship marathon runners have levels of body fat as low as 4%, whereas most casual runners have levels ranging from 12 to 15%. Why would marathoners be at greater risk during prolonged fasting?

12. Assume that a typical 70-kg man expends about 2000 kcal of energy per day. If the energy for his activities were all derived from ATP, how many grams of ATP would have to be generated on a daily basis? How many grams of glucose would be required to drive the formation of the needed amount of ATP? The molecular weight of ATP is 500, and that of glucose is 180. One mole of glucose generates 686 kcal of energy when completely oxidized, and 7.3 kcal are required to drive the synthesis of one mole of ATP. Assume that 40% of the energy from the oxidation of glucose can be used for ATP synthesis.

ANSWERS TO SELF-TEST

Strategy and Regulation of Metabolism

1. b, d, and e. Answer (e) is correct because phenylalanine hydroxylase requires O_2 and NADPH in the conversion of phenylalanine to tyrosine.

2. (a) 1, 2, 4 (b) 6 (c) 2 (d) 2, 4 (e) 3, 5. Regarding the answer to (a), (1) is correct because the synthesis of phosphatidylcholine in bacteria requires S-adenosylmethionine.

3. (a) 1, 4 (b) 5 (c) 2 (d) 1 (e) 3

4. d

Major Metabolic Pathways and Key Intermediates

5. (a) 5 (b) 1 (c) 4 (d) 3 (e) 2, 5

6. The different effects of glucagon and epinephrine in the liver and in muscle are due to the different properties of the kinase and phosphatase enzymes that catalyze the synthesis and degradation of fructose 2,6-bisphosphate in these organs. In the liver, the cAMP cascade leads to inhibition of the kinase and activation of the phosphatase. This decreases fructose 2,6-bisphosphate levels and inhibits glycolysis. In muscle, the phosphorylation of a homologous enzyme activates the kinase, thus stimulating the formation of fructose 2,6-bisphosphate, the activity of phosphofructokinase, and glycolysis.

7. Because NADPH and NADH are distinct compounds that bind specifically to different enzymes and are not readily interconverted, it is possible to have high $NADPH/NADP^+$ ratios and low $NADH/NAD^+$ ratios in the same compartment of a cell.

8. b

9. (a) 3, 4, 7 (b) 2, 6, 9, 11 (c) 1, 5, 8, 10

Metabolic Profiles of the Major Organs, Hormonal Regulators, and the Control of Blood Glucose Levels

10. b, c

11. a, c. The conversions of pyruvate into lactate or alanine mainly occur in the muscles and red cells. In the liver pyruvate is mostly used in gluconeogenesis or for lipid synthesis.

12. a, e

13. b

14. The synthesis of triacylglycerols requires glycerol 3-phosphate, which is derived from glucose. The glycerol that is released during triacylglycerol hydrolysis cannot be reutilized in adipose cells because they lack glycerol kinase. Thus, externally supplied glucose is required.

15. a, b, e

16. (a) 2 (b) 4, 6 (c) 1 (d) 3, 5, 7 (e) 2, 4

17. c

18. The selection of the pathway depends on whether or not the fatty acids enter the mitochondrial matrix, the compartment of β-oxidation and ketone-body formation. When citrate and ATP concentrations are high, as in the fed state, the activity of acetyl CoA carboxylase is stimulated. The resulting malonyl CoA, which is a precursor for fatty acid synthesis, inhibits carnitine acyl transferase I, which translocates fatty acids from the cytosol into the mitochondria for oxidation.

19. (a) 2, 6, 7, 9 (b) 2, 3, 4, 5, 8, 10 (c) 1, 4, 6, 10

20. (a) I (b) S (c) S (d) I (e) S (f) S (g) S

21. Phosphorylase *a* binds glucose, which makes this enzyme susceptible to the action of phosphatase. The resulting phosphorylase *b* is inactive; therefore, glycogen degradation is decreased. Since phosphorylase *b* does not bind phosphatase, phosphatase is released and activates glycogen synthase by dephosphorylating it, leading to the production of glycogen.

22. (a) The increase in the glucose level from A to B is due to the absorption of dietary glucose.

    (b) The decrease in the glucose level from B to C is due to the effects of insulin, which is secreted in response to increased blood glucose. Glucose is removed from the blood by the liver, which synthesizes glycogen, and by muscle and adipose tissue, which store glycogen and triacylglycerols.

    (c) The leveling off of the glucose level from C to D is due to the increased secretion of glucagon and the diminished concentration of insulin. Glucagon maintains blood glucose levels by promoting gluconeogenesis and glycogen degradation in the liver and by promoting the release of fatty acids, which partially replace glucose as the fuel for many organs.

    (d) The slight overshoot that is sometimes observed at C can be explained by the continued effects of insulin, which are not yet balanced by the metabolic effects of glucagon.

Metabolic Adaptations in Starvation and in Diabetes

23. The entire protein is converted by proteolysis to Ala molecules, which by transamination give rise to pyruvate. Gluconeogenesis from two pyruvate molecules yields one molecule to glucose. The triacylglycerol contains 10% of glycerol by weight, which is the only part of the lipid that can give rise to glucose. Glycerol enters glycolysis as dihydroxyacetone phosphate, and two molecules of this triose phosphate yield one molecule of glucose. Therefore, the glucose produced from 1 g of protein is

$$\frac{1\ g}{72\ g/mol} \times \frac{1}{2}$$

and the glucose produced from 1 g of triacylglycerol is

$$\frac{0.1\ g}{89\ g/mol} \times \frac{1}{2}$$

The ratio of the glucose from protein to the glucose from triacylglycerol is thus

$$\frac{(1\ g)/(72\ g/mol)}{(0.1\ g)/(89\ g/mol)} = \frac{1}{0.1} \times \frac{89}{72} = 12.4$$

24. (a) 2, 4 (b) 1 (c) 3, 5

25. b, c, d

26. b, d, and e. Answer (a) is incorrect because lipolysis remains essentially constant.

27. See Figure 26-2. In diabetic patients, the level of insulin is too low and that of glucagon is too high, so after a meal the glucose levels will reach higher values than in a normal individual. Also the removal of glucose from blood will be slower, and the fasting glucose levels in the blood may remain higher.

28. a, c, e

29. False. Although ketone-body concentrations in blood may become high in diabetes, glucose is even more plentiful. Therefore, the brain continues to use glucose as its major fuel.

30. d, e

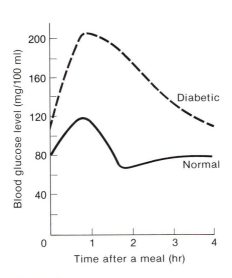

Figure 26-2
Blood glucose levels for a diabetic patient and a normal individual after a meal rich in carbohydrate.

1. (a) Under normal conditions, heart muscle consumes acetoacetate and β-hydroxybutyrate, both of which are converted to two molecules of acetyl CoA. Oxygen is required for terminal oxidation of acetyl CoA in the citric acid cycle.

   (b) Cardiac cells deprived of oxygen are unable to generate metabolic energy by oxidizing acetyl CoA. An alternative source of energy is glucose, which can be converted to lactate under anaerobic conditions, with the subsequent generation of ATP. Thus, under anaerobic conditions, glucose uptake by heart muscle increases; the concentration of glucose in cardiac veins will decrease, relative to glucose levels in coronary arteries. The concentration of lactate in the coronary veins is also elevated, compared to the levels in the coronary arteries.

   (c) Under anaerobic conditions, the limited amount of ATP produced by the glycolytic pathway is insufficient to maintain normal cell functions, and many of the cells in anaerobic or ischemic tissues will die. Cell lysis permits release of cytoplasmic enzymes into the circulation. Lactate dehydrogenase and the aminotransferases are relatively easy to assay and are therefore used as indicators of cellular damage. During recovery, as heart tissue undergoes some degree of repair, the release of cytoplasmic enzymes slows, so there is a concomitant decrease in the plasma concentrations of these enzymes.

2. (a) The organic acids that are elevated in the blood and urine are those that are normal intermediates in the degradation of leucine. Examination of the pathway for leucine degradation (see Stryer's text, pp. 510 and 511) shows that the deficient enzyme is probably the enzyme that converts β-hydroxy-β-methylglutaryl CoA (HMG CoA) to acetoacetate and acetyl CoA. This enzyme is known as HMG CoA lyase, and the reduced activity of this enzyme in the infant leads to a reduction in the rate of leucine catabolism.

   (b) The production of acetoacetate and β-hydroxybutyrate is also dependent upon the activity of HMG CoA lyase, which is deficient in the infant. Thus, the production of ketone bodies from HMG CoA in response to increases in the concentration of acetyl CoA is impaired because of the enzyme deficiency.

   (c) Fasting causes an increase in acetyl CoA production through the increased rates of lipolysis that occur in the attempt to generate sources of metabolic energy. Any increase in acetyl CoA concentration stimulates HMG CoA synthesis, and the inability to convert HMG CoA to acetoacetate leads to an increase in concentrations of β-hydroxy-β-methylglutarate, which is excreted by the cells into the plasma.

   (d) Consumption of fats such as triacylglycerols leads to the increased production of acetyl CoA, from the β-oxidation of fatty acids. As noted in the answer to 2(c), elevation in the level of acetyl CoA stimulates the production of HMG CoA, leading to an increase in the concentration of β-hydroxy-β-methylglutarate.

   (e) Most proteins contain leucine, and the fraction of leucine that is not required for the synthesis of other proteins will undergo oxidation. As noted in the answer to 2(a), the catabolism of leucine generates a number of intermediates that accumulate as a result of the deficiency of HMG CoA lyase.

(f) Because the liver is unable to generate normal levels of ketone bodies, those tissues that normally utilize them are required to use other substrates as metabolic fuels. Since glucose is the substrate of choice in such situations, increased demand for blood glucose results in hypoglycemia. The administration of β-hydroxybutyrate provides an alternative source of metabolic energy for organs such as the heart, and glucose is thereby conserved.

3. The limited amount of glucose is given to prevent the hydrolysis of muscle protein during fasting. It serves as a source of energy for the brain and blood cells; otherwise, during fasting, body proteins are hydrolyzed to provide carbon atoms for the generation of glucose by gluconeogenesis in the liver and kidney.

4. A high-fat diet will stimulate ketone-body formation because the oxidation of fatty acids causes an increase in acetyl CoA concentration, which in turn stimulates the production of ketone bodies. The overall profile of ketone-body production may resemble that found during starvation because a source of carbon for glucose production will be lacking. However, dietary fats will serve as a source of energy instead of the triacylglycerols stored in body tissue.

5. Under normal conditions, glucokinase acts to phosphorylate glucose when concentrations of the hexose are high. The failure to synthesize sufficient quantities of glucokinase means that the liver cannot control the levels of glucose in blood, compounding the other difficulties caused by insulin deficiency described on pages 641–642 Stryer's text.

6. (a) During the first few days of starvation, the brain continues to utilize glucose. Glycogen stores are exhausted, so the primary source of carbon atoms for gluconeogenesis is amino acids. Because the concentration of free amino acids in the tissues is limited, body proteins are broken down to provide the amino acids to support gluconeogenesis. Nitrogen excretion increases because the amino groups of those amino acids are eliminated as urea.

(b) After several weeks of fasting, the brain adapts to the utilization of ketone bodies as a source of energy, so less glucose is required. The resulting reduction in gluconeogenesis means a reduction in the rate of oxidation of amino acids and in the production of ammonia and urea.

(c) When triacylglycerol stores are depleted, the body relies on body proteins not only as a source of glucose for the brain but also as a source of energy for all other tissues. These requirements cause a great increase in the rate of body protein catabolism, with corresponding increases in amino acid oxidation and nitrogen excretion. Often more than a kilogram per day in weight is lost, indeed causing a threat to life.

7. Bacteria that lack a glyoxylate pathway are unable to carry out gluconeogenesis if leucine, a ketogenic amino acid, is the sole source of carbon. They can utilize alanine for the production of pyruvate, which in turn can be used as a source of carbon for the synthesis of glucose.

Bacteria having the enzymes of the glyoxalate pathway can use either leucine or alanine for the de novo synthesis of glucose.

8. One of the difficulties caused by fasting is that body proteins are hydrolyzed to provide a source of carbon atoms for gluconeogenesis. Normally, fatty acids cannot serve as a source of glucose because acetyl CoA is not gluconeogenic. The enzymes of the glyoxalate pathway provide a means for acetyl CoA to be used for the synthesis of glucose. Such a system in human cells could conceivably be used to prevent the hydrolysis of body proteins during fasting.

9. Most of the vitamins and cofactors discussed in previous chapters of Stryer's text would be needed during starvation because many of the essential metabolic pathways must continue to operate. Among the most obvious vitamins needed for those pathways are pyridoxal phosphate (for the transamination of amino acids), niacin and riboflavin (for electron transport), thiamin (for the oxidative decarboxylation of pyruvate, α-ketoglutarate, and the branched-chain amino acids), biotin (for the carboxylation of pyruvate), and cobalamin (for the conversion of methylmalonyl CoA to succinyl CoA).

10. (a) In the cytoplasm, the acyl chain of palmitoyl CoA can be esterfied to glycerol as part of the process of triacylglycerol formation in the liver cytosol, or the palmitoyl chain can be transferred to carnitine for subsequent transport to the mitochondria. In mitochondria, palmitoyl CoA is oxidized to CO_2 and H_2O.

(b) In mitochondria, acetyl CoA is oxidized to carbon dioxide and water, or it can undergo carboxylation to form oxaloacetate. It can also be used for the synthesis of ketone bodies. In the cytosol acetyl CoA serves as a precursor of malonyl CoA, as well as a precursor of HMG CoA and cholesterol.

(c) In the mitochondria, carbamoyl phosphate combines with ornithine to form citrulline in the urea cycle, whereas in the cytosol, it serves as a precursor of pyrimidines when it condenses with aspartate to yield carbamoyl aspartate.

(d) In mitochondria, NADH is reoxidized to NAD^+ in the electron-transport chain. In the cytosol, it can be reoxidized to NAD^+ through the action of lactate dehydrogenase or NADPH dehydrogenase or by means of the glycerol-phosphate or malate-aspartate shuttles.

(e) In mitochondria, glutamate undergoes oxidative deamination, yielding ammonia and α-ketoglutarate. It can also serve as a source of amino groups for a number of aminotransferase enzymes. In the cytosol, it can also serve as an amino donor for aminotransferases. Glutamate can also be used in the cytosol as a precursor of glutamine, and it can of course be incorporated into newly synthesized protein.

(f) Malate is converted to oxaloacetate in the citric acid cycle, which takes place in the mitochondria. In the cytoplasm, as a component of the malate-aspartate shuttle, it serves as an electron carrier to transfer electrons from NADH to the inner mitochondrial membrane. Malate can also be used as a source of electrons for the generation of NADPH in the reaction catalyzed by malic enzyme.

11. The greater the percentage of body fat, the larger the reserves of triacylglycerols, which are the primary source of metabolic energy dur-

ing fasting. Once these reserves are depleted, the body accelerates the breakdown of muscle protein as an energy source. Extensive hydrolysis of muscle tissue threatens many vital body functions and can lead to death.

12. Since 7.3 kcal are required to drive the synthesis of one mole of ATP, the caloric value of the energy in each mole of ATP is 7.3 kcal. Therefore, the amount of ATP needed per day is

$$\frac{2000 \text{ kcal/day}}{7.3 \text{ kcal/mol ATP}} = 274 \text{ mol ATP per day}$$

274 mol ATP/day $\times$ 500 g/mol = 137 kg ATP per day

If the oxidation of one mole of glucose yields 686 kcal of energy, and if 40% can be used to drive the synthesis of ATP, glucose yields (0.4)(686 kcal/mol) = 274 kcal/mol of usable energy. Therefore, the amount of glucose needed per day is

$$\frac{2000 \text{ kcal/day}}{274 \text{ kcal/mol glucose}} = 7.3 \text{ mol glucose per day}$$

7.3 mol glucose/day $\times$ 180 g/mol = 1.31 kg glucose per day

Normal fuel stores available to the typical 70-kg man include about 250 g of glycogen and approximately 25 g of glucose. These figures make it evident that humans can rely on stores of carbohydrate for only a short time.

CHAPTER 27

DNA Structure, Replication, and Repair

In Part V of the text, Stryer returns to the topic of the flow of genetic information and considers the detailed biochemical mechanisms underlying this complex process. Chapters 3 and 4 introduced you to DNA and RNA and outlined the storage, duplication, and expression of genetic information. In Chapter 6, the techniques used to analyze, construct, and clone DNA were presented. You should review these chapters to prepare for studying Chapter 27. Pay particular attention to DNA structure, the supercoiling of DNA, and DNA polymerase I in Chapter 4 and to the restriction endonucleases and DNA ligase in Chapter 6.

Chapter 27 covers the biochemistry of the flow of genetic information from parent to progeny by describing DNA and the enzyme systems that replicate and maintain it. Stryer expands his earlier coverage of DNA with a more thorough description of the A-, B-, and Z-structures it can assume and the underlying chemical determinants of these structural variations. The mechanisms by which proteins interact with DNA are exemplified by the DNase I and EcoRI nucleases. The mechanisms of DNA ligase are also given. Stryer then provides a detailed description of the topology of covalently-closed circular DNAs and the topoisomerases that modulate their linking numbers. He expands upon the previous description of the biochemistry of DNA polymerase I from *E. coli*. He then describes replication initiation; bidirectional, semiconservative, and RNA-primed semidiscontinuous DNA elongation; and the termination of replication in *E. coli*. The special problems of replication due to the structure of eucaryotic chromatin will be covered in Chapter 33. Stryer concludes the chapter with a description of mutations—their nature, causes, consequences, and repair. He describes the pathological consequences of a repair deficiency in humans, states the relationship of mutation to carcinogenesis, and describes a test system for detecting potential carcinogens through their mutagenic action on bacteria.

LEARNING OBJECTIVES

When you have mastered this chapter, you should be able to complete the following objectives.

DNA Structure and Protein-DNA Interactions (Stryer, pp. 649–659)

1. Contrast the structural information provided by the *X-ray diffraction analysis* of *DNA fibers* and *DNA crystals*.

2. Explain why *local DNA structure* depends upon *base sequence*. Contrast deformations in the DNA helix due to *smooth bending* and *kinking*.

3. Describe the *major* and *minor grooves* of the *B-DNA helix*. Distinguish among the four possible base pairs (AT, TA, CG, and GC) in terms of the unique arrays of *hydrogen-bond acceptors* and *donors* and methyl groups they present in the grooves of the DNA.

4. Indicate the role of the *puckering* of deoxyribose in determining the structural differences between *A-DNA* and B-DNA helices. Compare the structure of *double-stranded RNA* and *RNA-DNA hybrids* with that of the A-DNA helix.

5. Describe *Z-DNA* in terms of the *handedness* of the helix, the sequence determinants, and the *syn* and *anti conformations* of its constituent deoxyribonucleotides. Account for the effects of supercoiling and the methylation of cytosine residues at position C-5 on the formation of Z-DNA.

6. Describe the *electrostatic interactions* that give rise to the relatively *nonspecific interactions* between *deoxyribonuclease I* and DNA. Similarly, describe the *specific interactions* between *EcoRI endonuclease* and its *palindromic recognition sequence*.

7. List the substrates and outline the reaction mechanisms of the *DNA ligases*.

DNA Topology and Topoisomerases (Stryer, pp. 659–665)

8. Define the *linking number* of a circular DNA molecule and relate *supercoiling* to the *electrophoretic* and *centrifugal mobility* of the molecule.

9. Write the equation relating the linking number to the *twisting number* and *writhing number* of a DNA *topoisomer*. Give the equation for the *specific linking difference* (λ) of a topoisomer, and explain how a negative *superhelix density* can facilitate the unwinding of the helix.

10. Give the general reaction of the *topoisomerases*, distinguish between *type I* and *type II* topoisomerases, and describe the substrates, products, and mechanisms of *topoisomerase I* and *DNA gyrase*.

DNA Polymerases (Stryer, pp. 665–669)

11. Outline the key features of the reactions catalyzed by *DNA polymerase I* of *E. coli*. Relate its $3' \rightarrow 5'$ and $5' \rightarrow 3'$ *exonuclease activities* to the *fidelity* of the polymerization reaction and the physiological functions of the enzyme. Distinguish between the small and large fragments produced by limited proteolysis of DNA polymerase I.

12. Recount the significance of the discovery of a *mutant* of *E. coli* that was deficient in DNA polymerase I, and describe the reactions catalyzed by *DNA polymerase III*.

DNA Replication (Stryer, pp. 669–675)

13. Draw a *replication fork*, and describe the reactions and the movements of the DNA strands that occur during replication. Recount the

evidence for a unique *origin* of replication and *bidirectional replication* in *E. coli.*

14. Define *continuous replication* and *discontinuous replication* and relate these terms to the *leading* and *lagging strands.* Describe an *Okazaki fragment.*

15. Describe the features of the nucleotide sequence of *oriC.* List the proteins that interact with the DNA in this region of the chromosome, and give the reactions they catalyze.

16. Explain the roles of *RNA* in DNA replication. Describe the *primosome,* and name the enzyme that removes *RNA primers* from the genome.

17. List the distinctive features of *DNA polymerase III holoenzyme,* and describe how a dimer of the enzyme, along with other proteins, synthesizes the leading and lagging strands of the daughter duplexes.

18. Describe the reactions and identify the proteins that complete DNA replication to give intact DNA molecules.

Mutations—Their Nature, Consequences, and Repair (Stryer, pp. 675–681)

19. Distinguish among *substitution, insertion,* and *deletion mutations,* and relate them to changes in DNA. Distinguish between *transversion* and *transition* substitution mutations, and explain the origin of mutations that alter the reading frame in translation.

20. Relate an enzyme deficiency in the proofreading, $3' \rightarrow 5'$ exonuclease activity of DNA polymerase III holoenzyme to the *mutation rate* in *E. coli.*

21. Give examples of some *chemical mutagens* and their mechanisms. Describe the structure of the *pyrimidine dimer* formed by *ultraviolet light.* Outline the mechanisms for the enzymatic repair of damaged DNA.

22. Describe *xeroderma pigmentosum,* its cause, and its pathological consequences.

23. Explain why *thymine* rather than *uracil* is used in DNA.

24. Relate *mutagens* and *carcinogens.* Outline the *Salmonella mutagen assay.*

SELF-TEST

DNA Structure and Protein-DNA Interactions

1. Which of the following statements about the double-helical structure of DNA are *correct?*

(a) It has adenine paired with thymine and cytosine paired with guanine.
(b) It can assume at least three different forms: A-DNA, B-DNA, and Z-DNA.
(c) In B-DNA, all the hydrogen-bonded base pairs lie in a plane perpendicular to the helix axis.
(d) It can be deformed by both smooth bends and kinks.
(e) It is rigid and static.

2. Figure 27-1 shows the structures of the adenine and thymine
(A : T) and guanine and cytosine (G : C) base pairs in B-DNA. Use the
figure to answer the following questions:

Figure 27-1
Watson-Crick base pairs.

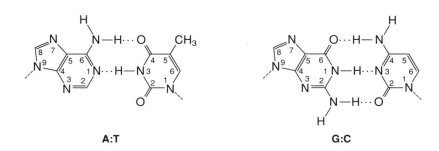

A:T   G:C

(a) Which hydrogen-bond donors or acceptors of the A : T base
pair are in the major groove?

_____

(b) Which hydrogen-bond donors or acceptors of the G : C base
pair are in the minor groove?

_____

(c) Using n to represent heterocyclic-ring nitrogen atoms, o for oxy
groups, and h for protons, give the patterns of hydrogen-bond
acceptors and donors in the major groove for a G : C and a C : G
base pair.

_____

(d) Using the same symbols as in (c), give the patterns of hydrogen-
bond acceptors and donors in the minor groove for an A : T
and a T : A base pair.

_____

(e) Explain the possible biological significance of the unique arrays
of hydrogen-bond donors and acceptors in the grooves of
B-DNA.

_____

_____

3. X-ray diffraction studies of DNA held at low relative humidity revealed the existence of A-DNA. Since such arid conditions presumably
never occur in the cell, what is the significance of the structure of this
DNA?

_____

_____

4. Match the features or characteristics in the right column with the
type of DNA helix in the left column.

(a) A-DNA _____   (1) Pyrimides are in the *anti* config-
                                uration.
(b) B-DNA _____
                           (2) Phosphates in the backbone are
(c) Z-DNA _____   zigzagged.

(3) The formation is favored by
negative supercoiling.
(4) Has a relatively wide and deep
major groove.
(5) Has a right-handed helix.
(6) Has 10.4 base pairs per turn.
(7) Has a structure similar to that of
double-stranded RNA.
(8) The strands in the helix have
opposite polarities.

5. Which of the following statements about DNase I are *correct?*

(a) It interacts with the bases of DNA in the enzyme-substrate complex.
(b) It makes contact with DNA by electrostatic interactions between glutamate and aspartate side chains and the DNA phosphates.
(c) It interacts with approximately one turn of the DNA helix.
(d) It requires a divalent metal ion for activity.
(e) It forms a hydroxyl ion that attacks the phosphorous of the scissile bond to initiate the hydrolytic reaction.

6. How does EcoRI endonuclease recognize the sequence it cleaves? What are the symmetry relationships in the interaction between the enzyme and the substrate?

_____

_____

_____

_____

_____

7. Which of the following statements about DNA ligase are *correct?*

(a) It forms a phosphodiester bond between a 5′-hydroxyl and a 3′-phosphate in duplex DNA.
(b) It requires a cofactor, either NAD^+ or ATP depending upon the source of the enzyme, to provide the energy to form the phosphodiester bond.
(c) It catalyzes its reaction by a mechanism that involves the formation of a covalently-linked enzyme adenylate.
(d) It catalyzes its reaction by a mechanism that involves the activation of a DNA phosphate through the formation of a phosphoanhydride or pyrophosphate bond with AMP.
(e) It is involved in DNA replication, repair, and recombination.

DNA Topology and Topoisomerases

8. The topological features of DNA may affect which of the following?

(a) The electrophoretic mobility of the DNA
(b) The sedimentation properties of the DNA
(c) Its affinities toward proteins that bind to the DNA
(d) The susceptibility of the strands of the DNA to unwinding
(e) The susceptibility of the DNA to the action of DNA ligase

9. Which of the following statements about DNA molecules that are topoisomers are *correct*?

   (a) They are bound to topoisomerases.
   (b) They differ from one another topologically only in that they have different linking numbers.
   (c) The values of their specific linking differences are unequal.
   (d) They have identical superhelix densities.
   (e) They are topological isomers.

10. If two covalently-closed circular DNA molecules with the same number of base pairs have the same linking number (*L*), can they contain different numbers of Watson-Crick base pairs? Explain.

_____

_____

_____

11. Which of the following statements about topoisomerases are *correct*?

   (a) They alter the linking numbers of topoisomers.
   (b) They break and reseal phosphodiester bonds.
   (c) They require NAD^+ as cofactor to supply the energy to drive the conversion of a supercoiled molecule to its relaxed form.
   (d) They form covalent intermediates with their DNA substrates.
   (d) They can, in the case of a particular topoisomerase, use ATP to form negatively supercoiled DNA from relaxed DNA in *E. coli*.

DNA Polymerases

12. Which of the following statements about DNA polymerase I are *correct*?

   (a) It adds deoxynucleotide units to the 3'-hydroxyl of a primer.
   (b) It uses the template strand to select which deoxynucleotide unit to add to the growing DNA chain.
   (c) It contains a 3'→5' nuclease that cleaves phosphodiester bonds to yield 3'-dNMPs and 3'-phosphate–terminated DNA.
   (d) It contains two nuclease activities in the same polypeptide chain that contains the polymerase active site.
   (e) It can be cleaved with a protease into two fragments, each of which has a nuclease activity.

13. Match the properties or functions in the right column with the DNA polymerase in the left column.

   (a) DNA polymerase I _____
   (b) DNA polymerase II _____
   (c) DNA polymerase III _____

   (1) Involved in replication
   (2) Requires a primer and a template
   (3) Physiological function is unknown
   (4) Makes most of the DNA during replication
   (5) Removes the primer and fills in gaps during replication

14. Which of the following statements about DNA replication in *E. coli* are *correct*?

 (a) It occurs at a replication fork.
 (b) It starts at a unique locus on the chromosome.
 (c) It proceeds with one replication fork per replicating molecule.
 (d) It is bidirectional.
 (e) It involves discontinuous synthesis on the leading strand.
 (f) It uses RNA transiently as a template.

15. Which of the following statements about DNA polymerase III holoenzyme from *E. coli* are *correct*?

 (a) It elongates a growing DNA chain approximately 100 times faster than does DNA polymerase I.
 (b) It associates with the parental template, adds a few nucleotides to the growing chain, and then dissociates prior to initiating another synthesis cycle.
 (c) It maintains a high fidelity of replication, in part by acting in conjunction with a subunit containing a $3' \rightarrow 5'$ exonuclease activity.
 (d) When replicating DNA, it is a molecular assembly composed of at least eight different kinds of subunits.

16. Why is RNA synthesis essential to DNA synthesis in *E. coli*?

_____ —

_____

_____

_____

17. Match the functions or features related to DNA replication in *E. coli* listed in the right column with the molecules or structures in the left column.

 (a) Replication fork _____

 (b) *OriC* _____

 (c) Lagging strand _____

 (d) Leading strand _____

 (e) Okazaki fragments _____

 (f) DnaB helicase _____

 (g) Single-strand binding _____

 (h) DNA gyrase _____

 (i) Primase _____

 (1) Synthesis direction is opposite that of replication fork movement
 (2) Unwinds strands at the origin of replication in association with dnaA and dnaC proteins
 (3) Is synthesized continuously
 (4) Synthesizes most of DNA
 (5) Is synthesized discontinuously
 (6) Relieves positive supercoiling

(j) DNA polymerase III
holoenzyme _____

(k) Rep protein _____

(l) DNA polymerase I _____

(m) DNA ligase _____

(7) Is the locus of
DNA unwinding
(8) Hydrolyzes ATP to
reduce the linking
number of DNA
(9) Binds dnaA, dnaB,
and dnaC proteins
(10) Fills in gaps where
RNA existed
(11) Is the point of ini-
tiation of synthesis
(12) Joins lagging strand
pieces to one
another
(13) Contains a $5' \rightarrow 3'$
exonuclease that
removes RNA primers
(14) Is the RNA polymerase
that makes primers
(15) Is a helicase that
functions at the
elongating replica-
tion fork
(16) Stabilizes unwound
DNA
(17) Uses NAD^+ to form
phosphodiester bonds

Mutations—Their Nature, Consequences, and Repair

18. Which of the following nucleotide substitutions are transition mutations?

(a) G for A
(b) A for C
(c) C for T
(d) T for G

19. Which of the substitutions in Question 18 are transversion mutations?

_____

20. How could the tautomerization of a keto group on a guanine residue in DNA to the enol form lead to a mutation in the DNA?

_____

_____

_____

21. Explain why most nucleotides that have been misincorporated during DNA synthesis in *E. coli* do not lead to mutant progeny.

_____

_____

_____

22. Match the type of mutation or physiological consequence in the right column with the appropriate mutagen in the left column.

    (a) 5-Bromouracil _____

    (b) 2-Aminopurine _____

    (c) Hydroxylamine _____

    (d) 9-Aminoacridine _____

    (e) Nitrous acid _____

    (f) Ultraviolet light _____

    (1) Transversion
    (2) Transition
    (3) Insertion or deletion
    (4) Translational frameshift
    (5) Block in replication

23. What property of DNA allows the repair of some residues damaged through the action of mutagens?

_____

_____

24. How does the repair machinery of *E. coli* identify a DNA strand that has recently misincorporated a noncomplementary nucleotide during replication in order to repair it?

_____

_____

_____

25. Since U pairs with A as well as T does, why is T used in DNA?

_____

_____

_____

26. Explain how some strains of *Salmonella* are used to detect carcinogens. How is an extract from human liver involved in this test?

_____

_____

_____

PROBLEMS

1. The following are regional nucleotide sequences on one of the two strands of B-DNA. Which sequence might induce kinking of the helix? Briefly explain your answer.

    (a) ACTGGUTT
    (b) GGTAAAGT
    (c) CTTTTTAG
    (d) GAACTCGC
    (e) GATCGATC

2. As discussed by Stryer on page 651, the major and minor grooves of B-DNA contain groups that are potential hydrogen-bond donors and acceptors and hence might be important in specific interactions between proteins and DNA. Suppose that a given region of B-DNA is GC rich. Will the number of potential hydrogen-bond–forming groups in the major groove and the minor groove differ from the number that would be present if the region were AT rich? Explain.

3. Which of the following nucleotide sequences might allow for the formation of Z-DNA? Explain your answer.

    (a) CCTAGTTA
    (b) TATATATA
    (c) CGTGTACA
    (d) TGAATTCA

4. Many proteins that bind to DNA consist of two α-helical recognition units that are related by a twofold axis of symmetry and are separated by a distance of 34 Å.

    (a) Are these proteins likely to bind to double-stranded or to single-stranded DNA? Explain.
    (b) Would the proteins bind preferentially to A-DNA, B-DNA, or Z-DNA? Explain.
    (c) Describe how these proteins likely interact with DNA.

5. Either ATP or NAD^+ may serve as the energy source for the reaction catalyzed by DNA ligase (see Stryer, p. 659). However, the energy potentials of ATP and NAD^+ differ greatly. Which will produce more energy? Why?

6. Suppose that two polynucleotide chains are joined by DNA ligase in a reaction mixture to which ATP labeled with ^{32}p in the α-phosphoryl group has been added as an energy source. What products of the reaction would be expected to carry the radioactive label? Explain.

7. Suppose that negatively supercoiled DNA with $L = 23$, $T = 25$, and $W = -2$ is acted upon by topoisomerase I. After one catalytic cycle, what would be the values of L, T, and W?

8. Suppose that negatively supercoiled DNA with $L = 23$, $T = 25$, and $W = -2$ is acted upon by DNA gyrase and ATP. After one catalytic cycle, what would be the values of L, T, and W?

9. Suppose that DNA synthesis involved the addition of nucleotides from 3'- rather than 5'-deoxynucleoside triphosphates. What properties of the DNA polymerases would be different? Explain briefly.

10. What property of DNA polymerase I leads to the observation that polA1 mutants are more sensitive to ultraviolet light than are their wild-type counterparts? (See p. 668 of Stryer.)

11. Suppose that a single-stranded circular DNA with the base composition 30% A, 20% T, 15% C, and 35% G serves as the template for the synthesis of a complementary strand by DNA polymerase.
    (a) Give the base composition of the complementary strand.
    (b) Give the overall base composition of the resulting double-helical DNA.

12. Suppose that a plasmid containing only genes A, B, C, and D begins to replicate rapidly at time $t = 0$. At $t = 1$, there are twice as many copies of genes B and C as there are copies of genes D and A. Is it possible to establish the order of the four genes on the plasmid? Explain.

13. Suppose that a bacterial mutant is found to replicate its DNA at a very low rate. Upon analysis, it is found to have entirely normal activity of DNA polymerases I and III, DNA gyrase, and DNA ligase. It also makes normal amounts and kinds of dnaA, dnaB, dnaC, and SSB proteins. The *oriC* region of its chromosome is found to be entirely normal with respect to nucleotide sequence. What defect might account for the abnormally low rate of DNA replication in this mutant? Explain briefly.

14. Which of the following mutations in a polypeptide chain could have been induced by a single hit of the mutagen 5-bromouracil on DNA? (See Stryer, p. 677. Also see the genetic code on p. 107 of Stryer.)

   (a) Phe $\longrightarrow$ Glu
   (b) Asp $\longrightarrow$ Ala
   (c) Phe $\longrightarrow$ Leu
   (d) Met $\longrightarrow$ Lys

15. Which of the following mutations could result from the action of a single hit of the mutagen hydroxylamine on DNA?

   (a) Gln $\longrightarrow$ Asn
   (b) Glu $\longrightarrow$ Lys
   (c) His $\longrightarrow$ Tyr
   (d) Gly $\longrightarrow$ Asp

16. The drug fluorouracil is widely used as an anticancer agent (see Stryer, p. 615). It irreversibly inactivates the enzyme thymidylate synthase. Explain how this treatment retards the growth of tumor tissue. Will the growth of normal cells be affected as well?

17. Mammalian cells of two differing genotypes can be fused together, usually in the presence of Sendai virus, to form multinucleate cells (heterokaryons) containing nuclei of both genotypes (see Stryer, p. 294). When fibroblasts from two patients suffering from xeroderma pigmentosum were fused, the resulting heterokaryons showed no deficiency in DNA repair. What conclusions can be drawn from this observation? Explain.

ANSWERS TO SELF-TEST

DNA Structure and Protein-DNA Interactions

1. a, b, and d. Answer (c) is incorrect because B-DNA has local variations from the average structure. The hydrogen-bonded base pairs are often twisted and tilted out of the plane that is perpendicular to the helix axis.

2. (a) In the major groove, N-7 of A is an acceptor, H-6 on the 6-exocyclic (i.e., not in the ring) amino group of A is a donor, and 0-4 on T is an acceptor.

(b) In the minor groove, N-3 of G is an acceptor, H-2 on the 2-exocyclic amino group of G is a donor, and 0-2 on C is an acceptor.

(c) The patterns in the major groove are *noh* for G : C and *hon* for C : G.

(d) The patterns in the minor groove are *no* for A : T and *on* for T : A.

(e) The patterns of hydrogen-bond donors and acceptors in the major and minor grooves of B-DNA can be recognized by complementary hydrogen-bond donors and acceptors on the amino acid side chains of proteins and thus facilitate the sequence-specific interaction of proteins and DNA. The methyl groups on T may also participate in this process through hydrophobic interactions with proteins.

3. Both double-stranded RNA and RNA-DNA hybrids in solution have structures like that of A-DNA. Thus, knowledge about the helical structure of A-DNA contributed to an understanding of two similar helices that are of physiological importance.

4. (a) 1, 5, 7, 8 (b) 1, 4, 5, 6, 8 (c) 1, 2, 3, 8

5. c, d, and e. Answer (a) is incorrect because DNase I interacts with the negatively charged phosphodiester bonds of the backbone and not with the bases. Answer (b) is incorrect because *positively* charged side chains on the enzyme interact with negatively charged phosphates of the DNA.

6. The enzyme interacts specifically with the palindromic DNA sequence GAATTC by forming 12 specific hydrogen bonds between side chains on the enzyme and the groups exposed in the major groove of the DNA (see Question 2). Only the sequence GAATTC offers the correct array of hydrogen-bond acceptors and donors for the 12 hydrogen bonds to form. The DNA is kinked by the enzyme to make room for the insertion of the four α-helices into the major groove of the DNA. In addition, there are electrostatic interactions between the enzyme and the DNA. The enzyme is a dimer and interacts symmetrically with the palindromic sequence.

7. b, c, d, and e. Answer (a) is incorrect because the enzyme joins a 3′-hydroxyl to a 5′-phosphate.

DNA Topology and Topoisomerases

8. a, b, c, and d. Regarding answer (e), supercoiling is a property of covalently-closed circular DNA—that is, of DNA in which there are no discontinuities in either strand of the helix. Hence, these molecules are not substrates for DNA ligase because they lack ends.

9. b, c, and e. The specific linking difference, or superhelix density, $\lambda = (L - L_0)/L_0$, where L is the linking number of the topoisomer and L_0 is the linking number of the relaxed molecule. Since, by definition, two topoisomers will differ in L values but will have identical L_0 values under the same conditions, λ will necessarily be different for the two topoisomers.

10. Yes. The linking number (*L*) of a given topoisomer equals the twisting number (*T*) plus the writhing number (*W*), and the value of *L* can be apportioned between *T* and *W* in nonintegral values. Thermal energy will cause an equilibrium distribution of *W* and *T* values in a population of identical topoisomers. If there is more writhing in a given topoisomer—that is, if *W* is larger—then *T* must be smaller, and there will be fewer Watson-Crick base pairs and more unwound bases. There are, of course, limits to the variabilities in the values of *T* and *W* for a given value of *L* that are set by the energetics of the system. Native DNA is more stable than denatured DNA, so base pairs are favored, and the helix is not infinitely flexible, so writhing is constrained by the stiffness of the DNA.

11. a, b, d, and e. Although all topoisomerases break and reseal phosphodiester bonds, an external energy source is not always required. Relieving the torsional stress in a negatively supercoiled DNA molecule by relaxing it with topoisomerase I is exergonic and requires no energy input, whereas introducing negative supercoils with DNA gyrase is endergonic and is coupled to ATP hydrolysis. The particular catalytic mechanisms of given topoisomerases determine whether they are coupled to ATP hydrolysis.

DNA Polymerases

12. a, b, d, and e. The $3' \rightarrow 5'$ exonuclease cleaves a 5'-dNMP from the 3'-hydroxyl of the primer, leaving a 3'-hydroxyl group on the DNA. The two exonucleases of DNA polymerase I can be separated from one another by cleaving the enzyme with a protease.

13. (a) 1, 2, 5 (b) 2, 3 (c) 1, 2, 4

DNA Replication

14. a, b, and d. Answer (c) is incorrect because the replicating *E. coli* chromosome has two replication forks that synthesize the DNA bidirectionally. Answer (e) is incorrect because only the lagging strand is synthesized discontinuously. Answer (f) is incorrect because RNA serves as a primer and not as a template.

15. a, c, and d. Answer (b) is incorrect because DNA polymerase III holoenzyme is a highly processive enzyme that synthesizes extensively before dissociating from its template.

16. Because DNA polymerases are unable to initiate DNA chains de novo and because they require a primer with a 3'-hydroxyl group, short RNA chains are used as primers to start DNA replication at the origin of replication and to initiate the Okazaki fragments of the lagging strand. RNA polymerases can start RNA chains by adding a nucleotide to an initiating NTP, but they do so with relatively low accuracy. RNA-initiated DNA chains facilitate high-fidelity replication at the beginning sequences of new chains because they allow DNA polymerase I to replace the ribonucleotides with DNA using its editing $3' \rightarrow 5'$ exonuclease by a nick translation reaction (see Stryer, p. 672).

17. (a) 7 (b) 9, 11 (c) 1, 5 (d) 3 (e) 1 (f) 2 (g) 16 (h) 6, 8 (i) 14 (j) 4 (k) 15 (l) 1, 10, 13 (m) 10, 12, 17. Answer (5) is not a correct match with (e) because each Okazaki fragment is synthesized continuously.

18. a and c. Transition mutations are substitutions of a purine for a purine or a pyrimidine for a pyrimidine.

19. b and d. Transversion mutations substitute a purine for a pyrimidine or vice versa.

20. The rare enol tautomer of G could base pair with T to allow its incorporation into a growing DNA strand during replication. If the proofreading process missed this erroneous incorporation, the resulting daughter DNA duplex would contain a GT base pair. During the next round of replication, the T would direct the incorporation of an A into its complementary daughter strand. The final results would be the substitution of an AT base pair for the original GC base pair.

21. The proofreading $3' \rightarrow 5'$ nuclease of the ϵ subunit of DNA polymerase III holoenzyme removes most of the misincorporated nucleotides that do not form a base pair with the template, thus giving the polymerase activity of the enzyme a second chance to incorporate the correct nucleotide.

22. (a) 2 (b) 2 (c) 2 (d) 3, 4 (e) 2 (f) 5

23. Since DNA is double stranded, damage to one strand of the DNA can often be repaired because the opposite strand contains the undamaged complementary nucleotide, which can direct the reincorporation of the correct nucleotide by acting as a template for a DNA polymerase.

24. The authentic parental template, which has not been used faithfully to direct the incorporation of the correct nucleotide, has its GATC sequences methylated, whereas the newly synthesized strand has not yet had time for its GATC sequences to become methylated. The methyl groups on the parental strand serve as "tags" to direct the repair enzymes to correct the newly synthesized, unmethylated strand.

25. In DNA, T but not U, facilitates the repair of a product of a spontaneous reaction of DNA that could lead to a mutation. C spontaneously deaminates to form U, a change that could lead to a mutation during replication since the U would pair with A. If U were the normal partner of A, the mutagenic product of a C deamination would not be a unique constituent of DNA and thus could not be so readily repaired by an enzyme that recognizes and removes it.

26. Special strains of *Salmonella* have been developed to detect substitution, insertion, and deletion mutations in their DNA as a result of exposure to exogenously supplied chemicals. The strains monitor the capacity of mutagens to alter their DNA and thus convert them from auxotrophs, which are unable to grow in the absence of histidine, to prototrophs. The revertants can grow and are detected with high sensitivity. Since there is a correlation between mutagenicity and carcinogenicity, these strains are used as an inexpensive initial test of the carcinogenic potential of a compound. Because animals sometimes metabolize innocuous compounds and convert them to carcinogens, incubation of a suspect chemical with a human liver extract before using the bacterial test can sometimes mimic what would happen to the chemical in vivo. This adjunct to the test expands its capacity to detect potential human carcinogens.

1. A run of at least four adenine residues can induce kinking (see Stryer, p. 651). Therefore, sequence (c) might induce kinking, since a run of five adenines would be present on the complementary strand.

2. Both GC and AT base pairs of B-DNA have one hydrogen-bond donor and two hydrogen-bond acceptors in the major groove. However, a GC pair has one donor and two acceptors in the minor groove, whereas an AT pair has only two acceptors. Thus, a GC rich region on DNA differs from an AT rich region with respect to the number of hydrogen-bond–forming groups in the minor groove.

3. Both (b) and (c) are alternating sequences of purines and pyrimidines and are therefore sites of possible Z-DNA formation (see Stryer, p. 653).

4. (a) They bind to a region of double-stranded DNA that also has a twofold axis of symmetry, like a palindromic region.
   (b) B-DNA. The distance between adjacent major grooves is 34 Å. Also these grooves are wide enough to accommodate the recognition helix of the binding protein.
   (c) The helical recognition regions bind to two adjacent major grooves of B-DNA.

5. When NAD^+ serves as the energy source, a pyrophosphate bond linking the adenosine portion and the nicotinamide portion of the molecule is broken, yielding approximately -7.3 kcal/mol of energy under standard conditions. When ATP serves as the energy source, approximately -7.3 kcal/mol of energy is provided by the cleaving of the ATP to AMP and pyrophosphate. Another -7.3 kcal of free energy may be obtained from the hydrolysis of pyrophosphate to two molecules of inorganic phosphate by pyrophosphatase. Therefore, ATP will produce more energy.

6. In the overall reaction, ATP is hydrolyzed to AMP and pyrophosphate. Only AMP would be labeled. The phosphates involved in the formation of phosphodiester bonds are furnished by the incoming polynucleotide chains and do not arise from ATP.

7. $L = 24$, $T = 25$, and $W = -1$. Topoisomerase I increases the linking number of DNA by 1 each catalytic cycle. (see Stryer, p. 663.) This increase comes about at the expense of unwinding the negative supercoil.

8. $L = 21$, $T = 25$, and $W = -4$. DNA gyrase catalyzes a reaction in which both DNA strands are broken, the linking number is decreased by 2, and the number of negative supercoils is correspondingly increased by 2.

9. The DNA polymerases would read in the $5' \rightarrow 3'$ direction and would synthesize new DNA in the $3' \rightarrow 5'$ direction. Under these conditions the incoming nucleotide residues would be added to a free $5'$-OH group. One would also expect DNA polymerase III to have $5' \rightarrow 3'$ rather than $3' \rightarrow 5'$ exonuclease activity to allow it to proofread and replace incorrectly matched nucleotides. DNA polymerase I would have both $5'$ and $3'$ exonuclease activity. However, the $3'$ exonuclease activity would be responsible for its role in DNA repair.

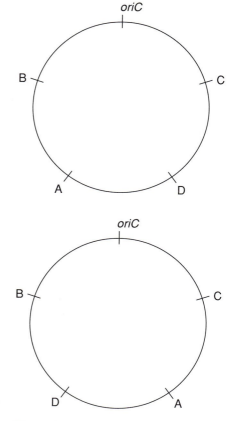

Figure 27-2
Two possible gene arrangements for Problem 12.

10. The polA1 mutants are deficient in DNA polymerase I, having only 1 percent of the activity of their wild-type counterparts. They can, however, replicate their DNA at normal rates because DNA polymerase III is the enzyme that is primarily responsible for DNA replication. They are extraordinarily sensitive to ultraviolet irradiation because they lack the $5' \rightarrow 3'$ exonuclease activity of DNA polymerase I and are therefore deficient at DNA repair.

11. (a) 30% T, 20% A, 15% G, 35%
    (b) 25% A, 25% T, 25% C, 25% G. The base composition of the double strand is the average of that of the two single strands.

12. The order cannot be unambiguously established. Two possibilities are shown in Figure 27-2.

13. A decrease in the activity of primase would account for the low rate of DNA replication. DNA replication requires the prior synthesis of RNA primers.

14. The mutagen 5-bromouracil changes AT pairs to GC pairs or GC pairs to AT pairs. The mutation in (c) could be induced by 5-bromouracil. For example, the DNA sequence AAA, which codes for phenylalanine, could be changed to the sequence AAG, which codes for leucine. The other mutations could not arise from treatment with 5-bromouracil. Remember that the genetic code presented in Stryer is expressed in terms of RNA. The sequence UUU on RNA corresponds to the sequence AAA on the informational strand of DNA. Leucine is encoded by the sequence CUU on RNA, which corresponds to the sequence AAG on the informational strand of DNA. Remember also that, unless otherwise specified, nucleotide sequences are written in the $5' \rightarrow 3'$ direction.

15. Hydroxylamine is a mutagen that causes the unidirectional change of CG pairs to TA pairs. The mutation in (a) cannot result from the action of hydroxylamine. Those in (b), (c), and (d) might. In (b), TTC (Glu) could change to TTT (Lys). In (c), ATG (His) could be converted to ATA (Tyr). In (d), ACC (Gly) could change to ATC (Asp) or GCC (Gly) could change to GTC (Asp).

16. Because thymidylate synthase is inactivated, the supply of 5'-dTTP is insufficient to support the synthesis of DNA at normal rates. If DNA synthesis is suppressed, so will be the rate of division of the tumor cells. This type of treatment takes advantage of the fact that tumor cells divide more rapidly than do normal cells. The dosage of the drug is adjusted so that it will primarily affect more rapidly dividing cells. However, the division of some normal cells may be retarded as well.

17. The fibroblasts from the two patients show complementation, so it is likely that the two patients suffer from different genetic variants of xeroderma pigmentosum. The action of several genes is likely responsible for the excision and subsequent repair of damaged DNA. One patient, for example, could have produced a normal nuclease that excises damaged DNA but could have been deficient in a ligase. The other patient could have produced normal ligase but could have been deficient in nuclease activity. There are at least seven different complementation groups among xeroderma patients.

CHAPTER 28

Gene Rearrangements: Recombination and Transposition

New sequences of nucleotides can be generated in DNA not only by mutation (Chapter 27) but also by recombination between two different DNA molecules or by the transposition of a segment of DNA within a chromosome. The breakage and reunion of fragments of DNA also rearranges gene order within the chromosome and is a mechanism for repairing damaged DNA and regulating gene expression. Review the material on plasmids and bacteriophages in Stryer (pp. 127–130) in preparation for studying this chapter.

Procaryotic gene rearrangements are presented in Chapter 28. Stryer first describes general recombination, which requires extensive sequence similarity between the interacting DNA duplexes. He describes the DNA strand movements that occur and explains how the recA and recBCD enzymes and a single-stranded DNA-binding protein effect the recombination. Stryer then introduces recombination systems that do not require extensive regions of sequence similarity. Plasmids are discussed in terms of their roles as mobile genetic elements. F-factor plasmids, which allow the exchange of DNA between bacteria, are described, as is their use to map the gene order of the bacterial genome when they are integrated into the bacterial chromosome to form an Hfr cell. Specialized R-factor plasmids, which contain genes that confer antibiotic resistance upon bacteria that harbor them, are presented next. Stryer concludes the chapter with a general discussion of the nature of transposons and of the kinds of changes they induce in DNA sequences as a result of their movements. Site-specific recombination, which occurs at unique sites not necessarily determined by sequence similarities between the reacting duplexes, will be covered in later chapters.

When you have mastered this chapter, you should be able to complete the following objectives.

Introduction (Stryer pages 687–688)

1. List the different kinds of *gene rearrangements*.

2. Distinguish between *general recombination* and *transposition* in terms of the requirement for *sequence similarity* between the reacting DNA molecules.

3. List the *biological functions* of gene rearrangement and recombination.

General Recombination: DNA Strand Mechanics (Stryer pages 688–691)

4. Draw a *generalized recombinant DNA molecule* and indicate the *parental origins* of each strand. Draw a *lap-joint* and list the enzymes that convert it to an intact duplex.

5. Explain how ^{32}P and *bromouracil* were used to elucidate the nature of the products resulting from general recombination.

6. Describe the *Holliday model* for general recombination. Explain how the *Holliday recombination intermediate* can form two distinct recombinant products.

7. Draw the *chi form*, and relate it to the Holliday model for recombination.

General Recombination: Enzymology (Stryer pages 691–695)

8. Outline the *strand-assimilation* reaction of the *recA protein* that gives rise to a *D loop*. List the substrates, intermediates, and products of the reaction and explain the roles of ATP.

9. Describe the consequences of the vectorial assembly of the *recA-single-stranded DNA filament* on recombination.

10. Relate the functioning of the *recBCD complex* to the generation of single-stranded DNA. Explain the roles of ATP and the *chi sequence* in this process.

11. Explain the functions of *single-strand-binding protein (SSB), topoisomerase I,* and *DNA ligase* in general recombination.

12. Outline the *SOS response* to *DNA damage*. Describe the role of the *protease* activity of *recA* and the role of the *lexA* repressor protein in this process.

F-Factor and R-Factor Plasmids (Stryer pages 695–698)

13. Define *allele* and distinguish between the rearrangement of alleles and the rearrangement of genes. Assign these events to either general recombination or to the action of *mobile genetic elements (transposons)*.

14. List the properties of *plasmids*. Describe *conjugation,* and explain the roles of *sex pili, pili receptors,* and *F factors* in this process.

15. Describe how an F factor can create an *Hfr cell*. Describe gene mapping using Hfr cells. Distinguish between an F factor and an *F' factor*.

16. Explain how the F factor and *lysogenic bacteriophages* facilitate *bacterial evolution*.

17. Describe the acquisition of *simultaneous multiple-drug resistance* by a bacterial population. Explain the role of *R-factor plasmids* in this process, and indicate the functions of the *resistance transfer factor (RTF)* and *r-genes*. Distinguish between *simple* and *complex* R-factor plasmids.

Transposons (Stryer pages 698–700)

18. Describe the structure of *insertion sequences (IS)*, and recount the functions of their *inverted terminal repeats* and *gene products*. Outline the events that occur during the insertion of a *transposon* into a chromosome.

19. Explain how transposons can lead to *gene activation* or *inactivation* and, depending upon *IS orientation*, to the *inversion, deletion,* or *duplication* of genes.

20. Distinguish between *simple* and *complex transposons*.

SELF-TEST

Introduction

1. Which of the following statements about gene rearrangements are *correct?*

   (a) They generate new combinations of genes.
   (b) They can move a segment of DNA from one chromosome to another.
   (c) They are mediated by the breakage of DNA and the rejoining of the resulting fragments.
   (d) They generate genome sequence variability, upon which natural selection can act.
   (e) They can regulate gene expression.
   (f) They always require extensive regions of sequence similarity between the interacting DNA molecules in order to juxtapose the recombining sites.

General Recombination; DNA Strand Mechanics

2. DNA was isolated from *E. coli* that had been infected, in the presence of a DNA-synthesis inhibitor, with a mixture of bromouracil-labeled and ^{32}P-labeled T4 bacteriophages.

   (a) If this DNA were centrifuged to equilibrium in a CsCl gradient, how would you expect it to distribute in the density gradient?

(b) Would the density distribution of the radioactivity and bromouracil be different if the DNA were heated prior to the centrifugation? Why?

_____

_____

_____

(c) What enzymes and substrates would you need to treat the isolated DNA prior to centrifugation to seal the lap-joints so that the recombinant molecules would be converted to intact duplex DNA?

_____

_____

_____

3. Place the following events in the order in which they would occur during general recombination between two DNA molecules.

(a) Strand exchange occurs between duplexes *via* branch migration at the crossover point.
(b) A pair of strands with similar sequences is cleaved in each duplex.
(c) Two duplexes align at a region of sequence similarity.
(d) Each invading strand becomes covalently joined to its corresponding strand in the other duplex.
(e) Strands of the recombinant intermediate are cleaved at or near the crossover point and are joined to their corresponding strands in each duplex.,
(f) The end of each single strand invades the other duplex and forms base pairs with its complementary strand.

4. Which of the following statements about structures like those at the nodes or crossover points of chi-form DNA molecules are *correct*?

(a) They are formed in high yields when bacteria bearing certain plasmids are grown in the presence of chloramphenicol, a protein-synthesis inhibitor.
(b) They are intermediates that are formed during plasmid replication.
(c) They are formed from plasmids in *E. coli* strains that have mutations in the rec genes.
(d) When they are derived from plasmids, they are composed of two duplexes that are connected *via* four single strands of DNA.
(e) When they are derived from plasmids, they can be resolved in two different ways; one resolution yields two unit-length circular molecules, and the other produces a single, two-unit long circular molecule.

5. When two circular plasmids containing a single restriction endonuclease site recombine and the recombination intermediates are isolated and cut with an endonuclease specific for that site, χ-shaped products with pairs of arms of equal length are formed. What do the structures

of the products reveal about the sequences at the site of the crossover that forms the node of the χ form?

_____

_____

_____

General Recombination: Enzymology

6. Which of the following statements about the recA protein of *E. coli* are *correct?*

(a) It is an ATP-dependent nuclease that generates single-stranded DNA.
(b) It catalyzes an ATP-dependent strand-assimilation reaction in which a single-stranded DNA molecule associates with duplex DNA.
(c) It hydrolyzes ATP to promote branch migration.
(d) It binds to single-stranded DNA to form a filament.
(e) It facilitates the search of duplex DNA for regions with sequence similarity to the invading single-stranded DNA.

7. Considering that duplex DNA is a stable molecule—that is, it ordinarily exists as a duplex in the cell—explain how the recA protein is capable of catalyzing the net insertion of a single-stranded DNA molecule into the duplex.

_____

_____

_____

8. Explain how the single-stranded DNA required for general recombination is generated.

_____

_____

_____

9. General recombination is likely to require which of the following?

(a) DnaB protein
(b) Topoisomerase I
(c) RecA protein
(d) RecBCD complex
(e) ATP
(f) NAD$^+$

(g) Single-strand-binding protein
(h) DNA-dependent RNA polymerase
(i) DNA polymerase I
(j) DNA ligase
(k) dATP

10. Which of the following statements about the lexA protein of *E. coli* are *correct?*

(a) It is cleaved by the recBCD complex.
(b) It represses the synthesis of recA protein.
(c) It is a protease as well as a repressor.
(d) It represses the synthesis of more than 15 different proteins.
(e) It is involved in the SOS response to DNA damage.

F-Factor and R-Factor Plasmids

11. Which of the following are transposable elements in *E. coli?*

   (a) Insertion sequences
   (b) Bacteriophage mu (μ)
   (c) R-factor plasmids
   (d) RecBCD complex
   (e) Bacteriophage lambda (λ)
   (f) Chi sequences
   (g) F-factor plasmids
   (h) F'-factor plasmids

12. Which of the following pairs of polypeptides have genes that are alleles?

   (a) Hemoglobin α chain and hemoglobin β chain
   (b) Hemoglobin β chain and hemoglobin β^S chain (found in Hb S, sickle-cell hemoglobin)
   (c) Hemoglobin α chain and hemoglobin γ chain (found in Hb F, fetal hemoglobin)
   (d) Hemoglobin α chain and myoglobin
   (e) Hemoglobin β chain and myoglobin
   (f) Lactic dehydrogenase (LDH) isozyme M (found in muscle) and LDH isozyme H (found in the heart)

13. Which of the following statements about plasmids are *correct?*

   (a) They can serve as accessory chromosomes in a bacterium.
   (b) They can be essential for bacterial chromosomal replication when the *oriC* sequences of both the plasmid and the bacterium are nearly identical.
   (c) They can sometimes recombine with the chromosome of the bacterium harboring them.
   (d) They can replicate independently of the chromosome of their bacterial host.
   (e) They can confer drug resistance upon the bacterium harboring them.
   (f) They can promote the transfer of genetic information between bacteria.

14. Which of the following are necessary for two bacteria to conjugate successfully?

   (a) Both must have pili.
   (b) Both must have pili receptors.
   (c) One must contain an F-factor plasmid.
   (d) DNA synthesis must occur in the donor.
   (e) DNA synthesis must occur in the acceptor.

15. What is an Hfr cell?

   _____

   _____

16. Which of the following statements about the F' factor are *correct?*

   (a) Its formation requires the integration of an F factor into the chromosome of a bacterium.
   (b) It can give rise to a bacterium that is diploid for some genes.
   (c) It is an F factor without a gene specifying a pilus protein.

(d) It is functionally equivalent to the transducing phages in that chromosomal DNA can associate with its DNA and be transferred between cells.

(e) It can result in a bacterium becoming auxotrophic for an amino acid.

17. A transmissible or infectious drug-resistance plasmid can be created when

(a) an *r*-gene integrates into an R-factor plasmid bearing a resistance transfer factor (RTF).

(b) an R-factor plasmid loses several *r*-genes.

(c) a RTF plasmid integrates into the chromosome of its host bacterium.

(d) a simple R-factor plasmid excises from the host bacterial chromosome.

Transposons

18. Which of the following statements about transposons are *correct*?

(a) They contain insertion sequences.

(b) They contain inverted terminal repeat sequences.

(c) They contain one or more genes specifying one or more enzymes that catalyze the transposition event.

(d) They require the rec gene products to complete their movements between or within genomes.

(e) They can lead to the duplication of short sequences of DNA in the recipient genome.

19. How could the insertion of a transposon into a bacterial chromosome enhance the expression of a bacterial gene?

_____

_____

_____

PROBLEMS

1. Suppose that one parental DNA has the genes A, B, and C, in that order, and that the other parental DNA has the alleles a, b, and c.

(a) Give all the recombinant genotypes that may be formed.

(b) Which recombinant genotype(s) will be produced least frequently? Why?

2. Suppose that an Hfr strain of *E. coli* that is sensitive to streptomycin (str^s) and is able to synthesize amino acids A and B is mixed together in a culture medium with an F$^-$ strain that is resistant to streptomycin and is unable to synthesize amino acids A and B. After some time, mating is interrupted by placing the mixture in a Waring blender and then transferring samples of the bacteria to other culture media to test whether or not conjugation has occurred.

(a) When the bacteria are transferred to a medium containing streptomycin but lacking amino acids A and B, some bacterial growth occurs. Explain this result.

(b) Suppose that you carry out a similar experiment but place the bacterial mixture into the blender after a shorter time. Explain why some bacteria might grow in a medium containing streptomycin and A but not B.

(c) Studies of conjugation in bacteria always involve following more than one marker. In this experiment, for example, two markers, A and B, are followed. Why do you think this is necessary? What possibility could not be excluded if only gene A were followed? (Hint: Recombination in bacteria is a rare event.)

3. Suppose that in a bacterium having a single circular chromosome there is a region in which there are 26 genes in alphabetical order. Several Hfr strains of this bacterium are known. When Hfr_1 is used in mating, gene D is the first of the genes in this region to be transferred to the F^- recipient. When Hfr_2 is used, gene L is the first to be transferred. When Hfr_3 is used, gene X is the first to be transferred. Explain these results.

4. An early and very attractive mechanism proposed for genetic recombination was the copy-choice model. It suggested that recombination between two parental DNA duplexes occurs during DNA replication when DNA polymerase switches from one parental duplex to the other so as to produce a recombinant daughter DNA duplex. The copy-choice model is now known to be incorrect. Cite experimental evidence involving T4 bacteriophage that shows the copy-choice model is not the mechanism of genetic recombination.

5. When one thinks about genetic recombination from a classical point of view, one may envision that the further apart two genes are from one another on a chromosome, the greater the likelihood of recombination between them will be. However, the recombination frequency between two given nucleotides is not always a function of the internucleotide distance separating them. Briefly explain how recombination may not occur, for example, between two sites in the *E. coli* chromosome that are 3000 nucleotide residues apart.

6. When transposons having long, inverted repeat sequences are denatured and are allowed to self-anneal, they form a characteristic structure that may be seen through electron microscopy. Sketch the structure that would be expected for a transposon having the sequence ABCDEFGHIJKLGFEDCBA.

7. Sketch the product(s) that would be expected when recombination between A and B occurs in the insertion sequence ABCWXYZcba. What happens to the sequence WXYZ?

8. Sketch the product(s) that would be expected when recombination between B and C occurs in the insertion sequence ABCDEFGHabcd. What happens to the sequence EFGH?

9. The kind of recombination displayed by transposons, sometimes called recA-independent recombination, has been regarded as having far greater evolutionary significance than does general, or recA-dependent, recombination. Why do you think this may be the case?

Introduction

1. a, b, c, d, and e. Answer (f) is incorrect because transposition, unlike general recombination, doesn't require extensive sequence similarity between the regions of DNA that react with one another.

General Recombination: DNA Strand Mechanics

2. (a) In addition to heavy, bromouracil-labeled DNA and light, ^{32}P-labeled DNA derived from unaltered parental genomes, you would expect to find recombinant molecules of intermediate density containing both labels.
   (b) Yes. Heating would disrupt the base pairing of the lap-joints of the recombinant molecules, so no intermediate-density molecules would be present.
   (c) DNA polymerase I, dNTPs, Mg^{2+}, T4 DNA ligase, and ATP (or NAD$^+$ and bacterial DNA ligase) would fill in the gaps in the lap-joints and form intact phosphodiester backbones in the DNA of the recombinant molecules.

3. c, b, f, d, a, and e. (See Figure 28-4, p. 689 in Stryer.)

4. a, d, and e. Answer (b) is incorrect because χ-form nodes are intermediates in recombination, not replication. Answer (c) is incorrect because their formation depends on the rec gene products. For answer (e), see Figure 28-6 on page 690 of Stryer.

5. Pairs of arms of equal length show that the crossover occurs at regions of sequence homology between the two joined molecules. Crossover at regions of sequence similarity will give rise to intermediates in which the single restriction site is uniquely located with respect to the joint. If the crossover occurred randomly, χ forms would be likely to have arms of unquel length because the unique restriction sites would be randomly distributed with respect to the node. Because the pairs of arms on χ structures can have any length, recombination can occur at almost any location on the DNA. (See Figure 28-7, p. 691 in Stryer.)

General Recombination: Enzymology

6. b, c, d, and e. Answer (a) is incorrect because the recA protein lacks nuclease activity.

7. The recA protein binds both single-stranded and duplex DNA and facilitates interactions between them at regions that share sequence similarity. ATP plays several roles in the overall recombination reaction. It is necessary for the binding of recA to single-stranded DNA, and its hydrolysis is required for the release of the DNA. ATP hydrolysis is coupled to the extension of the recA-DNA filament in the $5' \rightarrow 3'$ direction only, leading to unidirectional branch migration and invasion of the single-stranded DNA into the recipient duplex. In addition, ATP hydrolysis allows the invading single-stranded DNA to proceed past

regions of duplex DNA that do not share sequence similarity. In summary, ATP provides the energy to destabilize the DNA helix and facilitate the uptake of an invading single strand of DNA into the duplex to initiate recombination. It powers the branch migration and prepares the recombination intermediate for completion by the other required enzymes.

8. The products of the recB, recC, and recD genes form the recBCD complex, an ATP-drive, and site-specific nuclease and helicase that unwinds duplex DNA and hydrolyzes it to create free, single-stranded DNA. The *rec*BCD complex binds to the end of a linear duplex and acts as a helicase to unwind the DNA as it invades the duplex. The unwound DNA rewinds behind the moving complex more slowly than it is unwound, leading to the formation of two single-stranded loops. When the complex encounters a chi site, which has a specific sequence, it cuts the strand. The complex moves on, thus forming single-stranded DNA with a free end (see Figure 28-13, p. 693 in Stryer). The single-stranded DNA can then interact with the recA protein to initiate a strand-assimilation reaction with a homologous duplex.

9. b, c, d, e, f, g, i, j, and k. ATP is required for the actions of recA and the recBCD complex; dATP and the other 3 dNTPs, for DNA synthesis by DNA polymerase I; and NAD^+, for the DNA-ligase reaction.

10. b, d, and e. The lexA protein is a repressor that regulates several genes whose products are involved in DNA repair via the SOS pathway. Single-stranded DNA, which arises from the damaged DNA, binds to recA protein, and the resulting ssDNA-recA complex cleaves lexA proteolytically, allowing the proteins involved in the repair of DNA to be synthesized.

F-Factor and R-Factor Plasmids

11. a, b, c, e, g, and h. Answer (f) is incorrect because chi sequences promote general homologous recombination, which does not ordinarily result in the rearrangement of the order of genes on the chromosome. See Table 28-1 on page 695 of Stryer.

12. b. Only in choice (b) are the two proteins derived from two different or alternative forms of the same gene. For all the other pairs, the two proteins are derived from different genes. General homologous recombination generates new combinations of alleles, whereas transposition can reorder the positions of genes.

13. a, c, d, e, and f.

14. c, d, and e. Answers (a) and (b) are incorrect because the male bacterium must have pili and the female must have pili receptors.

15. Bacteria that harbor an F-factor plasmid in their chromosomes are called Hfr cells because they exhibit a high frequency of recombination. An Hfr cell is capable of transferring its entire chromosome into a female cell by conjugation. The F factor contains the genes for the formation of the sex pili and the other components required for conjugation.

16. a, b, d, and e. The integration of an F factor is a necessary prerequisite to the formation of an F' factor. When an F' factor carries a chromosomal gene into a bacterium that duplicates one already present, the recipient cell becomes diploid for that gene. If an F' factor integrates

into the bacterial chromosome such that a chromosomal gene encoding an essential enzyme needed to synthesize an amino acid is disrupted, the bacterium would then have a nutritional requirement, that is, it would be auxotrophic, for that amino acid.

17. a. An infectious plasmic bearing genes that confer drug resistance on a bacterium can be formed by recombination between two plasmids: an R-factor plasmid containing an RTF and a plasmid bearing one or more drug resistance genes. The excision of a simple R-factor plasmid would not produce a transmissible plasmid because it would lack the RTF genes.

Transposons

18. a, b, c, and e. A transposon can be composed of as little as an insertion sequence (IS) and of as much as the IS plus several genes, including those necessary for transposition. That is, a transposon may contain only a transposase or it may contain a transposase plus a resolvase plus additional genes that specify drug resistance as well as other functions. Transposition does not require the rec gene products, but it does use DNA polymerase I and DNA ligase.

19. A promoter on the transposon might start a transcript that extends into a neighboring bacterial gene. This transcript would lead to enhanced expression of the neighboring gene with respect to the level occurring prior to the transposition.

ANSWERS TO PROBLEMS

1. (a) The possible recombinant genotypes are Abc and aBC, resulting from recombination between A and B only; ABc and abC, resulting from recombination between B and C only; and AbC and aBc, resulting from simultaneous recombination between A and B and between B and C.

   (b) The genotypes AbC and aBc will be produced least frequently because they involve the simultaneous occurrence of rare events.

2. (a) Neither parent strain could grow in such a medium. The Hfr parent could not survive because the medium contains streptomycin, and the F⁻ parent would be killed because the medium lacks amino acids A and B. The result is explained by conjugation followed by genetic recombination. The Hfr strain could have donated DNA containing information for the production of enzymes that synthesize amino acids A and B to the F⁻ strain during mating. Provided that there was recombination between the newly-introduced DNA and the DNA of the F⁻ chromosome, the resulting progeny would be able to synthesize amino acids A and B and would be resistant to streptomycin.

   (b) This would occur if B were the first gene to be transferred during mating and A were the second. If mating were interrupted after the transfer of B but before the transfer of A, the result would be F⁻ bacteria that are streptomycin resistant but require the addition of A for growth.

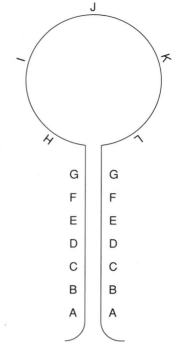

Figure 28-1
Structure of a transposon with long, inverted repeats following denaturation and self-annealing.

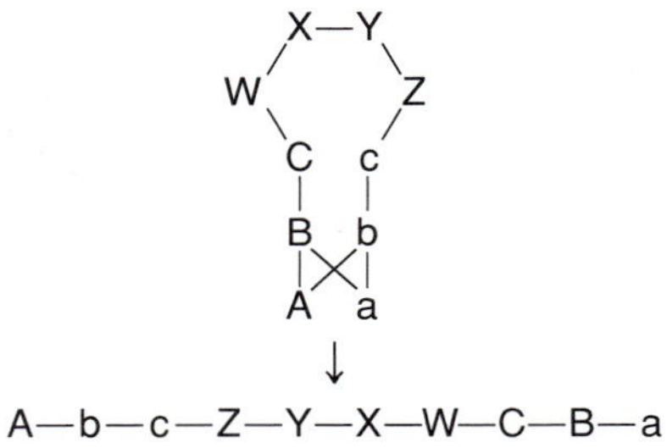

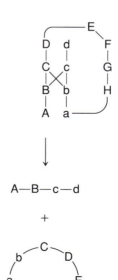

Figure 28-2
Solution to Problem 7.

Figure 28-3
Solution to Problem 8.

(c) In an F⁻ strain that is unable to synthesize A and B, it is always possible that regaining the ability to make either A or B could result from mutation and not from conjugation. (The frequencies of occurrence of mutation and recombination are similar.) However it is highly improbable that mutation would simultaneously reactivate the genes for producing both A and B.

3. The F factor in these strains has different sites of integration into the chromosome of the F⁻ strain. In Hfr_1 the F factor lies between genes C and D; in Hfr_2 it lies between K and L; and in Hfr_3 it lies between W and X.

4. In the copy-choice model, the parental DNA duplexes remain intact. An alternative to the copy-choice model is the breakage-reunion model, which requires that homologous parental DNA duplexes be broken and rejoined. One line of experimental evidence that favors the breakage-reunion model is described on page 688 of Stryer. One sample of T4 bacteriophage was labeled with bromouracil, and another was labeled with ^{32}P. When *E. coli* cells were infected with a mixture of these two labeled phages, recombinant DNA was found that contained both ^{32}P and bromouracil in the absence of any new DNA synthesis. The breakage and reunion of the parental DNA can account for the results, but the copy-choice model cannot.

5. The mechanism of recombination mediated by the recA protein requires that breaks be introduced into one strand of a DNA duplex by the recBCD complex (see Figure 28-13 on p. 693 of Stryer). The break occurs near a specific sequence called *chi*, which occurs on the average of once for each 4000 nucleotide residues in *E. coli* (see p. 694 of Stryer).

6. The structure is a "lollipop," as shown in Figure 28-1.

7. The new sequence is AbcZYXWCBa, as shown in Figure 28-2. The original sequence WXYZ is inverted.

8. The two products are shown in Figure 28-3. The sequence EFGH is deleted from one of the products.

9. RecA-dependent recombination can only occur between homologous segments of chromosomes. RecA-independent recombination allows for rather drastic rearrangements of chromosomes, including the insertion, deletion, and duplication of genes, as shown by the illustrations on page 699 of Stryer. Without recA-independent recombination, evolutionary possibilities would be far more limited.

RNA Synthesis and Splicing

An early step in the expression of genetic information is the conversion of DNA nucleotide sequences into RNA sequences. DNA-dependent RNA polymerases catalyze this reaction, and the product RNA transcript must sometimes be cleaved and modified in various ways before it becomes functional. You were introduced to these topics in Chapter 5; in this chapter Stryer describes them in detail in both procaryotes and eucaryotes. Reread Chapter 5, paying particular attention to the nomenclature conventions, structures, and kinds of RNA found in cells; RNA polymerases, promoters, and terminators; and introns and RNA splicing. Also refresh your understanding of the relationship between the mRNA nucleotide sequence and the protein amino acid sequence, which is outlined in Chapter 5. Review the material on DNA polymerase I on pages 84–85 and 666–667 of Stryer in order to be better able to contrast its properties with those of the RNA polymerases. In addition, reread pages 214–215 in Stryer on catalytic RNA prior to studying the section on self-splicing RNA in this chapter.

Stryer begins this chapter by describing RNA polymerase from *E. coli*. He represents the subunit structure of the enzyme and divides the RNA synthesis reaction into initiation, elongation, and termination stages. The nature of procaryotic promoters is given and the function of the σ subunit of RNA polymerase in specific transcript initiation is explained. Stryer next describes the structure of the ternary elongating complex consisting of the template DNA, RNA polymerase, and product RNA. Finally, two mechanisms of transcription termination, one of which requires the ρ protein, are detailed. Stryer outlines the cleavage and modification reactions involved in the maturation of ribosomal and transfer RNAs. In addition, he describes two antibiotics, rifamycin and actinomycin D, that inhibit transcription by different mechanisms.

Stryer next turns to transcription in eucaryotes. He describes the three RNA polymerases that carry out this process and relates them to the kinds of RNA they synthesize. Eucaryotic promoters and enhancers are then presented, as are the reactions that modify the 5′ and 3′ ends of typical eucaryotic mRNA transcripts to cap and add a poly A tail to them. In the last part of the chapter, the reactions that remove RNA segments from the interior of eucaryotic RNA precursors, that is, the splicing reactions that remove introns, are outlined. The mechanistic and possible evolutionary relationships among three different splicing mechanisms, two of which do not require the participation of proteins, are also given.

LEARNING OBJECTIVES

When you have mastered this chapter, you should be able to complete the following objectives.

RNA Polymerase from *E. coli* (Stryer pages 703–708)

1. Describe the *subunit structure* of *RNA polymerase* from *E. coli*. Where possible, assign functions to the individual subunits.

2. Name the three *stages* of *RNA synthesis,* and list the functions of RNA polymerase in these processes.

3. Define *promoter,* and recount how RNA polymerase protects promoters from hydrolysis by DNase. Explain how *footprinting* can be used to locate promoters.

4. Recognize the convention for numbering the nucleotides in the DNA template with regard to the *transcription start site*. Distinguish between the *template* (or *antisense*) *strand* and the *coding* (or *sense*) *strand* of the duplex DNA template.

5. Note the *consensus sequences* around the −35 and −10 positions of *E. coli* promoters. Contrast the rates of transcript initiation on *strong* and *weak promoters* in *E. coli.*

6. Explain how the σ *factor* enables RNA polymerase to *recognize promoters*. Describe the effect of the *unidimensional random walk* of RNA polymerase along the DNA template on the rate constant for the binding of RNA polymerase to the promoter.

7. Distinguish between the σ^{70} and σ^{32} subunits, contrast the sequences of *standard promoters* and *heat-shock promoters,* and provide examples of how σ factors can determine which genes are expressed.

8. Describe how topoisomerase I was used to determine the number of promoter *DNA base pairs that are unwound* upon the binding of RNA polymerase. Relate *negative supercoiling* to promoter efficiency. Distinguish between *closed promoter* and *open promoter complexes.*

Procaryotic Transcription (Stryer pages 708–715)

9. Draw the structure of the *5′ end* of an RNA transcript in *E. coli.* Recount an experiment showing that RNA polymerase elongates RNA chains in the 5′→3′ direction.

10. Describe the model for the *transcription bubble.* State the *number of base pairs in the RNA-DNA hybrid.* Appreciate the *rate of RNA chain elongation* in terms of both the nucleotides added and the distance on the template traversed by RNA polymerase.

11. Contrast the *fidelity of polymerization* of RNA polymerase with that of DNA polymerase, and explain the reason for this difference.

12. Sketch a *ρ-independent transcription termination structure* and state the roles of the *GC-rich hairpin* and *oligo-U stretch* of the RNA structure in the termination of transcription.

13. Contrast *ρ-dependent* and *ρ-independent* transcription termination. Outline the mechanisms of the *ρ* protein and explain the role of *ATP hydrolysis* in its function.

14. Compare the degree of *modification* of mRNA with that of *ribosomal* and *transfer RNA* (*rRNA* and *tRNA*) in procaryotes. Outline the mechanisms of *intron removal* from some *eucaryotic tRNA precursors*.

15. Describe the mechanisms of *inhibition of transcription* by *rifamycin* and *actinomycin D*.

Eucaryotic Transcription (Stryer pages 716–722)

16. Note the spatial and temporal differences in transcription and translation between procaryotes and eucaryotes. Consider the regulatory implications of these differences.

17. Contrast *mRNA processing* in procaryotes and eucaryotes. List the major reactions that process the *heterogeneous nuclear RNA* (hnRNA) of eucaryotes.

18. Describe the *RNA polymerases of eucaryotes* and list the kinds of RNA they synthesize. Compare the structural and catalytic features they share with the RNA polymerase of *E. coli*.

19. Account for the toxic effects of *α-amanitin* and describe how it can be used to differentiate the three eucaryotic RNA polymerases.

20. List the salient *sequence elements* of *eucaryotic promoters*.

21. Describe the functions of *transcription factors*.

22. Describe *enhancers*, list their functions, and present a molecular model that accounts for their activity.

23. Draw the structure of the *5′ end* of a typical *eucaryotic mRNA*, and distinguish *caps 0, 1,* and *2*. Outline the reactions required to cap the primary transcript.

24. Contrast eucaryotic and procaryotic transcription termination. Describe the events leading to the production of mRNA with a *poly A tail*.

RNA Splicing (Stryer pages 722–729)

25. Describe the *splicing of eucaryotic mRNA*, give the *consensus sequences* at the *splice site junctions*, and designate the other nucleotide *sequence elements* involved in the process.

26. Provide an example of *aberrant splicing* that leads to a form of *thalassemia* in humans.

27. Draw the structure of the nucleotide at the *branch site* of an mRNA *lariat intermediate*.

28. Describe the role of *transesterification reactions* in splicing, and compare the number of phosphodiester bonds broken and formed during mRNA splicing.

29. Describe the *spliceosome,* and detail the involvement of *small nuclear ribonucleoprotein particles (snRNPs)* in mRNA splicing.

30. Contrast the *group I* and *group II self-splicing* introns. Compare *spliceosome-catalyzed* splicing to self-splicing.

SELF-TEST

RNA Polymerase from *E. coli*

1. Give the subunit composition of both the core and holoenzyme for the RNA polymerase of *E. coli.*

---

2. Match the subunit of the RNA polymerase of the *E. coli* in the left column with its putative function during catalysis from the right column.

(a) α _____   (1) Binds the DNA template
(b) β _____   (2) Uncertain
                        (3) Binds NTPs and catalyzes bond
(c) β' _____   formation
(d) σ^{70} _____   (4) Recognizes the promoter and
                            initiates synthesis

3. Which of the following statements about *E. coli* promoters are *correct?*

(a) They may exhibit different transcription efficiencies.
(b) For most genes they include variants of consensus sequences.
(c) They specify the start sites for transcription on the DNA template.
(d) They have identical and characteristic sequences.
(e) They are activated when C or G residues are substituted into their -10 regions by mutation.
(f) Those that have sequences that correspond closely to the consensus sequences and are separated by 17 base pairs are very efficient.

4. The sequence of a duplex DNA segment in a DNA molecule is

5'-ATCGCTTGTACGGA-3'
3'-TAGCGAACATGCCT-5'

When this segment serves as a template for *E. coli* RNA polymerase, it gives rise to a segment of RNA with the sequence

5'-UCCGAACAAGCGAU-3'

Which of the following statements about the DNA segment are *correct?*

(a) The top strand is the coding strand.
(b) The bottom strand is the sense strand.
(c) The top strand is the template strand.
(d) The bottom strand is the antisense strand.

5. Which of the following statements about the σ subunit of RNA polymerase are *correct?*

   (a) It enables the enzyme to transcribe asymmetrically.
   (b) It confers the ability to initiate transcription at promoters upon the core enzyme.
   (c) It decreases the affinity of RNA polymerase for regions of DNA that lack promoter sequences.
   (d) It facilitates the termination of transcription by recognizing hairpins in the transcript.

6. The rate constant for the binding of RNA polymerase holoenzyme to a promoter on a long DNA molecule is greater than that for the collision of two small molecules in solution. Since small molecules diffuse through solutions more rapidly than large ones, how can this be true?

_____

_____

_____

7. When growing *E. coli* is subjected to a rapid increase in temperature, a new and characteristic set of genes is expressed. Explain how this alteration in gene expression occurs.

_____

_____

Procaryotic Transcription

8. A restriction fragment of bacterial DNA is transcribed *in vitro* by *E. coli* RNA polymerase in the presence of different ^{32}P-labeled ribonucleoside triphosphates (NTPs). In NTPs the innermost phosphate is designated α, the middle phosphate is β, and the outermost phosphate is γ. For which of the following pairs of radiolabeled NTPs would a mixture containing *either* of the radiolabeled NTPs along with the other three NTPs (unlabeled) necessary to make a complete set of all four NTPs result in a transcript that contains ^{32}P?

   (a) $[\alpha\text{-}^{32}P]ATP$ or $[\alpha\text{-}^{32}P]GTP$
   (b) $[\beta\text{-}^{32}P]ATP$ of $[\beta\text{-}^{32}P]GTP$
   (c) $[\gamma\text{-}^{32}P]ATP$ or $[\gamma\text{-}^{32}P]GTP$
   (d) $[\alpha\text{-}^{32}P]CTP$ or $[\alpha\text{-}^{32}P]UTP$
   (e) $[\beta\text{-}^{32}P]CTP$ or $[\beta\text{-}^{32}P]UTP$
   (f) $[\gamma\text{-}^{32}P]CTP$ or $[\gamma\text{-}^{32}P]UTP$

9. Which of the following are components or structures that lie within the transcription bubble formed by *E. coli* RNA polymerase?

   (a) The polymerization site of the enzyme
   (b) The σ subunit
   (c) Approximately 17 nucleotides of the DNA coding strand in the single-stranded form
   (d) Approximately 12 nucleotides of the 5' end of the elongating RNA strand
   (e) Approximately 12 base pairs of RNA-DNA duplex

10. Match the regions of a ρ-independent transcription termination signal in a DNA template in the left column with the structures or the functions performed by the encoded transcript segments in the right column.

(a) GC-rich palindromic region _____

(b) AT-rich region _____

(1) Oligo-U stretch in RNA
(2) Hairpin in RNA
(3) Promotes the dissociation of RNA-DNA hybrid helix
(4) Causes the enzyme to pause

11. Which of the following statements about the ρ protein of *E. coli* are correct?

(a) It is an ATPase that is activated by binding to single-stranded DNA.
(b) It recognizes specific sequences in single-stranded RNA.
(c) It recognizes sequences in the DNA template strand.
(d) It causes RNA polymerase to terminate transcription at template sites that are different from those that lead to ρ-independent termination.
(e) It acts as a RNA-DNA helicase.

12. To what extent are mRNA, rRNA, and tRNA modified in procaryotes? List some of the modifications that occur.

_____

_____

_____

13. Match the functions in the right column with the antibiotic inhibitors of *E. coli* transcription in the left column.

(a) Rifamycin _____

(b) Actinomycin D _____

(1) interacts with the template.
(2) interacts with RNA polymerase.
(3) prevents initiation.
(4) prevents elongation.
(5) intercalates into RNA hairpins.

14. Explain how a mutation might give rise to an *E. coli* that is resistant to the antibiotic rifamycin.

_____

_____

_____

Eucaryotic Transcription

15. Which of the following statements about eucaryotic mRNAs are *correct*?

(a) They are derived from hnRNA precursors.
(b) They result from extensive processing of their primary transcripts before serving as translation components.

(c) They usually have poly A tails at their 5′ ends.
(d) They have a "cap" at their 3′ ends.
(e) They are often encoded by noncontiguous segments of template DNA.

16. Match the descriptions in the right column with the appropriate eucaryotic DNA-dependent RNA polymerases in the left column.

(a) RNA polymerase I
_____

(b) RNA polymerase II
_____

(c) RNA polymerase III
_____

(1) is located in the nucleolus.
(2) is located in the nucleoplasm.
(3) makes hnRNA.
(4) makes tRNA.
(5) makes rRNA.
(6) makes rRNA precursors.
(7) is strongly inhibited by α-amanitin.
(8) syntheses RNA in the 5′→3′ direction.
(9) is composed of several subunits.

17. List the major sequence features of promoters for eucaryotic mRNA genes.

_____

_____

_____

18. Describe a biochemical model that explains how the upstream promoter elements referred to in Question 17 might act to facilitate transcription.

_____

_____

_____

19. Which of the following statements about enhancers are *correct?*

(a) They may function as promoters.
(b) They may function when in either orientation in the DNA.
(c) They may function when on either side of the activated promoter.
(d) They may function even when located many base pairs away from the promoter.
(e) They may function only in specific types of cells and tissues.

20. Which of the following are substrates or products in the capping reactions that form the structure at the 5′ end of mature mRNAs in eucaryotes?

(a) PP$_i$
(b) P$_i$
(c) GTP
(d) GDP
(e) GMP
(f) 5′-Triphosphate-terminated RNA
(g) 5′-Diphosphate-terminated RNA
(h) 5′-Monophosphate-terminated RNA
(i) 5′-Hydroxyl-terminated RNA
(j) *S*-Adenosylmethionine (SAM)
(k) *S*-Adenosylhomocysteine (SAH)
(l) $N^5 N^{10}$-Methylenetetrahydrofolate

21. Which of the following statements about the poly A tails that are found on most eucaryotic mRNAs are *correct*?

    (a) They are added as preformed polyriboadenylate segments to the 3' ends of mRNA precursors by an RNA ligase activity.

    (b) They are encoded by stretches of polydeoxythymidylate in the template strand of the gene.

    (c) They are added by RNA polymerase II in a template-independent reaction using ATP as the sole nucleotide substrate.

    (d) They are added by poly A polymerase using dATP as the sole nucleotide substrate.

    (e) They are cleaved from eucaryotic mRNAs by a sequence-specific endoribonuclease that recognizes the RNA sequence AAUAAA.

RNA Splicing

22. Which of the following are important sequence elements in the splicing reactions that produce eucaryotic mRNAs?

    (a) Exon sequences located between 20 and 50 nucleotides from the 5' splice site

    (b) Exon sequences located between 20 and 50 nucleotides from the 3' splice site

    (c) Intron sequences located between 20 and 50 nucleotides from the 5' splice site

    (d) Intron sequences located between 20 and 50 nucleotides from the 3' splice site

    (e) Intron sequences at the 5' splice site

    (f) Intron sequences at the 3' splice site

23. Eucaryotic mRNA splicing involves which of the following?

    (a) The formation of $2' \rightarrow 5'$ phosphodiester bonds

    (b) A sequence-specific endoribonuclease to hydrolyze the phosphodiester bond at the junctions of the intron with the exon

    (c) The spliceosome

    (d) The coupling of phosphodiester bond formation to ATP hydrolysis

    (e) The formation of lariat intermediates

24. What are snRNPs, and how are they involved in the eucaryotic mRNA splicing reaction?

_____

_____

_____

25. Place the following events in the order in which they occur during the formation of mature rRNA in *Tetrahymena*.

_____

    (a) The 3'-hydroxyl of a guanine nucleoside attacks the phosphodiester bond at the 5' splice site leaving the 5' (upstream) exon with a free 3'-hydroxyl and attaching the guanine nucleoside to the 5' end of the intron.

(b) The transcript from the rRNA gene folds into a specific structure.

(c) The rRNA transcript specifically binds a guanine nucleoside or nucleotide.

(d) Self-splicing occurs within the intron to form L19 RNA.

(e) The 3'-hydroxyl of the 5' exon attacks the bonds at the 3' splice junction to form the spliced rRNA and eliminate the intron.

26. Match the descriptions in the right column with the type of splicing in the left column.

(a) Group I splicing

———

(b) Group II splicing

———

(c) Nuclear mRNA splicing

———

(d) Eucaryotic tRNA splicing

———

(1) A 2'→5' phosphodiester bond is involved.

(2) Nuclease and ligase activities are required.

(3) Transesterification reactions break and form bonds.

(4) A guanine nucleoside or nucleotide is required.

(5) Spliceosomes are required.

(6) Lariat intermediates are involved.

(7) snRNPs are involved.

PROBLEMS

1. The technique of footprinting described by Stryer on page 705 involves, in part, the digestion of unprotected DNA by DNase I.

(a) Would you expect a restriction endonuclease such as EcoRI to be useful in footprinting experiments? Explain.

(b) Why is it necessary for digestion with the cleavage enzyme to be brief and not extensive?

2. One would expect an analog of 5'ATP that lacks an oxygen at the 3' position of its ribose—(3'-deoxy-5'ATP (see Figure 29-1)—to interrupt RNA formation since it cannot form phosphodiester bonds at its 3' position. Could such a compound be used to establish the direction of chain growth in RNA synthesis? Explain. (Refer to Figure 29-10 on p. 709 of Stryer.)

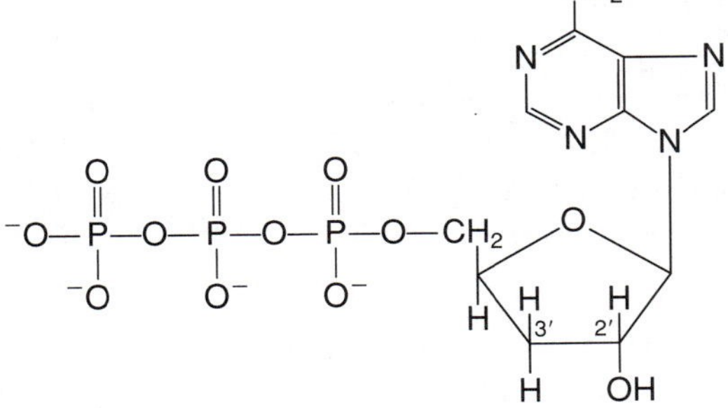

Figure 29-1
Structure of 3'-deoxy-5'-ATP.

3. The following is a partial nucleotide sequence of a recombinant plasmid containing a T7 promoter region and sites for cleavage by two restriction enzymes, EcoRI and BamHI.

5′-ATAGGGAGGA|GGATTACCCCTCGAATTCGAGCTCGCCCGGGGATCCTC-3′
3′-TATCCCTCCT|CCTAATGGGGAGCTTAAGCTCGAGCGGGCCCCTAGGAG-5′

T7 promoter

The top chain is the sense strand and the bottom strand is the template strand, and the first nucleotide to be transcribed is designated +1. It is possible to cleave the DNA with one of the restriction enzymes and then use the resulting cleavage fragment as a template for the preparation of "run-off" transcripts by T7 RNA polymerase.

(a) Give the RNA sequence that would be expected when the DNA is first cleaved with EcoRI and is then transcribed with T7 RNA polymerase. The cleavage pattern for EcoRI is 5′-G|AATTC-3′.

(b) Give the sequence expected when BamHI is used to prepare the template. The cleavage pattern for BamHI is 5′-G|GATCC-3′.

4. The technique of nearest-neighbor analysis was a valuable tool in the characterization of RNA transcripts in the days before sequencing techniques for nucleic acids became available. The technique involves the addition of a nucleotide triphosphate labeled with ^{32}P in the α-position to a reaction mixture in which RNA polymerase is active. Under such conditions, the label will be incorporated into the growing polynucleotide chain because phosphodiester bond formation involves the α-phosphoryl group of the incoming nucleotides. The newly-formed RNA can then be cleaved with a mixture of RNA-hydrolyzing enzymes that cleave RNA completely to 3′-mononucleotides. Separation of these nucleotides and the determination of their ^{32}P content can be used to establish the identity of nucleotide residues that were on the 5′ side of the labeled nucleotides in the RNA.

(a) Suppose that α-labeled UTP is used to prepare the transcripts in Problem 3(a) and that the resulting RNA is hydrolyzed to 3′-mononucleotides. Identify the 3′-mononucleotides that would be labeled and give their relative frequencies.

(b) Give the kinds and proportions of labeled 3′-mononucleotides that would be expected if α-labeled ATP had been used instead.

5. State whether each of the following cleavage fragments would be a major product of the digestion of the ribonucleotide pApG by alkali:

(a) pAp
(b) pG

(Hint: Alkaline digestion of RNA produces a mixture of 2′- and 3′-nucleoside monophosphates. Cyclic 2′-3′-phosphodiesters are intermediates in the digestion, and they are randomly cleaved by a base at both the 2′ and 3′ positions to yield 3′- and 2′-monophosphates respectively.)

6. The ciliated protozoan *Tetrahymena* contains an enzyme that can synthesize 5′-pseudouridine monophosphate from a mixture of PRPP

and uracil. For a time it was thought that this enzyme was instrumental in the synthesis of transfer RNAs in *Tetrahymena*. Explain why this is not the case.

7. When mammalian genes are cloned, a strategy that is frequently followed involves the isolation of mRNA rather than DNA from a cell and the preparation of a complementary DNA (cDNA) by the enzyme reverse transcriptase (see p. 132 of Stryer). Suppose that mRNA isolated from a cell specialized for the production of protein X is used as a template for the production of cDNA. What major difference(s) would you expect to find between the structure of that cDNA and genomic DNA for protein X?

8. Rifampicin specifically inhibits the initiation of transcription in procaryotes and may therefore be used in humans as a therapeutic antibacterial agent. Would you expect actinomycin D to be useful in antibacterial therapy? Why or why not?

9. The mRNAs produced by mammalian viruses undergo modification at the 5′ and 3′ ends in a fashion similar to that of eucaryotic mRNA. Why do you think this is the case?

10. Sketch the most stable secondary structure that could be assumed by the oligonucleotide AAGGCCCUACGUGGGCCG.

11. Suppose that human DNA is cleaved into fragments approximately the size of a given mature human messenger RNA and that RNA-DNA hybrids are then prepared. The corresponding procedure is then carried out for *E. coli*. When the RNA-DNA hybrids from each species are examined using an electron microscope, which will show the greater degree of hybridization? Explain.

12. In an attempt to determine whether a given RNA was catalytically active in the cleavage of a synthetic oligonucleotide, the following experimental results were obtained. When the RNA and the oligonucleotide were incubated together, cleavage of the oligonucleotide occurred. When either the RNA or the oligonucleotide was incubated alone, there was no cleavage. When the RNA was incubated with higher concentrations of the oligonucleotide, saturation kinetics of the Michaelis-Menten type were observed. Do these results demonstrate that the RNA has catalytic activity? Explain. (You may wish to refer to the discussion of catalytic RNA on pp. 214–215 of Stryer.)

ANSWERS TO SELF-TEST

RNA Polymerase from *E. coli*

1. The holoenzyme has the subunit composition $\alpha_2\beta\beta'\sigma$. The core enzyme lacks the σ subunit.

2. (a) 2 (b) 3 (c) 1 (d) 4

3. a, b, c, and f. The promoters of most *E. coli* genes include variants of characteristic consensus sequences that are centered at about the -35 and -10 positions. The nearer the sequences of a promoter are to the consensus sequences and the nearer the separation between them is

to the optimal 17-base-pair spacing, the more efficient the promoter. The -10 consensus sequence is TATAAT. The substitution of a C or G into the sequence would be expected to lower the efficiency of a promoter.

4. b, c

5. a, b, and c. The σ subunit recognizes promoter sites, decreases the affinity of the enzyme for general regions of DNA, and facilitates the specific initiation of transcription in one direction. This results in only one of the two DNA strands functioning as a template for RNA transcription; that is, it gives rise to asymmetric transcription.

6. The affinity of the RNA polymerase holoenzyme for nonspecific DNA sequences is less than its affinity for promoter sequences. This nonspecific affinity allows the enzyme to bind to DNA and then "slide" along the molecule in a random walk until it encounters a promoter sequence, for which its binding affinity is higher. Diffusion in one dimension is much faster than diffusion in three, which explains the rapid observed rate constant for the binding of RNA polymerase holoenzyme to promoter sequences. If one measured the encounter of the polymerase with the nonspecific regions of the DNA rather than with promoter sequences, the value of the rate constant would be much lower and would fit our expectations for a diffusion-limited reaction.

7. The temperature increase induces the synthesis of a new σ factor, σ^{32}, that directs RNA polymerase to promoters that have -10 and -35 sequences that are different from those recognized by σ^{70}. Transcription from these promoters gives rise to characteristic heat-shock proteins.

Procaryotic Transcription

8. a, b, c, and d. The β and γ phosphates of NTPs will be incorporated only with the nucleotide at the 5′ end of the transcript, which bears a 5′-triphosphate terminus. Since almost all transcripts start with purine nucleotides, only choices (b) and (c) are correct for the labeled β and γ phosphates. The α phosphates of all the NTPs will be incorporated as phosphodiester bonds in the transcript, with the β and γ phosphates of these NTPs being eliminated as PP$_i$.

9. a, c, and e. Answer (b) is incorrect because the σ subunit dissociates from the transcription bubble shortly after initiation. The enzyme unwinds approximately 17 base pairs of DNA and approximately 12 base pairs of the template strand are in a duplex with the 3′ end of the growing RNA chain leaving approximately 17 base pairs of the coding strand unpaired. (see Figure 29-11 on p. 710 in Stryer.)

10. (a) 2, 4 (b) 1, 3

11. d and e. The ρ protein recognizes and specifically binds stretches of RNA that are devoid of hairpins and are at least 72 nucleotides long. It acts to hydrolyze ATP and unwind the RNA-DNA hybrid in the transcription bubble.

12. In procaryotes, tRNA and rRNA are modified fairly extensively, whereas mRNA undergoes little, if any, alteration. Primary transcripts are cleaved by enzymes such as RNase III and have their ends trimmed by other enzymes, exemplified by RNase P. The tRNA transcripts are

also modified by the addition of nucleotides to their 3′ ends; the conversion of many of their bases to modified forms, for example, U to T or U to ψU; and the methylation of some of their ribose residues. Base and ribose modifications to the rRNA of procaryotes also occur.

13. (a) 2, 3 (b) 1, 4. Actinomycin D intercalates only into duplex DNA.

14. Rifamycin must bind to the β subunit of RNA polymerase to inhibit the enzyme. A mutation in the gene encoding this subunit that would interfere with the binding of the antibiotic but not with polymerization would produce a rifamycin-resistant cell.

Eucaryotic Transcription

15. a, b, and e. Answers (c) and (d) are incorrect because the poly A tail is found at the 3′ end and a "cap" is found at the 5′ end of typical eucaryotic mRNA. Segments of the primary transcript are discarded by RNA splicing to bring RNA sequences encoded by noncontiguous regions of the template together; that is, introns are removed.

16. (a) 1, 5, 8, 9 (b) 2, 3, 6, 7, 8, 9 (c) 2, 4, 5, 8, 9. Both RNA polymerases I and III are involved in rRNA synthesis, with polymerase I synthesizing the 18S, 5.8S, and 28S species and polymerase III synthesizing the 5S rRNA.

17. A TATA box centered at about −25 and consisting of a variant of the consensus sequence TATAAA is essential for promoter activity. In addition, many genes have a CAAT box located between −40 and −110. This sequence can be in either orientation, that is, in either DNA strand; it acts as an activating sequence. Genes that are expressed constitutively also have GC boxes in either orientation. Activating sequences upstream of the TATA box are necessary for most promoters.

18. The upstream activating sequences are recognized and bound by specific stimulatory proteins called transcription factors. The transcription factor-DNA complex could stimulate transcription by increasing the activity of RNA polymerase II or its affinity for the promoter.

19. b, c, d, and e. Answer (a) is incorrect because enhancer sequences do not serve as promoters; that is, they will not, by themselves, cause RNA polymerase II to initiate a transcript.

20. a, b, c, f, g, j, and k. PP$_i$ comes from GTP, and P$_i$ comes from the triphosphate terminus of the mRNA precursor. SAM provides the methyl groups that are transferred to the base and the sugars, and SAH is the reminant of SAM (see p. 582 in Stryer).

21. None of the statements are correct. Poly A polymerase uses ATP, not dATP, to add a stretch of A residues to the 3′-hydroxyl formed by the cleavage of an mRNA precursor.

RNA Splicing

22. d, e, f

23. a, c, and e. Answers (b) and (d) are incorrect because there is no hydrolysis of phosphodiester bonds during mRNA splicing up to the point of intron removal and lariat formation; only transesterification reactions occur, so consequently, no ATP is required for the synthesis of phosphodiester bonds.

24. snRNPs are small ribonucleoprotein particles that occur in the nucleus. Each is composed of a small RNA molecule and several characteristic proteins, some of which are common to different snRNPs. Distinct snRNPs recognize and bind to splice junctions and the branch site and are involved in assembling the spliceosome. They are requisite components of the splicing apparatus. The RNAs of some snRNPs form hydrogen bonds with sequences within introns and help to juxtapose properly the reacting splice junctions.

25. b, c, a, e, d

26. (a) 3, 4 (b) 1, 3, 6 (c) 1, 3, 5, 6, 7 (d) 2. The splicing of eucaryotic tRNAs requires protein nucleases to cut the phosphodiester bonds at the intron-exon junctions and protein RNA ligases to form the bonds joining the exons.

ANSWERS TO PROBLEMS

1. (a) EcoRI would not be very useful in footprinting experiments because it is highly specific for the rather rare nucleotide sequence GAATTC. DNase I, however, is rather nonspecific and will cleave duplex DNA into short, random, fragments. Thus potential cleavage sites would exist on either side of the protecting protein.

   (b) In principle, one wants to time the digestion with DNase so that there is, on the average, one cut per DNA molecule. Since there are many potential cleavage sites on each DNA molecule, the result will be a population of molecules of varying lengths (see Figure 29-3 on p. 705 of Stryer). Extensive digestion with DNase would produce a population of very short fragments.

2. A 3′-deoxy analog of ATP could be used to establish the direction of chain growth. In 5′→3′ growth the analog would donate a nucleotide containing 3′-deoxyadenosine and the polynucleotide chain would be terminated as a result. No additional nucleotides could be added because of the lack of a 3′-OH group on the terminal 3′-deoxyadenosine. In 3′→5′ growth the nucleotide could not be added to the growing polynucleotide chain because of the lack of the 3′-OH group. These possibilities could be distinguished by using an isotopically-labeled nucleotide.

3. (a) The following sequence would result:

   5′-GGGAGGAGGAUUACCCCUCGAAUU-3′

   Remember that EcoRI puts a staggered cut into DNA so that the template strand is longer than the coding strand in this instance.

   (b) The following sequence would result:

   5′-GGGAGGAGGAUUACCCCUCGAAUUCGAGCUCGCCCGGGGAUC-3′

4. (a) Each nucleotide to the left of U in the sequence will be labeled by this technique. The kinds and proportions of labeled mononucleotides are 2A:2U:1C. The U-containing segment of the

transcript, with the phosphates indicated, is as follows:

$$pAp\overset{*}{}Up\overset{*}{}UpApCpCpCpC\overset{*}{}pUpCpGpApAp\overset{*}{}Up\overset{*}{}U—OH$$

The phosphates marked with asterisks were donated by UTP when the RNA was synthesized, but they will depart with the neighbor on the left (the 5′ side) when the RNA is cleaved to a mixture of 3′-mononucleotides.

(b) The kinds and proportions of labeled mononucleotides are 4G:1A:1U.

5. (a) The fragment pAp would be produced by the cleavage of pApG between p and G. Guanosine is also formed.

(b) The fragment pG (5′-GMP) would not be produced because the cyclic 2′,3′-phosphodiester could not be an intermediate.

6. Nascent polynucleotides formed by RNA polymerases contain only the four usual bases. Subsequently, some of the bases are chemically modified. Were unusual nucleotides to be incorporated into a growing RNA chain, this would in turn require the presence of unusual bases on DNA. The pseudouridine found in transfer RNAs is formed by breaking the nitrogen-carbon bond linking uracil to ribose and forming a carbon–carbon bond instead. Only certain uracils are modified in this manner, owing to their position in the three-dimensional structure of the RNA and to the specificity of the enzymes that carry out the modification.

7. The cDNA prepared from mRNA would have a long poly T tail, unlike genomic DNA. Remember that the poly A tail is added to the 3′ end of mammalian mRNA and that there is no counterpart on DNA. A second striking difference would be that the cDNA would contain no intervening sequences (introns) and would therefore be much shorter than the corresponding sections of genomic DNA. Remember that most mammalian genes are mosaics of introns and exons.

8. In order to be useful as a therapeutic antibacterial agent, a compound must selectively inhibit processes in procaryotes but leave the corresponding processes in eucaryotes (including those in mitochondria) largely unaffected. Because rifampicin selectively inhibits the initiation of transcription in procaryotes but not in eucaryotes, it is useful as an antibacterial agent. Actinomycin D is an intercalating agent that binds to DNA duplexes and inhibits both DNA replication and transcription, although it has a greater inhibiting effect on transcription than on replication. It cannot discriminate between the duplex DNA of bacteria and that of humans, however, and will therefore bind to both. Because it disrupts eucaryotic as well as procaryotic processes, it is not very useful as an antibacterial agent. It is sometimes used as an anticancer agent, however, because of its ability to slow the replication rate of human DNA.

9. Viruses use the host's enzyme system to replicate their DNA and to synthesize their proteins. Since eucaryotic translation systems must synthesize viral protein, the structure of viral mRNAs must mimic that of the host mRNA.

10. The structure is a stem-and-loop ("lollipop") hairpin structure, as shown in Figure 29-2.

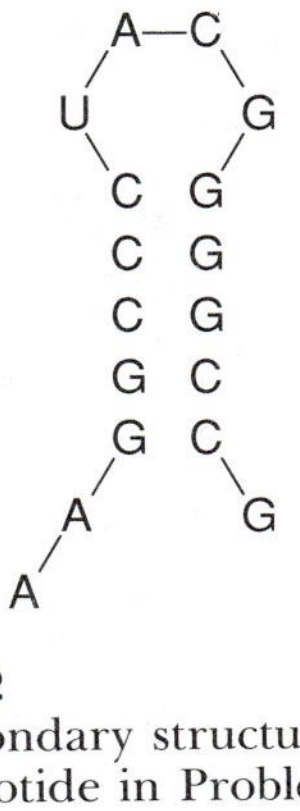

11. The RNA-DNA hybrid of *E. coli* will show greater hybridization because it is produced continuously from a DNA template without pro-

Figure 29-2
Stable secondary structure for the oligonucleotide in Problem 10.

cessing. Because of the presence of intervening sequences in human DNA, there will be regions in the human RNA-DNA hybrids where no base pairing occurs.

12. These results alone do not establish that the RNA has catalytic activity. A catalyst must be regenerated. It is entirely possible that the results observed could be accounted for by a stoichiometric, as opposed to a catalytic, interaction between RNA and the oligonucleotide in which the RNA may "commit suicide" as the oligonucleotide is cleaved. In such an interaction, a portion of the RNA would participate chemically in the cleavage of the oligonucleotide, but it would also be cleaved itself as a part of the reaction. Four reaction products would accumulate, two resulting from the cleavage of RNA and two from the cleavage of the oligonucleotide. To show that this particular RNA was catalytic, it would be necessary to demonstrate that it turns over and is regenerated in the course of the reaction.

CHAPTER 30

Protein Synthesis

Genetic information is stored as sequences of nucleotides, and for much of this information to be used it must be converted into sequences of amino acids. In Chapter 30, Stryer details the biochemistry of the translation from the language of nucleic acids to that of proteins. You should review the introduction to this process that was given in Chapter 5 of *Biochemistry*. Focus on the roles of the different kinds of cellular RNAs on pages 92–93, on mRNA as an intermediate in the conversion of nucleotide sequences to amino acid sequences on pages 93–95, on tRNA as an adaptor in the decoding of mRNA sequences on pages 99–101, and on the major features of the genetic code on pages 106–109.

Before describing the initiation, elongation, and termination stages of polypeptide synthesis, Stryer explains how amino acids are activated for the subsequent formation of peptide bonds through their attachment to tRNAs by the aminoacyl-tRNA synthetases. The mechanisms of both the activation of amino acids to form the acyl-tRNA esters and the hydrolysis of an aminoacyl-adenylate intermediate, which ensures the accuracy of attachment, are explained. The catalytic mechanism of the first step of the tyrosyl-tRNA synthetase reaction is considered in detail. The structure of the enzyme in a cocrystal with tyrosyl-AMP is known to atomic resolution and kinetic studies of genetically-engineered enzymes have been performed. The enzyme provides a striking example of delocalized catalysis in which amino acid side chains stabilize the transition state of the reaction, and thus catalyze it, by binding atoms of the substrate that are distant from the bonds being broken and formed. The detailed structures and conformations of the tRNAs are given next. Stryer presents the wobble hypothesis to explain the lack of strict one-to-one Watson-Crick base-pairing interactions between the nucleotides of the tRNA anticodons and the mRNA codons. He also explains how mutant tRNAs can act as suppressors of mutations in other genes. A description of the ribosome completes the prelude to the discussion of peptide bond formation.

The experiments that showed the polarities of polypeptide formation and the translation of mRNA are presented next. Then initiation is described, and the roles of a specialized initiator tRNA, the mRNA start codon and 16S rRNA sequences, and the protein initiation factors are outlined. The spatial and functional relationships of the sites on the ribosome that bind aminoacyl-tRNAs and peptidyl-tRNAs, the elongation factors, the peptide bond forming reaction, and the mechanism of the translocation of the peptidyl-tRNA from site to site on the ribosome are presented next in the description of the elongation stage of protein synthesis. The termination of translation is outlined, and the role of release factors that recognize translation stop codons is described. Stryer also explains the mechanisms of several potent inhibitors of translation.

The chapter closes with a brief overview of translation in eucaryotes that emphasizes its major contrasting features with respect to translation in procaryotes. Differences in the initiator tRNA, the selection mechanism of the initiator codon, the ribosomes, and the overall complexity of the process are highlighted. An example of eucaryotic regulation at the level of translation initiation is given. Lastly, the mechanism of the bacterial toxin that causes diphtheria is presented.

LEARNING OBJECTIVES

When you have mastered this chapter, you should be able to complete the following objectives.

Introduction (Stryer pages 733–734)

1. Provide an overview of protein synthesis that includes the roles of the *amino acids*, the *tRNAs*, the *amino acid activating enzymes, mRNA*, the *ribosomes*, and *ATP* and *GTP*.

Amino Acid Activating Enzymes (Stryer pages 734–739)

2. Write the two-step reaction sequence of the *aminoacyl-tRNA synthetases*. Enumerate the high-energy phosphate bonds that are consumed in the overall reaction.

3. Draw the *mixed-anhydride* and *ester linkages* that occur in the *aminoacyl-adenylate* intermediate and the *aminoacyl-tRNA* product of the activation reaction.

4. Describe the mechanisms of substrate selection and proofreading that contribute to the accuracy of the attachment of the appropriate amino acid to the correct tRNA.

5. Explain the *delocalized catalytic mechanism* of the *tyrosyl-tRNA synthetase* reaction.

6. Draw the *cloverleaf structure of a tRNA*, and identify the regions containing the *anticodon* and the *amino acid attachment site*.

7. List the common features of all tRNAs.

8. Relate the two-dimensional cloverleaf representation of the tRNA structure to its three-dimensional configuration.

9. Recount the experiment that showed that the tRNA rather than the amino acid in an aminoacyl-tRNA recognizes an mRNA codon.

10. Explain how some *codons are recognized by more than one anticodon*, that is, how they interact with more than one species of aminoacyl-tRNA. List the base-pairing interactions allowed according to the *wobble hypothesis*.

11. List the different classes of *suppressor mutations*, and describe how mutations in the genes that encode tRNAs give rise to *missense, nonsense*, and *frameshift suppressors*.

Ribosomes (Stryer pages 746–750)

12. Provide an overview of the roles of the *ribosomes* in protein synthesis. List the kinds and numbers of macromolecular components of the procaryotic ribosome. Give the *mass, sedimentation coefficient*, and *dimensions* of the *ribosome* of *E. coli*.

13. Describe the processing of the tRNA and rRNA primary transcripts to form the mature RNAs of *E. coli*.

14. State the significance of the *reconstitution* of a *functional ribosome subunit* from its constituent proteins and RNAs.

Polypeptide Formation (Stryer pages 750–760)

15. Recount the experiment of Dintzis that established the *polarity of polypeptide synthesis*.

16. Recount the experiment that related the *polarity* of polypeptide synthesis to that *of mRNA decoding*.

17. Define the *polysome* and outline its assembly. Correlate the polarity of ribosome movement with the polarity of the growing polypeptide chain.

18. Distinguish between *monocistronic* and *polycistronic mRNAs*, and specify the minimum sequence length found between the first nucleotide transcribed and the initiation codon.

19. Note the conventions for designating tRNAs and aminoacyl-tRNAs. Distinguish between $tRNA_f^{Met}$ and $tRNA_m^{Met}$ and outline the conversion of methionine into *formylmethionyl-tRNA$_f^{Met}$*.

20. Name the major *initiator codon* and the amino acid it encodes. Explain the roles of the nucleotide sequences in *16S rRNA*, mRNA, and tRNA in *selecting the initiation codon* rather than the identical codon that encodes an internal amino acid.

21. List the components of the *70S initiation complex*, and indicate the roles of the *initiation factors* (IF) and GTP in its formation.

22. Outline the elongation stage of protein synthesis, and describe the roles of the *elongation factors* (EFs) and GTP in the process. Locate the *aminoacyl-tRNAs* and *peptidyl-tRNAs* in the *A or P sites of the ribosome* during one cycle of elongation.

23. Explain the role of *EF-Tu* in determining the accuracy of protein synthesis.

24. Name the *translation stop codons*, describe the termination of translation, and explain the roles of the *release factors* (RFs) in the process.

25. Provide examples of *antibiotics that inhibit translation*, and describe their mechanisms of action.

26. Describe how *puromycin* was used to define the A and P sites of the ribosome.

Eucaryotic Translation (Stryer pages 760–763)

27. Contrast eucaryotic and procaryotic ribosomes with respect to composition and size.

28. Contrast the mechanisms of translation initiation in procaryotes and eucaryotes. Note the different initiator tRNAs, AUG codon selection mechanisms, and numbers of IFs and RFs.

29. Explain the role of the *protein kinases* in the regulation of the initiation of eucaryotic translation.

30. Describe the mechanism by which the *diphtheria toxin* inhibits protein synthesis in eucaryotes.

SELF-TEST

Introduction

1. Which of the following statements about translation are *correct*?

   (a) Amino acids are added to the amino terminus of the growing polypeptide chain.
   (b) Amino acids that are activated by attachment to tRNA molecules are used as acyl esters.
   (c) A specific initiator tRNA along with specific sequences of the mRNA ensure that translation begins at the correct codon.
   (d) Peptide bonds form between an aminoacyl-tRNA and a peptidyl-tRNA positioned in the A and P sites, respectively, of the ribosome.
   (e) Termination involves the binding of a terminator tRNA to a stop codon on the mRNA.

2. Which of the following statements about procaryotic translation are *true*?

   (a) The formation of a peptide bond is an exergonic reaction.

(b) The protein grows in the amino-to-carboxyl direction.
(c) The movement of peptidyl-tRNA on the ribosome is directly powered by the hydrolysis of ATP.
(d) Translation involves the conversion of genetic information in single-stranded DNA into sequences of amino acids.

Amino Acid Activating Enzymes

3. Which of the following statements about the aminoacyl-tRNA synthetase reaction are *correct?*

(a) ATP is a cofactor.
(b) GTP is a cofactor.
(c) The amino acid is attached to the 2'- or 3'-hydroxyl of the nucleotide cofactor.
(d) The amino group of the amino acid is activated.
(e) A mixed-anhydride bond is formed.
(f) An acyl-ester bond is formed.
(g) An acyl-thioester bond is formed.
(h) A phosphoamide (P—N) bond is formed.

4. The $\Delta G^{\circ\prime}$ of the reaction catalyzed by the *aminoacyl-tRNA synthetases* is

(a) $\cong 0$ kcal/mol.
(b) < 0 kcal/mol.
(c) > 0 kcal/mol.

5. Considering the correct answer to Question 4, explain how aminoacyl-tRNAs can be produced in the cell.

_____

_____

_____

6. Given the similarity of valine and isoleucine (there is one extra methylene group in the side chain of Ile) and the five-fold higher concentration of Val than Ile in the cell, calculations indicate that isoleucyl-tRNA synthetase should mistakenly incorporate valine into proteins in place of isoleucine once in every 40 reactions. In fact, however, this happens only once in every 3000 reactions. Explain.

_____

_____

_____

7. How can some aminoacyl-tRNA synthetases achieve accurate reactions even though they lack a hydrolytic proofreading mechanism?

_____

_____

_____

Transfer RNAs and Codon-Anticodon Interactions

8. Which of the following statements about functional tRNAs are *correct?*

(a) They contain many modified nucleosides.
(b) About half of their nucleosides are in base-paired helical regions.
(c) They contain fewer than 100 ribonucleosides.
(d) Their anticodons and amino acid accepting regions are within 5 Å of one another.
(e) They consist of two helical stems that are joined by loops to form a U-shaped structure.
(f) They have a terminal AAC sequence at their amino acid accepting end.

9. Explain why tRNA molecules must have both unique and common structural features.

___

___

10. In an experiment, it was found that Cys-$tRNA^{Cys}$ can be converted to Ala-$tRNA^{Cys}$ and used in an in vitro system that is capable of synthesizing proteins.

(a) If the Ala-$tRNA^{Cys}$ were labeled with ^{14}C in the amino acid, would the labeled Ala be incorporated in the protein synthesized in the place of other Ala residues? Explain.

___

___

___

(b) What does the experiment indicate about the importance of the accuracy of the aminoacyl-tRNA synthetase reaction to the overall accuracy of protein synthesis?

___

___

___

11. The wobble hypothesis

(a) accounts for the conformational looseness of the amino acid acceptor stem of tRNAs that allows sufficient flexibility for the peptidyl-tRNA and aminoacyl-tRNA to be brought together for peptide bond formation.
(b) accounts for the ability of some codons to be recognized by more than one anticodon.
(c) explains the occasional errors made by the aminoacyl-tRNA synthetases.
(d) explains the oscillation of the peptidyl-tRNAs between the A and P sites on the ribosome.
(e) assumes steric freedom in the pairing of the first (5′) nucleotide of the codon and the third (3′) nucleotide of the anticodon.

12. Assuming that each nucleoside in the left column is in the first position of an anticodon, with which nucleoside or nucleosides in the right column could it pair during a codon-anticodon interaction if each of the nucleosides on the right is in the third position (3′ position) of a codon?

(a) Adenosine _____   (1) Adenosine
                           (2) Cytidine
(b) Cytidine _____   (3) Guanosine
                           (4) Uridine
(c) Guanosine _____

(d) Inosine _____

(e) Uridine _____

13. According to the wobble principle, what is the *minimum* number of tRNAs required to decode the six leucine codons—UUA, UUG, CUU, CUC, CUA, and CUG? Explain.

_____

_____

_____

14. A suppressor mutation compensates for the effect of another mutation. Explain how tRNA could be involved in this process.

_____

_____

_____

15. Match the alterations in a DNA sequence that encodes an mRNA listed in the right column with the kind of mutation, listed in the left column, it produces.

(a) Missense _____   (1) The alteration of a codon for an
                             amino acid to a stop codon
(b) Nonsense _____   (2) The insertion of a nucleotide
                             into a codon
(c) Frameshift _____   (3) The deletion of a nucleotide
                             from a codon
                         (4) The alteration of a codon for one
                             amino acid to a codon for another
                             amino acid

Ribosomes

16. Which of the following statements about an *E. coli* ribosome are *correct?*

(a) It is composed of two spherically symmetrical subunits.
(b) It has a large subunit comprising 34 kinds of proteins and two different rRNA molecules.
(c) It has a sedimentation coefficient of 70S.
(d) It has two small subunits, one housing the A site and the other the P site.
(e) It has an average diameter of approximately 200 Å.
(f) It has a mass of approximately 270 kd, one-third of which is RNA.

17. What is the significance of the reconstitution of a functional ribosome from its separated components?

_____

_____

_____

Polypeptide Formation

18. If labeled amino acids are added to an in vitro system synthesizing protein under the direction of a single mRNA species and samples are withdrawn at different times, the following labeling patterns in the *completed* polypeptide chains would be observed. The dash (-) represents unlabeled amino acids, the X represents labeled amino acids, and A and B represent the ends of the intact protein.

Time 1 (early)	A---------------XXB
Time 2	A----------XXXXB
Time 3	A------XXXXXXB
Time 4 (late)	A--XXXXXXXXB

Which of the following statements about these proteins are *correct?*

(a) The labeled amino acids are added in the B-to-A direction.
(b) The labeled amino acids are added in the A-to-B direction.
(c) A is the amino terminus of the protein.
(d) B is the amino terminus of the protein.

19. Given an in vitro system that allows protein synthesis to start and stop at the ends of any RNA sequence, answer the following questions:

(a) What peptide would be produced by the polyribonucleotide 5'-UUUGUUUUUGUU-3'? (See the table with the genetic code on p. 107 of Stryer.)

_____

(b) For this peptide, which is the *N*-terminal amino acid and which is the *C*-terminal amino acid?

_____

20. What is the role of the vitamin folic acid in procaryotic translation?

_____

_____

21. The methionine codon AUG functions both to initiate a polypeptide chain and to direct methionine incorporation into internal positions in a protein. By what mechanisms are the AUG start codon selected?

_____

_____

22. Match the functions or characteristics of procaryotic translation in the right column with the appropriate translation components in the left column.

(a) IF1 _____

(b) IF2 _____

(c) IF3 _____

(d) EF-Tu _____

(e) EF-Ts _____

(f) EF-G _____

(g) Peptidyl
transferase _____

(h) RF1 _____

(i) RF2 _____

(1) moves the peptidyl-tRNA from the A to the P site.

(2) delivers aminoacyl-tRNA to the A site.

(3) binds to the 30S ribosomal subunit.

(4) recognizes stop codons.

(5) forms the peptide bond.

(6) delivers fMET-tRNA$_f^{Met}$ to the P site.

(7) cycles on and off the ribosome.

(8) binds GTP.

(9) prevents the combination of the 50S and 30S subunits.

(10) is involved in the hydrolysis of GTP to GDP.

(11) associates with EF-Tu to release a bound nucleoside diphosphate.

(12) hydrolyzes peptidyl-tRNA.

(13) modifies the peptidyl transferase reaction.

23. Which of the following statements about occurrences during translation are *correct*?

(a) The carboxyl group of the growing polypeptide chain is transferred to the amino group of an aminoacyl-tRNA.

(b) The carboxyl group of the amino acid on the aminoacyl-tRNA is transferred to the amino group of a peptidyl-tRNA.

(c) Peptidyl-tRNA may reside in either the A or P site.

(d) Aminoacyl-tRNAs are shuttled from the A to the P site by EF-G.

24. For each of the following steps of translation, give the nucleotide cofactor involved and the number of high-energy phosphate bonds consumed.

(a) Amino acid activation _____

(b) Formation of the 70S initiation complex _____

(c) Delivery of aminoacyl-tRNA to the ribosome _____

(d) Formation of a peptide bond _____

(e) Translocation _____

25. After the formation of the 70S initiation complex, how many high-energy phosphate bonds are expended for each amino acid that is added to the growing polypeptide chain?

_____

_____

26. Which of the following statements about release factors are *correct*?

(a) They recognize terminator tRNAs.

(b) They recognize translation stop codons.

(c) They cause peptidyl transferase to use H_2O as a substrate.

(d) They are two proteins in *E. coli*, each of which recognizes two mRNA triplet sequences.

27. Many antibiotics act by inhibiting protein synthesis. How can some of these be used in humans to counteract microbial infections without causing toxic side effects due to the inhibition of eucaryotic protein synthesis?

_____

_____

28. Increasing the concentration of which of the following would most effectively antagonize the inhibition of protein synthesis by puromycin?

   (a) ATP
   (b) GTP
   (c) Aminoacyl-tRNAs
   (d) Peptidyl-tRNAs
   (e) eIF3

Eucaryotic Translation

29. Which of the following statements about eucaryotic translation are _correct?_

   (a) A formylmethionyl-tRNA initiates each protein chain.
   (b) It occurs on ribosomes containing one copy each of the 5S, 5.8S, 18S, and 28S rRNA molecules.
   (c) The correct AUG codon for initiation is selected by the base pairing of a region on the rRNA of the small ribosomal subunit with an mRNA sequence upstream from the translation start site.
   (d) It is terminated by a release factor that recognizes stop codons and hydrolyzes GTP.
   (e) It involves proteins that bind to the 5′ ends of mRNAs.

30. Which of the following statements about the diphtheria toxin are _correct?_

   (a) It is cleaved on the surface of susceptible eucaryotic cells into two fragments, one of which enters the cytosol.
   (b) It binds to peptidyl transferase and inhibits protein synthesis.
   (c) It reacts with ATP to phosphorylate eIF2 and prevent the insertion of the Met-tRNA$_i^{Met}$ into the P site.
   (d) It reacts with NAD^+ to add ADP-ribose to eEF2 and prevent movement of the peptidyl-tRNA from the A to the P site.
   (e) One toxin molecule is required for each translation factor inactivated; that is, it acts stoichiometrically.

PROBLEMS

1. The template strand of DNA known to encode the _N_-terminal region of an _E. coli_ protein has the following nucleotide sequence:

   GTAGCGTTCCATCAGATTT

   Give the sequence for the first four amino acids of the protein. (Refer to the genetic code on p. 107 of Stryer.)

2. Suppose that the sense strand of the DNA known to encode the amino acid sequence of the *N*-terminal region of a mammalian protein has the following nucleotide sequence:

CCTGTGGATGCTCATGTTT

Give the amino acid sequence that would result.

3. The nucleotide sequence on the sense strand of the DNA known to encode the carboxyl terminus of a long protein of *E. coli* has the following nucleotide sequence:

CCATGCAAAGTAATAGGT

Give the resulting amino acid sequence.

4. List the major components that must be present in an in vitro system for the synthesis of an *E. coli* protein from a mixture of amino acids.

5. The synthesis of a protein molecule requires a considerable expenditure of energy.

   (a) Calculate the minimum number of high-energy phosphate bonds that must be expended to synthesize a single procaryotic protein consisting of 100 amino acid residues from a mixture of the components listed in your answer to Problem 4.
   (b) Explain why the number of necessary high-energy phosphate bonds you have calculated in part (a) is a minimum.

6. Suppose that a particular aminoacyl-tRNA synthetase has a 10 percent error rate in the formation of aminoacyl-adenylates and a 99 percent success rate in the hydrolysis of incorrect aminoacyl-adenylates. What percentage of the tRNAs produced by this aminoacyl-tRNA synthetase will be faulty?

7. Students of biochemistry are frequently distressed by "Svedberg arithmetic," that is, for instance, by the fact that the 30S and 50S ribosomal subunits form a 70S particle rather than an 80S particle. Why don't the numbers add up to 80? (See pp. 49–50 of Stryer.)

8. In the chain of events leading from a nucleotide sequence on DNA to the production of protein by ribosomes, where precisely does the process of translation occur? Explain.

9. Studies on random copolymers of nucleotides have been important in research on the genetic code and on the role of tRNAs in protein synthesis. (See Stryer pp. 102 and 743.) Give the kinds and proportions of amino acids that you would expect to be incorporated into a polypeptide chain when a random copolymer of A and C in the ratio of 3:1 is added as a template to an in vitro system for protein synthesis. (Protein synthesis in vitro may be initiated at random sites along a template lacking AUG providing that high concentrations of Mg^{2+} are present.)

10. The possible codons for valine are GUU, GUC, GUA, and GUG.

   (a) For each of these codons write down all the possible anticodons with which it might pair. (See Stryer, p. 744.)
   (b) How many codons could pair with anticodons having I as the first base? How many could pair with anticodons having U or G as the first base? How many could pair with anticodons beginning with A or C.

11. What amino acid will be specified by a tRNA whose anticodon sequence is IGG?

12. Chain terminating codons frequently occur in tandem in *E. coli*. (See Problem 3 for an example.) Explain how such repetition may ensure that suppression is directed toward nonsense mutations and does not prevent the proper termination of a completely synthesized polypeptide chain.

13. Suppose that the probability of inserting an incorrect amino acid at each position in a protein is 1 percent. Calculate the probability that a protein consisting of 100 amino acid residues is entirely correct.

14. Laboratory studies of protein synthesis usually involve the addition of a radioactively-labeled amino acid and either natural or synthetic mRNAs to systems containing the other components that you specified in your answer to Problem 4. To observe the formation of protein, advantage is taken of the fact that proteins, but not amino acids, can be precipitated by solutions of trichloroacetic acid. Thus, one can observe the extent to which radioactivity has been incorporated into "acid precipitable material" as a function of time to estimate the rate of formation of protein. In one such experiment, poly U is used as a synthetic mRNA in an in vitro system derived from wheat germ.

  (a) What labeled amino acid would you add to the reaction mixture?
  (b) What product will be formed?

   For each of the following procedures, explain the results observed. Assume that in a complete system 3000 cpm (counts per minute) are found in acid precipitable material at the end of 30 minutes.

  (c) RNase A is added to the complete system; 85 cpm are found at the end of 30 minutes.
  (d) Chloramphenicol is added to the complete system; 2900 cpm are found at the end of 30 minutes.
  (e) Cyclohexamide is added to the complete system; 300 cpm are found at the end of 30 minutes.
  (f) Puromycin is added to the complete system; 640 cpm are found at the end of 30 minutes.
  (g) Puromycin and extra wheat germ tRNA are added to the complete system; 1518 cpm are found at the end of 30 minutes.
  (h) Poly A is used instead of poly U; 120 cpm are found at the end of 30 minutes.

ANSWERS TO SELF-TEST

Introduction

1. b, c, and d. Answer (a) is incorrect because the incoming activated aminoacyl-tRNA, in the A site of the ribosome, adds its free amino group to the activated carboxyl of the growing polypeptide on a pep-

tidyl-tRNA in the P site. Answer (e) is incorrect because termination does not involve tRNAs that recognize translation stop codons but rather protein release factors that recognize these signals and cause peptidyl transferase to donate the growing polypeptide chain to H_2O rather than to another aminoacyl-tRNA.

2. b. Answer (a) is incorrect because energy is required to form a peptide bond. Answer (c) is incorrect because the shuttling of peptidyl-tRNA on the ribosome is powered by the hydrolysis of GTP. Answer (d) is incorrect because translation converts genetic information from nucleotide sequences in *RNA* into sequences of amino acids.

Amino Acid Activating Enzymes

3. a, e, and f. The carboxyl group of the amino acid is activated in a two-step reaction *via* the formation of an intermediate containing a mixed-anhydride linkage to AMP. The amino acid is ultimately linked by an ester bond to the 2'- or 3'-hydroxyl of the tRNA.

4. a. Since the standard free energy of the hydrolysis of an aminoacyl-tRNA is nearly equal to that of the hydrolysis of ATP, the reaction has a $\Delta G^{\circ\prime} \cong 0$; that is, it has an equilibrium constant near 1.

5. In the cell, the hydrolysis of PP_i by pyrophosphatase shifts the equilibrium toward the formation of aminoacyl-tRNA.

6. Isoleucyl-tRNA synthetase has a proofreading mechanism. The hydrolysis of any Val-AMP that is mistakenly formed in the first step of the reaction gives the enzyme a second chance to bind and react with the correct amino acid to form Ile-AMP. The "decision" to hydrolyze the aminoacyl-adenylate intermediate probably depends upon the size of the amino acid substituent. If it is as small as Val, the intermediate fits into the hydrolytic site and is destroyed. If it is as large as isoleucine, it does not fit, is not destroyed, and is added to the tRNA.

7. With these enzymes, the appropriate amino acid may not have to compete with an amino acid that closely resembles it. Consequently, the activating enzyme can easily distinguish the correct amino acid from all other amino acids through simple binding interactions. For example, the hydroxyl on tyrosine allows it to be unambiguously distinguished from phenylalanine.

Transfer RNAs and Codon-Anticodon Interactions

8. a, b, and c. The molecules consist of two helical stems, each of which is made of two stacked helical segments. However, the molecules are L-shaped, and the anticodon and amino acid accepting regions are some 80 Å from one another. Functional tRNAs have a CCA sequence, not an AAC sequence at their 3' termini.

9. Transfer RNAs need common features for their interactions with ribosomes and elongation factors but unique features for their interactions with the activating enzymes.

10. (a) No, the labeled Ala would not be incorporated. The tRNA, acting as an adaptor between the amino acid and mRNA, would associate with Cys codons in the mRNA through base pairing between codon and anticodon. The labeled alanine would incorporate only at sites encoded by the Cys codons and not at those encoded by Ala codons.

(b) The experiment demonstrates that the tRNA and not the amino acid reads the mRNA. Thus, if the activating enzyme mistakenly attaches an incorrect amino acid to a tRNA, that amino acid is destined to be incorporated erroneously into the protein.

11. b. Answer (e) is incorrect because the ambiguities in base pairing occur between the third nucleotide of the codon and the first nucleotide of the anticodon.

12. (a) 4 (b) 3 (c) 2, 4 (d) 1, 2, 4 (e) 1, 3. (See Table 30-3 on p. 744 of Stryer.)

13. A minimum of three tRNAs would be required. One tRNA having the anticodon UAA could decode both UUA and UUG. For the other four codons, which have C in the first position and U in the second, there are two different combinations of two tRNAs each that could decode them. The first combination would be two tRNAs that have anticodons with A in the second position and G in the third, one with I in the first position to decode CUU, CUC, and CUA, and the other with U or C in the first position to decode CUG. The second combination would be two tRNAs that have anticodons with A in the second position and G in the third, one with G in the first position to decode CUU and CUC, and the other with U in the first position to decode CUA and CUG.

14. The DNA encoding the tRNA could change so that the anticodon of the tRNA recognizes a different mRNA codon. For example, a tRNA^{Tyr} gene might mutate so that the anticodon it specifies would supress a nonsense mutation by reading a translation stop codon as a signal to incorporate Tyr (see pp. 745–746 of Stryer).

15. (a) 4 (b) 1 (c) 2, 3

Ribosomes

16. b, c, and e. Answer (f) is incorrect because two-thirds of the 2700-kd mass of a ribosome is rRNA.

17. It shows that the components themselves contain all the information necessary to form the structure and that neither a template nor any other factors are involved. Thus, the ribosome serves as model from which we might learn the general principles involved in self-assembly. Reassembly allows systematic study of the roles of the individual components through the determination of the effects of substitutions of mutant or altered individual proteins or rRNAs.

Polypeptide Formation

18. b and c. Although longer incubation times result in completed proteins that have labeled polypeptides closer to their amino terminus, the chains actually grow in the amino-to-carboxyl direction. When the labeled amino acid is introduced into the system, it begins adding to the carboxyl ends of the growing chains that are already present in all stages of completion. Thus, the completed chains in samples withdrawn after a short time will have labeled polypeptides only near their car-

boxyl end. As time passes, more and more polypeptides that began adding labeled polypeptides near their amino terminals will become complete chains.

19. (a) Phenylalanyl-valyl-phenylalanyl-valine
    (b) Phe is the *N*-terminal amino acid and Val is the *C*-terminal amino acid.

20. After folic acid is converted to N^{10}-formyltetrahydrofolate, it acts as a carrier of formyl groups and is a substrate for a transformylase reaction that converts Met-tRNA$_f^{Met}$ to fMet-tRNA$_f^{Met}$—the initiator tRNA. (See Table 13-3, p. 323, and pp. 580–582 in Stryer.)

21. A purine-rich mRNA sequence, three to nine nucleotides long (called the Shine-Dalgarno sequence), which is centered about ten nucleotides upstream or to the 5′ side of the start codon, base pairs with a sequence of complementary nucleotides near the 3′ end of the 16S rRNA of the 30S ribosomal subunit. This interaction plus the association of fMet-tRNA$_f^{Met}$ with the AUG in the P site of the ribosome sets the mRNA reading frame.

22. (a) 3, 7 (b) 3, 6, 7, 8, 10 (c) 3, 7, 9 (d) 2, 7, 8, 10 (e) 11 (f) 1, 7, 8, 10 (g) 5, 12 (h) 4, 7, 13 (i) 4, 7, 13

23. a and c. The aminoacyl-tRNA in the A site becomes a peptidyl-tRNA when it receives the carboxyl group of the growing polypeptide chain from the peptidyl-tRNA in the P site. After the free tRNA leaves, the extended polypeptide on its new tRNA is then moved to the P site by EF-G. Answer (d) is incorrect because transfer RNAs bearing amino acyl derivatives with free amino groups are never found in the P site.

24. (a) ATP, 2 (b) GTP, 1 (c) GTP, 1 (d) None (e) GTP, 1. With regard to the answer for (d), the formation of a peptide bond per se doesn't require a cofactor. The energy for the exergonic reaction is supplied by the activated aminoacyl-tRNA.

25. Four high-energy phosphate bonds are expended: two to form the aminoacyl-tRNA, one to insert the aminoacyl-tRNA into the A site, and one to move the peptidyl-tRNA from the A to the P site.

26. b, c, and d. Each of the two release factors of *E. coli* recognizes two of the three translation stop codons and interacts with the synthesis machinery such that peptidyl transferase donates the growing polypeptide chain to H_2O and thus terminates synthesis by hydrolyzing the ester linkage of the protein to the tRNA.

27. The inhibition of the procaryotic translation and not that of the eucaryote can result from differences between their respective ribosomes. Some antibiotics interact with components that are unique to bacterial ribosomes and, consequently, can inhibit bacterial growth without affecting the human cells.

28. c. Puromycin is an analog of aminoacyl-tRNA. It inhibits protein synthesis by binding to the A site of the ribosome and accepting the growing polypeptide chain from the peptidyl-tRNA in the P site and thus terminating polymer growth. Because aminoacyl-tRNAs compete with the puromycin for the A site, increasing their concentration would lessen the extent of inhibition.

Eucaryotic Translation

29. b, d, and e. Eucaryotic ribosomes usually scan the mRNA from the 5′ end for the first AUG codon, which then serves to initiate the synthesis. Answer (e) is correct because proteins that bind to the cap of the mRNA are involved in the association of the ribosome with the mRNA.

30. a and d. Answer (e) is incorrect because the toxin acts catalytically and is thus extremely deadly; one toxin molecule can inactivate many translocase molecules by modifying them covalently.

ANSWERS TO PROBLEMS

1. The sequence of the first four amino acids of the protein is (formyl)Met-Glu-Arg-Tyr. As the name implies, the template (antisense) strand of DNA serves as the template for the synthesis of a complementary mRNA molecule. (Remember that by convention nucleotide sequences are always written in the 5′ to 3′ direction unless otherwise specified). The template strand of DNA and the mRNA synthesized are as follows:

DNA template strand: 5′-GTAGCGTTCCATCAGATTT-3′

mRNA: 3′-CAUCGCAAGGUAGUCUAAA-5′

Remember that the codons of an mRNA molecule are read in the 5′ to 3′ direction. Because this particular nucleotide sequence specifies the N-terminal region of an *E. coli* protein, the first amino acid must be (formyl) methionine, which may be encoded by either AUG or GUG. Because there is no GUG and only a single AUG in the mRNA sequence, the location of the initiation codon can be established unambiguously. The portion of the mRNA sequence encoding protein and the first four amino acids it encodes are:

mRNA: 5′-AUG-GAA-CGC-UAC-3′

Amino acid sequence: (Formyl)Met-Glu-Arg-Tyr

2. The expected amino acid sequence is Met-Leu-Met-Phe. The nucleotide sequences on DNA and mRNA are

Sense strand of DNA: 5′-CCTGTGGATGCTCATGTTT-3′

mRNA: 5′-CCUGUGGAUGCUCAUGUUU-3′

In eucaryotes the first triplet specifying an amino acid is almost always the AUG that is closest to the 5′ end of the mRNA molecule. In this example there are two AUGs, so there will be two Met residues in the polypeptide that is produced. The reading frame and the resulting amino acids are as follows:

mRNA: 5′-CCUGUGG-AUG-CUC-AUG-UUU-3′

Amino acid sequence: Met-Leu-Met-Phe

3. The sequence is His-Ala-Lys. The DNA and mRNA sequences are:

Sense strand of DNA: 5'-CCATGCAAAGTAATAGGT-3'

mRNA: 5'-CCAUGCAAAGUAAUAGGU-3'

Since this sequence specifies the carboxyl end of the peptide chain, it must contain one or more of the chain termination codons: UAA, UAG, or UGA. UAA and UAG occur in tandem in the sequence, so we can infer the reading frame. The mapping of the amino acid residues to the mRNA is as follows:

mRNA: 5'-C-CAU-GCA-AAG-UAA-UAG-GU-3'

Amino acid sequence: His-Ala-Lys

4. The major components are as follows: amino acids, tRNAs, activating enzymes, ATP, GTP, ribosomes, mRNA, N^{10}-formyltetrahydrofolate, transformylase, IF1, IF2, IF3, EF-G, EF-Tu, RF1, RF2, and Mg^{2+}.

5. (a) The minimum number of high-energy phosphate bonds required would be 399. Each of the 100 amino acids must be attached to a tRNA molecule. Since ATP is cleaved to one AMP and two P_i in this process, a total of 200 high energy phosphates are required for the synthesis of the aminoacyl-tRNAs. In the synthesis of a protein of 100 residues 99 elongation and 99 translocation steps are involved, each requiring one GTP, for a total of 198 GTPs. Finally, the formation of the single initiation complex requires a GTP. Thus, there are 200 high energy phosphate bonds from ATP and 199 GTP molecules required for a total of 399 high-energy phosphates.

(b) The number of high-energy phosphate bonds calculated in (a) is a minimum because it reflects only the direct expenditure of energy. It does not take into account the energy required for the synthesis of ribosomes, activating enzymes, initiation factors, elongation factors, release factors, and so forth.

6. The percentage of tRNAs that will be faulty is 0.11%. For every 1000 aminoacyl-adenylates that are produced, 100 are faulty and 900 are correct. The 900 correct intermediates will be converted to correct aminoacyl tRNAs because the intermediates are tightly bound to the active site of the aminoacyl-tRNA synthetase. Of the 100 incorrect aminoacyl-adenylates, 99 will be hydrolyzed and will therefore not form aminoacyl tRNAs. Only 1 will survive to become an incorrect aminoacyl tRNA. The fraction of incorrect aminoacyl tRNAs is therefore 1/901, or 0.11%.

7. The Svedberg unit (S) is a sedimentation coefficient, which is a measure of the velocity with which a particle moves in a centrifugal field. It represents a hydrodynamic property of a particle, a property that depends on, among other factors, the size and shape of the particle. When two particles come together, the sedimentation coefficient of the resulting particle should be less than the sum of the individual coefficients because there is no frictional resistance between the contact surfaces of the particles and the centrifugal medium.

8. Translation involves conversion of the language of nucleotides to that of proteins. The agent of translation is the appropriate aminoacyl-

tRNA synthetase, which must recognize a particular amino acid and link it to a tRNA containing an anticodon for that amino acid.

9. The eight possible codons on the template together with their frequencies and the amino acids that they specify are as follows:

Codon	Frequency	Amino Acid
AAA	$3/4 \times 3/4 \times 3/4 = 27/64$	Lys
AAC	$3/4 \times 3/4 \times 1/4 = 9/64$	Asn
ACA	$3/4 \times 1/4 \times 3/4 = 9/64$	Thr
CAA	$1/4 \times 3/4 \times 3/4 = 9/64$	Gln
ACC	$3/4 \times 1/4 \times 1/4 = 3/64$	Thr
CAC	$1/4 \times 3/4 \times 1/4 = 3/64$	His
CCA	$1/4 \times 1/4 \times 3/4 = 3/64$	Pro
CCC	$1/4 \times 1/4 \times 1/4 = 1/64$	Pro

The total fraction of each amino acid incorporated is as follows:

Amino Acid	Proportion
Lys	27/64
Thr	$9/64 + 3/64 = 12/64$
Asn	9/64
Gln	9/64
His	3/64
Pro	$3/64 + 1/64 = 4/64$
Total	64/64

10. (a) The possible anticodons with which the codons might pair are as follows:

Codon	Possible Anticodon
GUU	AAC, GAC, IAC
GUC	GAC, IAC
GUA	UAC, IAC
GUG	CAC, UAC

(b) Three codons could pair with the anticodon beginning with I; two codons could pair with an anticodon beginning with U or G; and only one codon could pair with an anticodon beginning with A or C.

11. Proline. The three codons that will pair with IGG—CCU, CCC, and CCA—all specify proline.

12. In a nonsense mutation, a codon that formerly specified the incorporation of an amino acid has mutated into one of the chain termination codons—UAA, UAG, or UGA. As a result, incomplete polypeptide chains are produced. Intergenic suppression of nonsense mutations

usually occurs when mutant tRNAs recognize a termination codon and insert an amino acid. The result is a completed polypeptide chain that may be entirely functional even though it may contain an altered amino acid residue. If two different legitimate stop codons are present in tandem, it would be extremely improbable that mutant tRNAs would exist for both and would simultaneously bind to each of them and thereby prevent proper chain termination.

13. The probability of any one residue being the correct amino acid is 0.99. The probability that the entire protein consisting of 100 amino acid residues will be correct is therefore $(0.99)^{100}$ or 0.366.

14. (a) Poly U codes for the incorporation of phenylalanine. Therefore, labeled phenylalanine must be added to the reaction mixture.

   (b) Polyphenylalanine will be formed.

   (c) RNase A will digest poly U almost completely to 3'-UMP, thus destroying the template for polyphenylalanine synthesis. Also the tRNA will be digested and the ribosomes damaged.

   (d) Chloramphenicol inhibits the peptidyl transferase activity of the 50S ribosomal subunit in procaryotes but has not effect on eucaryotes. (Wheat is a eucaryote.)

   (e) Cyclohexamide inhibits the peptidyl transferase activity of the 60s ribosomal subunit in eucaryotes.

   (f) Puromycin mimics an aminoacyl-tRNA and causes premature polypeptide chain termination.

   (g) The addition of the extra wheat germ tRNA reduces the inhibiting effect of puromycin since they both compete for the A site on ribosomes.

   (h) Poly A directs the synthesis of polylysine; since there is no lysine (either labeled or unlabeled) in the system, no product can be detected. The 120 cpm is not significantly above background.

Protein Targeting

Proteins are not uniformly distributed throughout a cell. In Chapter 31, Stryer describes the biochemistry responsible for the delivery of proteins to their appropriate locations. In addition, he describes the receptor-mediated import of specific proteins into certain eucaryotic cells and a programmed pathway for protein degradation.

The translocation path for proteins that are synthesized on the rough endoplasmic reticulum (ER) and targeted for the lysosomes, for the plasma membrane, or for export is described first. Stryer explains the role of signal sequences within the nascent protein in initiating a series of events, involving a small cytoplasmic ribonucleoprotein particle and integral membrane proteins, that results in the translocation of the polypeptide across the membrane into the lumen of the ER. The ensuing events, in which the protein may be covalently modified by cleavage and glycosylation and is transported through the Golgi complex into transfer vesicles that will deliver it to its final destination, are also outlined. Stryer explains how the molecular marker that targets proteins to the lysosomes was revealed by a human disorder in which an entire class of enzymes is directed to the wrong intracellular location because its members lack a particular carbohydrate substituent.

Proteins synthesized on free cytosolic ribosomes may also be targeted to specific intracellular sites. Mitochondrial, chloroplast, and nuclear proteins are synthesized in the cytosol and are translocated through or into the membranes of these organelles. Stryer describes how chimeric proteins, produced by recombinant DNA techniques, have been useful in detecting the targeting signals responsible for directing proteins to specific sites.

The mechanism by which cells import specific proteins and their associated ligands is described next. Receptor-mediated endocytosis provides a way for cells to recognize particular extracellular proteins through interactions with receptors located in cell-surface structures called coated pits. Some viruses and toxins have appropriated this mechanism to gain entry into the cell. Stryer closes the chapter with a description of a mechanism cells use to designate proteins for degradation.

In preparation for studying this chapter, you should review the structure of the mitochondrion on page 398 and the structure of the chloroplast on page 581 in *Biochemistry.* In addition, reread pages 343–346 to refresh your memory of the relevant carbohydrate structures and the roles of oligosaccharide substituents on proteins. Review the role of nucleotide sugars in the synthesis of glycogen on pages 455–456 to contrast this mechanism of condensing carbohydrate residues with the one involving dolichol phosphate sugar derivatives described in this chapter. The material on pages 302–304 will remind you of the linkages by which carbohydrates can be attached to proteins and of the hydrophobic nature of the regions of integral proteins that span membranes. Membrane proteins should be reviewed by reading pages 292–294. Finally, review the introduction to receptor-mediated endocytosis that was given on pages 561–563 during the presentation of the metabolism of cholesterol.

LEARNING OBJECTIVES

When you have mastered this chapter, you should be able to complete the following objectives.

Protein Translocation across the Endoplasmic Reticulum Membrane (Stryer pages 767–773)

1. Summarize the role of *protein sorting* in the delivery of proteins to their cellular locations.

2. Distinguish between *ribosomes* that are *free in the cytosol* and those that are *bound to the endoplasmic reticulum (ER) membrane system* in terms of the destinations of the proteins they synthesize.

3. List the three major classes of proteins that are synthesized on the *rough ER.*

4. State the *signal hypothesis,* and list the common features of *signal sequences.* Explain how *chimeric proteins* have been used to show the crucial role of signal sequences in protein targeting.

5. Describe the composition of the *signal recognition particle* (SRP), and outline its role in translocating proteins across the ER membrane. Sketch the *SRP cycle,* and relate it to the *ribosome cycle.* Describe the roles of the *SRP-receptor, ribophorin I* and *II,* and *signal peptidase* in the translocation mechanism.

6. Explain the roles of the ribosome and SRP in maintaining a nascent polypeptide as an optimal substrate for translocation. Note that some proteins may be translocated across membranes after they have folded, and list the required components for this process.

7. Describe the function of the *stop-transfer sequence.*

Protein Modifications during Translocation and the Golgi Complex (Stryer pages 773–782)

8. List the types of *covalent modifications* that proteins undergo in the *lumen of the ER.*

9. Describe the structure of *dolichol phosphate*, and outline its role in the synthesis of the *pentasaccharide core* of N-*linked oligosaccharides*. Relate the effects of *bacitracin* and *tunicamycin* to dolichol phosphate metabolism.

10. Describe the *Golgi complex*, and list its major functions. Distinguish among the *cis*, *medial*, and *trans* compartments of the Golgi.

11. Describe the origin and destination of *transfer vesicles*.

12. Explain how *membrane budding* to form vesicles and *vesicle fusion* with the plasma membrane determine the orientation of integral membrane proteins.

13. Distinguish between *core* and *terminal glycosylation* of glycoproteins, and provide an overview of the reactions that occur in the three compartments of the Golgi.

14. State the molecular basis of *I-cell disease*. Explain how this disorder revealed the molecular signal that directs hydrolytic enzymes to the lysosome.

15. Describe the possible role of *signal patches* in protein targeting.

16. Summarize the functions of oligosaccharide substituents on proteins.

17. Distinguish between the *apical* and *basolateral plasma membranes* of polarized cells in terms of the targeting of integral membrane proteins.

18. Contrast the physiological roles of *constitutive* and *regulated secretory vesicles*. Indicate the destinations of soluble proteins in the lumen or integral membrane proteins in the membrane of the ER if they lack specific targeting signals.

19. List the potential destinations of *bacterial* proteins, and compare the translocation of bacterial membrane proteins with the translocation process in eucaryotes.

Mitochondrial, Chloroplast, and Nuclear Protein Targeting (Stryer pages 782–786)

20. Indicate the *origin* of most *mitochondrial proteins*, and list their potential *destinations* in the mitochondrion.

21. Describe the salient features of the *presequences* of proteins destined for various mitochondrial locations. Contrast the *energy source* for the translocation of proteins into the mitochondria with that used for the translocation of proteins across the ER.

22. Compare protein transport into *chloroplasts* and with that into mitochondria.

23. Describe the structural features of the *nuclear envelope* that are relevant to protein transport, and discuss the effects of *protein size* and *nuclear localization sequences* on the import of proteins into the nucleus.

Receptor-Mediated Endocytosis (Stryer pages 786–794)

24. List the roles of *receptor-mediated endocytosis*.

25. Describe the general features of *glycoprotein cell-surface receptors*, and name the cell-surface regions in which they are found.

26. Describe a *coated pit*, and outline the process of receptor-mediated endocytosis from *receptor binding* to the formation of the *endosome*. Describe the *triskelion* structure of *clathrin*.

27. List the possible fates of the internalized proteins of the endosome.

28. Explain the importance of pH changes during receptor-mediated endocytosis and during *lysosome functioning*.

29. Outline the roles of receptor-mediated endocytosis in the life cycle of some *membrane-enveloped viruses* and in the mechanisms of action of some *toxins*.

Programmed Protein Destruction (Stryer pages 794–795)

30. Note the different *half-lives of proteins* in the cell, and explain the role of *ubiquitin* in determining the stability of some proteins.

31. Relate the identity of the N-*terminal amino acid* of a protein to its *intracellular stability*.

SELF-TEST

Protein Translocation across the Endoplasmic Reticulum Membrane

1. What feature distinguishes the rough and smooth endoplasmic reticulum?

_____

_____

2. Membrane-bound ribosomes synthesize which of the following?
   (a) Integral membrane proteins of the mitochondria
   (b) Proteins spanning the ER membrane
   (c) Lysosomal proteins
   (d) Secretory proteins
   (e) Nuclear proteins

3. Ribosomes can be classified as "free in the cytosol" or "membrane bound." Describe the experiments that showed all ribosomes are intrinsically identical and differ only in their intracellular locations?

_____

_____

_____

4. When mRNA specifying a secretory protein was translated in vitro in the absence of ER membranes, the product was longer than the form secreted by living cells. Explain why.

_____

_____

_____

5. Which of the following changes in the gene encoding a lysosomal protein would be expected to impair the translocation of the protein across the ER membrane?

   (a) The deletion of the nucleotides specifying the 25 amino acids following the translation initiation codon

   (b) The conversion of the sole lysine codon after the translation initiation codon to an arginine codon

   (c) The conversion of the sole lysine codon after the translation initiation codon to an aspartic acid codon

   (d) The conversion of two leucine codons and one phenylalanine codon near the translation initiation codon to an isoleucine codon, an aspartic acid codon and a glutamic acid codon, respectively

6. Suppose you wish to engineer an enzyme that ordinarily resides in the cytosol such that it will be secreted by a human cell. How would you proceed?

_____

_____

_____

7. Match the functions or characteristics in the right column with the components in the left column that are involved in the mechanism that delivers ribosomes to the ER and leads to the translocation of proteins across the ER membrane.

   (a) 7SL RNA _____

   (b) Signal recognition particle (SRP) _____

   (c) SRP receptor _____

   (d) Ribophorin I and II _____

   (e) Signal peptidases _____

   (1) An α, β-heterodimer on the cytosolic face of the ER that is involved in the release of SRP from the ribosomes

   (2) Interferes with the entry of aminoacyl-tRNA into the ribosome and with the peptidyl transferase reaction

   (3) Ribosome receptors on the cytosolic face of the ER membrane

   (4) Binds to a ribosome bearing a polypeptide with a signal sequence

   (5) A component of SRP

   (6) Located on the lumenal face of the ER membrane

   (7) Releases SRP and, consequently, relieves translation arrest when it binds to SRP

8. If you could digest a nascent polypeptide associated with a ribosome during protein synthesis with a voracious and nonspecific protease, how big a fragment would survive? Explain.

_____

_____

_____

9. Which of the following statements about translocation across the ER membrane are *correct?*

(a) It is driven by peptidyl transferase.
(b) It requires the hydrolysis of ATP.
(c) It depends upon a proton-motive force across the membrane.
(d) It occurs only for unfolded proteins.
(e) It can be stopped by specific sequences of amino acids in the translocated protein.

Protein Modifications during Translocation and the Golgi Complex

10. Translocated proteins may undergo which of the following modifications in the lumen of the ER?

(a) Signal sequence cleavage
(b) The attachment of dolichol phosphate to form a lipid anchor
(c) Folding and disulfide bond formation
(d) The addition of oligosaccharides to their asparagine residues to form *N*-linked derivatives
(e) The addition of oligosaccharides to their tyrosine residues to form *O*-linked derivatives

11. Which of the following statements about dolichol phosphate are *correct?*

(a) It serves as an acceptor of monosaccharides.
(b) It serves as a donor of both monosaccharides and oligosaccharides.
(c) It acts as a lipid carrier to facilitate the transfer of sugar residues from the cytosol to the lumen of the ER.
(d) It is converted to dolichol pyrophosphate by a kinase that uses ATP as a phosphate source.
(e) It is produced from dolichol pyrophosphate by a phosphatase.

12. Which of the following statements about the Golgi complex are *correct?*

(a) It is a stack of flattened proteoglycan sacs.
(b) It carries out core glycosylation of the proteins being transported.
(c) It is the major protein-sorting center of the cell.
(d) It receives proteins from the ER by fusion with transfer vesicles.
(e) It forms secretory granules in its *trans* compartment.
(f) The cisternae of the *cis, medial,* and *trans* compartments are connected by pores.

13. If a protein that is targeted to become an integral plasma membrane protein begins its journey by being inserted into the ER membrane such that its amino-terminus is in the lumen, will the

amino-terminus reside inside or outside the cell when it reaches its destination? Explain.

_____

_____

_____

14. I-cell disease

    (a) results from the inability of lysosomes to hydrolyze glycosaminoglycans and glycolipids.

    (b) results from a chromosomal deletion of the genes specifying at least eight acid hydrolases ordinarily found in the lysosomes.

    (c) arises from a deficiency in an enzyme that transfers mannose 6-phosphate onto a core oligosaccharide that is normally found on lysosomal enzymes.

    (d) arises from the absence of a mannose 6-phosphate receptor in the _trans_ Golgi complex.

15. Carbohydrate substituents on glycoproteins may serve which of the following functions?

    (a) They hold the amino acids adjacent to the carbohydrate in the vicinity of a membrane.

    (b) They stabilize the protein against degradation by proteases.

    (c) They act as markers that can direct the protein to particular locations.

    (d) They increase the caloric content of storage proteins.

    (e) They act as labels that can signal the uptake and degradation of the protein.

16. Match each protein in the right column with the kind of secretory vesicle in the left column that is likely to translocate it.

    (a) Constitutive vesicle   (1) Serum albumin

    _____   (2) Trypsinogen

     (3) ACTH

    (b) Regulated vesicle   (4) Collagen

    _____   (5) Chymotrypsinogen

     (6) Prolactin

     (7) Integral plasma membrane protein

17. What is the fate of soluble proteins in the lumen of the ER if they do not contain signals directing them to specific locations?

_____

_____

18. Describe an experiment that indicated that some of the mechanisms involved in the translocation of proteins across membranes share common features in both eucaryotes and procaryotes.

_____

_____

_____

19. Which of the following statements about mitochondrial proteins are *correct*?

(a) They are mostly encoded by mitochondrial DNA.
(b) They are mostly synthesized on the rough ER.
(c) They are synthesized with *N*-terminal mitochondrial entry sequences that specify their mitochondrial locations.
(d) They require a proton-motive force across the inner mitochondrial membrane for their import into the mitochondrial matrix.
(e) Those that reside in the outer membrane do not have their *N*-terminal targeting sequences removed.

20. Which of the following statements about proteins are *correct*?

(a) If small (~15 kd), they can enter the nucleus through pores in the nuclear membrane.
(b) If large (> ~90 kd), they are excluded from the nucleus.
(c) Nuclear proteins are synthesized on the rough ER.
(d) They can be directed to the nucleus by engineering them to contain a short nuclear localization signal.

Receptor-Mediated Endocytosis

21. Which of the following statements about receptor-mediated endocytosis are *correct*?

(a) It delivers certain metabolites to cells.
(b) It is involved in the actions of some hormones and growth factors.
(c) It participates in the uptake and delivery of proteins to the lysosomes for degradation.
(d) It is used by some bacteriophages to gain entry into bacteria.
(e) When defective, it can cause human disorders.

22. Which of the following statements about clathrin are *true*?

(a) It is a component of the protein lattice of coated pits and coated vesicles.
(b) It has an extracellular domain containing carbohydrate substituents.
(c) It exists in different forms that are specific for each protein recognized and bound.
(d) It is a trimer comprising six polypeptide chains.
(e) It has a three-legged structure called a triskelion.
(f) It is recycled after being released from coated vesicles by an ATP-dependent uncoating protein.

23. Place the following probable initial steps in the receptor-mediated endocytosis of a protein in their proper order.

(a) Invagination of the coated pit
(b) Loss of clathrin
(c) Binding of the protein to its receptor in the coated pit
(d) Formation of the endosome (or receptosome)
(e) A decrease in pH leads to the dissociation of the protein-receptor complex
(f) Formation of the coated vesicle

24. What are the possible fates of the receptor and its protein ligand after they are internalized by receptor-mediated endocytosis?

_____

_____

25. Outline the steps in the infection of a mammalian cell by Semliki Forest virus, a membrane-enveloped virus.

_____

_____

_____

Programmed Protein Destruction

26. Which of the following statements about ubiquitin are *correct?*

    (a) It is a small and widely distributed protein.
    (b) It has been highly conserved during evolution.
    (c) Its *C*-terminal carboxyl group is activated as a thioester in an ATP-dependent reaction.
    (d) It is joined to other proteins, depending on the identity of their *C*-terminal amino acids.
    (e) It marks the proteins to which it attaches for destruction.
    (f) It joins to other proteins by isopeptide linkages.

PROBLEMS

1. In the classic study of the events involved in the processing of digestive enzymes for export by the endocrine cells of the pancreas, Jamieson and Palade incubated slices of guinea pig pancreas in a medium containing ^{3}H-leucine for 3 minutes (the "pulse"). After this period, the slices were transferred to an incubation medium containing unlabeled leucine (the "chase"). At varying intervals after the transfer the slices were prepared for autoradiography (see Stryer, p. 45 and p. 121) and electron microscopy.

    (a) Explain briefly what the experiment is designed to establish.
    (b) Predict the order in which "tracks" corresponding to silver halide exposed by β particles will appear in the following structures: (1) the *cis* side of the Golgi, (2) the RER, (3) the *trans* side of the Golgi, and (4) zymogen granules.
    (c) Explain why the chase with unlabeled leucine is indispensible to this experiment.
    (d) Identify the contents of the zymogen granules. What is their fate?

2. Trypsinlike enzymes are ubiquitous in cells and are frequently involved in important proteolytic processing events. Is it possible that signal peptidase is a trypsinlike enzyme? Explain by referring to Figure 31-2 on page 769 of Stryer.

3. In one process for the production of human insulin by recombinant DNA techniques, cDNA for insulin is ligated to *E. coli* DNA that carrys information for the synthesis of β-lactamase. The recombinant DNA is then added to *E. coli* cells, where it directs the synthesis of a fusion protein containing amino acid sequences for both lactamase and insulin. Despite the fact that the hybrid protein must be cleaved to release the insulin amino acid sequences, this technique offers considerable advantages in the quantitative isolation and purification of the hybrid protein product. Explain why this might be the case.

4. Explain the roles of (a) the phosphate group and (b) the long lipid chain of dolichol phosphate in the transport of polysaccharides across membranes.

5. What feature of the procaryotic plasma membrane allows proton-motive force, rather than the hydrolysis of high-energy phosphates, to power the translocation of proteins? Why can proton-motive force not serve as an energy source for translocation across the membranes of the ER in eucaryotes?

6. Suppose that glucose 1-phosphate labeled with ^{32}P is added to a cellular system designed to study the synthesis and processing of *N*-glycosylated proteins. When bacitracin is added to the system, a lipid-soluble intermediate labeled with ^{32}P accumulates. In the absence of bacitracin, the label appears in inorganic phosphate. Explain these results, and identify the lipid-soluble intermediate that accumulates. (Refer to Stryer, p. 774.)

7. What feature of the signal sequence likely prevents the proper identification of proteins targeted for the ER once significant protein folding has begun? Explain.

8. Suppose that you are trying to prepare a chimeric protein for the study of the events involved in the transport of proteins into mitochondria. You have two types of cDNAs that you can use to specify the *N*-terminus of the fusion protein, and you also have cDNAs for four proteins.

cDNAs for N-*terminus*

(1) MLRTSSLFTRRVQPSLFRNILRLQST

(2) MRTNSNNVQPNQNRTNQQLNVANTNT

cDNAs for proteins

(1) Cytochrome c_1

(2) Succinate dehydrogenase

(3) Glucose 6-phosphate dehydrogenase

(4) Malate dehydrogenase

Of the eight possible combinations of cDNAs, which one is most likely to give satisfactory results? Explain. (Refer to p. 782 of Stryer.)

9. Proteins are targeted for destruction by the formation of peptide bonds between the carboxy-terminal glycine of ubiquitin and the ε-amino group of lysine residues in the proteins destined for degradation. How many high-energy phosphates are required for the for-

mation of this peptide bond. Compare this number to the number required for the formation of peptide bonds by ribosomes. (Refer to Figure 31-39 on p. 794 of Stryer.)

10. Insulin is a polypeptide hormone that combines with a specific glycoprotein receptor on the surface of target cells. The insulin-receptor complexes then migrate to coated pits and are internalized by receptor-mediated endocytosis. The resulting vesicles may then fuse with lysosomes.

   (a) Because insulin is a small protein (MW 5700) investigators wondered for years whether it might diffuse into cells and directly modify cellular processes or whether it must combine with a cell surface receptor, causing the release of an intracellular second messenger that mediates its intracellular effects. Autoradiographic studies involving the use of radioactively labeled insulin appeared to indicate that it had entered cells. Does this weaken the notion that insulin may require a second messenger? Explain.

   (b) Divalent antibodies toward insulin receptors can be prepared. These are antibodies that combine with two molecules of insulin receptors. When a fluorescent group is covalently attached to the antibody and the fluorescently-labeled antibodies are employed in an attempt to map the location of insulin receptors on the cell surface, it is found that insulin must be added to the cells as well. Why do you think this is the case?

   (c) For years it has been known that liver tissue and kidney tissue have the capacity to remove circulating insulin from the blood and degrade it. Outline the steps that you think are probably involved in this process.

ANSWERS TO SELF-TEST

Protein Translocation across the Endoplasmic Reticulum Membrane

1. The rough ER has a studded appearance because of the ribosomes that are associated with it, whereas the smooth ER is devoid of ribosomes.

2. b, c, and d. Answers (a) and (e) are incorrect because the proteins of the mitochondrion and the nucleus are synthesized on cytosolic ribosomes.

3. Ribosomes of each class were isolated and mixed with appropriate mRNAs in an in vitro protein synthesizing system. Cytosol-derived ribosomes were capable of synthesizing secretory proteins and, conversely, ribosomes derived from the ER membrane were capable of synthesizing proteins that ordinarily reside in the cytosol.

4. Since the mRNA encoded a protein destined for secretion, it specified a signal sequence at the N-terminus of the protein. This sequence is usually cleaved from the protein by a signal peptidase during translocation across the ER membrane and is therefore not found on normally secreted mature proteins.

5. a, c, and d. The signal sequence, which immediately follows the translation initiation codon, is usually from 13 to 36 amino acids long and includes a positively charged amino acid near the N-terminus and a stretch of nearby hydrophobic residues. Either the removal of the positively charged amino acid near the N-terminus or the introduction of a charged residue into the hydrophobic region usually impairs translocation.

6. Insert DNA encoding the signal sequence from a secreted protein into the gene of the target enzyme by recombinant DNA techniques so that the protein will have an N-terminal signal sequence. If the inserted DNA doesn't disrupt expression of the gene, the encoded protein will be exported upon translation.

7. (a) 5 (b) 2, 4 (c) 1, 7 (d) 3 (e) 6

8. A fragment 35 to 40 amino acids in length should survive. The growing N-terminus of the polypeptide is extruded through a narrow tunnel in the large subunit of the ribosome, and the amino acids within the tunnel should be inaccessible to the protease.

9. b and e. Elongation and translocation are mechanistically distinct. Although unfolded proteins are probably the optimal substrates for the translocation apparatus, fully-folded proteins can be transported across the membrane using energy from the hydrolysis of ATP. Stop-transfer or membrane-anchor sequences can arrest translocation and stabilize a transmembrane peptide segment in the membrane.

Protein Modifications during Translocation and the Golgi Complex

10. a, c, and d. Answer (e) is incorrect because threonine and serine provide hydroxyls for the formation of O-linked oligosaccharides.

11. a, b, c, and e. Sugar-substituted dolichol phosphates serve both as acceptors of monosaccharides from nucleotide sugars and other dolichol phosphate sugars and as donors of monosaccharides and oligosaccharides to other dolichol phosphate sugar derivatives and proteins. As a result of glycosyl transfer by the dolicol oligosaccharide, dolichol pyrophosphate is formed, which must be hydrolyzed to dolichol phosphate by a phosphatase to regenerate the sugar carrier for continued use. (See Figure 31-10 on p. 774 in Stryer.)

12. c, d, and e. The Golgi complex carries out terminal glycosylation by modifying and adding to the core oligosaccharides that were constructed in the ER. Answer (f) is incorrect because the compartments of the Golgi are distinct, and components are transferred between them by vesicles.

13. Its amino-terminus will reside outside the cell. (See Figure 31-13 on p. 776 in Stryer.) All budding and fusion events with the surfaces of other closed membranous compartments within the cell will leave the N-terminus in the lumen of the resulting compartment. The final fusion is with the plasma membrane that encloses the entire cell. This fusion leads to an eversion of the fusing vesicle that places its contents and, in this case, the N-terminus of the integral membrane protein on the outside surface of the cell.

14. a. The disease results from a deficiency in a sugar phosphotransferase that initiates a two-step sequence leading to the formation of a

mannose 6-phosphate terminus on an oligosaccharide substituent of the eight or more affected lysosomal hydrolases. The phosphotransferase attaches a GlcNAc phosphate to a mannose residue of the oligosaccharide. Removal of the GlcNAc leaves the phosphate on the mannose. The enzymes lacking this mannose 6-phosphate "address" label are erroneously exported from the cell rather than being directed to the lysosomes. (See Figure 31-16 on p. 778 in Stryer.)

15. b, c, and e. Answer (a) is incorrect because carbohydrate substituents are polar; thus, they hold glycosylated regions of the protein in the aqueous phase.

16. (a) 1, 4, 7 (b) 2, 3, 5, 6. Serum albumin and collagen are likely to be secreted constitutively and not in response to specific hormonal signals. Production of these proteins is regulated, but not at the level of secretion.

17. They are exported from the cell by constitutive secretory vesicles.

18. Messenger RNA encoding the secreted bacterial enzyme β-lactamase was translated in an in vitro protein synthesis system derived from a plant and the protein was recognized by a mammalian SRP. In addition, determinations of the signal sequences and the specificities of mammalian and bacterial signal peptidases showed that they are similar.

Mitochondrial, Chloroplast, and Nuclear Targeting

19. c, d, and e. Answers (a) and (b) are incorrect because most mitochondrial proteins are encoded by nuclear DNA and are synthesized on cytosolic ribosomes. With regard to answer (e), outer membrane proteins, unlike those in the intermembrane space, the inner membrane, and the matrix, do not have their signal sequences proteolytically cleaved.

20. a and d. Answer (b) is incorrect because large proteins ($> \sim 90$ kd) can enter the nucleus if they contain a nuclear localization sequence. Answer (c) is incorrect because nuclear proteins are synthesized on free ribosomes.

Receptor-Mediated Endocytosis

21. a, b, c, and e. Answer (d) is incorrect because receptor-mediated endocytosis is a phenomenon of eucaryotes, not of procaryotes.

22. a, d, e, and f. With regard to answer (a), clathrin is located on the cytosolic face of coated pits and coated vesicles. Choices (b) and (c) are true for the extracellular receptors of coated pits but not for clathrin.

23. c, a, f, b, d, e

24. Depending upon the particular mode of receptor-mediated endocytosis (see Table 31-1 on p. 791 in Stryer), either component can be recycled, or transported to the lysosomes for degradation, or transported to another location in the cell.

25. Proteins on the surface of the virus are recognized by receptors in coated pits. These receptors probably recognize some other protein during the normal functioning of the cell. The virus is internalized by receptor-mediated endocytosis. In the endosome, the decreased pH

leads to the release of the nucleocapsid into the cytosol through fusion of the viral membrane envelope with the endosome membrane or with the membrane of the fused endosome-lysosome (see Figure 31-36 on p. 792 of Stryer). The viral RNA is exposed, translated, and replicated. The newly-made viral proteins form a nucleocapsid that exits the cell by budding from the plasma membrane at a site where glycoproteins encoded by the virus have clustered.

Programmed Protein Destruction

26. a, b, c, e, and f. Answer (d) is incorrect because it is the identity of the *N*-terminal amino acid of a protein that determines its tagging by ubiquitin and thus its destruction.

ANSWERS TO PROBLEMS

1. (a) The experiment is designed to allow a temporal sequence of processing events to be inferred.
   (b) The tracks indicate the locations of the labeled protein at the time the tissue was fixed and prepared for microscopy. It was found that label appeared in the RER after 3 minutes, in the *cis* side of the Golgi after 7 minutes, in the *trans* side of the Golgi after 37 minutes, and in zymogen granules after 117 minutes.
   (c) If unlabeled leucine were not added, one could not establish a temporal sequence with certainty. A temporal sequence can only be inferred when the label disappears from the precursor as it appears in the product. If labeled leucine were left in the incubation medium, the label would appear first in the RER and later in the Golgi. However, one could not be sure whether it appeared later in the Golgi because it was processed first by the RER and then entered the Golgi or because it entered the Golgi directly by a slower process.
   (d) The contents of the zymogen granules are the major zymogens that are manufactured by the pancreas, including trypsinogen, chymotrypsinogen, procarboxypeptidase, and prolipase, just to name a few. They are released by the pancreatic exocrine cells into the pancreatic duct and travel to the small intestine where they are activated by proteolytic cleavage.

2. Trypsin cleaves peptides on the carboxyl side of Arg and Lys residues. None of the cleavage sites depicted in Figure 31-2 satisfies these criteria. It is therefore reasonable to conclude that signal peptidase is not a trypsinlike enzyme.

3. β-Lactamase is a protein that is ordinarily exported by *E. coli* into the periplasmic space, the space between the inner and outer membranes. The fusion protein contains the amino acid leader sequence that directs β-lactamase to the periplasmic space where the fusion protein can be conveniently isolated by techniques that rupture the outer but not the inner membrane of the cell.

4. (a) The phosphate group serves as the site for the covalent attachment of sugar residues to the carrier.

   (b) The long lipid chain renders the carrier highly hydrophobic and thus membrane permeable.

5. The assemblies that carry out electron transport are part of the plasma membrane of procaryotes. Accordingly, a proton gradient is established across them. In eucaryotes the corresponding assemblies are found on the inner membrane of the mitochondrion and not on the membranes of the ER.

6. The lipid-soluble intermediate that accumulates is dolichol pyrophosphate, whose terminal phosphate comes from glucose 1-phosphate. (See Figure 31-10 on p. 774 of Stryer.) Bacitracin is an antibiotic that forms a 1:1 complex with dolichol pyrophosphate, preventing its hydrolysis to dolichol phosphate and inorganic phosphate. Thus, in the presence of bacitracin, the label will remain in dolichol pyrophosphate. In the absence of bacitracin, the terminal phosphate will be released as inorganic phosphate.

7. The signal sequence is rich in hydrophobic amino acids. Once folding has begun, these residues become buried on the inside of the protein structure, where they are stabilized by interactions among themselves and are no longer free to interact with SRP and thus be directed to the ER.

8. The cDNA that specifies amino acid sequence (1) for the *N*-terminus must be used. The transport of proteins into mitochondria requires this particular mitochondrial entry sequence, which is rich in threonine and serine. If our chimeric protein is to be used to demonstrate that the entry sequence is specific for the translocation of proteins into mitochondria, the balance of the protein must be derived from a protein that is not ordinarily present in mitochondria and is readily identifiable. Of the proteins on the list, only glucose 6-phosphate dehydrogenase is not ordinarily found in mitochondria. It has the additional advantage of being readily assayed.

9. Two high-energy phosphates are expended in the attachment of ubiquitin to lysine. An ATP is cleaved to AMP and pyrophosphate, and the pyrophosphate is subsequently cleaved to two molecules of P_i. A total of four high-energy phosphates are required in protein synthesis. (See the answer to Problem 5, p. 465.)

10. (a) No, it does not weaken the notion that a second messenger is required. Labeled insulin would be internalized by receptor-mediated endocytosis, but that insulin would likely be destroyed by the hydrolytic enzymes of lysosomes. It is important to realize that the interior of internalized vesicles is topologically equivalent to the exterior of the cell, so a case can be made for asserting that the labeled insulin really did not enter the cell at all.

   (b) When insulin combines with its receptor, the insulin-receptor complexes aggregate in coated pits. Such aggregates will combine with divalent antibody. Without the presence of insulin, the receptors remain dispersed on the surface of the cell.

(c) The insulin combines with receptors on the surface of the liver or kidney cells. The insulin-receptor complexes are then internalized by receptor-mediated endocytosis, and the resulting vesicles fuse with lysosomes, leading to the hydrolysis of the vesicle contents.

Control of Gene Expression in Procaryotes

Procaryotes regulate the activities and amounts of their enzymes and structural proteins in response to signals from their environments. The activities of enzymes present at any given moment are modulated by allosteric interactions with effectors and by covalent modifications. More fundamental regulation involves the expression of the genes that encode the proteins. This type of regulation occurs primarily at the level of transcription; that is, when a protein is needed, the gene that encodes it is transcribed and translated. In this chapter, Stryer describes the biochemistry of several mechanisms that control procaryotic gene expression.

The lactose *(lac)* operon is described in detail as an example of a mechanism in which the initiation of transcription is regulated. Negative control is exerted through the binding of a repressor protein to DNA carrying the *lac* operator. This blocks RNA polymerase from binding to the *lac* promoter, thereby preventing transcription. The repression is lifted when a lactose metabolite is bound to the repressor, thereby converting it to a form that is unable to bind to the operator DNA. Positive control of the *lac* operon is accomplished by a protein and cyclic AMP (cAMP) complex, which binds near the same promoter in the absence of the repressor and stimulates the activity of RNA polymerase at the *lac* promoter. The interplay of these mechanisms allows the amount of lactose-catabolizing enzymes to vary over a several-thousandfold range. The arabinose operon, another operon that encodes the enzymes of a catabolic pathway, illustrates how a single protein can serve as both a negative and a positive regulator, depending upon the levels of arabinose and cAMP in the cell. In addition, this operon exemplifies autoregulation—the control of the production of a protein by itself—and the ability of DNA sequences that are not directly adjacent to a promoter to affect the function of that promoter.

The regulation of bacteriophage λ in its alternate life cycles is explained because control of both the initiation and termination of transcription play important roles in determining which of the phage genes are expressed and, hence, the fate of the phage. Proteins called antiterminators allow genes downstream of transcription terminators to be expressed and thus enable the timed development of functions to occur during lytic expression of the phage. The λ repressor, which maintains the prophage in a nearly quiescent state by silencing transcription, is discussed because its structure has been determined to atomic resolution and its mechanism of interaction with a series of adjacent operator sequences is well understood. This repressor also regulates its own synthesis by either stimulating or depressing RNA polymerase activity at its promoter, depending upon which of the operator sites it occupies. The λ repressor provides an example of the helix-turn-helix motif, which several other DNA-binding proteins also use to attain specific association with their target sequences.

Another example of the regulation of gene expression at the level of transcription termination is provided by the attenuation phenomenon in the *trp* operon. This mechanism, which is used by several biosynthetic amino acid operons, depends upon the coupling of transcription and translation in procaryotes. The location of a ribosome on the mRNA determines whether a transcription termination structure will form in the RNA and, consequently, whether the structural genes of the operon will be silent or transcribed and thus expressed. The position of the ribosome on the mRNA is itself determined by the availability of the particular amino acids whose biosynthetic enzymes are encoded by the structural genes of the operon. This elegant molecular switch can provide complete control of the expression of some operons. Translational repression, in which proteins bind to the mRNAs that encode them and thus prevent protein synthesis, is exemplified by some ribosomal proteins. Stryer also describes the stringent response, in which the synthesis of ribosomal and transfer RNA is coordinated with protein synthesis. The chapter closes with a description of a system in which site-specific recombination acts as a switch to juxtapose a promoter and a sequence that encodes a repressor protein. This mechanism provides an absolute and reversible switch between two alternative DNA sequences that give rise to two distinct structural proteins in *Salmonella*.

A review of mRNA initiation and termination and protein synthesis initiation on pages 704–706, 710–713, and 750–753 in Stryer will help you to understand the control mechanisms described in this chapter. Reread pages 695–697 on F and F′ factors as an aid to comprehending the genetic experiments on the *lac* operon. Also, read about the life cycles of bacteriophage λ on page 128 and the formation, hydrolysis, and actions of cAMP on page 458 and pages 461–464. The material on page 651 describes some of the functional groups of DNA that can serve as determinants for specific interactions with proteins and will aid your understanding of how repressors interact with DNA. Pages 656–658 describe the binding of EcoRI endonuclease to DNA and will enable you to contrast its mode of binding with that used by the repressors described in this chapter. Finally, reread pages 698–699 on transposons as an aid to understanding the section on DNA inversion in *Salmonella*.

487

Chapter 32
CONTROL OF GENE EXPRESSION
IN PROCARYOTES

LEARNING OBJECTIVES

When you have mastered this chapter, you should be able to complete the following objectives.

The Lactose Operon and Catabolite Repression (Stryer pages 799–807)

1. List the mechanisms bacteria use to *regulate* their *patterns of metabolism.*

2. Specify the primary mechanism for the *regulation of gene expression* in bacteria.

3. Provide an overview of the metabolism of lactose in *E. coli.* Draw the structure of *lactose,* describe its entry into the cell, and write the equations for the reactions catalyzed by *β-galactosidase,* naming both the substrates used and the products formed.

4. Recount the genetic experiments that led to the concept of the *lac operon* and its regulation.

5. Draw the *genetic map* of the lac operon and outline the functions of the *promoter, repressor, operator,* and *inducer* in controlling the production of *polycistronic* lac *mRNA.*

6. Describe the role of the *lac* repressor in determining whether a bacterium is *inducible* or *constitutive for* β-galactosidase.

7. Describe the *subunit structure* of the *lac* repressor and relate it to the *symmetrical sequence* of the *lac* operator. Compare the affinities of the *lac* repressor for operator DNA and nonspecific DNA. Describe the effect of the binding of *allolactose* by the repressor on its affinity for the operator.

8. Explain the functions of *cyclic AMP* (cAMP) and the *catabolite activator protein* (CAP) in modulating the expression of the *lac* operon. Contrast the functions of cAMP in bacterial and mammalian cells.

9. Provide an overview of the *arabinose (ara) operon* and its expression. Explain the alternative functions of the ara*C protein.*

10. List four general principles of gene regulation exemplified by the *ara* operon.

Bacteriophage λ—The Lytic-Lysogenic Switch (Stryer pages 807–813)

11. Outline the alternative fates of *bacteriophage* λ after it has infected *E. coli.* List the stages of the *lytic pathway.*

12. Outline the roles of *antitermination factors N and Q* as *positive regulatory proteins* during lytic development.

13. List the three stages of the *lysogenic cycle,* and name the only gene expressed in the *prophage.*

14. Distinguish between the patterns of gene expression during the lytic and lysogenic states.

15. Describe the structure of the λ *repressor,* and outline the organization and salient sequence features of the operators to which it binds.

16. Describe the interactions of the λ repressor with its three target sites in the *left* and *right operators* (O_L and O_R). Explain the *autoregulatory mechanism* of the repressor.

17. Describe the release from lysogeny, focusing on the role of the *recA protein* in decreasing the concentration of λ repressor. Describe the function of *cro protein* in depressing λ repressor synthesis by contrasting the affinities of the λ repressor and cro protein for O_R sequences.

18. Describe the *helix-turn-helix motif* of some sequence-specific DNA binding proteins. Recount an experiment that indicated that amino acids in one of the helices determine the binding specificity of a repressor.

The Tryptophan Operon and Transcription Termination (Stryer pages 813–817)

19. Provide an overview of the regulation of the *tryptophan (trp) operon.* Describe the activation of the *trp repressor* by the *corepressor* tryptophan.

20. Sketch the secondary structures of the *trp attenuator* (3:4 hairpin) and *antiterminator* (2:3 hairpin) and relate them to the control of the *trp* operon. Explain the function of the *leader peptide* and the consequences of the *coupling of transcription and translation* in its synthesis.

21. Describe the mechanism of *translational repression* in the regulation of ribosomal protein synthesis, and note the role of the *stringent response* in the coordination of the synthesis of ribosomal and transfer RNA with the synthesis of other proteins.

Regulation by DNA Inversion (Stryer pages 817–818)

22. Outline *phase variation* in *Salmonella.*

23. Describe the role of *site-specific recombination* in determining which flagellin appears on the surface of the bacterium.

SELF-TEST

The Lactose Operon and Catabolite Repression

1. Which of the following are common mechanisms used by bacteria to regulate their metabolic pathways?

   (a) Control of the expression of genes
   (b) Control of enzyme activities through allosteric activators and repressors
   (c) Formation of altered enzymes by the alternate splicing of mRNAs
   (d) Deletion and elimination of genes that specify enzymes
   (e) Control of enzyme activities through covalent modifications

2. Which of the following statements about β-galactosidase in *E. coli* are *correct?*

    (a) It is present in varying concentrations depending upon the carbon source used for growth.

    (b) It is a product of a unit of gene expression called an operon.

    (c) It hydrolyzes the β-1,4-linked disaccharide lactose to produce galactose and glucose.

    (d) It forms the β-1,6-linked disaccharide allolactose.

    (e) It is allosterically activated by the nonmetabolizable compound isopropylthiogalactoside (IPTG).

    (f) Its levels rise coordinately with those of galactoside permease and thiogalactoside transacetylase.

3. Match each feature or function in the right column with the appropriate DNA sequence element of the *lac* operon in the left column.

(a) *i* _____	(1) Bound specifically by the *lac* repressor
(b) *p* _____	(2) Encodes a galactoside permease
(c) *o* _____	(3) Bound specifically by the gene activator cAMP-catabolite protein complex
(d) *z* _____	(4) Encodes a protein capable of interfering with the binding of RNA polymerase
(e) *y* and *a* _____	(5) Encodes a protein that binds allolactose
(f) CAP binding site _____	(6) Bound specifically by RNA polymerase
	(7) Encodes β-galactosidease
	(8) Is a regulatory gene
	(9) Is the *lac* promoter
	(10) Encodes thiogalactoside transacetylase
	(11) Is the *lac* operator
	(12) Encodes the *lac* repressor

4. Explain the direct correlation among the concentrations of β-galactosidase, galactoside permease, and thiogalactoside transacetylase upon the induction of the *lac* operon.

_____

_____

_____

5. When *E. coli* is added to a culture containing both lactose and glucose, which of the sugars is metabolized preferentially? What is the mechanism underlying this selectivity?

_____

_____

_____

6. What happens when the first sugar is depleted during the experiment described in Question 5?

_____

_____

_____

7. Which of the following statements about the C protein encoded by the *ara* operon are *correct?*

    (a) It binds specifically to DNA only when it is complexed with arabinose.

    (b) It binds to an operator (O_1) and shuts off its own synthesis.

    (c) It binds to an operator (O_2) and shuts off the expression of the structural genes of the operon.

    (d) When complexed to arabinose and when cAMP is present, it binds to O_1 and to a region of DNA *(araI)* adjacent to the promoter for the structural genes.

    (e) It brings noncontiguous segments of DNA into proximity by the simultaneous binding of DNA sequences at different sites.

8. It is postulated that the cAMP-CAP complex stimulates the *lac* operon because it creates an additional interaction site for RNA polymerase when it occupies the CAP binding site on the DNA. This site is directly adjacent to the *lac* promoter. Speculate about how the cAMP-CAP complex can accomplish the same positive activating role in the *ara* operon when its binding site on the DNA is separated from the *araBAD* RNA polymerase binding site by the *araI* sequence?

_____

_____

_____

Bacteriophage λ—The Lytic-Lysogenic Switch

9. Which of the following statements about bacteriophage λ are *correct?*

    (a) It synthesizes only the λ repressor in a lysogenic bacterium.

    (b) It escapes from the bacterial chromosome when the λ repressor interacts with single-stranded DNA produced by agents that damage the DNA.

    (c) It exists as a prophage during the lysogenic phase of its life cycle.

    (d) It maintains its lysogenic state, in the absence of an inducer, because the synthesis of the λ repressor is autoregulated.

    (e) In switching from the lysogenic to lytic phase of its life cycle, it turns off the synthesis of the λ repressor because cro protein binds to the λ operator O_R3.

10. Which of the following statements about bacteriophage λ during lytic development are *correct?*

    (a) It synthesizes three different classes of mRNA, which are designated by the times after infection they appear.

(b) It carries out a programmed synthesis of the proteins necessary to replicate its genome and produce its structural components.

(c) It forms the N and Q gene products, which act as positive regulatory proteins, leading to the sequential production of λ-encoded proteins.

(d) It initially produces two proteins; one acts as an inhibititor of λ repressor synthesis and the other acts as a transcription termination factor.

11. Which of the following statements about the λ repressor are *correct?*

(a) It exists in solution as a monomer with a $M_r = 26,000$.

(b) The monomer contains an N-terminal domain, which interacts specifically with DNA, and a C-terminal domain, which contains the allolactose binding site.

(c) It binds specifically to some of the same sites on DNA as does the cro protein.

(d) It binds specifically to O_R and O_L operators with a stoichiometry of up to six polypeptide chains per operator.

(e) It shows different affinities to subsites within O_R and O_L.

(f) It binds to a region of DNA that is 17 base pairs long and is imperfectly symmetrical.

(g) It is inactivated through the hydrolysis of one of its peptide bonds by recA.

12. Explain how the λ repressor regulates its own synthesis in a lysogenic cell.

_____

_____

_____

13. Which of the following statements about the helix-turn-helix motif are *correct?*

(a) It is an arrangement of secondary-structure elements in some proteins that bind specifically to DNA.

(b) It occurs in two symmetrically disposed arrays in dimeric proteins to facilitate interaction with DNA sequences of inverted symmetry.

(c) It positions the positively charged N-terminus of its recognition α helix into the major groove of the DNA to promote electrostatic interactions with the negatively charged phosphodiester backbone of the DNA.

(d) It is arranged in dimers so that two nearly parallel α helices can be positioned 34 Å apart.

(e) It is found in the λ and tryptophan repressors, the cro protein, and the EcoRI restriction endonuclease structures.

(f) It contains a recognition helix of invariant amino acid composition.

14. Describe the experiment that showed unambiguously that amino acids on one face of the recognition helix of a helix-turn-helix structure

in a particular repressor were responsible for its sequence-specific binding to DNA.

_____

_____

_____

_____

The Tryptophan Operon and Transcription Termination

15. Regulation of the *trp* operon involves which of the following?

   (a) Controlling the amount of polycistronic mRNA formed at the level of transcription initiation
   (b) Controlling the amount of polycistronic mRNA at the level of transcription termination
   (c) The sequential and coordinate production of five enzymes of tryptophan metabolism from a single mRNA
   (d) The sequential and coordinate production of five enzymes of tryptophan metabolism from five different mRNAs produced in equal concentrations
   (e) The production of transcripts of different sizes, depending upon the level of tryptophan in the cell

16. With regard to the *trp* operon, why is tryptophan called a corepressor?

_____

_____

17. Which of the following statements concerning the *trp* operon leader RNA, which has 162 nucleotides preceding the initiation codon of the first structural gene of the operon, are *correct?*

   (a) A deletion mutation in the DNA encoding the 3′-region of the leader RNA gives rise to increased levels of the biosynthetic enzymes forming Trp.
   (b) A short open reading frame, containing Trp codons among others, exists within the leader RNA.
   (c) The leader RNA encodes a "test" peptide whose capacity to be synthesized monitors the level of Trp-tRNA^{Trp} in the cell.
   (d) The leader RNA contains one Shine-Dalgarno ribosome-binding site.
   (e) The leader RNA may form two alternate and mutually-exclusive secondary structures.
   (f) The structure of the leader RNA in vivo depends upon the position of the ribosomes translating it.

18. Explain how the synthesis of ribosomal RNA and the synthesis of other proteins are coordinated.

_____

_____

_____

_____

19. Which of the following statements about site-specific recombination are *correct?*

(a) When it occurs between two repeated sequences that are inverted with respect to one another on the same DNA molecule, it leads to the deletion of the DNA between the sequences.

(b) It can move a promoter for a gene such that it is "aimed" away from the DNA sequences that encode the mRNA it ordinarily produces.

(c) In the *Salmonella* phase-variation phenomenon, it disrupts an operon such that the repressor it made before the recombination is no longer produced.

(d) In the *Salmonella* phase-variation phenomenon, it selects genes for transcription by two mechanisms—operator-repressor interaction and operon disruption.

(e) It provides an absolute switch that turns on the synthesis of one flagellin or another at any given time in *Salmonella.*

20. Match each feature or requirement in the right column with the appropriate regulated element or mechanism of regulation of gene expression listed in the left column.

(a) *Lac* repressor ———

(b) λ repressor ———

(c) *Trp* repressor ———

(d) cAMP-CAP complex
———

(e) *Ara*C protein ———

(f) N and Q proteins of λ ———

(g) *Trp* operon attenuation ———

(h) Ribosomal proteins
———

(i) Ribosomal and transfer RNAs ———

(j) *Hin* recombinase ———

(1) Requires concomitant protein synthesis to function

(2) Controlled by translational repression

(3) Inverts the orientation of a promoter with respect to a repressor gene

(4) Interferes with the functioning of RNA polymerase by binding to DNA

(5) Stimulates transcription of several catabolic operons

(6) Binds to DNA sequences with regions of twofold symmetry

(7) Is inactivated by recA protein

(8) Its regulatory effects are antagonized by cro protein

(9) Breaks and rejoins DNA phosphodiester bonds

(10) Depends upon the tight coupling of translation and transcription

(11) Regulates its own synthesis

(12) Involves ribosomes

(13) Stimulates transcription by allowing continued transcription elongation

(14) Its concentration depends on glucose levels

(15) Stimulates transcription initiation

(16) Interacts with repeated sequence variants within its operators

(17) Acts as both a positive and negative regulator

(18) Participates in global regulation

(19) Is affected by the level of aminoacyl-tRNAs in the cell

(20) Involves ppGpp

PROBLEMS

1. What property of enzymes makes them more suitable than, say, structural proteins for studies of the genetic regulation of protein synthesis? Explain.

2. When lactose is used as an inducer, there is a lag before the enzymes of the lactose operon are synthesized. With IPTG there is no lag. Explain this observation. (See Figure 32-1.)

3. Some of the known constitutive mutations of the lactose operon occur in the operator gene rather than the regulator gene.
   (a) Would you expect such an o^c mutant to be dominant or recessive to its wild-type o^+ allele? Explain.
   (b) Is a constitutive mutation in an operator *cis* or *trans* in its effects? Explain.
   (c) Design an experiment involving the genes i^+, o^c, o^+, and z^+ that would confirm your answer to part (b). Assume that it is possible to detect whether enzymes are produced in diploid (++) or haploid (+) amounts.

4. Since the permease required for the entry of lactose into *E. coli* cells is itself a product of the lactose operon, how can the first lactose molecules enter uninduced cells? Explain.

5. Design an experiment to show that lactose stimulates the synthesis of new enzyme molecules in *E. coli* rather than fostering the activation of preexisting enzyme molecules.

6. The three known enzymes of the lactose operon in *E. coli* are not produced in equimolar amounts following induction. Rather, more ga-

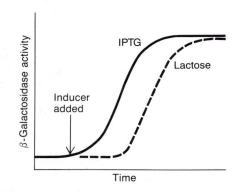

Figure 32-1
Kinetics of β-galactosidase induction by lactose and IPTG. (Assume that each inducer has been removed after an appropriate period.)

lactosidase than permease is produced, and more permease than trans-acetylase is produced. Propose a mechanism to account for this that is consistent with known facts about the lactose operon.

7. Assume that the following allelic possibilities exist for the i genes and o genes of the lactose operon of *E. coli:*

i^+ = wild-type regulator gene

i^c = regulator constitutive mutation, makes inactive repressor

i^s = repressor substance insensitive to inducer

o^+ = wild-type operator gene

o^c = operator consitutive mutation

In addition, assume that the mutations z^-, y^-, and a^- lead to nonfunctional enzymes Z, Y, and A, respectively.

For each of the following, predict whether active enzymes Z, Y, and A will or will not be produced. For partially diploid cells, assume semidominance; that is, the enzyme activity in a diploid cell will be twice that found in a haploid cell. Use the following answer code:

O = active enzyme absent

+ = active enzyme present in haploid amounts

++ = active enzyme present in diploid amounts

	Without IPTG			With IPTG		
	Z	Y	A	Z	Y	A
(a) $i^+o^+z^+y^+a^+$						
(b) $i^co^+z^+y^+a^+$						
(c) $i^so^+z^+y^+a^+$						
(d) $i^+o^cz^+y^+a^+$						
(e) $\dfrac{i^+o^+z^+y^+a^+}{i^co^+z^+y^+a^+}$						
(f) $\dfrac{i^+o^+z^+y^+a^+}{i^co^+z^-y^+a^+}$						
(g) $\dfrac{i^+o^+z^+y^+a^+}{i^+o^cz^+y^+a^+}$						
(h) $\dfrac{i^so^+z^+y^+a^+}{i^+o^+z^+y^-a^-}$						
(i) $\dfrac{i^so^+z^+y^+a^+}{i^+o^cz^-y^+a^+}$						
(j) $\dfrac{i^+o^+z^+y^+a^+}{i^co^cz^+y^+a^-}$						

8. Assume that the structural genes x and y, which code for the repressible enzymes X and Y, are subject to negative control by a regulator gene r, which produces a substance S whose ability to bind to the operator gene o is modified by cosubstance T.

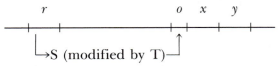

Assume that the following allelic possibilities exist for genes r and o:

o^+ = wild-type operator gene

o^o = operator gene unable to bind S (with or without T)

r^+ = wild-type regulator gene

r^o = inactive regulator gene product (with or without T)

r^s = regulator gene product always active (with or without T)

In addition, assume that the mutations x^- and y^- in the two structural genes lead, respectively, to nonfunctional enzymes X and Y.

(a) Are enzymes X and Y likely to be biosynthetic or degradative? Justify your answer briefly.

(b) Is substance S active or inactive in the presence of T? Explain.

(c) For each of the following, predict whether active enzymes X and Y will or will not be produced under the specified conditions. For partially diploid cells, assume semidominance; that is, the enzyme activity in a diploid cell will be twice that found in a haploid cell. Use the following answer code:

0 = active enzyme absent

+ = active enzyme present in haploid amounts

++ = active enzyme present in diploid amounts

	Without T		With T	
	X	Y	X	Y
(1) $r^+o^+x^+y^+$	——	——	——	——
(2) $r^+o^+x^-y^+$	——	——	——	——
(3) $r^so^+x^+y^+$	——	——	——	——
(4) $r^so^ox^+y^+$	——	——	——	——
(5) $r^+o^ox^+y^+$	——	——	——	——
(6) $r^oo^+x^+y^+$	——	——	——	——
(7) $\dfrac{r^+o^+x^+y^+}{r^+o^+x^-y^+}$	——	——	——	——
(8) $\dfrac{r^+o^+x^+y^+}{r^oo^+x^+y^+}$	——	——	——	——
(9) $\dfrac{r^so^+x^-y^-}{r^oo^+x^+y^+}$	——	——	——	——
(10) $\dfrac{r^+o^ox^+y^-}{r^oo^+x^+y^+}$	——	——	——	——

9. The kinetics of induction of enzyme X are shown in Figure 32-2. What percentage of total cellular protein is due to enzyme X in induced cells?

10. Assume that the dissociation constant K for the repressor-operator complex is 10^{-13} M and that the rate constant for association of operator and repressor is 10^{10} M^{-1} s^{-1}. Calculate the rate constant k_{diss} for the dissociation of the repressor-operator complex. What is the $t_{1/2}$ (half-time of dissociation or half-life) of the repressor-operator complex?

11. In systems of genetic regulation involving positive control, a regulatory gene produces a substance that enhances rather than inhibits transcription. Are there elements of positive control in the lactose operon of *E. coli*? Explain.

12. Suppose that a mutant C protein of the arabinose operon of *E. coli* were found that could not bind arabinose. Predict the results on (a) the level of C protein and (b) the rate of transcription of the mRNA for *BAD* enzymes in the presence of cAMP-CAP complex, explaining your reasoning.

13. What is the minimum number of promoter regions that are necessary for the transcription of the genes for λ repressor, cro protein, and N protein. Where are these regions located? See Figure 32-18 on page 808 of Stryer.

14. Suppose that the relative affinities of the three O_R genes and the three O_L genes for the λ repressor were reversed and that O_R3 and O_L3 have the greatest affinity for a repressor. What would be the likely effect on the mechanism of infection by λ? Explain.

15. An operon for the biosynthesis of amino acid X in a certain bacterium is known to be regulated by a mechanism involving attenuation. What can one confidently predict about the amino acid sequence in the leader peptide for that operon? Explain.

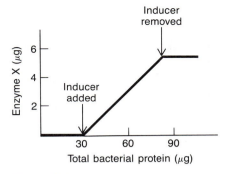

Figure 32-2
Kinetics of induction of enzyme X.

ANSWERS TO SELF-TEST

The Lactose Operon and Catabolite Repression

1. a, b, and e. Answer (c) is incorrect because the splicing of mRNA is not common in bacteria.

2. a, b, c, d, and f. Although IPTG is an inducer for the synthesis of β-galactosidase, it is neither a substrate nor an allosteric activator of the enzyme, so answer (e) is incorrect.

3. (a) 4, 5, 8, 12 (b) 6, 9 (c) 1, 11 (d) 5, 7 (e) 2, 10 (f) 3. For (d), 5 is correct because allolactose is the product of a reaction catalyzed by β-galactosidase and, as a product, it binds to the enzymes.

4. The genes encoding these three enzymes are transcribed as a polycistronic mRNA, so there are equal numbers of copies of the mRNA sequences specifying each of the enzymes. See problem 6.

5. Glucose is metabolized preferentially because it results in a decrease in the synthesis of cAMP by adenylate cyclase. The lack of cAMP

prevents the formation of the cAMP-CAP complex, which is necessary for the efficient transcription of the *lac* operon and other catabolite-repressible operons.

6. When glucose is depleted, the concentration of cAMP rises. The cAMP-CAP complex forms and binds to the CAP binding site just upstream of the RNA polymerase binding site in the *lac* promoter. At the same time, some lactose has entered the cell, has been converted to allolactose by β-galactosidase, and is bound by the *lac* repressor so that it no longer binds to the *lac* operator. RNA polymerase now binds to the *lac* promoter even more effectively because of the presence of the cAMP-CAP complex. The enzymes and permease of the *lac* operon are expressed fully; consequently, lactose readily enters the cell and is efficiently metabolized.

7. b, c, d, and e. Answer (a) is incorrect because the *ara*C protein (C protein) binds to specific sequences of DNA both by itself and when it is complexed to arabinose—but with different outcomes. The binding of arabinose converts the *ara*C protein from a repressor to an activator of transcription.

8. The *ara*C protein and arabinose complex binds to *araI* and mediates the action of the cAMP-CAP complex. The binding of the cAMP-CAP complex upstream of the binding site for the *ara*C protein and arabinose complex probably affects the way the complex interacts with the adjacent RNA polymerase so that the transcription is more efficient.

Bacteriophage λ—The Lytic-Lysogenic Switch

9. a, c, d, and e. Answer (b) is incorrect because the prophage is excised from the lysogen by a series of events that are initiated by the proteolytic cleavage of the λ repressor by the recA protein. The activation of the proteolytic activity of recA results from the production of single-stranded DNA in the cell, often as a result of agents that damage DNA. Both Cro and λ repressors bind to O_R3 but with different affinities; Cro binds more tightly and, consequently, blocks repressor synthesis. See Figures 32-21 and 32-22 on page 810 of Stryer.

10. a, b, and c. Answer (d) is incorrect because, in addition to the cro protein, which suppresses λ repressor synthesis, the N protein is an *anti*terminator, not a terminator.

11. c, d, e, f, and g. Answers (a) and (b) are incorrect because the λ repressor exists as a dimer in solution, with the *C*-terminal domains interacting with one another to maintain the dimer. Each dimer interacts with a partially symmetrical sequence that is repeated three times in both O_R and O_L. Because the sequence of each of the subsites is different, the repressor displays different affinities for each.

12. Alternative interactions between the λ repressor and the subsites within O_R can either stimulate or repress the activity of RNA polymerase at the promoter for λ repressor synthesis during lysogeny and thus autoregulate its synthesis. At lower levels of repressor, the distal O_R1 and O_R2 subsites are filled and, because they are adjacent to the promoter and allow interaction with RNA polymerase, stimulate it to transcribe the repressor gene. Conversely, at higher repressor concentrations, the proximal O_R3 subsite is also occupied by the repressor and RNA polymerase is then precluded from binding to the promoter.

13. a, b, and d. Answer (c) is incorrect because, of the proteins mentioned in the question, only the EcoRI restriction endonuclease inserts the *N*-termini of its recognition helices into the major groove of the DNA at the sequence recognized (see p. 658 in Stryer). Answer (f) is incorrect because the sequences of the recognition helices of those proteins that utilize the helix-turn-helix motif vary, specifying the different sequences they recognize.

14. Recombinant DNA techniques were used to construct a modified bacteriophage 434 gene encoding a repressor protein in which five amino acids on one face of its recognition helix were replaced with those found in the bacteriophage P22 repressor. The modified 434 repressor recognized and bound the same operator as did the P22 repressor and, in addition, no longer bound to the 434 operator.

The Tryptophan Operon and Transcription Termination

15. a, b, c, and e. When the RNA is not terminated at the attenuator, a single polycistronic mRNA, which encodes five enzymes, is produced, so answer (d) is incorrect.

16. The *trp* repressor does not bind DNA specifically unless it is complexed to Trp. The binding of Trp to the repressor is necessary to induce a conformational change that forms a specific DNA binding surface on the protein.

17. a, b, c, e, and f. The discovery of a deletion mutation in front of the first structural gene of the operon, and not in the operator, was the first clue that a control mechanism operating at the level of transcription termination was involved in regulating the expression of the *trp* biosynthetic enzymes. This deletion altered potential mRNA structures so that the rho-independent transcription-termination structure could no longer form. Answer (d) is incorrect because the leader RNA contains two Shine-Dalgarno ribosome binding sites: one in front of the initiation codon of the leader peptide and one preceding the first structural gene of the operon.

18. Ribosomes are synthesized rapidly during periods of growth when nutrients are in good supply. If amino acids become limiting in the cell, tRNAs cannot be aminoacylated, and uncharged tRNAs bind to the A site of the ribosomes. This event, coupled with the presence of stringent factor, gives rise to the synthesis of guanosine tetraphosphate (ppGpp) through the transfer of the terminal pyrophosphoryl group of ATP to the 3'-hydroxyl of GDP. ppGpp inhibits the synthesis of rRNA at the level of transcription initiation. If the synthesis of ribosomal proteins exceeds that of rRNAs, the excess of particular ribosomal proteins bind to the mRNAs that encode them and the other ribosomal proteins in the same operons. The repression continues until enough rRNAs, which bind ribosomal proteins more tightly that the mRNAs that encode these proteins, are available to allow the formation of ribosomes. Through this interplay of controls at the levels of transcription and translation, the cell maintains proper ratios of ribosomal proteins and rRNAs.

Regulation by DNA Inversion

19. b, c, d, and e. Answer (a) is incorrect because site-specific recombination between inverted repeats on the same molecule leads to the inversion of the DNA between them (see Stryer, p. 699). In the *Salmonella*

phase-variation phenomenon, the inversion occurs in a piece of DNA containing the promoter for a polycistronic mRNA that encodes one type of flagellin and a repressor for a gene that encodes a second type of flagellin. Inverting the DNA disrupts the expression of this operon and, because of the absence of the repressor it encodes, allows the expression of a new mRNA that encodes the second type of flagellin molecule.

20. (a) 4, 6 (b) 4, 6, 7, 8, 11, 15, 16, 17. The λ repressor acts both as a repressor and stimulator of its own transcription, depending upon its concentration in the cell. In acting as both a positive and negative regulator, it is like the *ara*C protein. (c) 4, 6 (d) 5, 6, 14, 15, 18 (e) 4, 11, 15, 17 (f) 13 (g) 1, 10, 12, 19 (h) 2, 12, 19 (i) 12, 19, 20. The ribosomes are necessary for the formation of ppGpp. (j) 3, 6, 9. The *hin* recombinase binds and reacts with inverted repeats (insertion sequences) to flip the orientation of the intervening DNA (see Stryer, pp. 699–700).

ANSWERS TO PROBLEMS

1. Enzymes are catalysts, and thus small amounts can be readily detected. An enzyme having a turnover number of 300,000 s^{-1} will provide an assay that is 300,000 times as sensitive as that for a structural protein, which must be assayed stoichiometrically. Many cellular proteins are produced in amounts that are too small to be detected by direct chemical methods.

2. The actual inducer of the lactose operon in vivo is 1,6-allolactose (see Stryer, p. 800). The lag represents the time it takes for lactose to be converted into 1,6-allolactose. IPTG itself directly acts as an inducer. Therefore, no lag is observed.

3. (a) Imagine a partial diploid that has one o^+ and one o^c gene. The o^+ gene will bind a repressor, so the structural genes on its chromosome will not be expressed. The o^c gene will not bind a repressor, so the structural genes on its chromosome will always be expressed. Thus, an o^c mutant would be dominant to its wild-type o^+ allele.

   (b) Repressors are not bound by o^c genes. Only the structural genes on the same chromosome as the o^c mutant will be affected, a *cis* effect.

   (c) One could prepare a partial diploid with the following genotype

   $$\frac{i^+o^+z^+}{i^+o^cz^+}$$

   If the effect of the mutation is *cis*, the haploid amount of enzyme Z will be produced in the absence of inducer. (Its synthesis will be specified by the chromosome containing $i^+o^cz^+$.) If the effect is *trans*, the diploid amount of Z will be produced in the absence of inducer.

4. Very low levels of lactose operon enzymes are synthesized even in the absence of an inducer (see Stryer, p. 800).

5. Add IPTG to *E. coli* cells growing in a medium containing a carbon source other than lactose in both the presence and the absence of an inhibitor of procaryotic protein synthesis, like chloramphenicol. If zymogen activation is involved, chloramphenicol will not inhibit induction. If the synthesis of new protein is involved (as it is), induction will not be observed in the presence of chloramphenicol.

6. Differential expression of the three structural genes in the lactose operon must be at the level of translation and not transcription. Following induction, mRNA transcripts containing genetic information for all three genes are produced. Some ribosomes might drop off the messenger at the end of the structural genes, with a smaller number reading through.

7. (a) 0 0 0 +   +   +
  (b) + + + +   +   +
  (c) 0 0 0 0   0   0
  (d) + + + +   +   +
  (e) 0 0 0 ++ ++ ++
  (f) 0 0 0 +   ++ ++
  (g) + + + ++ ++ ++
  (h) 0 0 0 0   0   0
  (i) 0 + + 0   +   +
  (j) + + 0 ++ ++ +

8. (a) The enzymes will be biosynthetic. The clue to this is provided by the statement that the enzymes are repressible. (Degradative enzymes are inducible, whereas biosynthetic enzymes are repressible.)
  (b) Because this operon is under negative control, substance S must act as a repressor. Substance S is active as a repressor in the presence of T, so T functions as a corepressor.
  (c) (1) +   +   0 0
      (2) 0   +   0 0
      (3) 0   0   0 0
      (4) +   +   + +
      (5) +   +   + +
      (6) +   +   + +
      (7) +   ++ 0 0
      (8) +   ++ 0 0
      (9) 0   0   0 0
     (10) ++ +   + 0

9. From the graph in Figure 32-2, we see that 3 μg of enzyme X is present when the total bacterial protein present is equal to 60 -μg. Thus, the percentage of enzyme X is $100\% \times \frac{3}{60} = 5\%$.

10.
$$[RO] \rightleftharpoons [R] + [O]$$

$$\frac{[R][O]}{[RO]} = K = 10^{-13} \text{M} = \frac{k_{diss}}{10^{10} \text{ M}^{-1} \text{ s}^{-1}}$$

$$k_{diss} = 10^{-3} \text{ s}^{-1}$$

$$t_{1/2} = \frac{0.693}{k_{diss}} = \frac{0.693}{10^{-3} \text{s}^{-1}} = 693 \text{ s} \cong 11.6 \text{ min.}$$

11. Catabolite activator protein (CAP) is a positive control element. When the level of glucose in cells is low, the level of cAMP is high, leading to the formation of cAMP-CAP complex. The cAMP-CAP complex binds to DNA in the promoter region, creating an entry site for RNA polymerase. The result is the transcription of the lactose operon (providing that no repressor is present). (See Figure 32-9 on p. 804 of Stryer.)

12. (a) The level of C protein would be unaffected because C protein binds to $araO_1$ in the absence of arabinose, preventing synthesis of C mRNA.
    (b) The rate of transcription of mRNA for *BAD* enzymes would be reduced. The maximal transcription rate is only achieved when both cAMP-CAP complex and C protein with bound arabinose are present. (See Figure 32-14 on p. 806 of Stryer.)

13. Three promotor regions must be involved. One within the O_R1 region for the transcription of the *cro* gene, one within O_L1 for the transcription of the *N* gene, and one within O_R3 for the transcription of the *cI* gene.

14. If O_R3 has the greatest affinity for the λ repressor, the result would be abnormally low levels of the λ repressor itself since it represses its own synthesis by binding to O_R3. Also, transcription of the mRNA for both the N protein and the Q protein would proceed because O_L1 and O_R1 would be unoccupied by the repressor. These effects would lead to lytic, rather than temperate, infection by λ.

15. The leader sequence should be rich in amino acid X. If there is already sufficient X present in the cell, there will be sufficient X-tRNA^X for the synthesis of the leader peptide (as well as for the synthesis of other X-containing proteins in the cell). Therefore, there will be no need to biosynthesize the enzymes needed to produce X.

CHAPTER 33

Eucaryotic Chromosomes and Gene Expression

In Chapter 33, Stryer elaborates three aspects of the biochemistry of nucleic acids that reflect both the problems and the opportunities afforded by the profusion of DNA in eucaryotic cells. He describes the structure of nucleosomes, which are the basic, repeating units of chromatin. Each nucleosome consists of DNA wrapped about an octameric protein core composed of histones. Individual nucleosomes are linked together by intervening stretches of DNA. The nucleosomes are packed, along with other proteins, to form a highly condensed chromosome. The replication and repair of eucaryotic DNA are also described, with special emphasis on multiple initiations, the DNA polymerases involved, the deposition of newly synthesized histones on only one of the two daughter duplexes, and the formation of the telomeres.

The organization of chromosomal DNA is introduced with an explanation of how the kinetics of hybridization of fragmented and denatured eucaryotic DNA revealed the existence of repeated sequences. Stryer provides examples of different arrangements of repeated genes and uses the silk fibroin gene to illustrate the capacity of single-copy genes to produce abundant proteins. Examples of gene amplification and the organization of a developmentally regulated gene are also presented. Stryer next describes the modifications of chromatin structure that occur during transcription. He also describes the structure of a transcription factor that uses the zinc-finger motif to bind DNA specifically and activate the 5S rRNA gene for transcription. The chapter concludes with a brief description of the role of the homeo box in the development of the basic architectural plan of higher eucaryotic organisms.

In preparation for studying this chapter, you should review Chapters 27 and 29, focusing on the sections dealing with DNA replication, eucaryotic RNA synthesis and transcription factors, and tRNA and rRNA processing. Also review pages 81–82 on the melting and annealing of DNA as an introduction to the section on the kinetics of DNA hybridization and pages 150–151 on the changing patterns of hemoglobin synthesis during embryonic and fetal development to prepare yourself for the section on the organization of the hemoglobin gene.

LEARNING OBJECTIVES

When you have mastered this chapter, you should be able to complete the following objectives.

The Eucaryotic Chromosome (Stryer pages 823–830)

1. State the relative amounts of *genomic DNA* in the *haploid genomes* of *E. coli, yeast,* the *fruit fly,* and *humans.*

2. Relate the number of chromosomal DNA molecules to the number of chromosomes in a haploid cell, and discuss the empirical bases for arriving at this number.

3. Describe the composition of *chromatin.* List the types of *histones* and describe their general characteristics. Note the evolutionary stability of the sequences of the H3 and H4 histones.

4. Describe the composition and structure of the *nucleosome,* and relate the nucleosome to the proposed structure of the *chromatin fiber.*

5. Describe the evidence provided by electron microscopy, X-ray and neutron diffraction analyses, nuclease digestion, and reconstitution studies that led to the current model of the structure of the nucleosome.

6. Distinguish between the DNA associated with the *nucleosome core* and that in the *internucleosome linker.*

7. Assign the types of histones to their locations within the nucleosome.

8. Explain how the *packing ratio* of 7 arises from the structure of the nucleosome, state the packing ratios that have been estimated for the *metaphase* and *interphase chromosomes,* and outline a molecular model that partially accounts for these latter ratios.

9. Compare *DNA replication* in procaryotes and eucaryotes. Contrast the rates of movement of the *replication fork* in procaryotes and eucaryotes, and explain how *multiple origins* allow for the rapid duplication of genomes in eucaryotes.

10. List the enzymatic properties and functions of the *DNA polymerases* of eucaryotes.

11. Explain the difficulty in replicating the end of a linear DNA duplex, and outline a model for *telomere replication*.

12. Account for the *conservative segregation of parental histones*. Identify the daughter DNA duplex on which newly synthesized histones assemble during replication.

13. Describe the phenotype of *petite mutants* in yeast, and explain the significance of their independent segregation.

14. Name the *organelles* in plants and animals that contain functional DNA.

15. Describe the human *mitochondrial genome,* and contrast its features with those of the nuclear genome.

Repetitive DNA and Gene Organization (Stryer pages 834–839)

16. Write the chemical reaction for the *renaturation of denatured DNA* and the equation that relates the fraction of single-stranded DNA present to the total DNA concentration and the time of hybridization.

17. Draw a C_0t *curve*, and describe the significance of the $C_0t_{0.5}$ value with respect to *genome size* and the content of *repeated DNA sequences*.

18. Provide examples of the repeated DNA sequences that are found in mammals.

19. Describe *satellite DNA,* locate it in the chromosome, and explain the properties that allow it to be separated from the remaining genomic DNA.

20. Describe the intracellular location and spatial organization of the 18S, 5.8S, and 28S *rRNA genes* and their *spacer elements*.

21. Outline the biochemistry of the expression and maturation of the *rRNA gene cluster*.

22. Explain how the hybridization of histone mRNA to genomic DNA revealed the repetitions of the *histone genes*.

23. List the distinctive features of the histone genes.

Eucaryotic Gene Expression (Stryer pages 839–846)

24. Explain, how a *single-copy gene* can give rise to an *abundant protein,* using the silk fibroin protein from the silkworm as an example.

25. Relate the *developmental expression* of α-globin and β-globin to the chromosomal location and organization of their genes.

26. Relate the globin gene sequences and chromosomal organization to their evolution. Distinguish between genes and *pseudogenes*.

27. State the fraction of mammalian DNA that is likely to encode proteins.

28. Relate *chromosomal puffs* to gene expression.

29. Provide the experimental bases for the correlations between *nuclease susceptibility* and the *degree of methylation* of the chromosome to *transcriptional activity* in mammals.

30. Outline the transcriptional activation of the 5S rRNA genes. Describe the *zinc-finger structure* of *transcription factor IIIA*, and appreciate the ubiquity of finger domains.

31. Outline the function of *homeotic genes* and the effect of *homeotic mutations*. Describe the relationship between the *homeo box* and the *homeo domain*.

SELF-TEST

The Eucaryotic Chromosome

1. Which of the following statements about DNA that has been isolated from a eucaryotic chromosome are *correct*?

   (a) It is resistant to breakage by shearing forces because of its large size.
   (b) It is linear and unbranched.
   (c) It can be sized by a viscoelastic technique, which measures the time the stretched and elongated molecule takes to relax to its normal conformation.
   (d) It is a single molecule.
   (e) It can be more than 100 Mb long.

2. Match the organism listed in the left column with the correct amount of DNA in its haploid genome from the right column.

   (a) *E. coli* (bacterium) _____   (1) 4000 Kb
   (b) *S. cerevisiae* (yeast) _____   (2) 170 Mb
                                             (3) 3900 Mb
   (c) *D. melanogaster* (fruit fly) _____ (4) 14 Mb
   (d) *H. sapiens* (human) _____

3. Which of the following statements about histones are *correct*?

   (a) They are highly basic because they contain many positively charged amino acid side chains.
   (b) They are extensively modified after their translation.
   (c) They can be fractionated by ion-exchange chromatography.
   (d) They can be removed from DNA with dilute acid.
   (e) They account for approximately one-fifth of the mass of a chromosome.

4. Which of the following statements about nucleosomes are *correct*?

 (a) They constitute the repeating units of a chromatin fiber.
 (b) Each contains a core of eight histones.
 (c) They contain DNA that is surrounded by a coating of histones.
 (d) They occur in chromatin in association with approximately 200 base pairs of DNA, on average.

5. What did each of the following techniques contribute to our understanding of chromatin structure?

 (a) Electron microscopy

 _____

 _____

 (b) X-ray and neutron diffraction

 _____

 _____

 (c) Nuclease digestion

 _____

 _____

 (d) Reconstitution

 _____

 _____

6. Describe the model of the nucleosome.

_____

_____

_____

_____

7. Does the formation of nucleosomes account for the observed packing ratio of human metaphase chromosomes? Explain.

_____

_____

_____

Chromosome Replication and Organelle DNA

8. Which of the following statements about eucaryotic DNA are *correct*?

 (a) It is replicated conservatively.
 (b) It is replicated bidirectionally.
 (c) Its replication originates at numerous initiation points on each chromosome.
 (d) Its replication requires primers.
 (e) Its replication is template independent.
 (f) It is synthesized by chain growth in the $3' \rightarrow 5'$ direction.

9. Match the DNA polymerase in the left column with the function or property with which it is associated from the right column.

(a) DNA polymerase α _____

(b) DNA polymerase β _____

(c) DNA polymerase γ _____

(1) A single polypeptide
(2) Increases more than tenfold during the S-phase of the cell cycle
(3) Involved in DNA repair
(4) Involved in chromosomal DNA replication
(5) Involved in mitochondrial DNA replication

10. Describe the replication of the DNA at the end of linear chromosomes.

_____

_____

_____

11. Which of the following duplex DNA segments is the product of one round in the replication of a eucaryotic chromosome in the absence of any mitotic events, that is, in the absence of recombination. The parental strand is represented by ---, and the newly synthesized daughter strand is represented by ***.

(a) -------

(b) *******
    ****
(c) -------
    ****

12. Which of the following statements about histones during DNA replication are *correct?*

(a) Their number must double.
(b) Those that are present in the G1 phase of the cell cycle initially associate with both daughter DNA duplexes at the replication fork.
(c) They segregate semiconservatively.
(d) Those that are newly synthesized associate with the daughter duplex containing the lagging strand.

13. Why do the petite mutations of yeast segregate independently of mutations located on the nuclear genome?

_____

_____

_____

14. Which of the following statements about human mitochondrial DNA are *correct?*

(a) It is a linear, single-stranded molecule.
(b) It encodes both protein and RNA products.

    (c) It is responsible for the synthesis of several subunits of proteins that are involved in electron transport and oxidative phosphorylation.

    (d) It encodes all the tRNA molecules that are required for mitochondrial protein synthesis.

    (e) It contains noncoding spacer DNA between many of its genes.

Repetitive DNA and Gene Organization

15. When the DNA from bacterium is broken into small fragments, thermally denatured, and renatured under defined conditions, which of the following occur?

    (a) The formation of duplex DNA follows a time course that is characteristic of a bimolecular reaction.

    (b) The fraction of the strands that are renatured is a function of the ratio of the time of renaturation to the concentration of the DNA.

    (c) The $C_0t_{0.5}$ value, which corresponds to the reassociation of half of the DNA, can be determined.

    (d) The rate of reassociation is a function of both the concentration of the DNA fragments and the time of reaction.

16. Under a standard set of conditions and with equimolar concentrations of DNA nucleotides, the *E. coli* genome gives a higher $C_0t_{0.5}$ value than does the genome of bacteriophage T4. Explain why.

_____

_____

_____

17. When DNA from a mouse cell, which contains approximately 1000 times more DNA than an *E. coli* cell, was subjected to a denaturation and renaturation protocol similar to that outlined in Question 16, it did *not* yield a $C_0t_{0.5}$ value that was 1000 times that of *E. coli*. Explain why.

_____

_____

_____

18. Which of the following statements about repeated DNA are *correct?*

    (a) It is abundant in all eucaryotic organisms.

    (b) It can sometimes be separated from unique DNA sequences.

    (c) It is sometimes transcribed.

    (d) It can be found at centromeres.

    (e) It may have arisen, in some cases, by reverse transcription of RNA sequences followed by the incorporation of the cDNA copies into the chromosome.

19. Which of the following statements about ribosomal RNA genes are *correct?*

    (a) They are localized in the nucleolus.

    (b) They are present in the chromosome as single copies behind the most efficient promoters known so that they can be transcribed rapidly and repeatedly.

(c) They are present in the chromosome as repeated units that are located at highly variable distances from one another.

(d) They exist as units comprising the 18S, 5.8S, and 28S genes plus a spacer region.

(e) They are amplified approximately 40,000 times during oogenesis in the African clawed toad.

20. Which of the following statements about the transcript of the ribosomal RNA gene cluster are *correct?*

(a) It is produced by RNA polymerase II in mammals.

(b) It is cleaved into the final products, 18S, 5.8S, and 28S rRNAs, by nucleases.

(c) It is modified by the methylation of some of its nucleotides at the 2′-hydroxyl groups of their ribose residues.

(d) It is modified by the isomerization of some of its uridine residues into ribothymidine.

21. The rate of hybridization of excess sea urchin histone mRNA to sea urchin DNA revealed what property of histone genes? Explain why.

_____

_____

_____

22. Which of the following statements about histone genes are *correct?*

(a) The numbers of copies present per genome correlates with the need of the organism for rapid histone synthesis.

(b) They exist as clusters of tandemly repeated sets of H1, H2A, H2B, H3, and H4 sequences in some organisms.

(c) They exist as dispersed H1, H2A, H2B, H3, and H4 sequences in some organisms.

(d) They contain introns.

(e) They produce mRNAs without poly A tails.

Eucaryotic Gene Expression

23. Which of the following statements about the silk fibroin protein from *Bombyx mori* are *correct?*

(a) It is abundant protein during certain stages of development.

(b) It is produced from a gene that is amplified during certain stages of development.

(c) It can be produced in large amounts because its gene exists in multiple copies.

(d) It is an example of an abundant protein that is produced by a single-copy gene.

24. Explain the role of methotrexate, an inhibitor of dihyrofolate reductase (DHFR), in the induced amplification of the mammalian gene that encodes DHFR.

_____

_____

_____

25. Which of the following statements about α-globin and β-globin gene clusters are *correct?*

   (a) They are on different chromosomes.
   (b) They are evolutionarily related to one another and to myoglobin.
   (c) Each contains globin genes that are arranged in the order the encoded globins appear during the development of the zygote.
   (d) Each contains α-globin or β-globin pseudogenes, which are among the first genes to be expressed after the fertilization of the egg.

26. The central dogma of molecular biology summarizes the flow of information in a cell by stating that DNA makes RNA, which makes protein. What proportion of human DNA participates in this pathway? What are the functions of the remaining DNA?

_____

_____

_____

27. Which of the following statements about transcriptionally active regions of chromosomes are *correct?*

   (a) They are found in the puffed regions of the polytene chromosomes of insects.
   (b) They are less sensitive to DNase I digestion than are inactive regions.
   (c) They vary with the stage of development of the organism.
   (d) They are tissue specific.
   (e) They are undermethylated in comparison to transcriptionally inactive regions of DNA in the same organism.

28. Which of the following statements abut transcription factor IIIA are *correct?*

   (a) It is a sequence-specific DNA-binding protein.
   (b) It potentiates the transcription of the genes encoding the 5S rRNA molecule.
   (c) It binds to the upstream activating sequence (UAS) of an rRNA gene.
   (d) It remains bound to its target gene during transcription.
   (e) It contains several domains comprising 30-residue sequences of amino acids with two Cys and two His side chains disposed such that they chelate a zinc ion.

29. Which of the following statements about the homeo box are *correct?*

   (a) It is found only in procaryotes and lower eucaryotes.
   (b) It is a group of similar sequences of DNA that are associated with the genes that control development.
   (c) It encodes proteins that probably bind to DNA with sequence specificity.
   (d) It encodes proteins that contain a homeo domain and are localized in the nucleus.

1. Would you expect the interaction between protamines and DNA to be enhanced or diminished in solutions that are highly ionic? Explain the basis for your answer.

2. One measure of the evolutionary divergence between two proteins is the number of amino acid differences between them. It can be argued that a better measure would be the minimum number of mutational events that must have occurred to result in those differences.

   (a) The differences in the amino acid sequences of histone H4 between calf thymus and pea seedlings are as follows:

AA position	Pea seedlings	Calf thymus
60	Ile	Val
77	Arg	Lys

   What minimum number of mutational events are necessary to account for these differences. Give the changes that must occur on both mRNA and DNA. (Refer to the genetic code, p. 107 of Stryer.)

   (b) Comment on the nature of the changes based on your knowledge of amino acid chemistry. What conclusion follows about the function of H4?

3. Norbert Numbskull, a not-too-eminent biochemistry student, and his laboratory partner, Lucy Luckless, decide to fractionate a mixture of chromatin into its component histones. Norbert adds salt to the mixture and pours the resulting solution onto a gel-filtration column. Lucy takes the original mixture and adds it to an ion-exchange chromatography column. Comment on the prospects of success for either of these experiments, giving your reasoning. If necessary, suggest an alternative experiment.

4. Chromatin that is transcriptionally active *(euchromatin)* is disperse in structure, whereas chromatin that is transcriptionally inactive *(heterochromatin)* is compact. When the nuclei of chick globin-producing cells were treated briefly with pancreatic DNase, the adult globin genes were selectively destroyed, but the genes for embryonic globin and ovalbumin remained intact. When the nuclei of oviduct cells were treated with DNase, the ovalbumin genes were destroyed. Explain these results.

5. The chromatin of globin-producing cells can be treated with a micrococcal nuclease that cleaves DNA in the linker regions between nucleosomes. When the resulting nucleosomes are isolated and their DNA is examined, it is found to contain the DNA sequence for the synthesis of globin. Are these results consistent or inconsistent with the explanation for the results in Problem 4? Explain.

6. Suppose that the following duplex sequence lies at the end of a eucaryotic chromosome:

   5'-ATAGGCC-

   3'-CGGTATCCGG-

Suppose further that two blocks of the sequence 5'-TTCCCGG-3' are added to the overhanging 3' end. The 3'-OH of the terminal G then loops around and serves as a primer for the synthesis of a complementary 3'→5' sequence. The resulting loop of five nucleotides is then cleaved by a nuclease that excises five nucleotides. Give the sequence of the resulting DNA duplex.

7. Figure 33-1 depicts the replication of a region of a eucaryotic chromosome. Sketch the expected location of parental histones on the daughter duplexes, using circles to depict nucleosomes.

8. The antibiotic erythromycin binds to the 50S subunit of procaryotic ribosomes, blocking chain elongation, but it does not bind to the 60S subunit of eucaryotic ribosomes. The tetracyclines bind to the 30S subunit of procaryotic ribosomes, blocking the A site, but they have no corresponding effect on the 40S subunit of eucaryotic ribosomes. Although erythromycin and the tetracyclines are useful in human antibiosis, one is best advised to be cautious in their administration. Give one good reason why this is the case.

9. Predict whether each of the following manipulations will cause an increase or decrease in the rate at which two separated, complementary DNA sequences reassociate under conditions typical of those used to establish C_0t curves. Briefly state the basis for your prediction.

    (a) An increase in salt concentration
    (b) An increase in incubation temperature
    (c) A decrease in DNA concentration

10. Calculate the relationship that exists between $C_0t_{0.5}$ and k from the following relationship, which is given at the bottom of page 834 of Stryer:

$$f = \frac{1}{1 + kC_0t}$$

11. As can be seen in Figure 33-21 on page 835 of Stryer, mouse satellite DNA has a $C_0t_{0.5}$ value of approximately 10^{-3} M s. Would you expect a mixture of poly G and poly C to have a $C_0t_{0.5}$ value that is smaller than, approximately the same as, or greater than that of mouse satellite DNA? Explain your answer.

12. Suppose that a DNA sample is subjected to C_0t analysis with the results depicted in Figure 33-2. How would you interpret these results?

13. The nucleotide sequence TCACCCTATAGCCATCG is found on the template (-) strand of mitochondrial DNA known to specify the amino terminal sequence of a mitochondrial protein. Give the amino acid sequence that is specified. (See pp. 107 and 834 of Stryer.)

14. Figure 33-31 on page 839 of Stryer depicts the histone gene cluster of the sea urchin. Do each of the five genes in the cluster produce equal amounts of protein products? Why or why not?

15. Suppose that the DNA specifying the synthesis of an enzyme having a turnover number of 10^4 s^{-1} is transcribed into 10^3 molecules of mRNA and that each of the mRNA molecules is translated into 10^5 molecules of enzyme protein. How many molecules of substrate are converted into product per second for each transcription of the DNA?

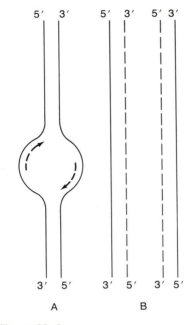

Figure 33-1
A eucaryotic chromosome (A) at the initiation of replication and (B) after replication. The daughter strands are shown as dashed lines.

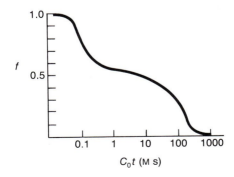

Figure 33-2
C_0t results for Problem 12.

16. Would you expect a zinc deficiency in eucaryotes to be associated with any sort of developmental abnormality? Explain.

ANSWERS TO SELF-TEST

The Eucaryotic Chromosome

1. b, c, d, and e. Answer (a) is incorrect because DNA molecules more than a few kilobases long are sensitive to fragmentation by the shearing forces developed when solutions are stirred.

2. (a) 1 (b) 4 (c) 2 and (d) 3. Note that the length of the bacterial chromosome is expressed in kilobases, not megabases.

3. a, b, c, and d. Histones make up nearly half of the mass of a chromosome, so answer (e) is incorrect.

4. a, b, and d. A nucleosome core consists of 140 base pairs of DNA wrapped around a histone octamer. The nucleosome cores are connected by linker DNA, which contains from fewer than 20 to more than 100 base pairs, the exact length depending upon the organism and the tissue.

5. (a) Electron microscopy showed that chromatin was a linear array of particles (nucleosomes), with diameters of approximately 100 Å, joined by flexible DNA linkers.
   (b) X-ray diffraction analyses also showed 100-Å diameter particles. Neutron diffraction analyses indicated that the DNA was near the surface of the nucleosome and not buried in its interior.
   (c) Limited nuclease digestion with DNase I followed by gel-electrophoresis of the products produced a ladder of discrete fragments. The fragments contained multiples of a basic DNA unit of about 200 base pairs. Digestion with micrococcal nuclease revealed that 140 base pairs of DNA were closely associated with the nucleosome core.
   (d) When purified histones and purified DNA were mixed, they associated with a stoichiometry of two each of histones H2A, H2B, H3, and H4 and approximately 200 base pairs of DNA to form characteristic beaded chromatin fibers. Histone H1 was not required.

6. The nucleosome core has a disc shape with an average diameter of 100 Å and a thickness of about 55 Å. The octameric core of histones has 140 base pairs of DNA wound about it in approximately $1\frac{3}{4}$ turns of a left-handed supercoil. Two copies each of histones H2A, H2B, H3, and H4 are on the inside of the toroidal coil, whereas histone H1 is associated with the DNA that links the cores.

7. No. Each nucleosome is associated with approximately 200 base pairs of DNA. If this DNA were coiled into a sphere with a diameter of approximately 100 Å, it would be condensed from 200 base pairs × 3.4 Å per base pair = 680 Å of linear DNA to 100 Å, which is a packing ratio of about 7. The chromatin fiber must be wound into a helical array and the resulting 360-Å coils must themselves be looped and

folded. Scaffolding proteins, topoisomerases, and small basic molecules, such as the polyamines, also contribute to the ultimate compaction of 10^4 that is observed.

Chromosome Replication and Organelle DNA

8. b, c, and d. Answers (a) and (e) are incorrect because eucaryotic DNA is replicated semiconservatively in a template-dependent reaction that is analogous to the replication process in procaryotes.

9. (a). 2, 4 (b) 1, 3 (c) 5

10. Since DNA polymerases require primers and can synthesize only in the $5' \rightarrow 3'$ direction, the ends of linear duplex DNA represent a challenge. Semiconservative replication of such a molecule results in a protruding, single-stranded 3' tail. (See Figure 33-17 and the text on p. 832 in Stryer.) This problem is solved, in essence, by looping the extended tail bearing the 3'-hydroxyl group around so that it hybridizes to a region of complementary sequence and serves as a primer to complete the telomere. Characteristic sequences with short repeated elements are added to the 3' ends to facilitate this event.

11. c. Since DNA synthesis is semiconservative, the daughter DNA duplex will contain one strand of parental DNA and one newly synthesized strand.

12. a and d. The old histones remain with the daughter DNA duplex, which is composed of a parental strand and a newly synthesized leading strand. New histones are deposited on the daughter DNA duplex that is formed from the lagging strand and the parental template. The initial association of histones with only one daughter DNA duplex is termed conservative segregation and contrasts with the semiconservative segregation of the parental DNA template strands (see Question 11).

13. Petite mutations segregate independently because they occur in the mitochondrial DNA, not the nuclear chromosomes.

14. b, c, and d. Human mitochondrial DNA is a circular duplex that is densely packed with encoded information for both proteins and RNAs, including the repertoire of tRNAs needed for mitochondrial protein synthesis.

Repetitive DNA and Gene Organization

15. a, c, and d. Answer (b) is incorrect because the fraction of strands that are renatured is a function of the product, not of the ratio, of the DNA concentration and the time of renaturation.

16. The DNA of both *E. coli* and bacteriophage T4 is composed almost entirely of unique sequences, with the bacterial DNA being approximately 25 times longer than the phage DNA. When equal amounts, in terms of total nucleotides, of bacterial and phage DNA are fragmented, the concentrations of individual fragments will be approximately 25 times lower for the bacterial DNA than for the T4 DNA. Consequently, reassociation for the bacterial DNA will be slower, which causes the $C_0t_{0.5}$ to be higher.

17. Unlike procaryotic DNA, eucaryotic DNA contains repeated sequences. Consequently, when a given concentration of eucaryotic DNA is fragmented, the concentrations of the repeated sequences are higher than those of the unique sequences, so the repeated sequences will renature more rapidly. Since there are multiple classes of complexity in eucaryotic DNA, the C_0t curve will not be a simple sigmoid as it is with DNA of a single complexity. The curve will be a complex mixture of individual sigmoidal components, each with a characteristic $C_0t_{0.5}$ value. Since some 30 percent of mouse DNA is repeated, the $C_0t_{0.5}$ value was lower than the approximately 15,000 M s that would be expected for DNA containing all unique sequences (see p. 835 in Stryer).

18. a, b, c, d, and e. The repeated sequences of rRNA, tRNA, protein encoding genes (e.g., the histone genes) and some others (e.g., Alu sequences) are transcribed. Other repeated sequences (e.g., satellite sequences) are not transcribed.

19. a, d, and e. The ribosomal RNA genes are repeated and exist in tandem clusters. In the toad, the number of copies increases some 40,000 times during the formation of the egg so that there are sufficient ribosomes to support protein synthesis during the initial stages of development.

20. b and c. Answer (a) is incorrect because RNA polymerase I transcribes the rRNA gene cluster to form a long (45S) transcript. The other component of eucaryotic ribosomes, 5S RNA, is formed by RNA polymerase III, which also transcribes the tRNA genes (see p. 717 in Stryer). Some of the uridines of rRNAs are isomerized to pseudouridines. Thymine contains a methyl group on the 5 position of the pyrimidine ring and is not an isomer of uridine, so answer (d) is also incorrect.

21. The kinetics of hybridization of RNA to DNA reveals information about the number of copies of genes, just as the kinetics of hybridization of DNA to DNA does. Because the rate of hybridization of mRNA to DNA was hundreds of times faster than could be accounted for by the concentration of sea urchin DNA assuming a single copy of the histone gene per genome, it was concluded that histone genes were reiterated 300 to 1000 times. The use of excess mRNA ensures that the rate of hybridization is a function of the DNA, not the RNA, concentration.

22. a, b, c, and e. Histone genes are atypical in that they lack introns and their mRNAs don't have poly A sequences at their 3′ termini.

Eucaryotic Gene Expression

23. a and d. Single-copy genes encode most of the proteins produced by eucaryotes.

24. DHFR is required for the synthesis of DNA because it catalyzes a step in formation of TTP. An inhibitor of this enzyme would be expected to retard cell growth. Cells grown in the presence of methotrexate are thus under selective pressure to overproduce DHFR. Cells containing spontaneous duplications of the DHFR gene produce more of this enzyme and thus grow better, as do their progeny. Further duplications of the gene lead to more cells with even better survival potentials. The net result of this selection is the amplification of the DHFR gene.

25. a, b, and c. Pseudogenes are not expressed because they lack one or more of the features that are required for transcription, so answer (d) is incorrect.

26. Probably only approximately 2 percent of human DNA encodes proteins. With the exception of the DNA that encodes known RNA products—for example, the ribosomal and snRNP RNAs and telomere sequences—the function of most of the remaining DNA is unknown.

27. a, c, d, and e. Answer (b) is incorrect because regions of the chromosome that extend for considerable distances on both sides of transcriptionally active genes are sensitive to digestion by DNase I in vitro. Some regions, usually at the 5′ ends, become hypersensitive to the nuclease; that is, they are more digestible than is naked DNA. With respect to answer (d), tissue specificity of transcription reflects the functional differences among differentiated cells. Common regions of transcriptional activity represent the common functions of cells.

28. a, b, d, and e. Transcription factor IIIA binds to a specific sequence within the 5S RNA gene, thus enabling the subsequent binding of other transcription factors and, ultimately, the transcription of the gene by RNA polymerase III.

29. b, c, and d. The homeo box sequences are widely distributed among eucaryotes, including mammals, so answer (a) is incorrect.

ANSWERS TO PROBLEMS

1. One would expect the interaction between protamines and DNA to be looser in highly ionic solutions. Protamines are arginine-rich proteins whose positively charged guanidinium groups extend outward from the long axis of the α helix (see p. 830 of Stryer). The protamines are bound tightly to DNA by electrostatic interactions between the positively charged guanidinium groups and the negatively charged phosphoryl groups of DNA. The positively and negatively charged ions that result from the addition of a salt to a solution compete with these interactions and, hence, weaken them.

2. (a) There would be a minimum of two mutations involved.
The possible codons for Ile and Val are as follows:

Ile	Val
AUU	GUU
AUC	GUC
AUA	GUA
	GUG

A single change from A to G in the first position of the codon would give a substitution of Val for Ile. This corresponds to a change from an AT to a GC base pair on DNA.

The possible codons for Arg and Lys are as follows:

Arg	*Lys*
CGU	
CGA	
CGG	
CGC	
AGA	AAA
AGG	AAG

Again a single change, from G to A in the second position of the codon, would account for the amino acid difference. This corresponds to a change from a GC to an AT base pair on DNA.

(b) These are conservative changes. Both Lys and Arg have positively charged side chains, and both Ile and Val are hydrophobic. Therefore, we would expect virtually no structural or functional difference in the H4 of calf thymus and pea seedlings. However, the two tissues are surely divergent in other respects.

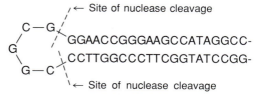

Figure 33–3
Cleavage sites on the telomeric nucleotide sequence given in Problem 6.

3. The prospects of success in either case are dim. Histones are of sufficiently similar size that a gel-filtration column, which is essentially a molecular sieve, would not be the best choice for separating them. An ion-exchange column would be a better choice for separating the histones, but *only* if the histones were first dissociated from DNA by the addition of salt.

4. In cells that are actively synthesizing adult globins, chromatin is dispersed so that the globin genes may be transcribed into mRNA. Accordingly, this region is sensitive to DNase. In an adult cell specialized for the production of globins, neither embryonic globins nor ovalbumin is produced to any significant extent. Accordingly, the regions of DNA carrying the information for these genes are not dispersed and are therefore not sensitive to DNase.

5. The results are consistent. The genes for globin synthesis are contained within the nucleosomes. When these genes become transcriptionally active, the chromatin becomes dispersed.

6. The cleavage points are shown in Figure 33-3. The duplex sequence that will result is as follows:

5′-GGAACCGGGAAGCCATAGGCC-
3′-CCTTGGCCCTTCGGTATCCGG-

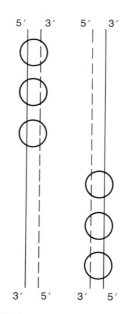

Figure 33–4
Distribution of parental histones on daughter strands for the answer to Problem 7.

7. The histones will be found on the leading strands of the daughter duplexes, as shown in Figure 33-4. Remember that DNA polymerase reads in the 3′→5′ direction and synthesizes polynucleotide chains in the 5′→3′ direction.

8. The mitochondria of eucaryotes have ribosomes that resemble those of procaryotes. Thus, both erythromycin and tetracycline can inhibit protein synthesis in mitochondria as well as in procaryotes.

9. (a) Increasing the salt concentration will decrease the repulsion between negatively charged phosphoryl groups and, hence, will increase the rate of reassociation.

(b) An increase in the incubation temperature will decrease the stability of the hydrogen bonds that must form between complementary bases and, hence, will decrease the rate of reassociation.

(c) A decrease in DNA concentration will decrease the rate of reassociation. Reassociation involves the coming together of two molecules in a bimolecular process, and the rate is proportional to the product of the concentrations of (+) and (−) strands.

10. When reassociation is half complete, $f = 0.5$ and $C_0t = C_0t_{0.5}$. First, we substitute these values into the original equation:

$$f = \frac{1}{1 + kc_0t}$$

$$0.5 = \frac{1}{1 + kC_0t_{0.5}}$$

We then cross multiply:

$$0.5(1 + kC_0t_{0.5}) = 1$$

$$0.5 + 0.5kC_0t_{0.5} = 1$$

$$0.5kC_0t_{0.5} = 0.5$$

$$C_0t_{0.5} = \frac{1}{k}$$

Thus $C_0t_{0.5}$ is the reciprocal of k, the second-order rate constant for the association of complementary fragments.

11. One would expect the $C_0t_{0.5}$ value for a mixture of poly G and poly C to be substantially lower than that for mouse satellite DNA. Although mouse satellite DNA is highly repetitive when compared to DNA from other sources, the sequence of duplex consisting exclusively of G on one strand and C on the other is even more repetitive.

12. At least two frequency classes of DNA must be present in the sample, one of which anneals more quickly than the other.

13. First, the nucleotide sequence of the template strand is used to determine the mRNA sequence:

Template DNA   3′-GCTACCGATATCCCACT-5′

mRNA   5′-CGAUGGCUAUAGGGUGA-3′

The reading frame is established by the occurrence of AUG.

mRNA   5′-CG AUG GCU AUA GGG UGA-3′

Thus, the amino sequence of the protein is

Met-Ala-Met-Gly-Trp

Note that AUA codes for Met rather than Ile and UGA codes for Trp rather than chain termination in mitochondria.

14. In nucleosomes there are two copies each of histones H2A, H2B, H3, and H4 but only a single copy of histone H1. Therefore, one would

expect only half as much protein product to be produced from histone H1 as from the other four.

15. For each transcription, the number of molecules of substrate that are converted into product per second is given by

$$10^4 \text{ s}^{-1} \times 10^3 \times 10^5 = 10^{12} \text{ s}^{-1}$$

16. The transcription of many eucaryotic genes is activated by proteins containing "zinc fingers," which are zinc-containing regions that bind to DNA and promote transcription. A zinc deficiency would therefore lead to developmental abnormalities.

Viruses and Oncogenes

Stryer concludes the part of the text covering the flow of genetic information with a chapter on viruses. These intracellular parasites provide paradigms for the duplication and expression of chromosomes because viruses are little more than coated packets of genetic information. The genomes of viruses are composed of either DNA or RNA. They are surrounded by a protein capsid or a membranous envelope. The genomes of viruses enter cells, interdict the normal flow of genetic information, express their own information, and, as a result, reproduce themselves. Because of their simplicity, the study of viruses has been crucial to the growth of our understanding of molecular biology. Viruses are also of fundamental practical importance because of their roles as etiologic agents. In addition, the oncogenic viruses have contributed to our knowledge of cancer and the control of normal cell growth and development. In this chapter, Stryer presents selected examples that typify important viral features and mechanisms.

Tobacco mosaic virus provides a simple model of the self-assembly of a biologically active macromolecular complex. Its single-stranded RNA genome is encapsidated by multiple copies of a single protein. Bacteriophage T4 is a more complex virus that employs enzymes and scaffolding proteins to aid its self-assembly. Site-specific recombination between bacteriophage lambda (λ) and the chromosome of *E. coli* shows how viral DNA can become a hitchhiker on a cell's DNA. Tomato bushy stunt virus serves as an example of how the strictly symmetrical capsid-protein interactions that form an icosahedral surface can be relaxed so that more nucleic acid can be packed into a quasi-icosahedral capsid.

The genome of RNA viruses is either single- or double-stranded RNA. The single-stranded RNA genome may act directly as an mRNA, called (+) RNA, or it may be the complement, called (−) RNA, of mRNA and thus require conversion to its mRNA complement through template-directed transcription. Furthermore, some RNA viruses replicate through DNA intermediates that integrate into the genomes of cells. Stryer provides examples of viruses in each of these four classes. The genomes of small, single-stranded RNA bacteriophages act directly as mRNAs. Poliovirus RNA acts directly as an mRNA to produce a single, long polypeptide that is subsequently cleaved by proteases into separate proteins. This strategy allows the production of more than one protein from a single mRNA—a requirement imposed by the mechanism of translation initiation in eucaryotes (see p. 761 in Stryer). In contrast, vesicular stomatitis virus is an example of a (−) RNA virus that must carry an RNA replicase in its virion so that a mRNA can be made to initiate its expression and replication. Stryer uses reovirus as the example of how double-stranded RNA genomes, called (±) RNA, are expressed and replicated.

The last part of the chapter deals with viruses that can cause cancer in susceptible hosts. Stryer lists the properties of transformed cells. He describes how SV40 and polyoma, both double-stranded DNA viruses, either replicate productively in permissive host-cells or become integrated at low frequencies into the chromosomes of nonpermissive host-cells and thereby convert the cells into the transformed (cancerous) state. The retroviruses, which carry a reverse transcriptase in their virion, are described in detail because their biology reveals so much about the functioning of both cancerous and normal cells. These RNA viruses are converted into DNA sequences that are transcribed to form (+) RNA genomes only when they are integrated into the cellular genome as DNA proviruses. Some retroviruses have picked up cellular genes during their replication, and the overexpression of some of these genes (oncogenes), or modified versions of them, can transform cells infected by the retroviruses carrying them. Since oncogenes are closely related to normal cellular genes, it is likely that some normal cellular genes (proto-oncogenes) that are involved in the control of growth and differentiation can be mutated to forms that cause cancer. Stryer also describes the most complicated retrovirus known to date—the human immunodeficiency virus that causes acquired immune deficiency syndrome (AIDS). The chapter ends with a description of the mechanisms of antiviral agents called interferons.

You should review the other chapters of Part V of *Biochemistry* in anticipation of studying this chapter. Pay particular attention to the descriptions of the DNA and RNA polymerases in Chapters 27 and 29 and the role of mRNA in translation in Chapter 30.

LEARNING OBJECTIVES

When you have mastered this chapter, you should be able to complete the following objectives.

Introduction (Stryer page 851)

1. Provide an overview of the major *structural and functional features of viruses.*

2. Define *virion,* and distinguish between a *capsid* and an *envelope.*

3. List some reasons for studying viruses.

Viral Assembly, DNA Restriction and Modification, and Viral Integration into a Host Chromosome (Stryer pages 852–862)

4. Describe the two major forms of viral capsids, and explain why all small viruses are either *rods* or *spheres* or a combination of these shapes.

5. Describe the composition, structure, and size of *tobacco mosaic virus (TMV)*. Provide an overview of the *self-assembly* of TMV.

6. Distinguish between the structures of the *two-layered protein disc* and the *two-turn helix (lockwasher)* formed by TMV coat subunits, and describe their roles in the formation of the virion.

7. Sketch the structure of *bacteriophage T4* and note its genetic complexity. Outline the infective cycle of the T4 phage, and note the distinctive aspects of its gene expression and genome composition.

8. Outline the assembly of T4 phage and contrast it with that of TMV. Note the roles of self-assembly, *scaffolding proteins*, and *assembly enzymes* in *T4 morphogenesis*. Compare the lengths and volumes of T4 DNA and the T4 phage head, and describe the insertion of the DNA into the head.

9. Describe the *restriction* and *modification phenomena* displayed by some bacterial cells when they encounter foreign DNA. Identify the enzymes responsible for the phenomena, and describe their activities. Contrast the properties of the three major types of *restriction endonucleases*.

10. Distinguish between *lytic* and *temperate bacteriophages*. Outline the pathways by which *bacteriophage lambda* (λ) is inserted into and excised from the chromosome of *E. coli*. Identify the proteins required for the *site-specific recombination* of λ.

11. Explain how the relaxation of the *symmetry requirements* for interactions among identical capsid proteins allows *tomato bushy stunt virus* to encapsidate its genome.

RNA Viruses (Stryer pages 862–870)

12. Distinguish between viruses that have *(+) RNA genomes* and those that have *(−) RNA genomes*. Note that there are also viruses that have *(±) RNA genomes*.

13. Provide an overview of the mechanisms of replication of the four classes of RNA viruses, and give examples of viruses in each class. Note the roles of *RNA-directed RNA polymerases (RNA replicases)* and *RNA-directed DNA polymerases (reverse transcriptases)* in the replication of RNA viruses.

14. Describe the salient features of the structure of the genome of RNA phases such as *bacteriophage Qβ*. Explain how translation and transcription are coordinated in these RNA phages and how the coat protein acts as a *translational repressor*.

15. Outline the experiment that demonstrated *evolution in a test tube*. Explain the relationship between the mutation rate of RNA viruses and the size of their genomes.

16. Describe the *viroids*, and state a hypothesis explaining their origin.

17. Outline the replication of *poliovirus*, and describe the role of *polyproteins* in the life cycle of poliovirus.

18. Outline the replication of *vesicular stomatitis virus (VSV)*. Contrast the mechanism used by this virus to produce discrete protein products with that used by the *picornaviruses*.

19. Explain how viral-encoded proteins in the envelope of *influenza virus* enable it to interact with susceptible cells. Explain how the *segmented genome* of the virus contributes to its ability to evade the immune response of its host.

20. Describe the genome structure of *reovirus*, and outline the mechanisms used to express and replicate it.

Viruses and Cancer (Stryer pages 870–876)

21. Note that *retroviruses* are the only *RNA tumor viruses* and that *oncogenic DNA viruses* also exist.

22. Describe the process and the consequences of *transforming tissue-culture cells* with an oncogenic DNA or RNA virus. Appreciate that transformed cells form a tumor when introduced into a susceptible animal.

23. Distinguish between the *lytic* and *transforming responses* of cells infected with *SV40* or *papilloma viruses*. Outline the expression and replication of SV40 virus.

24. Describe the experimental findings that suggested the involvement of DNA synthesis in the replication of RNA tumor viruses.

25. Describe the virion and genome structures of the retroviruses. Outline the life cycle of a retrovirus.

26. List the reactions catalyzed by reverse transcriptase.

27. Describe the roles of *tRNA* and *terminally redundant nucleotide sequences (long terminal repeats)* in the replication of retroviruses.

28. Note the requirement for integration of the DNA transcript into the chromosome of the host during the life cycle of a retrovirus. Contrast this feature of retroviral replication with the alternate life cycles of oncogenic DNA viruses.

29. State the functions of the four genes of the *avian sarcoma virus*.

30. Contrast the effects of infection by an oncogenic DNA virus and infection by a retrovirus on cell viability.

Oncogenes, AIDS, and Interferons (Stryer pages 876–881)

31. Describe the experiments with a temperature-sensitive avian sarcoma virus that led to the concept that a single gene product could effect cellular transformation. Describe the activity of the *src protein*.

32. Define *oncogene*. Describe the five major classes of oncogenes. Summarize the salient features of oncogenes and their normal cellular counterparts—the *proto-oncogenes*.

33. Describe *acquired immune deficiency syndrome (AIDS)*, and explain its cause.

34. Describe the structure and composition of the virion and genome of the *human immunodeficiency virus (HIV)*. Note that HIV has the most complex retroviral genome known to date and that it is genetically unstable.

35. Describe the kinds, origins, and properties of the *interferons*. Write the reactions that are catalyzed by the two enzymes elicited by interferons.

Introduction

1. Which of the following statements about viruses are *correct?*

   (a) Viruses contain either DNA or RNA.
   (b) Viruses are self-reproducing intracellular parasites.
   (c) Viruses are composed minimally of a capsid or envelope and a genome.
   (d) Viruses introduce energy-generating systems into the cells they infect.
   (e) Viruses contain single-stranded or double-stranded nucleic acid genomes.
   (f) Viruses are good models for the study of the flow of genetic information in uninfected cells.

Viral Assembly, DNA Restriction and Modification, and Viral Integration into a Host Chromosome

2. Tobacco mosaic virus (TMV) has over 300,000 amino acids in its capsid, but its genome consists of fewer than 7000 nucleotides as single-stranded RNA. How can this small genome produce its capsid?

_____

_____

_____

3. Which of the following statements about TMV are *correct?*

   (a) The virion of TMV exhibits icosahedral symmetry.
   (b) TMV is a macromolecular assembly of single-stranded DNA and protein.
   (c) Under suitable conditions, TMV spontaneously assembles from its capsid proteins and genome.
   (d) The capsid proteins of TMV initially form a two-layered disc structure in the absence of the genome.
   (e) The genome of TMV contains a sequence that is recognized by the capsid proteins.
   (f) After the initial specific complex between capsid proteins and nucleic acid is formed, a change in protein conformation occurs that entraps the nucleic acid and initiates virion formation.

4. Which of the following statements about bacteriophage T4 are *correct?*

   (a) T4 phage is a complex virus having more than 150 genes.
   (b) T4 phage infects *E. coli* by binding with its tail fibers to specific sites on the outer membrane of the bacterium.

(c) T4 phage injects its DNA into the cytoplasm of the bacterium via a process involving an energy-dependent change in the shape of the virion.

(d) Infection by T4 phage leads to the hydrolysis of the bacterial chromosome.

(e) The assembly of T4 phage proceeds in a manner analogous to that of TMV virus through an ordered association of capsid proteins with the genome.

(f) T4 phage forms an empty head structure into which DNA from a concatameric replication intermediate is inserted.

5. Why doesn't the DNase encoded by the T4 phage that hydrolyzes the bacterial DNA also degrade the T4 DNA produced during phage replication?

_____

_____

_____

6. When foreign DNA, that is, DNA from a different organism, enters some bacteria, it is often hydrolyzed (restricted). Explain the enzymatic basis of this restriction phenomenon.

_____

_____

_____

7. Explain how foreign DNA sometimes escapes the hydrolysis referred to in Question 6.

_____

_____

_____

8. How does the chromosome of a bacterium with a restriction-modification system escape cleavage during replication, considering that the newly synthesized DNA is unmethylated?

_____

_____

_____

9. Match the type of restriction endonuclease in the left column with its characteristics from the right column.

(a) Type I _____

(b) Type II _____

(c) Type III _____

(1) Requires ATP

(2) Hydrolysis of DNA blocked by methylation of the recognition site

(3) Recognizes symmetric sequences

(4) Requires S-adenosylmethionine

(5) A multisubunit complex containing methylase

(6) Hydrolyzes DNA at specific sites

(7) Movement along DNA is energy-dependent and processive
(8) Hydrolyzes DNA randomly, thousands of base pairs away from the recognition site

10. Which of the following statements about bacteriophage λ are *correct?*

(a) Bacteriophage λ is a temperate phage.
(b) Bacteriophage λ has two modes of replication.
(c) During lytic growth, the linear genome of bacteriophage λ is converted into a covalently closed circular form.
(d) The genes of bacteriophage λ recombine with those of *E. coli* by a recA-mediated reaction.

11. The site-specific recombination reaction between bacteriophage λ and *E. coli* may be symbolized as B-O-B′ + P-O-P′, where B and B′ represent bacterial DNA sequences, P and P′ represent phage sequences, and 0 represents a shared sequence. Which of the following statements concerning the reactants or the reaction are *correct?*

(a) The products of the recombination reaction may be symbolized as B-O-P and P′-O-B′.
(b) The B-O-B′ sequence is also called *attB* and the P-O-P′ sequence is called *attP*.
(c) The exchanges of DNA strands between the reacting duplexes are catalyzed by the phage-encoded enzyme integrase.
(d) The reaction results in a lysogenic bacterium.
(e) Two phage-encoded enzymes are required to free the prophage from the bacterial chromosome, that is, to reverse the reaction.

12. No more than 60 identical proteins can assemble into an isometric icosahedral shell to form a precisely symmetrical viral capsid. What is the molecular basis for the ability of tomato bushy stunt virus (TBSV) to form a quasi-symmetrical capsid of 180 proteins, each of which is encoded by the same gene? What is the value to the virus of using quasi-equivalent interactions to form its capsid?

_____

_____

_____

RNA Viruses

13. What is the relationship of a (−) RNA virus genome to the mRNA used in its replication and expression? What does this relationship imply about the enzymes that must be present in (−) RNA virus virions?

_____

_____

14. Which of the following statements about the small RNA bacteriophages MS2 and Qβ are *correct?*

(a) Bacteriophage MS2 contains a single molecule of (+) RNA that is approximately 3600 nucleotides long.

(b) Each three successive nucleotides of MS2 RNA specifies only a single amino acid.

(c) The genome of Qβ serves both as messenger RNA and as a template for replication.

(d) The genome of Qβ encodes four proteins that associate to form the RNA transcriptase that replicates the RNA.

15. How does bacteriophage Qβ avoid collisions between the ribosomes that are translating its genome in the $5' \rightarrow 3'$ direction and the RNA replicases that are transcribing the RNA in the $3' \rightarrow 5'$ direction?

_____

_____

_____

16. How is more coat protein than RNA replicase formed during Qβ infection, given that there is only one copy of each gene in the genome?

_____

_____

_____

17. In an experiment in Spiegelman's laboratory, a solution containing bacteriophage Qβ RNA, Qβ replicase, the four ribonucleoside triphosphates, and Mg^{2+} was incubated for a short time. The solution was then diluted into a new mixture containing the same components, except the Qβ RNA was missing. The dilution procedure was repeated at brief intervals. Which of the following statements concerning this experiment are *true?*

(a) The purpose of the dilutions was to replenish the replicase because it rapidly became inactive during the incubation.

(b) For each dilution, the amount of RNA that was synthesized decreased because the concentration of template molecules was decreased.

(c) "Evolution" occurred during the experiment because RNA mutants that were different from the wild-type Qβ and were better suited to their "environment" were formed.

(d) Longer RNA molecules were selected because they allowed the synthesis of more product per initiation event than did shorter molecules.

18. Which of the following statements about poliovirus are *correct?*

(a) The genome of poliovirus is a single-stranded, long (+) RNA molecule.

(b) The genome of poliovirus encodes several proteins, including those that comprise the coat and an RNA replicase.

(c) Poliovirus transcribes its genome into discrete segments of separate mRNAs, each of which can direct the synthesis of a protein.

(d) Ribosomes bind to several different sites on the poliovirus mRNA to initiate the translation of each of the different poliovirus-encoded proteins.

19. Both poliovirus and vesicular stomatitis virus (VSV) encode and produce several virus-specific proteins, but they do so by quite different mechanisms. Contrast these mechanisms.

_____

_____

_____

20. Which of the following statements about influenza virus are *correct?*

  (a) The function of the neuraminidase on the surface of influenza virus may be to alter the gangliosides of the surfaces of potential target cells to make them more susceptible to the virus.
  (b) Influenza virus is a ($-$) RNA virus with a lipid-bilayer envelope that is derived from the plasma membrane of the infected cell.
  (c) Antibodies formed against the hemagglutinin of influenza virus have allowed mass immunizations that have effectively eliminated influenza as a potential pandemic disease.
  (d) The outer surface of the influenza virion contains virus-encoded proteins.
  (e) The genome of influenza virus is segmented, which leads to the formation of discrete mRNAs upon transcription and to genetic reassortment during infection.

21. Place the following events of reovirus infection and replication in their correct order.

_____

  (a) The genome is asymmetrically transcribed in the core.
  (b) mRNAs are translated to form viral proteins.
  (c) The outer shell surrounds the core.
  (d) Capped mRNAs escape the core and enter the cytosol.
  (e) The virus enters the cell.
  (f) mRNAs serve as templates for the formation of ($\pm$) RNAs.
  (g) The outer shell comes off the core.
  (h) The virus exits the cell.
  (i) The core assembles around ($\pm$) RNAs.

Viruses and Cancer

22. What was the significance of the finding that a cell-free extract from a chicken with a sarcoma caused a sarcoma when it was injected into another chicken?

_____

_____

_____

23. Which of the following statements about retroviruses are *correct?*

  (a) Retroviruses can transform susceptible cells in a tissue culture.
  (b) Retroviruses carry a RNA transcriptase in their virions to convert their ($-$) RNA genomes into ($+$) mRNAs.
  (c) Retroviruses propagate themselves via a double-stranded RNA intermediate.
  (d) Retroviruses are encapsidated in a lipid-bilayer membrane.

24. Which of the following are properties of tissue-culture cells that have been transformed by DNA or RNA tumor viruses?

(a) They cause tumors when they are injected into susceptible animals.

(b) They can be formed from normal cells by oncogenic DNA or oncogenic RNA viruses.

(c) Viral-specific antigens appear on the surfaces of cells that have been transformed by tumor viruses.

(d) Viral nucleotide sequences are present in cells that have been transformed by tumor viruses.

(e) Growth is disordered and continuous.

(f) Genes are expressed that are ordinarily silent in the normal cells.

25. Which of the following statements concerning SV40 virus are *correct?*

(a) The virion of SV40 doesn't contain cellular proteins.

(b) SV40 either kills the cells it infects or becomes integrated into the chromosome of the cells and alters their growth properties.

(c) SV40 transforms infected cells at low ($\approx 10^{-5}$) frequencies.

(d) The SV40 virion contains a double-stranded DNA genome that is approximately 5000 base pairs long.

(e) The large T antigen produced by SV40 is needed only for the establishment and not for the maintenance of the transformed state.

(f) Lytic, or productive, infection by SV40 involves an ordered expression of its genome.

(g) Alternative mRNA-splicing patterns and translation in different reading frames are used to form the capsid proteins of SV40.

(h) During a nonproductive, or transforming, infection of nonpermissive cells, only the early genes of SV40 are expressed.

26. What was the significance of the observation that methotrexate and 5-fluorodeoxyuridine interfered with the infection of susceptible cells by the avian sarcoma virus?

_____

_____

_____

27. Which of the following statements about reverse transcriptase are *correct?*

(a) Reverse transcriptase uses DNA as a template to synthesize DNA.

(b) Reverse transcriptase uses RNA as a template to synthesize RNA.

(c) Reverse transcriptase uses DNA as a template to synthesize RNA.

(d) Reverse transcriptase is found in the virions of all RNA tumor viruses.

(e) Reverse transcriptase hydrolyzes RNA in DNA-RNA hybrid duplexes.

28. Place the following events in the order in which they occur during the infection of a cell by a retrovirus.

(a) Reverse transcriptase forms a duplex composed of the genome and a DNA complement.
(b) DNA-directed RNA polymerase II produces RNA transcripts from the DNA provirus.
(c) A portion of the plasma membrane carrying a viral-ribonuclear-protein complex buds to form the virion.
(d) The virion enters the cytosol of the cell through the plasma membrane.
(e) Processed and unprocessed RNA transcripts traverse the nuclear membrane and enter the cytosol.
(f) Some RNA transcripts are spliced to form mRNAs whereas others remain unspliced to form the genomic RNA.
(g) The virion binds to a specific receptor on the external surface of the plasma membrane.
(h) Ribonucleoprotein complexes of genomic RNA and viral proteins incorporate into the plasma membrane.
(i) Double-stranded DNA enters the nucleus and is incorporated into the chromosome of the infected cell.
(j) The virion is uncoated.
(k) Reverse transcriptase forms a double-stranded DNA copy of the viral genome.

Oncogenes, AIDS, and Interferons

29. Which of the following statements about oncogenes are *correct?*

(a) Oncogenes may encode proteins that are altered versions of similar cellular proteins.
(b) Oncogenes are nucleic acid sequences that bind essential cellular repressors, thereby decreasing their concentration and derepressing crucial cell regulatory genes.
(c) Temperature-sensitive v-oncogenes demonstrate that cellular transformation can be caused by a single protein.
(d) A cellular gene that can be converted into an oncogene by mutation, duplication, or translocation is called a proto-oncogene.
(e) Retroviral oncogenes may transform cells by overproducing a key protein that performs a crucial function in the regulation of normal cell growth.

30. Which of the following statements about the products of oncogenes are *correct?*

(a) Oncogenes may encode proteins that are tyrosine kinases.
(b) Oncogenes may encode proteins that are growth factors.
(c) Oncogenes may encode proteins that are receptors for growth factors.
(d) Oncogenes may encode proteins that alter the sequences of proto-oncogenes, that is, act as mutagens, so as to activate them.
(e) Oncogenes may encode proteins that are guanyl-nucleotide-binding proteins.
(f) Oncogenes may encode proteins that are nuclear proteins.

31. Which of the following statements about acquired immune deficiency syndrome (AIDS) or its etiologic agent, the human immunodeficiency virus (HIV), are *true?*

(a) Patients generally succumb to AIDS because HIV binds to and inactivates their IgG antibodies.
(b) HIV contains (+) RNA genomes.
(c) HIV is surrounded by a membrane containing viral-specific proteins.
(d) The HIV virion contains human tRNA molecules that are involved in its replication.
(e) The genome of HIV is more complex than those of the known oncogenic retroviruses.
(f) One of the surface glycoproteins in the coat of HIV has different amino acid sequences in different patients and sometimes in viruses formed at different times in the same patient.

32. Which of the following statements about the interferons or their physiological effects are correct?

(a) α Interferons interfere with DNA viruses, β interferons with (+) RNA viruses, and γ interferons with ($\pm$) RNA viruses.
(b) Interferons stimulate a protein kinase that phosphorylates a translation initiation factor and consequently interferes with protein synthesis.
(c) Interferons lead to the activation of RNase L, which cleaves RNAs near uridine residues.
(d) Interferons are synthesized and secreted by vertebrate cells that have been infected with a virus.
(e) Interferons stimulate a RNA-directed RNA polymerase that adds poly A tails onto viral RNAs.
(f) Double-stranded RNA molecules effectively stimulate the production of interferons.

PROBLEMS

1. Having a viral capsid composed of identical polypeptide subunits rather than a single large protein provides a safeguard against mistakes in protein synthesis.

(a) Suppose that the coat protein of tobacco mosaic virus were composed of a single, large protein molecule instead of 2130 protein molecules of 158 amino acid residues each. Calculate the probability that such a protein molecule would consist of entirely correct amino acid residues if an incorrect amino acid is inserted at any position 1 time out of 10,000.
(b) Now calculate the probability of getting an entirely correct protein molecule consisting of 158 amino acid residues assuming the same error rate. Compare this answer with your answer to part (a).

2. Some of the earliest experiments aimed at elucidating the nature of the genetic code were carried out on tobacco mosaic virus. In an effort to determine whether the code was overlapping or nonoverlapping, RNA from TMV was treated with nitrous acid such that there was, on

the average, one "hit" of nitrous acid per molecule of RNA. (The mutagenic effects of nitrous acid are discussed by Stryer on p. 677.) The treated RNAs were then added to tobacco leaves. Subsequently, viral progeny were harvested, coat protein was isolated, and its amino acid composition was examined.

(a) Predict the result that was obtained, explaining your answer.
(b) What would the result have been if the genetic code were completely overlapping?
(c) What features of TMV make it particularly well-suited to the purposes of this experiment?

3. What furnishes the nucleotides that are necessary for the synthesis of T4 DNA? What mechanisms enable T4 to take advantage of this source? (See the discussion of T4 DNA replication on p. 856 of Stryer.)

4. T4 DNA contains 5-hydroxymethylcytosine rather than cytosine as a base. tRNAs contain several unusual nucleotides. What fundamental difference exists between the mechanism for the incorporation of 5-hydroxymethylcytosine into T4 DNA and that for the incorporation of various modified bases into tRNAs? Explain.

5. Suppose that a DNA-containing bacterial virus has five genes, A, B, C, D, and E, in that order on its genome. The viral DNA is known to exist as a concatamer during infection. After replication, viral DNA is stuffed into preformed heads and is snipped by an endonuclease once the head is filled. The capacity of the head is a length of DNA that contains six genes. Although the DNA of this virus is a linear duplex, when genetic mapping studies are conducted, the DNA maps as a circle. Explain how this might come about.

6. A very small bacteriophage contains one molecule of single-stranded (+) circular DNA. Suppose that the single-stranded DNA is isolated and is used as a template for the synthesis of complementary (−) strands in an in vitro system containing, among other components, DNA and RNA polymerases. The resulting (−) strands are in turn isolated and are used as templates for the synthesis of additional (+) strands. This process is then repeated for 15 generations, after which bacteria infected with (+) strands of the viral progeny are indistinguishable from the starting bacteriophage.

(a) What conclusion follows about the fidelity of replication by DNA polymerase?
(b) Why are a large number of generations necessary for the purposes of this experiment?
(c) What is the role of RNA polymerase?

7. EcoRI is a restriction endonuclease that cleaves DNA as follows:

$$\text{5′-G|AATT C-3′}$$

$$\text{3′-C TTAA|G-5′}$$

Suppose that the following piece of DNA, which has been isolated from the *E. coli* chromosome, is found to be cleaved at only one site by EcoRI. (A* is methylated adenine.)

$$\text{5′-AACTCAGTGAA*TTCCCAAGTGAATTCAGAA-3′}$$

$$\text{3′-TTGAGTCACTTA*AGGGTTCACTTAAGTCTT-5′}$$

Indicate the location of the cleavage site, explaining your answer.

8. The use of overlapping codons by some RNA viruses allows them to maintain a compact genome, but it imposes some limitations on amino sequences.

(a) What amino acids must follow a Met (either *N*-terminal or internal) in a region of a protein known to be specified by a completely overlapping code? Explain your answer.

(b) What is the maximum number of possible amino acid sequences that could occur in this instance? What would be the number with a nonoverlapping code?

9. The viral-mediated transfer of host DNA from one bacterial cell to another is known as transduction. Bacteriophage P1 may carry a wide variety of *E. coli* genes from one bacterial cell to another. In contrast, bacteriophage λ carries only *gal* or *bio* genes from one cell to another. Explain the difference in the transducing behavior of the two phages.

10. (a) Suppose that you were trying to put together an in vitro system for the replication of genomic RNA from avian sarcoma virus. List the major components you would need, and briefly explain the role of each.

(b) What major difference would there be between the system you have described and an in vivo system? Explain.

11. Would you expect a synthetic double-stranded nucleotide, such as poly(I:C), to have any potential in antiviral therapy? Explain briefly.

12. Much research has been directed toward finding agents that will inhibit viral DNA replication but will have few toxic effects on the host. Generally, such agents are effective for viruses that have large genomes, but they are ineffective for viruses with very small genomes. Why do you think this is the case?

13. The drug arabinosyl thymine (ara-T), shown in Figure 34-1, can be phosphorylated by some thymidine kinases to form a derivative that inhibits DNA polymerase. One of the early genes of poxvirus encodes the synthesis of a viral thymidine kinase. What condition must be fulfilled for ara-T to be useful as an antiviral agent? Explain.

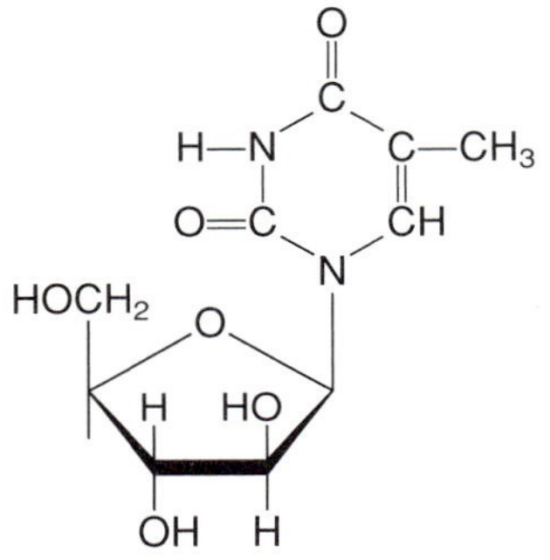

Figure 34-1
Arabinosyl thymine (ara-T).

ANSWERS TO SELF-TEST

Introduction

1. a, b, c, e, and f. Answer (d) is incorrect because viruses appropriate the energy-generating systems of the host cell.

Viral Assembly, DNA Restriction and Modification, and Viral Integration into a Host Chromosome

2. The capsid is composed of 2130 identical protein subunits, each containing only 158 amino acids. Fewer than 500 nucleotides are required to encode a protein of this size. By using identical capsid subunits, viruses coat themselves without having to devote a large fraction of their genomic sequences to the specification of coat proteins.

3. c, d, e, and f. Answers (a) and (b) are incorrect because TMV is a filamentous virus with a single-stranded RNA genome.

4. a, b, c, d, and f. Answer (e) is incorrect because the assembly process for T4 is different from that for TMV in that an empty phage head is formed before the coat proteins associate with the nucleic acid. Furthermore, scaffold proteins and proteases are needed for T4 morphogenesis.

5. T4 DNA lacks the cytosine residues found in the *E. coli* DNA and contains instead 5-hydoxymethylcytosine (HMC), a modified cytosine derivative. In addition, some of these HMC residues become glucosylated. As a result, the T4 DNA is immune to the activity of the phage-encoded DNase.

6. The bacteria produce endonucleases that recognize specific sequences in the foreign DNA and hydrolyze it to produce double-stranded breaks. Other bacterial nucleases further degrade the linear DNA to render it nonfunctional. The bacterial DNA itself is immune to the restriction endonucleases (see Question 7).

7. For their own survival, bacteria harboring restriction endonucleases also carry genes for methylase (methyl transferase) enzymes that recognize the same sequence as the corresponding nuclease. The methylases methylate a nucleotide within the sequence and thereby convert the DNA into a form that is immune to the nuclease. The modification methylases transfer a methyl group from S-adenosylmethionine to the N^6-amino group of A, the N^4-amino group of C, or the C-5 position of C within the particular recognition sequence. When unmethylated foreign DNA enters a cell, there is a competition between the endonuclease and the methylase to react with the DNA. If the methylase wins, which it does occasionally, the DNA will be immune to the nuclease and will survive.

8. DNA replication is semiconservative, so each daughter duplex will contain one parental and one new DNA strand. The parental strand will be methylated, whereas the new strand will not be. Fortunately for the bacterium, a single methyl group on the symmetric recognition sequence (hemi-methylation) of a type II restriction endonuclease is sufficient to render it refractory to cleavage. Shortly after the sequence is synthesized, the modification methylase will methylate the new strand so that replication can again occur safely.

9. (a) 1, 2, 4, 5, 7, 8 (b) 2, 3, 6 (c) 1, 2, 4, 5, 6, 7. Although not stated in the text, the type I and III restriction enzymes recognize asymmetric sequences. Type III enzymes, depending upon the particular enzyme, appear to cut DNA from 20 to 30 base pairs away from the recognition sequences.

10. a, b, and c. Answer (d) is incorrect because the phage and the bacterium do not ordinarily share genes or extensive regions of sequence similarity; consequently, there is little general (homologous) recombination.

11. b, c, d, and e. Answer (e) is correct because excisionase as well as integrase is required for the reaction of the B-O-P' and P-O-B' sites on the chromosome to form the free phage.

12. The interactions between TBSV capsid subunits are not perfectly symmetrical. The conformational flexibility of the capsid protein allows it to assume different conformations that can interact with one another quasi-equivalently. The resulting imperfectly symmetrical structure enables a larger genome to be encapsidated without more of the nucleic

acid sequence of the virus having to be used to encode different capsid proteins.

RNA Viruses

13. The sequence of a (−) RNA genome is complementary to that of the mRNA it will produce. Since the (−) RNA genome can't serve as an mRNA and the cell lacks the enzymes to make its complement, (−) RNA viruses must carry an RNA replicase in their virions and bring it into the cell upon infection.

14. a, b and c. For answer (b), note that some of the genes of MS2 overlap and a nucleotide can therefore be in two different codons simultaneously. Answer (c) is correct because the (+) RNA serves as a template for the formation of (−) RNAs, which are themselves templates for the production of more copies of (+) RNA. Answer (d) is incorrect because only one subunit of the transcriptase is phage encoded. The other three subunits—the protein elongation factors EF-Tu and EF-Ts and a ribosomal protein—are recruited from the cell.

15. Immediately after infection, RNA replicase is absent. Ribosomes bind to the RNA and translate it to form, among other products, the replicase. When sufficient RNA replicase accumulates, it binds to the RNA at ribosome entry sites near the 5′ end of the RNA, thereby blocking translation and precluding collisions with the approaching replicase. The transcribing replicase binds to its synthesis initiation site at the 3′ end of the (−) RNA to transcribe the entire molecule to form the (+) RNA.

16. Ribosomes bind more effectively to the initiation site for the translation of coat protein than to the other initiation sites on the RNA; consequently, more copies of coat protein than other gene products are made. In addition, the coat protein acts as a translational repressor by binding to the initiation site for the translation of the replicase gene to suppress the synthesis of the replicase. The result is that the coat protein gene is translated more efficiently than the replicase gene.

17. c. The purpose of the dilutions was to decrease the concentration of the original template RNA so that products that would replicate faster than Qβ RNA itself might be selected. The selection occurred, and the RNA continued to be made in spite of the dilution of the original template because it was made more efficiently. The faster-replicating templates were shorter than the original, for they had only to fulfill the criterion of rapid replication in their in vitro environment, which supplied them with enzyme and rNTPs. They could thus dispense with most of the original sequence, which had been selected to adapt the phage to in vivo conditions. (See p. 864 in Stryer.)

18. a and b. Answers (c) and (d) are incorrect because poliovirus produces individual proteins by translating its (+) RNA genome into a giant precursor polypeptide that is subsequently cleaved by proteases into separate proteins. The first cleavage of the polyprotein is autocatalytic.

19. As mentioned in the answer to Question 18, the poliovirus genome mRNA forms one precursor polypeptide, which is subsequently cleaved into several products. In contrast, the (−) RNA genome of VSV is first transcribed by a virus-encoded RNA replicase into five discrete mRNAs. Each mRNA leads to the formation of a polypeptide. To form

its genome, the (−) RNA is also transcribed into a long (+) RNA that contains all the genetic information of the virus and serves as the template for the progeny genomes.

20. b, d, and e. Answer (a) is incorrect because the likely function of the neuraminidase is to act on the sialic acid residues found in the mucus of the respiratory tract to lessen its ability to entrap the virus. Answer (c) is incorrect because the genetic reassortment of the RNA segments that encode the hemagglutinin gene leads to rapid changes in this viral surface protein. This thwarts the immune system of the host, maintains the virulence of the virus, and precludes effective mass-immunization programs.

21. e, g, a, d, b, f, i, c, and h

Viruses and Cancer

22. Since a sarcoma is a tumor, this discovery by Peyton Rous showed that cancer can be caused by an agent that was present in the cell-free extract. The agent turned out to be a virus, so the discovery indicated that cancer can be virally induced.

23. a and d. Answers (b) and (c) are incorrect because retroviruses are (+) RNA viruses that replicate by a mechanism involving double-stranded DNA.

24. a, b, c, d, e, and f

25. b, c, d, f, g, and h. Answer (a) is incorrect because the SV40 virion contains histones that are encoded by the host cell and then become bound to newly synthesized SV40 DNA. Answer (e) is incorrect because the large T antigen is required for both functions.

26. Since methotrexate and 5-fluorodeoxyuridine are primarily inhibitors of DNA synthesis and the avian sarcoma virus is an RNA virus, the observation indicated that DNA synthesis is involved in the replication of RNA tumor viruses.

27. a, d, and e. Answers (b) and (c) are incorrect because the enzyme doesn't make RNA. It uses both DNA and RNA as templates to form complementary DNA.

28. g, d, j, a, k, i, b, f, e, h, and c

Oncogenes, AIDS, and Interferons

29. a, c, d, and e

30. a, b, c, e, and f

31. b, c, d, e, and f. Answer (a) is incorrect because the HIV virus infects and compromises the functioning of T-cells, which are involved in the immune response. The virus doesn't interact with circulating IgG antibodies.

32. b, c, d, and f. Answer (a) is incorrect because interferons show little specificity toward viruses and are designated α, β, or γ on the basis of the kind of cell that synthesizes them. Answer (e) is incorrect because interferons stimulate a template-independent polymerase that synthesizes short oligoadenylates in which the AMP residues are joined by 2′, 5′ phosphodiester bonds.

1. (a) In the hypothetical "giant" coat protein for TMV, there would be $2130 \times 158 = 336{,}540$ amino acid residues. If there is 1 chance in 10,000 that a residue will be incorrect, the probability that any given residue will be correct $= 1 - 0.0001 = 0.9999$. The probability that all 336,540 residues will be correct is given by

$$(0.9999)^{336,540} = 2.42 \times 10^{-15}$$

Thus, the chances that entirely correct protein would be produced are virtually nil.

(b) In a protein of 158 amino acid residues, the probability that all will be correct is given by

$$(0.9999)^{158} = 0.984$$

Thus, almost all of the protein produced will be correct. If there is a way that faulty subunits will be rejected with considerable accuracy, there will be virtually no erroneous amino acids in the coat protein.

2. (a) Nitrous acid produces AT $\leftrightarrow$ GC transitions. One hit per molecule of RNA would produce a change in one base pair per molecule. A change in one base pair would give, at most, a change in one amino acid of each molecule of coat protein if the code were nonoverlapping. (There would be no amino substitution at all if the mutation was for an alternative codon for the same amino acid.)

(b) If the code were entirely overlapping, a given base would be a member of three adjacent codons, so up to three adjacent amino acids in the sequence could be changed by a mutation of a single base.

(c) There is only one molecule of nucleic acid and one type of coat protein in TMV, and the virus may be readily isolated. Moreover, the coat protein is amplified, there being 2130 copies per virion. The fact that protein is homogeneous and is produced in abundance greatly facilitated analytical studies.

3. The nucleotides necessary for T4 DNA synthesis are provided by the hydrolysis of host-cell DNA. One of the immediate-early gene products of T4 is a DNase that degrades the host-cell DNA. This DNase does not degrade T4 DNA because T4 DNA contains modified cytosine nucleotides.

4. In T4 DNA, 5-hydroxymethylcytosine completely replaces cytosine as a base, and hydroxymethylcytosine nucleotides furnished by 5-hydroxymethyl dCTP are incorporated into DNA chains during biosynthesis. Thus, the raw material for the synthesis of T4 DNA is a pool of 5'-dATP, 5'-dTTP, 5'-dGTP, and 5'-hydroxymethyl dCTP. In the case of tRNA synthesis, the "usual" four nucleotides are first incorporated into the newly biosynthesized RNA. Some of these "usual" nucleotides are subsequently altered by modification enzymes. In the biosynthesis of T4 DNA, some modification of bases at the polynucleotide level also takes place, inasmuch as some of the 5-hydroxymethylcytosine residues are glycosylated.

5. The concatameric DNA of this virus has the following gene sequence:

ABCDEABCDEABCDEABCDEABCDEABCDEABCDE . . . .

Because the viral DNA is stuffed into preformed heads, and because the maximum capacity of each head is six genes, the first virion will have the sequence ABCDEA; the second, BCDEAB; the third, CDEABC; and so forth. When genetic mapping experiments are conducted with these viruses, the genetic map will be circular because, for example, gene A will be equally close to genes B and E. Thus, concatameric linear DNA behaves genetically as if it were circular.

6. (a) DNA polymerase must replicate DNA with remarkable fidelity. For the fifteenth generation (+) strands to have full biological activity, few, if any, mistakes could have occurred.

   (b) The large number of generations ensures that the biological activity at the end of the experiment is due to replicated DNA and is not simply the result of the carryover of a few intact original (+) strands.

   (c) The RNA polymerase is necessary to form short primers that are extended by DNA polymerase. DNA polymerase can only extend preexisting polynucleotide chains.

7. The site of cleavage by EcoRI is

   5′-AACTCAGTGAA*TTCCCAAGTG|AATTCAGAA-3′

   3′-TTGAGTCACTTA*AGGGTTCACTTAA|GTCTT-5′

Cleavage does not occur at the site containing GAA*TTC because of the methylation of the second adenine nucleotide.

8. (a) The sequence is Met-(Cys or Trp)-(Val, Ala, or Gly). The Met must be specified by AUG. The second codon would then be UGX, where X denotes any nucleotide. The possibilities are UGU or UGC, which specify Cys, or UGG, which specifies Trp. (The second codon cannot be UGA because that would lead to termination.) The third codon would then be GUX (Val), GCX (Ala), or GGX (Gly).

   (b) Only three sequences are possible with a completely overlapping code. They are Met-Cys-Val, Met-Cys-Ala, and Met-Tyr-Cly. With a nonoverlapping code, the number is $1 \times 20 \times 20 = 400$. (Remember that the conditions of the problem specify that Met is the first amino acid, so there is only one possibility for the first position.)

9. Bacteriophage λ is integrated as a prophage at a specific site between the *gal* and *bio* genes on the *E. coli* chromosome. If it carries any host-cell DNA with it as it breaks out of the host chromosome, it will be the DNA immediately adjacent to its site of integration. In contrast, bacteriophage P1 is a lytic not a temperate phage. As with T4, preformed capsids are filled with DNA (see Stryer, p. 857). When a host-cell DNA fragment is packaged into a preformed capsid, the host-cell DNA will be carried to a cell subsequently infected by the virus particle. (A virus particle carrying host-cell DNA to the exclusion of its own DNA could not be infective.) The host DNA carried in the capsid may be a fragment of any part of the host chromosome.

10. (a) The first task involves making duplex DNA from the RNA. To do this you would need genomic RNA as a template; Trp-tRNA as a primer; 5′-dATP, 5′-dTTP, 5′-dGTP, 5′-dCTP as the raw materials for DNA synthesis; reverse transcriptase; and Mg^{2+} ion. Next you would need a system to transcribe the (+) RNA strands from the duplex DNA. To do this you would need duplex DNA as a template; RNA polymerase; 5′-ATP, 5′-UTP, 5′-GTP, and 5′-CTP; and Mg^{2+} ion. (See Figure 34-41 on p. 875 of Stryer.)

(b) Transcription of retroviral DNA in vivo demands that the DNA first be integrated into the host chromosome.

11. Poly(I:C) is a synthetic double-stranded RNA. Double-stranded RNA has antiviral effects (see Figure 34-47 on p. 881 of Stryer). It stimulates the degradation of mRNA and tRNA in interferon-treated cells, and it also inhibits protein synthesis. In fact, poly(I:C) has been found to be effective as an antiviral agent in cells in culture. Its use in human therapy is limited by the facts that it is toxic and is rapidly degraded by ribonucleases.

12. Viruses with small genomes lack the genetic information for the synthesis of unique enzymes and rely instead on host enzymes to carry out DNA replication. Therefore, any inhibitor of DNA replication will affect the replication of host DNA as well. Viruses with large genomes have enzymes that are uniquely involved in the replication of viral DNA.

13. Ara-T would be useful as an antiviral agent only if host-cell thymidine kinase fails to phosphorylate it. Otherwise, the administration of the drug would lead to the inhibition of DNA synthesis in all cells, not just those that are infected with poxvirus. Since ara-T is indeed not phosphorylated by host-cell thymidine kinase, it is effective in inhibiting DNA synthesis in cells infected by poxvirus.

CHAPTER 35

Molecular Immunology

In Part VI of *Biochemistry*, Stryer describes the molecular mechanisms of several physiological systems. In these descriptions, many of the general biochemical concepts you studied in the first five parts of the text, as well as peculiarities of the individual tissues, are presented as integrated systems that perform characteristic functions for the organism. There is little you can specifically review in preparation for studying this section of the text because it applies so much of what you have learned from earlier chapters. Instead, you should make a special effort to look up terms or concepts that were presented earlier and are unclear in their new contexts.

Chapter 35 deals with the biochemistry of the immune system. The cells and proteins of this system cooperate to detect and inactivate foreign (nonself) molecules, microorganisms, and viruses. The humoral immune response acts through the secretion into the circulatory system of soluble antibodies that bind antigens with high specificity and affinity. The cellular immune response acts through antibodies on the surface of specialized cells that bind to antigens that are themselves on the surface of cells.

After defining essential immunlogical terms, Stryer describes the kinetics of the appearance in the serum of immunoglobulin M (IgM) and immunoglobulin G (IgG) following the introduction of an antigen into an animal. He then describes the clonal selection theory that explains the origin of antibody specificity. The structure of IgG, including the antigen-combining site, is described in detail. Stryer also explains how hybridoma cells, which produce large amounts of homogeneous monoclonal antibodies, can be formed by the fusion of myeloma and spleen cells. The roles of the variable, hypervariable, and constant amino acid sequences of the light and heavy chains in the functions of immunoglobulins are given. The five classes of immunoglobulins with their characteristic polypeptide compositions and functions are also presented.

Stryer next relates the variable and constant regions of immunoglobulins to the organization of the genes that encode them. Numerous different variable-region genes for both light (L) and heavy (H) chains join to a much more limited set of constant-region genes. Joining and diversity genes expand the number of DNA segments that can be apposed in different combinations to generate a large number of different immunoglobulin sequences. Imprecision in the recombinations and the addition of extra nucleotides by the terminal deoxyribonucleotidyl transferase generate further diversity by introducing yet more sequence variability into the immunoglobulin genes. DNA recombination also plays a role in class switching; that is, in the change from producing IgM to another class with the same antigen-binding specificity. Recombination joins one of several different possible class-defining H chain constant regions to the unique H chain variable region that along with the L chains defines the specificity of the antibody. Alternate RNA splicing produces two different mRNAs that specify two versions of the initial immunoglobulin formed. One mRNA encodes an H chain containing a membrane-anchoring sequence that causes the immunoglobin to remain on the surface of the plasma cell. The other mRNA lacks the anchoring sequence and the immunoglobulin synthesized from it is secreted.

The chapter concludes with a description of the functions of T cells and the cellular immune response. The T cell receptor and the major histocompatability complex proteins are emphasized. These proteins display immunoglobulinlike binding specificities for foreign epitopes that are on the surface of cells. They are encoded by genes in the immunoglobulin gene superfamily and show characteristic structural and evolutionary relationships.

LEARNING OBJECTIVES

When you have mastered this chapter, you should be able to complete the following objectives.

Introduction and the Clonal Selection Theory (Stryer pages 889–892)

1. Define or describe the following immunological terms: *antibody (immunoglobulin), plasma cell, B lymphocyte (B cell), antigen (immunogen), antigenic determinant (epitope),* and *hapten.*

2. Describe the *humoral immune response* in terms of the immunoglobulins involved and the order of their appearance after *immunization.*

3. Explain how a small foreign molecule can give rise to an immune response.

4. Distinguish between the *instructive theory* and the *selective (clonal selection) theory* of antibody formation. Describe an important experiment that supports the clonal selection hypothesis.

5. List the salient features of the clonal selection theory. Describe how selection contributes to the descrimination between self and nonself macromolecules.

Immunoglobulin Structures, Monoclonal Antibodies, and Antibody-Antigen Interactions (Stryer pages 892–904)

6. Relate the intact structure of an *immunoglobulin G (IgG)* molecule to the F_{ab} and F_c fragments produced from it by proteolysis. Describe the functions performed by the regions of IgG contained in the fragments.

7. Sketch the polypeptide chains of an IgG molecule, and relate the *heavy chain (H)–light chain (L) subunit composition (H_2L_2)* to the F_{ab} and F_c fragments of the molecule. Appreciate the function of the *hinge region* of IgG.

8. List the similarities and differences between the *combining sites* of antibodies and the active sites of enzymes.

9. Explain why antibodies display a range of binding affinities for the antigen that elicits them.

10. Describe *multiple myeloma* and the nature of the antibodies produced by *myeloma cells*.

11. Outline the procedure for forming a *hybridoma cell* by fusing a myeloma cell with a lymphocyte. Describe *monoclonal antibodies*, list some of their uses, and explain how they are produced.

12. Describe the origin and structure of *Bence-Jones proteins*. Describe the *constant (C), variable (V)*, and *hypervariable* features of the amino acid sequences of the light chain of IgG molecules. Distinguish between the κ and λ L-chain *allotypes*. Note the Mendelian inheritance patterns of the L-chain allotypes.

13. Compare the regions of amino acid sequences of the H chain to those of the L chain. Relate the hypervariable sites of the H and L chains of IgG to the *antigenic combining sites* of the immunoglobulin. Describe experiments that helped to reveal this relationship. Describe the function of the constant regions of the H and L chains of IgG.

14. Describe the function of the *complement system* in the immune response.

15. Describe the *domain structure* of IgG, and note the presence of the *immunoglobulin fold* as a common structural feature.

16. Sketch the structure of an IgG molecule, and locate the antigen combining sites. Summarize the types of bonds that form complexes between immunoglobulins and haptens or antigens.

17. List the different classes of immunoglobulins and give their functions. Note the common occurrence of κ and λ L chains in all classes and the α, μ, δ, γ, or ϵ H chains that provide the structural bases for the function of each class.

Immunoglobulin Gene Structure and Expression and B Cell Function (Stryer pages 904–911)

18. Compare the relative spatial relationships of the *immunoglobulin genes* that encode the V and C regions of the L chains of IgG in *embryonic (germ-line) tissue* and *antibody-producing (plasma) cells.*

19. Describe the contributions of the number of immunoglobulin genes *(the germ-line repertoire), somatic recombination,* and *somatic mutation* in the generation of antibody diversity. Describe the roles of the *V, C, D,* and *J genes, DNA recombination in alternative reading frames,* and *terminal deoxyribonucleotidyl transferase* in these processes.

20. Note the functions of *RNA splicing* in joining a *signal sequence* mRNA to the immunoglobulin mRNA structural sequence by *intron removal,* and delineate the roles of RNA splicing and DNA recombination in immunoglobulin expression.

21. Provide a quantitative example of the diversity of immunoglobulin structures that arise from the *combinatorial association* of different genes and from somatic mutation.

22. Describe the phenomenon of *class switching,* and note its significance in maintaining constant recognition specificity among the immunoglobulin classes.

23. Outline the process by which *B lymphocytes* are triggered to develop into *plasma cells* and *memory B cells,* and state the functions of the two types of cells. Contrast the structures and antigen specificities of *membrane-bound* and *soluble* immunoglobulins. Describe the role of *alternate mRNA splicing* in producing these two forms of immunoglobulins.

T Cell Function and the Cellular Immune Response (Stryer pages 911–917)

24. Contrast the usual targets of the humoral immune response and the *cellular immune response.*

25. Compare the structures and features of *T cell receptors* with those of the immunoglobulins. Note the sizes and conformations of the epitopes that are recognized by each kind of protein.

26. List the kinds of *T cells,* and describe their functions.

27. Describe the functions of the three classes of products encoded by the *major histocompatibility complex (MHC) genes.* Note the cellular distribution and the *aggrotope* recognition variabilities of the *MHC proteins.*

28. Explain the origins of the diversity of T cell receptors.

29. Outline a model that accounts for the recognition of a *combined epitope* by a T cell receptor.

30. Explain how the humoral and cellular immune responses contribute to the distinction an organism makes between self and nonself.

31. Note that the domains of the proteins of the immune system suggest both structural and evolutionary relationships among the proteins. Distinguish between a *multigene family* and a *supergene family.*

Introduction and the Clonal Selection Theory

1. Match the immunological term in the left column with its description or definition in the right column.

    (a) Antigen _____

    (b) Antibody _____

    (c) Epitope _____

    (d) Hapten _____

    (1) A particular site on an immunogen to which an antibody binds

    (2) Protein synthesized in response to an immunogen

    (3) Macromolecule that elicits antibody formation

    (4) Small foreign molecule that elicits antibody formation

2. Small foreign molecules do not usually elicit the formation of soluble antibodies, and the cellular immune system also ordinarily responds only to macromolecules. Explain how an antibody can sometimes be directed against a small foreign molecule.

_____

_____

_____

3. Which of the following events usually follow the injection of an antigen into a rabbit?

    (a) IgG antibodies reach a relatively constant concentration in the serum approximately three weeks after injection.

    (b) IgG antibodies appear in large quantities approximately three days after antigen injection.

    (c) A second administration of the antigen a month or more after the initial injection elicits still higher levels of IgG antibodies.

    (d) IgM antibodies appear aproximately 10 days after antigen injection.

4. Describe a key experiment with a fragment of IgG that bound antigen that contradicted a prediction of the instructive theory of antibody formation. State the contradicted prediction of the instructive theory.

_____

_____

_____

5. Which of the following statements about the clonal selection theory of antibody formation are *correct?*

    (a) Each specific antibody is made by a single cell or its clonal descendents.

    (b) All the cells producing antibodies die when the antigen stimulating those cells is absent.

    (c) If the surface antibody of an immature, antibody-producing cell interacts with its antigen during fetal life, the cell is stimulated to divide and begin production of soluble antibodies.

    (d) Each antibody-producing clone contains *unique* sequences of DNA that encode the polypeptide chains of its antibodies.

Immunoglobulin Structures, Monoclonal Antibodies, and Antibody-Antigen Interactions

6. Match the structure or feature listed in the right column with the appropriate IgG fragment on the left.

(a) F_{ab} ———

(b) F_c ———

(1) Contains an antigen combining site
(2) One is formed per IgG molecule
(3) Contains an H-chain fragment
(4) Contains an intact L chain
(5) Mediates complement fixation in the intact IgG
(6) Two are formed from an IgG molecule
(7) Forms a precipitate upon binding an antigen

7. Explain why the antibodies produced in an animal in response to a hapten display a range of binding constants for the antigen eliciting them.

_____

_____

_____

8. What is the unique feature of the immunoglobulins produced in a patient with multiple myeloma?

_____

_____

9. Which of the following statements about the immunoglobulins produced by a clone of a myeloma cell are *correct*?

(a) Myeloma antibodies generally have shorter C regions than do normal IgG immunoglobulins.
(b) Myeloma antibodies display a range of binding constants for the antigen that elicits their synthesis.
(c) Myeloma antibodies are directed against several epitopes on the surface of the antigen that elicits their synthesis.
(d) Myeloma antibodies are monoclonal antibodies.

10. Because they are nearly homogeneous populations, myeloma immunoglobulins were useful for studying antibody structure. What rendered them less than ideal for the study of antigen-antibody interactions.

_____

_____

11. What seminal discovery did Milstein and Kohler make that overcame the difficulty in studying the antibody-antigen interactions referred to in Question 10?

_____

_____

_____

12. Match the component used in the production of a hybridoma cell listed in the left column with its function or property from the right column.

(a) Myeloma cells lacking HGPRTase _____

(b) Hypoxanthine _____

(c) Aminopterin _____

(d) Thymine _____

(e) Mouse spleen cells _____

(f) Polyethylene glycol _____

(1) Prevents de novo nucleotide synthesis

(2) Induces cell fusion

(3) Allows DNA synthesis in the absence of de novo dTMP synthesis

(4) Allows purine synthesis in cells containing HGPRTase

(5) Cells that require hypoxanthine for growth

(6) Antibody-producing cells

13. Describe how HAT medium selects for hybridoma cells when spleen cells are fused to myeloma cells lacking HGPRTase.

_____

_____

_____

14. Which of the following statements about monoclonal antibodies are *correct?*

(a) Monoclonal antibodies can be produced in large amounts by growing many mice, each of which has been immunized with the same antigen.

(b) Monoclonal antibodies that bind nearly any antigen can be produced.

(c) Monoclonal antibodies are useful for quantifying specific proteins in human blood.

(d) Immobilized monoclonal antibodies can be used to purify scarce proteins by affinity chromatography.

(e) Fluorescent-labeled monoclonal antibodies can be used in conjunction with fluorescence-activated cell sorting to isolate cells having a given cell-surface antigen.

15. Which of the following statements about the L and H chains of IgG are *correct?*

(a) The H chains of IgG molecules have variable and constant regions of amino acid sequences.

(b) Bence-Jones proteins provided the first information about the amino acid sequence of the L chains of IgG.

(c) The constant region of L chains exists in two allotypes (κ and λ), which have a single difference at one amino acid position.

(d) The variable region of the L chain has a counterpart of the same length and amino acid sequence in the variable region of the H chain.

16. Which of the following statements about IgG structure are *correct?*

    (a) Each of the two antigen-combining sites on an IgG molecule can bind to a structurally distinct epitope.

    (b) Both interchain and intrachain disulfide bonds stabilize IgG structure.

    (c) Unique and variable amino acid sequences in IgG are involved in transporting antibodies across the placental membrane.

    (d) Both the L and H chains of IgG contain domains with similar structures.

    (e) The F_c unit of IgG binds to a complement-pathway component to initiate the lysis of an infected cell.

    (f) The hypervariable regions of the L chain are the sole determinants for the binding of the IgG to the specific antigen.

17. Match the immunoglobulin class listed in the left column with its property or function from the right column.

    (a) IgA _____

    (b) IgD _____

    (c) IgE _____

    (d) IgG _____

    (e) IgM _____

    (1) Most prevalent soluble antibodies
    (2) Unknown function
    (3) First soluble antibodies to appear in serum after immunization
    (4) Protects against parasites
    (5) Major antibodies in tears, saliva, and mucus

Immunoglobulin Gene Structure and Expression and B Cell Function

18. If the mRNA encoding the L chain of an IgG molecule were isolated, radiolabeled, and hybridized to genomic DNA that has been isolated from either the plasma cell producing the antibody or from germline cells from the same organism, what would you observe with respect to the relative locations of the L-chain gene sequences on the two DNAs?

_____

_____

_____

19. Some antibody diversity arises from the combination of one *V* gene and one *C* gene from pools containing numerous different copies of each. Why doesn't this mechanism account completely for the observed diversity?

_____

_____

_____

20. Which of the following statements concerning the roles of mRNA splicing in immunoglobulin synthesis are *correct?*

    (a) The primary transcripts of both L and H chains contain introns.

    (b) The exons of immunoglobulin genes reflect the structural domains of the encoded proteins.

    (c) RNA splicing generates somatic mutations that expand the diversity of the antibody response.

(d) During the differentiation of B lymphocytes into plasma cells, splicing at alternative sites produces either membrane-bound or secreted immunoglobulins.

(e) RNA splicing is essential for immunoglobulin function because it attaches exon transcripts that encode a signal sequence to the L and H chains.

21. Which of the following statements about class switching are correct?

(a) RNA splicing joins the sequences that encode V_HDJ_H regions to sequences that encode different class C_H regions.

(b) Plasma cells that initially synthesize IgM switch to form IgG with the same antigen specificity.

(c) Class switching doesn't affect the variable region of the H chains.

(d) Class switching allows a given recognition specificity of an antibody to be coupled with different effector functions.

22. What are the physiological functions of plasma cells and memory B cells?

_____

_____

T Cell Function and the Cellular Immune Response

23. Which of the following statements about T cell receptors are *correct?*

(a) T cell receptors recognize soluble foreign molecules in the extracellular fluid.

(b) T cell receptors recognize T cells.

(c) For a T cell receptor to recognize a foreign molecule, the molecule must be bound by proteins encoded by the genes of the major histocompatability complex.

(d) The T cell receptor is structurally similar to an IgG immunoglobulin in that it has two H and two L chains.

(e) The T cell receptor is encoded by genes that arise through the recombination of a repertoire of V, J, D, and C DNA sequences.

(f) T cell receptors primarily recognize fragments derived from foreign macromolecules.

24. Which of the following statements about the major histocompatability complex (MHC) proteins are *correct?*

(a) MHC proteins play a role in the rejection of transplanted tissues.

(b) One class of MHC proteins is present on the surfaces of nearly all cells, binds fragments of antigens, and presents them to T cell receptors.

(c) MHC proteins are encoded by multiple genes.

(d) One class of MHC proteins provides components of the complement system.

(e) The genes encoding MHC proteins produce three classes of soluble proteins.

25. In what sense is the following analogy correct? Lamarck is to Darwin as Pauling is to Burnet.

PROBLEMS

1. Most antigens are polyvalent; that is, they have more than one antibody binding site. In the case of macromolecules that contain regular, repeating sequences, like polysaccharides, it is easy to understand how a molecule might have multiple binding sites. In the case of proteins with nonrepeating sequences, it is more difficult to envision how polyvalence might be accounted for. Yet proteins with single polypeptide chains are polyvalent as antigens. What feature of antibody production accounts for this behavior?

2. Assuming that antigen-antibody precipitates have lattice-like structures, draw simple sketches showing possible arrangements of antigen and antibody molecules in a precipitate in which the ratio of antibodies to antigens is (a) 1.14 and (b) 2.83.

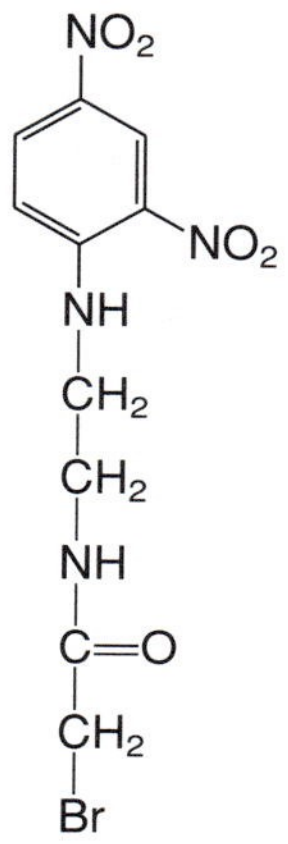

3. Suppose that dinitrophenol is attached to a protein with many potential DNP binding sites and that the resulting antigen is used to stimulate antibody production in rabbits. When serum is harvested and the gamma globulin fraction is purified and mixed together with antigen, no precipitate forms, yet fluorescence measurements reveal the presence of antigen-antibody complexes. Explain this seeming paradox.

4. In individuals who are heterozygous for two hemoglobin types, say HbA and HbS, a given red blood cell contains approximately a 50 percent mixture of the two types. What happens when individuals are heterozygous for immunoglobulin genes? Why is this process necessary?

5. When polyacrylamide-gel electrophosesis (Stryer, p. 44) of a monoclonal antibody preparation is conducted, a single sharp band appears. When the antibody preparation is treated with β-mercaptoethanol, two bands appear. Why is this the case?

6. Pepsin cleaves IgG molecules on the carboxyl-terminal side of the interchain disulfide bonds between heavy chains. How many physical pieces would result from the cleavage of IgG by pepsin? How many of the pieces derive from the F_c region of IgG?

7. The binding constant for the binding of a given hapten to an antibody is 10^{-9} M, and the rate constant for its binding is 10^8 M^{-1} s^{-1}. Calculate the rate constant for the dissociation of the hapten from the antibody.

8. The addition of a bifunctional DNP affinity-labeling reagent (one with two affinity-labeling groups) to myeloma protein produces light and heavy chains that are cross-linked through Tyr-34 and Lys-54, respectively. What conclusion is suggested by this observation?

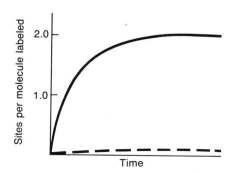

Figure 35-1
Affinity-labeling agent.

Figure 35-2
Time course for the labeling of the antibody in the presence (dashed line) and absence (solid line) of DNP-lysine.

9. Dinitrophenol is covalently attached to a protein, and the DNP-labeled protein is then used to produce a monoclonal antibody. A DNP affinity-labeling agent (Figure 35-1) is then added to the antibody preparation in the presence and absence of excess ϵ-DNP-lysine, giving the results depicted in Figure 35-2.

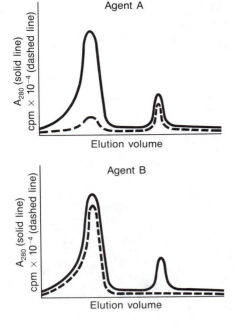

Agent A

Agent B

(a) Which group on the affinity-labeling agent is recognized by the hapten-binding sites on the antibody?
(b) Which group on the affinity-labeling agent is likely to be involved in covalently labeling the antibody?
(c) What is the role of the added ϵ-DNP-lysine?

10. An antibody that is prepared as described in Problem 9 is treated with two radiolabeled affinity-labeling agents, A and B. β-Mercaptoethanol is then added, and the mixture is poured over a molecular sieving column, giving the results depicted in Figure 35-3. Explain what conclusion likely follows from this experiment.

Figure 35-3
Results for Problem 10.

Introduction and the Clonal Selection Theory

1. (a) 3 (b) 2 (c) 1 (d) 4

2. If the small molecule (hapten) becomes attached to a macromolecule (carrier), it can act as an immunogen and serve as an epitope (haptenic determinant) to which an antibody can be selected to bind.

3. a, c Answer (d) is incorrect because IgM antibodies appear a few days after immunization and decrease after 10 days.

4. A fragment containing an antigen-binding site was obtained after the limited proteolysis of IgG. The fragment was denatured and tested for its ability to bind antigen. In the denatured state, it failed to associate with the antigen, but when it was renatured in the absence of the antigen it regained the ability to form a complex with the antigen. The instructive theory predicted that the antigen serves as a template upon which the antibody directed against it is folded. The experiment indicated that this was not the case.

5. a and d. Answer (b) is incorrect because some cells from the antibody-producing clone survive in the absence of the antigen and remain ready to begin antibody production immediately should the antigen reappear. Answer (c) is incorrect because immature antibody-producing cells are killed when their surface antibodies bind antigens. This process destroys the cells that are capable of forming antibodies against the organism's own macromolecules.

Immunoglobulin Structures, Monoclonal Antibodies, and Antibody-Antigen Interactions

6. (a) 1, 3, 4, 6 (b) 2, 3, 5. Answer (7) is inappropriate for either fragment because the F_c fragment lacks an antigen-binding site and the F_{ab} fragment contains only one. The insoluble lattice of antigen-anti-

body molecules forms because intact IgG molecules each have two antigen-binding sites and can therefore link several antigens together.

7. The hapten stimulates several B lymphocytes that bear different surface antibodies that recognize it to differentiate into plasma cells and secrete antibodies. Each plasma cell secretes a different kind of antibody that binds the hapten through a unique array of noncovalent interactions between the hypervariable regions of the antibody and the hapten.

8. Most of the antibodies in these patients arise from a clone of a single malignant plasma cell that has proliferated uncontrollably. These cells all produce a single molecular species of immunoglobulin.

9. d. Answers (a), (b), and (c) are incorrect because myeloma immunoglobulins have the same molecular architecture as normal IgG immunoglobulins and are directed against a single epitope on the surface of the antigen that elicits them. Since they are identical molecules they display a single association constant for the antigen they bind.

10. The antigen eliciting the immunoglobulin in a myeloma cell is generally unknown.

11. Milstein and Kohler discovered that any particular antibody-producing spleen cell could be fused to a strain of myeloma cell to form a hybridoma. The hybridoma was immortal and could produce large amounts of a monoclonal antibody to the antigen that had stimulated the spleen cell, thereby facilitating the study of antigen-antibody interactions. Note: Although not stated in *Biochemistry,* the reason that the hybridoma cell secretes the antibody specified by the spleen cell and not the one made by the myeloma cell is that a strain of nonsecreting myeloma cells is used.

12. (a) 5 (b) 4 (c) 1 (d) 3 (e) 6 (f) 2

13. Myeloma cells lacking HGPRTase cannot salvage hypoxanthine to form IMP and from it AMP and GMP, and in the presence of aminopterin they cannot synthesize nucleotides de novo. Consequently, they die unless they can fuse to a spleen cell with an active HGPRTase gene and thereby synthesize nucleic acids. The thymine in the medium allows dTMP to be formed and DNA synthesis to proceed in the presence of aminopterin, which indirectly inhibits dTMP synthetase and thus DNA synthesis (see p. 615 in Stryer).

14. b, c, d, e

15. a, b, and c. Answer (d) is incorrect because the variable sequences of the L and H chains in a given IgG are different from one another.

16. b and d. Answer (a) is incorrect because the two antigen combining sites on an IgG molecule are directed toward the same epitope. Thus, an IgG can bind only one kind of antigen. Answer (c) is incorrect because common (constant) sequences are involved in the transplacental transport of antibodies. Answer (e) is incorrect because it is the F_c unit of the IgG-antigen complex, not of IgG alone, that associates with a component of the complement pathway to trigger cell lysis. Finally, answer (f) is incorrect because the hypervariable regions of both H and L chains form the antigen-combining site.

17. (a) 5 (b) 2 (c) 4 (d) 1 (e) 3

18. The mRNA probe would hybridize to one region on the DNA from the plasma cell but to widely separated regions on the germ-line DNA. The gene on the plasma-cell DNA encodes the intact gene sequence for the V, J, and L regions of the L chain, including introns. The genes encoding the V, J, and L regions of the L chain are in distant locations in the germ-line DNA because the DNA has not yet rearranged to bring these regions into proximity.

19. There aren't enough unique *V* and *C* genes to provide a sufficient number of different sequences when they are recombined in all the possible combinations. Joining (*J*) and diversity (*D*) genes increases the number of possible combinations. (See Stryer, pp. 906–907).

20. a, b, d, e

21. b, c, and d. Answer (a) is incorrect because the sequence rearrangements of class switching take place through DNA recombination.

22. Plasma cells arise from antigen-triggered B lymphocytes, and they produce IgG antibodies. Memory B cells arise from the same clone of B lymphocytes, are long-lived, and can be stimulated to begin rapid IgG production should the antigen reappear in the organism.

T Cell Function and the Cellular Immune Response

23. c, e, and f. Answer (a) is incorrect because T cell receptors recognize fragments of foreign macromolecules only when the antigen is bound on the surface of a cell. Answer (d) is incorrect because the T cell receptor is composed of one α and one β chain, each having sequences that are homologous to the V regions of the chains of immunoglobulins.

24. a, b, c, and d. Answer (e) is incorrect because MHC proteins are all bound to the cell surface and are not soluble.

25. Just as Lamarck proposed environmentally directed changes whereas Darwin proposed selection from preexisting differences as being responsible for the evolution of species, Pauling proposed the antigen-directed formation of antibodies and Burnet championed the selection of preexisting antibodies to explain antibody diversity.

ANSWERS TO PROBLEMS

1. A given antigenic protein will stimulate the production of a mixed population of antibodies, with each type of antibody being specific for a different region in the tertiary structure of the antigenic protein.

2. The sketches for the two precipitates are shown in Figure 35-4. In (A) the Ab/Ag ratio is $8/7 = 1.14$. In (B) the ratio is $17/6 = 2.83$

3. The results would occur if the haptens were clustered so densely on the antigen that bivalent antibodies combined preferentially with two neighboring haptens on a given antigen molecule. Such behavior is actually found in some systems and is termed *monogamous bivalency*.

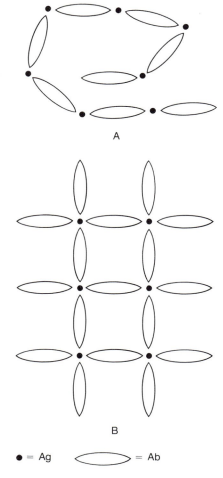

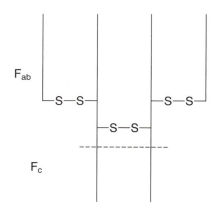

● = Ag   ⬭ = Ab

Figure 35-4
Possible lattice structures for Ab/Ag ratios of (A) 1.14 and (B) 2.83.

F_{ab}

F_c

Figure 35-5
Cleavage of IgG by pepsin.

4. Lymphocytes undergo allelic exclusion (Stryer, p. 907), so only one of two alleles is expressed in a given cell. This is necessary for all of antibodies produced by a single cell to be identical.

5. Polyacrylamide-gel electrophoresis separates proteins on the basis of size. β-Metcaptoethanol reduces the disulfide bonds that link the light and heavy antibody chains. Thus, the light and heavy chains are separated from one another on the gel.

6. Three pieces result from the cleavage of IgG by pepsin, as shown in Figure 35-5. One piece contains both F_{ab} units. The other two pieces derive from the bisection of the F_c region.

7. The mass-action expression for an antigen-antibody reaction is

$$\text{Ag-Ab} \underset{k_2}{\overset{k_1}{\rightleftharpoons}} \text{Ag} + \text{Ab}$$

Since

$$K = k_1/k_2$$

substituting in the given values yields

$$10^{-9} \text{ M} = \frac{k_1}{10^8 \text{ M}^{-1} \text{ s}^{-1}}$$

$$k_1 = 10^{-1} \text{ s}^{-1}$$

8. The observation suggests that both Tyr-34 on the light chain and Lys-54 on the heavy chain are involved in binding of hapten.

9. (a) In the preparation of the antibody, DNP served as the hapten. The DNP portion of the affinity-labeling agent will therefore be recognized by the hapten-binding site on the antibody.
   (b) The reagent alkylates the side chains of cysteine present in the hapten binding site of the antibody in a reaction involving the elimination of HBr. Thus, the end of the molecule opposite the ring is involved.
   (c) When DNP-Lys is present, the DNP ring binds to the hapten binding site of the antibody, thus competitively precluding the binding of the affinity agent. Because little or no affinity-labeling agent is bound to the hapten-binding site of the antibody, the amino acid side chains cannot be affinity labeled.

10. The β-mercaptoethanol reduces disulfide bonds in the antibody, producing a mixture of light and heavy chains, which is then resolved on the molecular sieving column. Affinity-labeling agent B apparently has labeled the heavy chains at the hapten-binding site, and agent A has labeled the light chains. (Remember that larger molecules elute first from a molecular sieving column.)

Muscle Contraction and Cell Motility

The transduction of chemical energy into mechanical energy—the use of ATP hydrolysis to drive the contraction of muscles or to move cells, for example—is the subject of Chapter 36. Stryer first describes the structure of vertebrate skeletal muscle by showing how thick and thin protein filaments give rise to its striated appearance in micrographs. He next explains that the thick filaments are primarily myosin and the thin filaments are primarily actin. He describes the ATPase and actin-binding activities of the globular domains of myosin and the structural basis for its filamentous domain. He shows the structure of the F-actin polymer, which is composed of a linear coiled array of G-actin monomers. The association of myosin and the actin polymer to form actomyosin is described. Actomyosin threads that have been formed in vitro shorten when they are incubated with ATP, Mg^{2+}, and K^+, which provides a hint of the molecular basis for the contraction of skeletal muscle. The opposite polarities of the thick and thin filaments within a sarcomere further indicate how molecular movement can result in the shortening of myofibrils.

The exact mechanism of the energy transduction remains obscure. However, the repeated association and dissociation of the S1 heads of myosin with actin and the conformational changes in myosin that are effected by the binding of ATP, its hydrolysis to ADP and P_i, and the release of the hydrolysis products suggest how the power stroke occurs. Stryer next describes how Ca^{2+} functions to regulate muscle contraction through its interaction with the troponin complex and the resulting effects on the location of tropomyosin on the actin polymer. Vertebrate smooth muscle provides an example of a Ca^{2+}-mediated mechanism that involves the covalent modification of myosin by phosphorylation as the means of regulating contraction. A discussion of the role of creatine phosphate as a reservoir of phosphoryl groups concludes the section of the chapter on muscle.

Stryer next considers the more universal roles of actin and myosin in cells. He explains a technique to disrupt a gene in vivo to determine the requirement for the product it encodes. For example, this technique has been used to demonstrate the need for actin for cell growth in yeast and the need for myosin for cell division in a slime mold. The structures and functions of microfilaments, intermediate filaments, and microtubules are considered next. Microfilaments, which are composed primarily of actin, interact with a variety of actin-binding proteins and are involved in several cytoskeletal functions. Intermediate filaments and microtubules are formed from other proteins. There are several types of intermediate filaments. They are constructed from proteins that are encoded by a large, multigene family, and they have two-strand or three-strand α-helical coiled-coil structures. Microtubules are found in nearly every cell and serve multiple structural and functional roles. They are composed primarily of α-tubulin and β-tubulin monomers that form tubular structures that have larger diameters than microfilaments or intermediate filaments. Microtubules also contribute the basic macromolecular assembly of the axoneme, which is the fundamental structural component of the cilia and flagella of eucaryotic cells. Dynein and kinesin are auxiliary proteins that interact with microtubules to bend cilia and flagella and to move vesicles along microtubules, respectively. Dynein and kinesin are ATPases. They transduce chemical energy into mechanical movement as does myosin. Microtubules also serve to separate the chromosomes during mitosis. The rapid association of GTP-tubulin with microtubules and the rapid dissociation of GDP-tubulin from the ends of tubulin polymers in conjunction with the GTPase activity of the tubulins explains the dynamic instability of microtubules. Stryer closes the chapter by describing how selection and stabilization of correctly targeted microtubules from among the many being formed and degraded give rise to the final structures that are observed.

LEARNING OBJECTIVES

When you have mastered this chapter, you should be able to complete the following objectives.

Microscopic Vertebrate Muscle Structure (Stryer pages 921–924)

1. Define *sarcolemma, sarcomere, sarcoplasm,* and *myofibril.*

2. Identify the *A band, H zone, M line, I band,* and *Z line* of a sarcomere in an electron micrograph of a myofibril.

3. Relate the locations of the *thick filaments* and *thin filaments* to the I band, the A band, and the H zone of a sarcomere.

4. Relate the structures of the thick and thin filaments to their compositions of *actin, myosin, tropomyosin,* and *troponin*. Note the *cross-bridges* between the thick and thin filaments.

5. Describe muscle contraction in terms of the *sliding-filament model,* and relate contraction to changes in the sizes of the A band, I band, and H zone.

Myosin, Actin, and Actomyosin in Vertebrate Skeletal Muscle (Stryer pages 924–927)

6. Sketch the general structure of the polypeptide backbone of a myosin molecule, and note its dimensions. Describe the subunit composition of myosin, and note its *α-helical coiled-coil structure* and its *dual globular head*.

7. Describe the amino acid sequence regularities of the regions of the myosin polypeptide chains that form the α-helical coiled coil.

8. Describe the fragmentation of myosin into *light meromyosin* (LMM) and *heavy meromyosin* (HMM) by proteolysis and the further fragmentation of HMM into its *S1* and *S2 subfragments*. Sketch the polypeptide structures of the fragments, and associate them with the *ATPase activity,* the *actin-binding sites,* and the *thick-filament–forming domains* of the intact molecule.

9. Note the dimensions of *G-actin,* and distinguish between G-actin and *F-actin*.

10. Describe *actomyosin,* and note the effects of adding ATP or ATP, K^+, and Mg^{2+} to a solution of actomyosin. Relate these effects to *muscle contraction*.

Muscle Structure at the Biochemical Level and Muscle Contraction (Stryer pages 928–936)

11. Note the *bipolarity* of the thick and thin filaments with respect to the Z lines of a sarcomere. Relate the opposing polarities of the molecules to the coordinate and macroscopic movement that occurs during contraction.

12. Provide an overview of a model for the mechanism of the *power stroke* during muscle contraction. Explain the roles of actin, the *hinges* of myosin, and the ATPase activity of myosin in the model.

13. Recount the experiment that demonstrated unidirectional movement of myosin molecules along an actin cable.

14. Describe the roles of Ca^{2+}, *tropomyosin,* and the *troponin complex* in the contraction of skeletal muscle. Note the resemblance of *troponin C* to *calmodulin,* and outline the flow of information from Ca^{2+} to myosin during the stimulation of contraction.

15. Distinguish between *smooth muscle* and skeletal muscle in terms of the mechanisms that are used for regulating contraction. Note the roles

of Ca^{2+}-calmodulin, *myosin light-chain kinase*, and the *cAMP-activated protein kinase* in the regulation of the contraction of smooth muscle.

16. Describe the roles of *creatine phosphate*, *creatine kinase*, and *adenylate kinase* in maintaining the ATP pool during muscle contraction.

Microfilaments, Intermediate Filaments, Microtubules, and the Cytoskeleton
(Stryer pages 936–940)

17. Describe some of the roles of actin and myosin in the movements of nonmuscle cells. Note the composition and diameters of *microfilaments*.

18. Outline the technique for demonstrating the requirement for a protein by disrupting the gene encoding it.

19. Describe *antisense RNA,* and explain how it is used to block the production of a specific protein.

20. Describe the *filamentous composition* and the functions of the *cytoskeleton*. Explain how *immunofluorescence microscopy* is used to visualize the cytoskeleton.

21. Outline the effects of ATP and *actin-binding proteins* on the *polymerization* and *depolymerization* of actin. Define *critical concentration*, and relate it to the polymerization of actin.

22. Describe the effects of *phalloidin* and *cytochalasin B* on the synthesis and degradation of actin filaments.

23. Describe the *intermediate filaments*. Note their diameters, common structural motifs, and cell and tissue locations.

Cilia, Flagella, and Microtubule Formation (Stryer pages 940–946)

24. Sketch the general structure of a *microtubule*, noting its diameter and showing its alternating subunits of *α-tubulin* and *β-tubulin*.

25. Distinguish between the *cilia* and *flagella* of eucaryotic cells in terms of their relative lengths and physiological functions. Describe the structure of the *axoneme*, and sketch the *9 + 2 array* of the microtubule doublets and singlets that form its basic ring motif.

26. Locate the *dynein* in an axoneme, and compare its biochemical activities to those of myosin. Explain the roles of *dynein cross-bridges, radial spokes,* and *nexin links* in causing the bending rather than the contraction of the axoneme.

27. List some of the causes and consequences of the *immotile-cilia syndrome*.

28. Describe the movement of *vesicles* and *organelles* along microtubules. Distinguish between the plus and minus ends of a microtubule, and relate this polarity to the direction of transport.

29. Compare *kinesin* to dynein and myosin.

30. Describe the dynamics of the polymerization and depolymerization of microtubules, noting the role of the *microtubule organizing-centers* (MTOC) on the *centrisomes* and *poles* of the *mitotic spindles* and the roles of *GTP* and *GDP* in these processes. Describe the effects of *colchicine* on the polymerization of microtubules. Note the role of selection in target location by growing microtubules.

Microscopic Vertebrate Muscle Structure

1. Energy is required to drive the contraction of striated muscle, the beating of flagella or cilia, and the intracellular transport of vesicles along microtubules. Which of the following statements about energy transduction in these systems are *correct*?

   (a) The proton-motive force across the plasma membrane surrounding the cell provides the energy for these movements.
   (b) The binding of ATP to proteins, which then undergo a conformational change, provides the energy for these processes.
   (c) The hydrolysis of ATP is coupled to the phosphorylation of tyrosine residues on the proteins of these systems to drive them.
   (d) The hydrolysis of protein-bound ATP and the release of $ADP + P_i$ lead to conformational transitions that complete a movement cycle.
   (e) The binding of GTP to oriented proteins at the interface of the moving assemblies drives these processes.

2. Which of the following statements about striated vertebrate muscle are *correct*?

   (a) The muscle cells (myocytes) are multinucleate and are surrounded by an electrically excitable plasma membrane.
   (b) Myocytes contain parallel myofibrils that are immersed in the sarcoplasm.
   (c) Each myofibril is composed primarily of thick and thin protein filaments.
   (d) The functional unit of a muscle cell that is directly responsible for contraction is called a sarcomere.
   (e) The sarcolemma separates individual myofibrils from one another.

3. The following figure is a schematic diagram of a longitudinal segment of a skeletal muscle myofibril. Label the structures indicated in the figure by matching them with the listed choices.

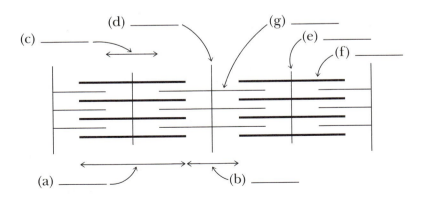

   (1) I band
   (2) A band
   (3) Thin filaments
   (4) M line

   (5) H zone
   (6) Thick filaments
   (7) Z line

4. During contraction, reversible cross-bridges form between which of the two structures listed in Question 3?

_____

5. Assign the proteins in the right column to the appropriate myofibrilar component in the left column.

(a) Thin filament _____

(b) Thick filament _____

(c) Z line _____

(d) Flexible mesh around filaments _____

(1) Titin
(2) α-Actinin
(3) Tropomyosin
(4) Myosin
(5) Actin
(6) Troponin complex
(7) Nebulin

6. Which of the following structures change their dimensions or alter their positions with respect to one another during contraction? Answer with *Yes* or *No*. (Refer to Question 3 for a diagram of a myofibril.)

(a) I band _____

(b) A band _____

(c) Thin filaments _____

(d) M line to M line _____

(e) H zone _____

(f) Thick filaments _____

(g) Z line to Z line _____

(h) Z line to edge of adjacent H zone _____

(i) Z line of M line _____

Myosin, Actin, and Actomyosin in Vertebrate Skeletal Muscle

7. Which of the following statements about myosin are *correct*?

(a) Myosin binds to polymerized actin.
(b) In vitro, myosin assembles spontaneously into the thin filaments of the I band of the sarcomere.
(c) Myosin is an ATPase.
(d) Myosin has domains that interact with one another to effect its physiological functions.
(e) Myosin is composed of two polypeptide chains, one of which forms an α-helical coiled coil and the other of which is a globular head.

8. Which of the following statements about heavy meromyosin (HMM) and light meromyosin (LMM) are *correct*?

(a) HMM and LMM are formed from myosin by tryptic cleavage.
(b) HMM can be cleaved by papain to yield two globular proteins that polymerize to form the thin filaments.
(c) LMM is an α-helical coiled coil composed of two polypeptide chains that can form filaments in vitro.
(d) HMM contains the two globular heads of the myosin molecule, hydrolyzes ATP, and binds actin in vitro.

9. Which of the following statements about actomyosin or its constituents are *correct*?

(a) The polymerization of G-actin is controlled by the binding of guanyl nucleotides.

(b) The addition of ATP to a solution of actomyosin leads to the contraction of the actomyosin threads.

(c) F-actin is composed of a helical array of two strings of G-actin monomers.

(d) Cross-bridges form between the globular heads of myosin and the F-actin polymer.

(e) Mixing solutions of F-actin and myosin leads to the formation of a viscous solution of actomyosin.

Muscle Structure at the Biochemical Level and Muscle Contraction

10. Which of the following statements concerning the contraction of vertebrate skeletal muscle are *correct?*

(a) Coherent movement occurs during contraction because actomyosin complexes are highly ordered and oppositely oriented within the sarcomere.

(b) Cross-bridges form between adjacent myosin molecules to stabilize the thick filaments.

(c) Whereas myosin is obviously an asymmetric molecule that forms an asymmetric filament, thin filaments are formed from globular actin (G-actin) subunits and are thus nondirectional or symmetric polymers.

(d) The polarity of the thick and thin filaments reverses at the Z line.

11. Which of the following statements concerning events related to the power stroke of muscle contraction are *correct?*

(a) The hydrolysis of ATP to ADP and P_i by myosin is fast relative to the release of the ADP and P_i from the protein.

(b) The binding of actin to myosin stimulates the ATPase activity of myosin by facilitating the release of ADP and P_i.

(c) Actin and myosin are joined by cross-bridges that are stabilized by the binding of ATP to the myosin head domains.

(d) Repeated cycles of ATP binding, ATP hydrolysis, and the resulting association and dissociation of cross-bridges and conformational changes in myosin contribute to the contractile process.

(e) In the region of overlapping thick and thin filaments of a sarcomere, the cross-bridges will either all be formed or all be dissociated, depending upon the phase of the power stroke.

12. Considering only the power stroke of skeletal muscle contraction and the events that precede and follow it, place the following states or processes that occur in going from the resting state to the contracted state and back again in their proper order.

(a) The thick filament moves with respect to the thin filament.

(b) The S1 heads of myosin interact with actin, P_i is released, and the interaction with actin strengthens.

(c) ATP binds to myosin.

(d) The S1 heads of myosin are dissociated from the thin filament and contain bound ADP and P_i.

(e) ATP is hydrolyzed to ADP and P_i.

(f) ADP dissociates from the myosin.

13. Which of the domains of myosin is primarily responsible for generating the force of skeletal muscle contraction?

   (a) The hinge between the S1 and S2 domains
   (b) The hinge between the S2 and the LMM domains
   (c) The LMM domain
   (d) The S1 globular head
   (e) The S2 domain that connects the S1 domain to the α-helical coiled coil
   (f) The α-helical coiled coil

14. Which of the following statements about the troponin complex and tropomyosin of skeletal muscle are *correct*?

   (a) Tropomyosin is a two-stranded α-helical rod that winds about the α-helical coiled-coil tail of myosin.
   (b) The troponin complex is composed of three subunits.
   (c) The subunits of the troponin complex bind actin, tropomyosin, or Ca^{2+}.
   (d) The conformational change induced in the troponin complex by the binding of Ca^{2+} is transmitted to tropomyosin.
   (e) Troponin C binds Ca^{2+} and is structurally similar to calmodulin.
   (f) The movement of tropomyosin allows the S1 heads of the thick filament to interact with the thin filaments.
   (g) The regulation of contraction by Ca^{2+} is mediated by proteins that are associated with the thin filaments.

15. Considering the events associated with the regulation of the contraction of skeletal muscle, place the following processes in the order in which they occur upon the stimulation of a muscle to contract.

   (a) A nerve impulse leads to the release of Ca^{2+} from the sarcoplasmic reticulum.
   (b) Tropomyosin changes position on the thin filament.
   (c) The formation of the complex between troponin C and Ca^{2+} causes the other components of the troponin complex to move.
   (d) An ATP-dependent pump lowers the Ca^{2+} concentration of the sarcoplasm.
   (e) Ca^{2+} binds to troponin C in the troponin complex.
   (f) The S1 heads of myosin bind to actin.

16. Contrast the regulation of contraction and the contractile properties of smooth muscle and skeletal muscle in vertebrates.

   _____

   _____

   _____

17. Which of the following statements about the phosphoryl-transfer potential of skeletal muscle are *correct*?

   (a) The ATP of muscle can sustain contraction for less than a second.
   (b) Creatine phosphate serves as a phosphoryl reservoir that replenishes the ATP pool.

(c) Creatine phosphate can support contraction for up to four minutes.

(d) The phosphoguanidino groups of creatine phosphate and arginine phosphate have large negative standard free energies of hydrolysis.

(e) Creatine phosphate is formed by a reaction between creatine and arginine phosphate.

(f) The ATP pool of muscle can be partially replaced by the activity of the enzyme adenylate kinase.

Microfilaments, Intermediate Filaments, Microtubules, and the Cytoskeleton

18. Which of the following statements about actin in nonmuscle cells are *correct*?

(a) The ratio of actin to myosin is typically much higher in nonmuscle cells than in muscle cells.

(b) The amino acid sequence of the actin in nonmuscle cells is highly diverged from that of the actin in muscle cells.

(c) The actin of yeast cells is dispensable if the cells are grown in liquid culture.

(d) Actin typically makes up approximately 10 percent of the protein of nonmuscle cells.

(e) Actin is the primary component of the microfilaments of nonmuscle cells.

(f) The assembly of actin filaments requires ATP hydrolysis.

19. Describe how you would use the technique of gene disruption to demonstrate that a particular protein is essential for the viability of a yeast. Assume the yeast contains only a single copy of the gene encoding the protein in question. State explicitly what materials you would use.

_____

_____

_____

_____

20. Which of the following statements about the cytoskeleton or its components are *correct*?

(a) A filamentous scaffolding composed primarily of proteins forms the cytoskeleton of eucaryotic cells.

(b) The cytoskeleton of eucaryotic cells is composed of three major classes of assemblies—microfilaments, intermediate filaments, and microtubules.

(c) The microfilaments of the cytoskeleton are stable once they are formed, and a given filament exists for the life of the cell.

(d) Cytochalasin B binds along the length of the actin filaments and prevents their contraction.

(e) Phalloidin blocks cell movement by preventing microfilaments from dissociating.

(f) Intermediate filaments are composed of two or three F-actin polymers that are coiled about one another.
(g) The keratins, neurofilaments, vimentin filaments, and lamnins are all intermediate filaments.

Cilia, Flagella, and Microtubule Formation

21. Which of the following statements concerning microtubules are *correct?*

(a) Microtubules are filaments composed of α-helical coiled-coil polypeptide chains.
(b) Microtubules contain α-tubulin and β-tubulin protomers that are disposed in a helical array around a hollow core to form a cylindrical structure.
(c) The cilia and flagella of eucaryotic cells contain nine microtubule doublets that surround a pair of microtubule singlets.
(d) The outer microtubules in an axoneme are linked together and to an ATPase called dynein.
(e) The powered movements of dynein in an axoneme shorten the structure.

22. When defective dynein causes the immotile-cilia syndrome, sometimes not only chronic respiratory disorders but also infertility occurs in males. Explain.

_____

_____

23. A neuron can move a vesicle approximately a meter from the central cell body to the end of an axon in a day. How are microtubules involved in this process, what provides the energy for the movement, and what protein is directly involved in the movement?

_____

_____

_____

24. Which of the following statements about microtubule assembly are *correct?*

(a) Nucleation centers are required to initiate the efficient formation of a new microtubule.
(b) Colchicine blocks the polymerization of microtubules and thus the movement of chromosomes.
(c) Microtubule-organizing centers serve as foci at which α-tubulin and β-tubulin polymerize to form microtubules.
(d) α-Tubulin and β-tubulin are potent ATPases.
(e) GTP-tubulin adds to growing microtubules more readily than does GDP-tubulin.
(f) GDP-tubulin rapidly dissociates from the ends of microtubules.
(g) GDP-tubulin remains in microtubules if it is in an internal position.
(h) Growing microtubules find their targets by forming along pre-existing microfilament tracts.

1. Decide whether each of the following will remain unchanged or will decrease upon muscle contraction. Assume that the sliding-filament model applies. Refer to page 923 of Stryer.

   (a) The distance between adjacent Z lines
   (b) The length of the A band
   (c) The length of the I band
   (d) The length of the H zone

2. The symmetry of thick and thin filaments in a sarcomere is such that six thin filaments ordinarily surround each thick filament in a hexagonal array. (See Figure 36-3 on p. 922 of Stryer.) In electron micrographs of cross sections of fully contracted muscle, the ratio between thin and thick filaments has been found to be double that of resting muscle.

   (a) Propose an explanation for this observation based on the sliding-filament model.
   (b) How might the explanation you have given in part (a) also account for the long appreciated fact that a fully contracted muscle is, paradoxically, "weaker" than a resting muscle.

3. In an early experiment on the energy requirements for muscle contraction, isolated muscle fibers were poisoned with iodoacetate to shut down the glycolytic pathway. These poisoned fibers would continue to contract upon stimulation, however, as long as a certain phosphorylated compound remained in abundance. Name that compound, and explain its role in providing the energy needed for muscle contraction.

4. In an early experiment designed to elucidate the role of creatine phosphate in muscle, an extract of muscle tissue was prepared and filtered so that it contained only large molecules like proteins. The filtrate could only hydrolyze creatine phosphate when another compound was added. What was that compound? Explain.

5. In his classic experiments on metabolism that led to his discovery of the citric acid cycle, Krebs studied oxygen consumption by actively respiring minced pigeon-breast muscle. Do you think his results would have been as striking if he had used minced breast muscle from the domestic turkey instead? Explain.

6. Predict whether each of the following substitutions in the amino acid sequence of the myosin tail would significantly affect the secondary structure of the myosin tail. Briefly explain your answers. (Refer to pp. 925 and 37 of Stryer.)

   (a) Ile for Leu
   (b) Met for Ala
   (c) Asp for Glu

7. Examine the model of the two-stranded α-helical coiled coil shown in Figure 36-10 on page 925 of Stryer.

   (a) Is the superhelix right-handed or left-handed? How can you tell?
   (b) Are each of the individual helices right-handed or left-handed? How do you know?

8. Ethylenediaminetetraacetate (EDTA), a heavy-metal chelator, acts as a muscle relaxant. Explain the basis for this effect.

9. The enzyme creatine kinase catalyzes the reaction

$$\text{Creatine phosphate} + \text{ADP} + \text{H}^+ \rightleftharpoons \text{creatine} + \text{ATP}$$

(a) The standard change in free energy for this reaction is −3.0 kcal/mol. Calculate the free-energy change for the reaction as it occurs in a typical muscle cell using the following concentrations: [ATP] = 4 mM, [ADP] = 0.013 mM, [creatine phosphate] = 25 mM, and [creatine] = 13 mM. Use the thermodynamic relationships given on page 182 of Stryer. Assume that the temperature is 25°C.

(b) What can you say about the direction of this reaction in resting muscle? How does this relate to the physiological role of creatine kinase?

10. The enzyme adenylate kinase catalyzes the following reaction, which is close to equilibrium in muscle cells:

$$2\,\text{ADP} \rightleftharpoons \text{ATP} + \text{AMP}$$

Give an explanation of the role that this reaction fulfills as it proceeds (a) in the direction written and (b) in the reverse direction.

11. Onion root-tip cells that have been incubated in solutions of colchicine for several days become polyploid rather than diploid. Propose an explanation for this observation.

12. The mold products vincristine and vinblastine interfere with the polymerization of microtubules as does colchicine. Vincristine and vinblastin are widely used in the treatment of rapidly growing cancers. Explain the basis for their effects.

13. When muscle tissue is completely depleted of ATP, the condition of rigor mortis develops. Explain the molecular basis for this.

ANSWERS TO SELF-TEST

Microscopic Vertebrate Muscle Structure

1. b and d

2. a, b, c, and d. Answer (e) is incorrect because the sarcolemma is the plasma membrane that surrounds the myocytes.

3. (a) 2 (b) 1 (c) 5 (d) 7 (e) 4 (f) 6 (g) 3

4. 3 and 6. Cross bridges form between thick filaments and thin filaments.

5. (a) 3, 5, 6 (b) 4 (c) 2 (d) 1, 7

6. (a) Yes (b) No (c) No (d) Yes (e) Yes (f) No (g) Yes (h) No (i) Yes

7. a, c, and d. Answer (b) is incorrect because myosin assembles to form the thick filaments. Answer (e) is incorrect because myosin is composed of six polypeptide chains. Two heavy chains intertwine their C-terminal portions to form the α-helical coiled-coil, with their N-terminal segments forming two globular heads. Four more polypeptide chains of two kinds associate with the globular heads.

8. a, c, and d. Answer (b) is incorrect because the two globular heads arising from papain digestion cannot polymerize for they lack α-helical coiled-coil structures.

9. c, d, and e. Answer (b) is incorrect because ATP dissociates the actomyosin complex.

Muscle Structure at the Biochemical Level and Muscle Contraction

10. a and d. Answer (b) is incorrect because cross-bridges form between the S1 heads of myosin and actin. Answer (c) is incorrect because thin filaments also have polarity or directionality (see Stryer, p. 928).

11. a, b, and d. Answer (c) is incorrect because the binding of ATP to actomyosin dissociates the cross-bridges. Answer (e) is incorrect because, at any instant, cross-bridges will be in all stages of forming and breaking because the process is dynamic and asynchronous (see Stryer, p. 930).

12. d, b, a, f, c, e, d

13. d. Although the whole molecule is required for muscle contraction, the activities of the S1 head domains provide the biochemical activity for the power generation. The cyclic changes in the conformation of the head domains and in their affinities for actin, ATP, ADP, and P_i are the bases for the energy transduction.

14. b, c, d, e, f, and g. Answer (a) is incorrect because tropomyosin is associated with actin, not myosin.

15. a, e, c, b, f, d

16. Smooth muscle contraction is regulated by a Ca^{2+}-mediated phosphorylation of its myosin light chains, whereas a troponin complex and tropomyosin respond directly to Ca^{2+} levels to control skeletal muscle contraction by affecting myosin-actin interactions. Because the reversal of the covalent modification by a phosphatase is relatively slow compared to the fluctuations in sarcoplasmic Ca^{2+} levels, smooth muscle undergoes more sustained contractions than does skeletal muscle. Furthermore, since a cAMP-activated protein kinase is involved in the regulation of the myosin light-chain kinase, smooth muscle contraction in cells containing β-type receptors can be modulated by epinephrine (see Stryer, p. 935).

17. a, b, d, and f. Answer (c) is incorrect because the amount of creatine phosphate is sufficient to maintain only a few seconds of intense contraction. Answer (e) is incorrect because creatine phosphate is formed from ATP and creatine in a reaction catalyzed by creatine kinase. The

ATP for this reaction is generated by glycolysis in anaerobic muscle or by respiration in muscle with sufficient oxygen.

Microfilaments, Intermediate Filaments, Microtubules, and the Cytoskeleton

18. a, d, e, and f. Answer (b) is incorrect because the actins have highly conserved amino acid sequences. Answer (c) is incorrect because actin is essential for the growth of yeast under any conditions. Note that, although not stated in the text, answer (f) is also true for the polymerization of G-actin to F-actin in muscle cells.

19. You would need the cloned gene so that you could delete a portion of it by recombinant DNA techniques to render it nonfunctional. You would then introduce the incomplete gene into the yeast on a plasmid that could not replicate in yeast. Homologous recombination between regions of shared sequences would disrupt the gene in the yeast (see Stryer, pp. 937–938). An antibiotic-resistance marker on the plasmid would facilitate the experiment; recombinants could be selected by growing the culture in the presence of the antibiotic.

20. a, b, e, and g. Answer (c) is incorrect because actin filaments are dynamic structures that are continuously being formed and degraded. Answer (d) is incorrect because cytochalasin B binds to the ends of the polymers and prevents their growth. Answer (f) is wrong because intermediate filaments don't contain actin.

Cilia, Flagella, and Microtubule Formation

21. b, c, and d. Answer (a) is incorrect because microtubules are assembled from relatively globular tubulin subunits and lack the α-helical coiled-coil structure of myosin and the intermediate filaments. Answer (e) is incorrect because dynein movements lead to the bending not to the contraction of the axoneme.

22. Defective dynein molecules not only immobilize the cilia of the respiratory tract, but they also render sperm immotile.

23. The microtubules, which form a network of fibers that traverse the cell, provide tracts along which vesicles and organelles can move. ATP hydrolysis by the protein kinesin acts as a molecular engine, in a manner analogous to the way myosin acts, to power the movements.

24. a, b, c, e, f, and g. Answer (h) is incorrect because microtubules form randomly, and only those that reach a target that stabilizes them are not rapidly depolymerized.

ANSWERS TO PROBLEMS

1. (a) The distance between adjacent Z lines will decrease. (b) The length of the A band will remain unchanged. (c) The length of the I band will decrease. (d) The length of the H zone will decrease. See Figure 36-1.

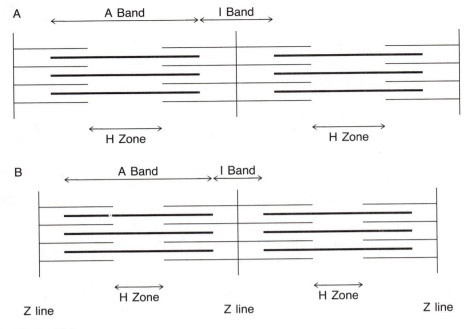

Figure 36-1
Schematic illustration of (A) uncontracted and (B) partially contracted sarcomeres.

2. (a) The number of thin filaments per thick filament doubles because the thin filaments override one another when the sarcomere is fully contracted. See Figure 36-2.

(b) Thin filaments and thick filaments must have the same polarity to interact with one another to form cross-bridges, but the polarity of the filaments reverses halfway between the Z lines. (See Figure 36-19 on p. 929 of Stryer.) Thus, in the region where the thin filaments override one another in a fully contracted sarcomere, the thin filaments have a polarity opposite that of the adjacent thick filament. Hence, a slightly smaller number of cross-bridges is formed, so the fully contracted muscle develops less tension; that is, it is "weaker."

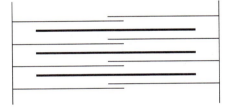

Figure 36-2
Schematic illustration of a fully contracted sarcomere.

3. The compound is creatine phosphate. It serves as a reservoir of high-potential phosphoryl groups in muscle, and it rephosphorylates ADP to ATP in the presence of the enzyme creatine kinase.

4. The compound was ADP, a coparticipant with creatine phosphate in the reaction catalyzed by creatine kinase:

$$\text{Creatine phosphate} + \text{ADP} + \text{H}^+ \longrightarrow \text{ATP} + \text{creatine}$$

5. Domestic turkey breast muscle would not have worked. Pigeon breast muscle is a red muscle. It is well supplied with oxygen and rich in myoglobin, and it generates most of its energy aerobically. It is therefore well endowed with mitochondria, the sites for the reactions of the citric acid cycle. Domestic turkey breast muscle is white. It is more poorly supplied with oxygen, has substantially less myoglobin, and generates most of its energy anaerobically by glycolysis. It is a muscle with substantially fewer mitochondria and would be a poor choice for studies on respiration.

6. The myosin tail is composed of a two-stranded α-helical coil. Therefore, the amino acid sequence must be compatible with the maintenance of the α helix.

(a) Whereas Leu is sterically compatible with the α helix, Ile is not. Ile branches at the β-carbon atom. This would disrupt the α helix and would therefore distort the myosin tail.

(b) Both Met and Ala are compatible with the α helix, so the substitution of Met for Ala should not affect the secondary structure of the myosin tail.

(c) Although Asp and Glu both have negative charges in their side chains at physiological pH, Asp has one less methylene group. The negative charge in its side chain is close enough to the hydrogen-bonding goups in the backbone of the α helix to interact with them and disrupt regular hydrogen bonding. Accordingly, the substitution of Asp for Glu would alter the secondary structure of the myosin tail.

7. (a) The superhelix is right-handed. An easy way of confirming that it is right-handed is to remember that common machine screws are right-handed and that tightening them involves the clockwise rotation of a screwdriver. Imagine that the superhelix of the myosin tail has been converted into a machine screw as illustrated in Figure 36-3. If you imagine placing one of your hands above the helix and holding a screwdriver in the other, impaling the hand with the myosin screw will require the clockwise rotation of the screwdriver. Therefore, the superhelix is right-handed.

(b) Figure 36-10 on page 925 of Stryer does not show the handedness of the helix, but we can confidently predict that it is right-handed because all α helices found in proteins are right-handed. (Note: There are many, many turns of each of the component α helices in each turn of the superhelix of the myosin tail.)

8. The cycle of muscle contraction is initiated by the release of calcium ion from the tubules of the sarcoplasmic reticulum. Relaxation must therefore involve the sequestration or removal of that calcium. EDTA binds calcium, thereby mimicking its removal by the sarcoplasmic reticulum by sequestering it.

9. We start with equation 6 on page 182 of Stryer:

$$\Delta G = \Delta G° + RT \log_e \frac{[C][D]}{[A][B]}$$

Substitution yields

$$\Delta G = -3.0 + 1.98 \times 298 \times \log_e \frac{[13 \times 10^{-3} \text{ M}][4 \times 10^{-3} \text{ M}]}{[25 \times 10^{-3} \text{ M}][0.013 \times 10^{-3} \text{ M}]}$$

$$= -3.0 + 0.00198 \times 298 \times \log_e (160)$$

$$= -3.0 + 2.99 = -0.01 \text{ kcal/mol}$$

(b) As seen in part (a), the free-energy change for this reaction in resting muscle is essentially zero, which means that the reaction is at equilibrium. It is for just this reason that creatine phosphate is effective as an energy storage form. When muscle contraction occurs and the concentration of ADP rises, the reaction goes toward the right, and ADP is

Figures 36-3
An imaginary myosin machine screw.

rephosphorylated to ATP by creatine phosphate. As ATP is made available from energy-yielding metabolic reactions, it and creatine combine to regenerate creatine phosphate.

We could have concluded that the creatine kinase reaction was at equilibrium in the cell by comparing the value for K calculated from the free-energy data with the experimentally observed value of Γ, the mass action constant, which is equal to the concentration term in the equation used in part (a). The calculated value of K is 162 (see Stryer, p. 935). The experimentally observed value for Γ is 160. Because these values are very nearly equal, the reaction is essentially at equilibrium in the cell.

10. (a) As muscle contraction occurs, the level of ADP tends to rise. The reaction catalyzed by adenylate kinase then takes two ADP and makes one ATP, which can be used to drive further contraction, and one AMP, which stimulates the glycolytic pathway (at the phosphofructokinase step) to produce even more ATP.

(b) At the end of vigorous exercise, energy-yielding metabolic reactions yield ATP, which rephosphorylates the AMP, the ultimate breakdown product in the adenine nucleotide system, to ADP. The ADP is rephosphorylated to ATP by energy-yielding metabolic reactions, and the adenine nucleotide system is thus completely recharged.

11. Colchicine inhibits the polymerization of microtubules (see Stryer, p. 945). When colchicine is added to mitotic onion root-tip cells, the formation of spindle fibers does not occur. Chromosomes divide at the centromere, but the daughters are not pulled to the opposite poles of the cell. Thus, the number of chromosomes per cell increases.

12. Cancer cells have a greater than normal rate of cell division. In the presence of vincristine or vinblastine, mitotic spindle fibers fail to form, so cell division is retarded.

13. ATP is necessary for the cycle of cross-bridge breakage and reformation during muscle contraction and relaxation. When ATP supplies are totally depleted, the cross-bridges become frozen and rigor mortis develops.

CHAPTER 37

Membrane Transport

This chapter describes the molecular mechanisms for the transport of small molecules and ions across cell membranes. These transport processes are essential for the maintenance of the intracellular ionic composition, the uptake of nutrients, the excretion of waste products, and the excitability of cells. Stryer introduces the concept of electrochemical potential by defining active and passive transport. He then describes several active transport systems that derive the energy required for transport from the hydrolysis of ATP, Na^+ gradients, proton gradients, phosphoenolpyruvate, or light. The Na^+-K^+ pump, powered by ATP, is described in detail because it is the best understood and most widely distributed transport system in animal cells. It maintains the ionic environment of cells and drives, indirectly, many other transport systems. The Ca^{2+}-pump, the cotransport of Na^+ with sugars or amino acids, the transport of sugars driven by a proton gradient, the phosphorylation of sugars during transport, and the generation of a proton gradient by light are discussed as additional examples of active transport systems. Finally, Stryer turns to passive transport, which involves the spontaneous movement of ions and molecules down their electrochemical gradients. Three systems are described: the anion exchange protein of erythrocytes, transport antibiotics, and gap junctions. Previous chapters that contain important background material include Chapter 12, on the structure of membrane proteins and the permeability of membranes; Chapter 17, on the synthesis of ATP driven by the electromotive force of the proton gradient generated by electron flow; Chapter 22, on the generation of ATP by light in photosynthesis; and Chapter 36, on the contraction of muscle and its dependence on the ability of sarcoplasmic reticulum to pump Ca^{2+}.

When you master this chapter, you should be able to complete the following objectives.

Introduction (Stryer pages 949–950)

1. List the functions of the *membrane transport systems*.

2. Distinguish between *active* and *passive transport*. Write the equation for the *electrochemical potential*, and explain the significance of its terms.

The Na$^+$-K$^+$ Pump (Stryer pages 950–955)

3. Discuss the physiological role of the *Na$^+$-K$^+$ pump* in animal cells. Summarize the evidence that the Na$^+$-K$^+$ ATPase is responsible for the transport of Na$^+$ and K$^+$.

4. Describe the *Na$^+$-K$^+$ ATPase* activity of the Na$^+$-K$^+$ pump, its orientation in the membrane, and its binding of ions and inhibitors and the coupling of ion transport and *phosphorylation*.

5. Describe the subunit structure of the Na$^+$-K$^+$ pump. List the possible functions of the subunits and their relationships to the membrane bilayer.

6. Outline the proposed model for the mechanism of the Na$^+$-K$^+$ pump. Distinguish between the E_1, E_1-P, E_2 and E_2-P *conformational states* of the Na$^+$-K$^+$ pump, and explain their properties.

7. Discuss the inhibition of the Na$^+$-K$^+$ pump by *cardiotonic steroids* and explain their clinical significance.

The Ca^{2+}-ATPase and Other ATPases (Stryer pages 955–958)

8. Describe the role of the *Ca^{2+}-ATPase* of the *sarcoplasmic reticulum* in muscle contraction.

9. Compare the structures, ATPase activities, and transport mechanisms of the Ca^{2+}-ATPase and the Na$^+$-K$^+$ pump.

10. Outline the experimental system that allows a Ca^{2+} gradient to generate ATP. Compare this system with the mitochondrial ATP synthase (Stryer, pp. 413–417).

11. Describe the control of Ca^{2+} transport by *phospholamban* and the binding of Ca^{2+} in the endoplasmic reticulum by *calsequestrin*.

12. Distinguish between the three classes of *ion-motive ATPases: P-type*, *V-type*, and *F-type*. Describe their functions in eucaryotic cells.

Transport of Sugars and Amino Acids in Animal Cells and Bacteria and Light Driven Pumps (Stryer pages 958–963)

13. Define the terms *cotransport*, *symport*, and *antiport*.

14. Describe the pumping of sugars, amino acids, and ions by animal cells, using symports and antiports driven by a *Na$^+$ gradient*. Note the indirect involvement of the Na$^+$-K$^+$ pump in these processes.

15. Explain the role of *proton gradients* in powering transport systems in bacteria. Describe the *lactose permease* of *E. coli,* and name other compounds that are transported by similar systems.

16. Describe the accumulation of sugars by the *phosphotransferase system (PTS)* of bacteria, which is driven by *phosphoenolpyruvate.*

17. List the components of the PTS system, and describe their roles in the translocation and phosphorylation steps, their generality or specificity for the transported sugars, and their regulation of the uptake of carbon sources.

18. Describe the *isomerization* of *retinal* by *light* in relation to the pumping of protons by *bacteriorhodopsin*. (See pp. 307–308 of Stryer for a review of the structure of bacteriorhodopsin.)

19. Summarize the various energy sources that drive active transport processes.

Examples of Passive Transport: The Anion-Exchange Protein, Transport Antibiotics, and Gap-Junctions (Stryer pages 963–971)

20. Describe the structure, physiological function, and proposed mechanism of the *anion-exchange protein* of erythrocytes. (See pp. 304–305 of Stryer for a review of the structure of this protein.)

21. Discuss the sources, general properties, and uses of the *transport antibiotics valinomycin* and *gramicidin A.*

22. Contrast the properties of *carrier* and *channel-forming* transport antibiotics. Note that all membrane transport proteins form transmembrane channels.

23. Describe the structural basis for the binding and translocation of ions by the transport antibiotics.

24. Describe the physiological roles of *gap junctions* between eucaryotic cells, and indicate the classes of compounds that flow through them.

25. Discuss the structure of the gap-junction protein and the closing of the channel by Ca^{2+} and H^+.

SELF-TEST

Introduction

1. Which of the following in *not* a physiological function of membrane transport?

   (a) It regulates cell volume.
   (b) It maintains intracellular pH and ionic composition.
   (c) It concentrates metabolic fuels and extrudes toxic substances.
   (d) It generates reducing power in the form of NADPH.
   (e) It generates ionic gradients that are essential for the excitability of certain cells.

2. An uncharged molecule is transported from side 1 to side 2 of a membrane.

(a) If its concentration is 10^{-3} M on side 1 and 10^{-6} M on side 2, will the transport be an active or a passive process? Explain your answer.

_____

_____

(b) If the concentration is 10^{-1} M on side 1 and 10^{-4} M on side 2, how will the free-energy change compare with that in question (a)? Explain.

_____

_____

(c) How will the rate of transport in (a) and (b) compare? Explain your answer.

_____

_____

The Na$^+$-K$^+$ Pump

3. The orientation of the Na$^+$-K$^+$ pump in cell membranes determines the side of the membrane where the various processes involved in the transport of Na$^+$ and K$^+$ will take place. Assign each of the steps or processes in the right column to the intracellular or extracellular side of the pump.

(a) Intracellular side _____

(b) Extracellular side _____

(1) Binding of vanadate
(2) Binding of cardiotonic steroids
(3) Hydrolysis of ATP and phosphorylation of the pump
(4) Binding of K$^+$
(5) Binding of Na$^+$

4. Explain why an electric current is generated during the transport of Na$^+$ and K$^+$ by the Na$^+$-K$^+$ pump.

_____

_____

5. Which of the following statements about the subunit structure of the Na$^+$-K$^+$ pump are *correct?*

(a) The pump contains α and β subunits, possibly associated as $\alpha_2\beta_2$ tetramers.
(b) The β subunits contain the ATPase activity.
(c) The α subunits contain the binding sites of cardiotonic steroids.
(d) The α subunits have large extracellular domains and contain oligosaccharide chains.
(e) The transmembrane channel is made by the interface between α and β subunits.

6. The proposed model for the mechanism of the Na^+-K^+ pump is based on the existence of four conformational states of this enzyme. Match each conformational state in the left column with the appropriate descriptions from the right column.

(a) E_1 _____

(b) E_1-P _____

(c) E_2 _____

(d) E_2-P _____

(1) High affinity for K^+
(2) High affinity for Na^+
(3) Phosphorylated by ATP
(4) Ion-binding sites face the cytosol
(5) Ion-binding sites face the extra-cellular space
(6) Dephosphorylated upon the binding of K^+
(7) Stabilized by cardiotonic steroids
(8) Stabilized by vanadate

The Ca^{2+}-ATPase and Other ATPases

7. Which of the following statements describe properties that are common to the Ca^{2+}-ATPase of the sarcoplasmic reticulum and the Na^+-K^+ pump?

(a) Both are very abundant membrane proteins in the sarcoplasmic reticulum.
(b) Both have homologous α subunits containing numerous transmembrane helices.
(c) Both contain an aspartate residue that is phosphorylated by ATP.
(d) Both translocate the same number of ions per transport cycle.
(e) Both probably have four major conformational states.

8. Explain how the Ca^{2+}-ATPase can be made to synthesize ATP in vitro.

_____

_____

9. Which of the following statements about phospholamban are *correct?*

(a) It is also called calmodulin.
(b) It binds Ca^{2+} in the lumen of the sarcoplasmic reticulum.
(c) It stimulates the Ca^{2+} pump after becoming phosphorylated.
(d) It inhibits the binding of Ca^{2+} by calsequestrin.
(e) It contributes to the feedback regulation of Ca^{2+} levels in the cytosol.

10. Match the classes of ion-motive ATPases found in eucaryotic cells in the left column with the appropriate properties from the right column.

(a) P type _____

(b) V type _____

(c) F type _____

(1) Found in lysosomes
(2) Found in the inner mitochondrial membranes
(3) Found in cell plasma membranes
(4) Contain a phosphorylated aspartate intermediate
(5) Carry out ATP synthesis driven by a H^+ gradient

(6) Carry out the active transport of ions such as Na^+, K^+, and Ca^{2+}

(7) Acidify the contents of endocytic and exocytic vesicles

(8) Cycle through four conformational states during ion pumping

Transport of Sugars and Amino Acids in Animal Cells and Bacteria and Light Driven Pumps

11. The transport of glucose into many animal cells occurs via which of the following?

  (a) Cotransport with Na^+
  (b) Cotransport with K^+
  (c) An antiport driven by a Na^+ gradient
  (d) An antiport driven by a H^+ gradient
  (e) A symport driven by a Na^+ gradient.

12. Which of the following statements about the transport of amino acids into animal cells via a symport with Na^+ are *correct*?

  (a) The input of energy is not required because the transport of amino acids is always a passive process.
  (b) The symport hydrolyzes ATP to drive the transport.
  (c) The energy for transport is derived from the Na^+ gradient.
  (d) The energy for transport is indirectly derived from the Na^+-K^+ATPase.

13. Which of the following statements about the lactose permease of *E. coli* under physiological conditions are *correct*?

  (a) It derives energy for transport from a Na^+ gradient.
  (b) It derives energy for transport from a H^+ gradient.
  (c) It derives energy for transport from the flow of electrons originating in NADH.
  (d) It is an antiport for lactose and H^+.
  (e) It is a symport for lactose and Na^+.

14. Group translocation systems in bacteria use which of the following molecules containing high-energy phosphate groups?

  (a) ATP
  (b) GTP
  (c) Phosphocreatine
  (d) Phosphoinositol
  (e) Phosphoenolpyruvate

15. Match the protein components of the bacterial phosphotransferase system (PTS) in the left column with their roles in sugar transport from the right column.

  (a) Enzyme I _____
  (b) Enzyme II _____
  (c) Enzyme III _____
  (d) HPr _____

  (1) participates in the transport of many sugars.
  (2) is specific for particular sugars.
  (3) forms the transmembrane channel.
  (4) becomes phosphorylated by phosphoenolpyruvate.

(5) catalyzes the transfer of the phosphoryl group from enzyme III to the sugar.

(6) has a His residue that becomes phosphorylated.

(7) becomes phosphorylated on a Ser residue in response to high sugar levels inside the cell.

(8) is a peripheral membrane protein.

16. Which of the following statements about the retinal prosthetic group of bacteriorhodopsin are *correct?*

(a) It is covalently bound to the protein via a Schiff base.

(b) It absorbs light with a peak at 350 nm.

(c) It becomes isomerized from an all-*trans* to an all-*cis* retinal upon the absorption of light.

(d) It dissociates from the protein upon the absorption of light.

(e) It undergoes a series of isomerizations in the time scale from picoseconds to milliseconds.

17. *Halobacterium halobium* cells synthesize ATP by oxidative phosphorylation in the presence of O_2, but they become photosynthetic in the absence of O_2. Explain this change in metabolism.

_____

_____

Examples of Passive Transport: Anion-Exchange Protein, Transport Antibiotics, and Gap Junctions

18. Which of the following statements about the anion-exchange protein of erythrocytes are *correct?*

(a) It alters the membrane potential.

(b) It exchanges Cl^- for HCO_3^- ions.

(c) It serves as a symport for the transported ions.

(d) It becomes phosphorylated by ATP during ion transport.

(e) It has at least four states during the transport process.

19. Match each of the transport antibiotics in the left column with the appropriate properties from the right column.

(a) Valinomycin _____

(b) Gramicidin A _____

(1) is a channel former.

(2) is an ion carrier.

(3) is a cyclic peptide.

(4) is a linear peptide.

(5) has alternating L and D amino acid residues.

(6) is active in transport as a dimer.

(7) has a transmembrane diffusion constant for ions of around $10^7 \, s^{-1}$.

(8) has a transmembrane diffusion constant for ions of around $10^3 \, s^{-1}$.

20. Explain the consequences of the alternating D and L amino acid residues on the structure of gramicidin A.

_____

_____

21. Why is the binding affinity of valinomycin much higher for K^+ than for Na^+?

_____

_____

22. Which of the following is *not* a characteristic of gap junctions between eucaryotic cells?

(a) They allow the exchange of ions and metabolites between cells.
(b) They allow the exchange of cytoplasmic proteins.
(c) They are essential for the nourishment of cells that are distant from blood vessels.
(d) They are possibly important in development and differentiation.
(e) They are controlled by Ca^{2+} and H^+ concentrations in cells.

23. Explain how relatively small conformational changes in the structure of connexin in response to Ca^{2+} can close a 20-Å channel.

_____

_____

PROBLEMS

1. The interior of most animal cells is electrically negative with respect to the exterior. Explain why the Na^+-K^+ ATPase can contribute to the membrane potential but the H^+-K^+ ATPase (see Stryer, p. 957) cannot.

2. In experiments to investigate the mechanism of transport of two substances, X and Y, across cell membranes, cells were incubated in media containing various concentrations of X and Y, and the initial rate of transport of each of the substances into the cell was determined. The results that were obtained are depicted in Figure 37-1. What conclusion is suggested by the results? Explain. It may be helpful to refer to the discussion of enzyme kinetics beginning on page 187 of Stryer.

3. Calculate the free-energy change for the transport of an uncharged species from a concentration of 5 mM outside a cell to a concentration of 150 mM inside. Assume that the temperature is 25°C.

4. Now repeat the calculation of Problem 3 for an ion with a charge of +1 that is crossing a membrane with a potential of −60 mV, with the interior negative with respect to the exterior.

5. In dog skeletal muscle, the extracellular and intracellular concentrations of Na^+ are 150 and 12 mM, and those of K^+ are 2.7 and 140 mM.

(a) Calculate the free-energy change as three Na^+ are transported out and two K^+ are transported in by the Na^+-K^+ pump. As-

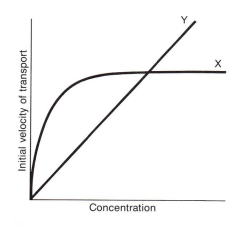

Figure 37-1
Initial velocity of transport versus concentration for substances X and Y.

sume that the temperature is 25°C and that the membrane potential is −60 mV.

(b) Does the hydrolysis of a single ATP provide sufficient energy for the process in part (a)? Explain.

6. In resting frog sartorius muscle, the concentration of K^+ outside the cell is 2.5 mM and that inside is 125 mM. Calculate the membrane potential that would exist in the resting muscle if the membrane were permeable only to K^+.

7. Synthetic lipid bilayers containing two newly-discovered transport antibiotics, X and Y, were prepared. When the ionic conductance across these bilayers was studied as a function of temperature, the results depicted in Figure 37-2 were obtained. Propose a mechanism for each of these transport antibiotics, briefly explaining your answer.

8. The maximum turnover number of the Na^+-K^+ ATPase is 100 s^{-1}. Does this limit the number of ions that can be transported per second? If so, what is that number?

9. Would you expect Li^+ to be transported by valinomycin? Why or why not?

10. You have been making hydropathy plots of two proteins. One is a cytosolic enzyme that you have been purifying, and the other is a transmembrane protein that you suspect is involved in a transport function. Unfortunately, you forgot to label which plot corresponds to which protein. Which of the two proteins depicted in Figure 37-3 is likely to be the transmembrane protein, and how many transmembrane helices does it have? Explain your answer. See page 304 of Stryer for a discussion of hydropathy plots.

11. Design an experiment using ATP labeled with ^{32}P in the γ-position that would suggest that the Na^+-K^+ ATPase reaction involves a stable enzyme-phosphate intermediate. (See Stryer, p. 952.)

12. Suggest an experiment involving a membrane preparation, ATP labeled in the β-position with ^{32}P, and unlabeled ADP that would lend additional support to the finding that the Na^+-K^+ ATPase reaction is a double-displacement. (Hint: Your experiment should involve the reisolation of ADP from the reaction mixture.)

13. Predict the effect of digitoxigenin on the force of contraction of an isolated muscle fiber when it is (a) added to a medium bathing the muscle fiber or (b) injected directly into the muscle fiber. Explain your answers.

14. In the three following experiments, erythrocyte ghosts are prepared and are then placed into medium A. After suitable equilibration, the erythrocytes are transferred to medium B. Predict the extent of hydrolysis of internal ATP in each of the experiments, explaining your answer briefly.

Experiment	Medium A	Medium B
1	ATP, high K^+	High Na^+
2	ATP, high Na^+	High K^+
3	ATP, only Na^+	Only Na^+

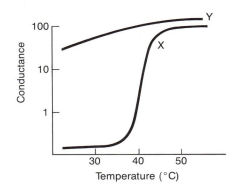

Figure 37-2
Conductance studies of two transport antibiotics.

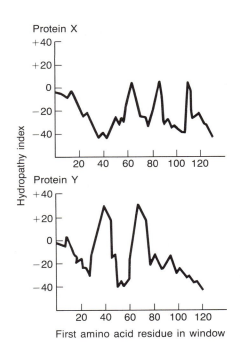

Figure 37-3
Hydropathy plots for proteins X and Y.

15. When erythrocyte ghosts are equilibrated in a medium containing ADP, phosphate, and K^+ and are then transferred to a medium containing high Na^+, ATP is formed within the erythrocytes. What is the significance of this observation?

ANSWERS TO SELF-TEST

Introduction

1. d

2. (a) The transport will be a passive process. Because the concentration on side 1 is higher than that on side 2, the molecule will move spontaneously down its concentration gradient. One can use the expression

$$\Delta G = 2.303\ RT \log \frac{c_2}{c_1}$$

$$= 2.303\ RT \log \frac{10^{-6}}{10^{-3}}$$

to show that ΔG has a negative value. Assuming a temperature of 25°C

$$\Delta G = 2.303 \times 1.98\ \frac{\text{cal}}{\text{mol} \cdot \text{K}} \times 298\ \text{K} \times (-3)$$

$$= -4.07\ \frac{\text{kcal}}{\text{mol}}$$

A negative ΔG value in the direction of movement is the hallmark of passive transport.

(b) Since the ratios of the concentrations in (a) and (b) are equal, then the free-energy change for the transport is the same in both.

$$\Delta G = 2.303 \times 1.98\ \frac{\text{cal}}{\text{mol} \cdot \text{K}} \times 298\ \text{K} \times \log \frac{10^{-4}}{10^{-1}}$$

$$= -4.07\ \frac{\text{kcal}}{\text{mol}}$$

See Figure 37-2 on page 950 of Stryer for a graphical representation of this relationship.

(c) The rate of a chemical process is always equal to a rate constant multiplied by the concentration of the chemical species undergoing the change (see Stryer, p. 188). Thus, the rate of transport will be equal to $k(c_1)$. Since the concentration of c_1 is greater in (b) than in (a), the rate of transport will be greater in (b).

The Na^+-K^+ Pump

3. (a) 1, 3, 5 (b) 2, 4

4. Since three Na^+ ions are transported out for every two K^+ ions that are transported in, there is a net efflux of one positively charged ion. The net movement of ions sets up an electric current.

5. a, c

6. (a) 2, 3, 4, 8 (b) 2, 4 (c) 1, 5 (d) 1, 5, 6, 7

The Ca^{2+} ATPase and Other ATPases

7. b, c, e

8. Ca^{2+}-ATPase can be reconstituted into phospholipid vesicles so that there is a high concentration of Ca^{2+} inside the vesicles and the phosphorylation site of the Ca^{2+}-ATPase is exposed to the outside of the vesicles. If P_i and ADP are added to the vesicle suspension, the passive transport of Ca^{2+} from the inside to the outside of the vesicles will cause the phosphorylation of the Ca^{2+}-ATPase and the synthesis of ATP.

9. c, e

10. (a) 3, 4, 6, 8 (b) 1, 7 (c) 2, 5

Transport of Sugars and Amino Acids in Animal Cells and Bacteria and Light Driven Pumps

11. a, e

12. c, d

13. b, c

14. e

15. (a) 1, 4 (b) 2, 3, 5 (c) 2, 8 (d) 1, 6, 7

16. a, e

17. In the presence of O_2 and reduced cofactors, the respiratory chain of *Halobacterium halobium* generates the proton-motive force that drives the synthesis of ATP by ATP synthase. Under anaerobic conditions, electron transfer and proton pumping are inhibited, and ATP synthesis also stops. Under these conditions, and in the presence of light, the bacteriorhodopsin can pump protons, thus restoring the proton gradient required for ATP synthesis. For a review of oxidative phosphorylation see Chapter 17 of Stryer.

Examples of Passive Transport: Anion-Exchange Protein, Transport Antibiotics, and Gap Junctions

18. b, e

19. (a) 2, 3, 8 (b) 1, 4, 5, 6, 7

20. The alternating sequence of D and L amino acid residues in gramicidin A allows the formation of a helical channel with all of the nonpolar side chains facing the hydrophobic interior of the membrane. The polar peptide groups face the interior of the channel and transiently bind the ions as they move through it.

21. The higher affinity of valinomycin for K^+ is due to the lower stability of the hydrated form of K^+ than that of Na^+. In other words, it is energetically more costly to displace the water surrounding Na^+ than that surrounding K^+.

22. b

23. After binding Ca^{2+}, the connexin subunits in each hemichannel rotate and slide relative to each other to adopt a more perpendicular orientation relative to the plane of the membrane. This effectively closes the channel. Because six subunits participate cooperatively in this process, small displacements in each subunit lead to the closure of the large channel.

ANSWERS TO PROBLEMS

1. The difference between the two ATPase systems lies in the stoichiometry of their exchange of ions. In the case of the Na^+-K^+ ATPase, three Na^+ ions are extruded and two K^+ ions are taken up during each pump cycle, making the interior of the cell more negative (less positive) for each pump cycle. The H^+-K^+ ATPase, in contrast extrudes one H^+ ion for each K^+ taken up, so its operation is electrically neutral.

2. The curve for X shows saturation, which would be expected if some membrane carrier is involved in the transport of substance X. The curve for Y shows no saturation, which is consistent with the notion that substance Y diffuses through the membrane without a carrier. Such behavior is shown by lipid soluble substances, which dissolve in the hydrophobic tails of membrane phospholipids and can thus enter cells without a carrier.

3. We use the equation at the top of page 950 of Stryer.

$$\Delta G = RT \log_e \frac{c_2}{c_1}$$

$$= 1.99 \times 298 \times \log_e \frac{150}{5}$$

$$= +2.0 \text{ kcal/mol}$$

4. The relevant equation is given on page 950 of Stryer:

$$\Delta G = RT \log_e \frac{c_2}{c_1} + ZF \Delta V$$

$$= 1.99 \times 298 \times \log_e \frac{150}{5} + (+1) \times 23.062 \times -0.060$$

$$= +2.0 - 1.4$$

$$= +0.6 \text{ kcal/mol}$$

Note that the membrane potential favors the entry of a positively charged ion.

5. (a) To solve this problem, we first calculate the free-energy change for transporting three Na^+ and then that for two K^+. The total free-energy change will be the sum of the two.

Again we use the following equation from page 950 of Stryer:

$$\Delta G = RT \log_e \frac{c_2}{c_1} + ZF \Delta V$$

Substituting in the values for Na$^+$ yields

$$\Delta G_{Na^+} = 1.99 \times 298 \times \log_e \frac{150}{12} + (+1) \times 23.062 \times 0.060$$

$$= +1.5 + 1.4$$

$$= +2.9 \text{ kcal/mol}$$

Note that when Na$^+$ is transported out of the cell, work must be done against both a concentration gradient and an electrical gradient.

Now we carry out the corresponding calculation for the K$^+$ ion.

$$\Delta G_{K^+} = 1.99 \times 298 \times \log_e \frac{140}{2.7} + (+1) \times 23.062 \times -0.060$$

$$= +2.3 - 1.4$$

$$= +0.9 \text{ kcal/mol}$$

Note that potassium ion is being transported against a concentration gradient but with an electrical gradient. Accordingly, the sign for the electrical term in the equation is negative.

To get the total energy expenditure, we must account for the stoichiometry of transport by summing the energy required for the transport of Na$^+$ and that required for the transport of K$^+$.

$$\Delta G = 3 \Delta G_{Na^+} + 2 \Delta G_{K^+}$$

$$= (3 \times 2.9) + (2 \times 0.9)$$

$$= 10.5 \text{ kcal/mol}$$

(b) Although the free-energy change for the hydrolysis of ATP under standard conditions is -7.3 kcal/mol, the free-energy change for the concentrations that exist in typical cells is approximately -12 kcal/mol. Thus, the energy furnished by the hydrolysis of a single ATP is sufficient.

6. If the membrane is permeable only to K$^+$, then K$^+$ would tend to diffuse out along its concentration gradient until its tendency to exit is just balanced by the tendency of the negative membrane potential to pull it back in, which means that an electrochemical equilibrium is established. We start with the equation on page 950 of Stryer:

$$\Delta G = RT \log_e \frac{c_2}{c_1} + ZF \Delta V$$

At equilibrium, $\Delta G = 0$, so

$$ZF \Delta V = -RT \log_e \frac{c_2}{c_1}$$

$$\Delta V = \frac{-RT \log_e c_2/c_1}{ZF}$$

$$= \frac{-1.99 \times 298 \times \log_e (125/2.5)}{23.062}$$

$$= -100 \text{ mV}$$

7. Antibiotic X promotes ion transport via a carrier mechanism and Y is a channel former. A carrier diffuses through the membrane, whereas a channel former does not. Antibiotic X shows a large, abrupt transition as the temperature increases, which is consistent with the idea that a carrier is involved that cannot traverse a "frozen" membrane. Channel formers do not show such abrupt transitions with increasing temperature because they do not diffuse through the membrane, so antibiotic Y must be a channel former.

8. The turnover number of an enzyme is the number of molecules of substrate that are converted to product per second per molecule of enzyme (see Stryer, p. 191). The stoichiometry of the Na^+-K^+ ATPase is that three Na^+ exit and two K^+ enter for each molecule of ATP that is cleaved. (The stoichiometry of the Na^+-K^+ ATPase is discussed on p. 952 of Stryer.) Thus, the maximum number of Na^+ that can be transported out per second is 300 per molecule of ATPase present. The maximum number of potassium ions that may be transported in is 200 per molecule of ATPase.

9. Li^+ should not be transported by valinomycin. Na^+ is bound to valinomycin a thousand time less tightly than is K^+. Since Li^+ has a smaller ionic radius than Na^+, the binding of it to valinomycin should be negligible.

10. Protein Y is most likely to be the transmembrane protein, and it could contain two transmembrane helical regions. A run of 20 hydrophobic amino acids is required to traverse the membrane core, and protein Y shows two such regions. Protein X has no 20-residue runs of hydrophobic amino acids, as evidenced by the lack of any peaks above +20 in the hydropathy plots.

11. The Na^+-K^+ ATPase reaction is a two-step process involving two independent half-reactions:

$$E + ATP \xrightleftharpoons{Na^+, Mg^{2+}} E\text{-}P + ADP \qquad (1)$$

$$E\text{-}P + H_2O \xrightleftharpoons{K^+} E + P_i \qquad (2)$$

In the first half-reaction, ATP and unmodified enzyme interact in the presence of Na^+ and Mg^{2+} to give phosphorylated enzyme and the first product of the overall reaction, ADP. In the second half-reaction, the phosphorylated enzyme is hydrolyzed by water in the presence of K^+ to give unmodified enzyme and inorganic phosphate. An overall reaction that consists of two independent half-reactions is known as a *double displacement*.

The following experiment would suggest that such an overall reaction occurs. Incubate a fragmented membrane preparation with K^+ ion to hydrolyze any phosphate that might be bound to the enzyme. Then wash the membrane preparation to remove all K^+, and transfer the membrane to a medium containing γ-labeled ATP, Na^+, and Mg^{2+}. After a suitable incubation period, wash the membrane preparation to remove any unreacted labeled ATP. Then carry out scintillation counting on the membrane preparation to detect the presence of labeled phosphate. The presence of radioactivity in the membrane fraction would suggest that a stable enzyme-phosphate intermediate had been formed. In fact, a covalent aspartyl phosphate derivative is formed at the active site of the ATPase. (See Figure 37-15 on p. 957 of Stryer.)

12. You could incubate the β-labeled ATP and the unlabeled ADP with the membrane preparation in the presence of Na^+ and Mg^{2+} but in the absence of K^+. After a suitable period, reisolate the ADP from the reaction mixture and assay it for radioactivity. If an E-P intermediate has been produced in this half-reaction (as it is), the ADP should be radioactive. A characteristic of double displacement reactions is that the half-reactions will occur separately.

13. (a) Extracellularly, digitoxigenin acts to inhibit the K^+-dependent dephosphorylation of the E-P intermediate of the Na^+-K^+ ATPase (step 2 in the solution to Problem 11). Therefore, it increases the intracellular concentration of Na^+, which diminishes the gradient that drives the extrusion of Ca^{2+} from the intracellular fluid. Thus, the intracellular concentration of Ca^{2+} increases as well, which increases the contractile force of muscle.

    (b) Direct intracellular injection of the digitoxigenin should have little or no effect on the force of contraction because the K^+ hydrolysis of the E-P intermediate takes place on the exterior surface of the plasma membrane.

14. Experiment 1 mimics the conditions that exist in a typical erythrocyte in the blood. There is relatively little hydrolysis of ATP because the ionic gradient is already established. In experiment 2, there will be significant hydrolysis of ATP within the erythrocyte because Na^+ and K^+ are actively pumped to establish a gradient that is the reverse of that originally imposed. In experiment 3, only Na^+ is present. Thus, there would be little hydrolysis of ATP because the hydrolysis of the E-P intermediate on the exterior surface of the membrane requires K^+.

15. The Na^+-K^+ pump is apparently reversible. The ionic gradient established across the membrane provides sufficient free energy to drive the synthesis of ATP, much as an ionic gradient across the inner membrane of mitochondria drives the synthesis of ATP in oxidative phosphorylation.

Hormone Action

Multicellular eucaryotes have evolved mechanisms to allow communication between distantly located cells so that their characteristic activities may be optimally coordinated to serve the needs of the whole organism. Two types of signals for the integration of the biochemical activities of separated tissues are electrical impulses, which are distributed via the nervous system, and chemicals, which are transported by the circulatory system. In Chapter 38 Stryer considers hormones, the chemical messengers, and focuses on the strategies by which they specifically alter the activities of cells in target tissues.

Stryer points out that many hormones initiate their physiological actions by binding specifically to integral membrane protein receptors. These hormones remain outside the cell, and their binding to the receptor activates it to initiate a series of intracellular events that lead to the modification of the activities of the targeted proteins. The history of the discovery of cyclic AMP (cAMP) serves as an introduction to the details of one general pathway common to many hormones. The hormone-initiated activation of adenylate cyclase, an integral membrane enzyme that forms cAMP, shows how an external signal supplied by a hormone can lead to a change in the level of an intracellular allosteric effector.

Cyclic AMP serves as an intracellular second messenger, the hormone itself being regarded as the first messenger, that delivers the stimulus from the hormone to the targeted enzymes. Stryer explains that other proteins, the guanyl-nucleotide-binding proteins (G proteins), are required to couple the signal from the specific hormone-receptor complex to the enzyme adenylate cyclase. The G proteins are a family of peripheral membrane proteins that are activated through binding to the cytosolic face of the activated hormone receptor. An activated stimulatory G-protein subunit bearing a GTP molecule diffuses to and interacts with adenylate cyclase to stimulate the production of cAMP. The stimulatory G-protein subunit has an intrinsic GTPase activity that hydrolyzes the bound GTP to GDP, thereby inactivating the subunit and stopping the production of cAMP. Some G proteins are inhibitory rather than stimulatory and lead to the sequestration of the stimulatory G-protein subunit so that it cannot promote the formation of cAMP. Thus, the G proteins either increase or decrease cAMP levels, depending on the identity of the triggering hormone. Stryer describes the structure of the seven-helix transmembrane β-adrenergic receptor and indicates how it might transmit a signal arising from a hormone that binds on the extracellular surface of the cell to the intracellular side of the plasma membrane. The description of the information pathway from hormone stimulus to G proteins to adenylate cyclase to the targeted protein activities is completed by a discussion of how cAMP activates specific protein kinases. The cAMP binds to the regulatory subunits of the inactive protein kinase tetramer, which leads to the dissociation of the active catalytic subunits. The catalytic subunits then phosphorylate serine and threonine residues in the target enzymes to activate or inactivate them. A small number of hormone molecules results in an amplified response because each activated enzyme in the cascade forms numerous products.

Stryer next describes an analogous hormone-stimulated system—the phosphoionositide cascade. In this system, the hormone activates, probably by means of G proteins, a specific phospholipase (phospolipase C) that cleaves a plasma membrane phospholipid, phosphatidyl inositol 4,5-bisphosphate (PIP_2), to form two second messengers. The inositol phosphate derivative, inositol 1,4,5-trisphosphate (IP_3), which is short-lived, triggers the opening of ion channels so that the Ca^{+2} concentration in the cytosol is increased. The remainder of the PIP_2 molecule, diacylglycerol (DAG), is another second messenger that activates protein kinase C. The increased Ca^{+2} levels and the activated protein kinase C affect a variety of biochemical reactions. Stryer describes the structure of Ca^{+2}-binding proteins and explains how the binding of the ion is highly specific and leads to a large conformational change in the protein, characteristics that are all desirable qualities for molecules serving as Ca^{+2} sensors and signal transducers.

The eicosanoid hormones that are formed from the polyunsaturated fatty acid arachidonate are described next. These hormones are short-lived. Because they are destroyed so rapidly that they cannot be widely distributed beyond the sites of their synthesis, they exert their effects locally rather than globally. The endorphins are peptide hormones that are analgesics, They are presented to illustrate how a single polypeptide, in this case pro-opiocortin, is cleaved proteolytically into a group of products having diverse biological activities. Some of the products share stretches of amino acids from the same region of the precursor. For example, β-lipotropin and β-endorphin share a region of identical sequence, but they have quite different physiological effects.

The pathway for the formation of insulin from its precursors is presented next. Stryer describes the mechanism of insulin action and the tyrosine kinase activity of the activated insulin receptor. By analogy to the effects of the hormonally activated protein kinase C, this activity suggests a mechanism by which insulin might elicit the phosphorylation of targeted proteins and thus affect their activities. The nerve-growth-factor receptor, the epidermal-growth-factor receptor, and the products of some oncogenes are also tyrosine kinases. The critical roles of these proteins in growth processes further indicate the importance of the tyrosine kinases in cell regulation.

Stryer closes the chapter with a description of the steroid and thyroid hormones. In contrast to the hormones that remain outside the cell, these hormones must enter the cell to exert their effects. The steroid and thyroid hormones interact with cytosolic receptors and travel to the nucleus, where they act as transcriptional enhancers. They modify cellular processes by stimulating the synthesis of proteins rather than by modifying the activities of preexisting proteins. Analyses of the structures of the receptors of these hormones and of analogous oncogene products reveal that they have zinc-finger domains and can bind to specific DNA sequences.

LEARNING OBJECTIVES

When you have mastered this chapter, you should be able to complete the following objectives.

Introduction (Stryer pages 974–975)

1. Provide an overview of the functional and structural classes of *hormones*.

Cyclic AMP—Its Formation and Its Role as a Second Messenger (Stryer pages 975–978)

2. Recount the history of the discovery of *adenosine 3',5'-monophosphate* (*cyclic AMP* or *cAMP*), noting the observations of enzyme regulation by *covalent modification* and hormone action in a cell-free extract.

3. Write the reactions catalyzed by *adenylate cyclase* and *cAMP phosphodiesterase*, and note the standard free-energy changes for each reaction.

4. List the salient features of the *second-messenger concept*. Appreciate that a hormone need not enter a cell to affect intracellular reactions.

5. Cite the evidence that implicates cAMP as a second messenger.

6. Appreciate the multiplicity of hormones that use cAMP as their intracellular effector and the diversities and similarities of the physiological responses elicited.

Guanyl-Nucleotide-Binding Proteins and Adenylate Cyclase (Stryer pages 978–985)

7. Describe the roles of the *guanyl-nucleotide-binding proteins* (*G proteins*) in coupling a *hormone-receptor complex* to adenylate cyclase and in amplifying the stimulus.

8. Locate G-proteins in the cell, and describe their catalytic characteristics and molecular mechanisms of activation and inactivation.

9. Describe the structure of the *β-adrenergic receptor,* which binds *epinephrine,* and explain the phenomenon of *desensitization* or *adaptation* in hormone function.

10. Explain the ability of cAMP to affect diverse cellular functions. Describe the mechanism by which cAMP modulates *protein kinases* in eucaryotes.

11. List three steps in the *cAMP cascade* that contribute to the amplification of the hormonal stimulus, and explain how the amplified response is achieved.

12. Describe the molecular basis of the toxicity of *cholera toxin.* Note that the v-*ras* oncogene product, which is the viral form of a cellular regulatory protein, has diminished GTPase activity that keeps it permanently activated as a *stimulatory G protein.*

13. Marshal the evidence that cAMP might be an *ancient molecular hunger signal* for cells, and list the properties of cAMP that suit it to this task.

14. Describe the molecular basis of the toxicity of *pertussis toxin.* Contrast the mechanisms of the *inhibitory G proteins* and stimulatory G proteins.

15. Appreciate that there are families of G proteins that enable hormones to effect a variety of physiological functions.

16. List the common structural and functional properties of the G proteins.

The Phosphoinositide Cascade and the Modulation of Ca²⁺ Concentrations
(Stryer pages 985–991)

17. Draw the structure of *phosphatidyl inositol 4,5-bisphosphate (PIP₂).* Write the reaction catalyzed by *phospholipase C* to produce the second messengers *inositol 1,4,5-trisphosphate (IP₃)* and *diacylglycerol (DAG).* Note that G proteins may be involved in the regulation of phospholipase C (see p. 984, of Stryer).

18. Outline the *phosphoinositide cascade,* and note the diversity of the elicited physiological responses.

19. Describe the biochemical fates of the second messengers produced from PIP_2. Note that the phosphoinositide cascade often produces *arachidonate.*

20. Describe the effects of IP₃ on *Ca²⁺ sequestration* by the endoplasmic reticulum and the sarcoplasmic reticulum, and list some biochemical processes affected by intracellular Ca^{2+} concentration.

21. Describe the effects of DAG on *protein kinase C* and the mechanistic and functional relationships between IP₃ and DAG. Describe the biochemical and physiological effects of *phorbol esters.*

22. Describe the *EF-hand structural motif* of *calcium-binding proteins,* and explain how it binds Ca^{2+}.

23. Describe the structure of *calmodulin* and its biochemical function. Note the value of calcium ionophores, calcium buffers, and fluorescent indicators in studying the functions of Ca^{2+} in cells.

24. Indicate the relationships between the *eicosanoid* arachidonate and the *prostaglandins, thromboxanes,* and *leukotrienes.*

25. Explain why the eicosanoid hormones are local in their actions. Describe the effects of prostaglandin PGE_1 on adipose tissue. Note the diversity of the physiological effects of the eicosanoid hormones.

26. Explain the biochemical basis for the *anti-inflammatory activity of aspirin.*

27. Describe the *endorphins* and their physiological functions. Describe the processing of *pro-opiocortin,* and appreciate this mechanism of generating several polypeptide hormones from a single polypeptide precursor.

Insulin and the Hormones that Modulate the Tyrosine Kinases (Stryer pages 994–1000)

28. List the physiological effects of *insulin* on both the cells and the organs of humans. Describe the polypeptide composition of insulin. Outline the synthesis of *preproinsulin* and *proinsulin,* and describe the export of insulin from the islet cells of the pancreas.

29. Describe the structure of the *insulin receptor,* and list the reactions it catalyzes when activated by insulin binding. Note the consequences on the *tyrosine kinase* activity of the receptor when it is *autophosphorylated* or phosphorylated on serine or threonine residues by protein kinase A or protein kinase C.

30. Discuss the major unknown features in the mechanism of insulin action.

31. Compare the mechanisms of action of *epidermal growth factor* and insulin. Appreciate that several oncogene-encoded proteins have tyrosine kinase activity.

The Steroid and Thyroid Hormones (Stryer pages 1000–1001)

32. Provide an overview of the mechanisms of action of the *steroid hormones* and the *thyroid hormones.* Note their primary intracellular site of action.

33. Note that both thyroid and steroid hormone-receptor complexes are DNA-binding proteins and act as *transcriptional enhancers.* Appreciate that some oncogene-encoded proteins are similar in structure and function to the thyroid receptor.

SELF-TEST

Introduction

1. Which of the following statements about hormones in mammals are *correct?*

    (a) Hormones are enzymes.
    (b) Hormones are synthesized in specific tissues.
    (c) Hormones are secreted into the blood.

(d) Hormones act as second messengers to conver the electrical signals of the nervous system into a chemical form.

(e) Hormones alter one or more activities in the cells to which they are targeted.

(f) Hormones display specificity toward the tissues with which they interact.

(g) Hormones are involved in biochemical amplification systems.

2. Which of the following are direct precursors of hormones?

(a) Arachidonate
(b) cAMP
(c) Cholesterol
(d) Amino acids
(e) Glucose

3. Match the hormonal precursors in the column on the left with the appropriate products from the column on the right.

(a) Amino acid _____   (1) Glucagon
                            (2) Cortisol
(b) Cholesterol _____   (3) Epinephrine
                            (4) Prostaglandin
(c) Arachidonic acid _____

Cyclic AMP—Its Formation and Its Role as a Second Messenger

4. Which the following statements about adenosine 3',5'-monophosphate (cyclic AMP or cAMP) are *correct*?

(a) ATP is converted to cAMP by the enzyme adenylate cyclase in one step.

(b) Cyclic AMP contains a phosphorous atom in a phosphodiester bond.

(c) If cAMP were incubated in the presence of a phosphomonoesterase, it would be hydrolyzed to P_i and the ribonucleoside adenosine.

(d) ATP reacts with adenosine to form cAMP and ADP in a reaction catalyzed by adenylate kinase.

(e) Cyclic AMP is converted to 5'-AMP by a phosphodiesterase-catalyzed reaction with H_2O.

5. Which of the following statements about cAMP and the second messenger mechanism of hormone function are *correct*?

(a) The hormonal stimulus leads to increased amounts of adenylate cyclase.

(b) The formation of a hormone-receptor complex leads to the activation of adenylate cyclase.

(c) Cyclic AMP acts as an allosteric modulator to affect the activities of specific protein kinases.

(d) Cyclic AMP interacts with a hormone-receptor complex to dissociate the hormone.

(e) The hormone-receptor complex enters the cell and affects the activities of target enzymes.

6. Why do thyroid cells respond to thyroid stimulatory hormone (TSH) whereas muscle cells do not?

7. Would you expect drinking coffee to have an affect on blood glucose levels? Explain.

_____

_____

_____

8. Suppose a patient is suffering from a disorder in which adenylate cyclase is deficient and, as a result, cAMP levels aren't readily increased by hormones. Explain why the infusion of cAMP probably won't remedy the problem.

_____

_____

_____

_____

Guanyl-Nucleotide-Binding Proteins and Adenylate Cyclase

9. Which of the following statements about GTP and its role in the cAMP-mediated hormone response system are correct?

(a) GTP is associated with the α-subunit of a guanyl-nucleotide-binding protein (G protein).

(b) GTP reduces the magnitude of the hormone response because it is converted to cyclic GMP (cGMP)—a compound that antagonizes the effects of cAMP.

(c) GTP maintains the steady-state level of cAMP by rephosphorylating AMP to ATP in a nucleotide-kinase catalyzed reaction.

(d) GTP activates G protein so that it interacts with adenylate cyclase.

(e) GTP couples the stimulus from a hormone-receptor complex or an activated receptor to a system that produces an allosteric effector.

(f) The effect of GTP on hormone response is antagonized by the GTPase activity of a subunit of the G protein.

(g) One GTP binding event with a stimulatory G protein leads to one cAMP molecule being formed.

10. Which of the following are _correct_ statements about stimulatory G proteins and their functioning in cAMP-mediated hormonal systems?

(a) G proteins bind hormones.

(b) G proteins are integral membrane proteins.

(c) G proteins are heterotrimers.

(d) G proteins bind adenylate cyclase.

(e) In their GDP form and in the absence of hormone, G proteins bind to hormone receptors and are converted to their GTP forms.

(f) When G protein in the GDP form binds to a hormone-receptor complex, GTP exchanges with GDP.

(g) The α-subunit of G proteins is a GTPase.

11. If cells with β-adrenergic receptors are exposed for extended times to epinephrine, a hormone that activates adenylate cyclase, the G pro-

tein fails to carry out the GDP-GTP exchange reaction and adenylate cyclase is no longer activated. What is this phenomenon called, what is its biological function, and how does it occur?

_____

_____

_____

12. Which of the following statements about cAMP and its functioning in hormone action are *correct?*

    (a) The effects of cAMP in eucaryotic cells are exerted through the activation of protein kinases.

    (b) Cyclic AMP binds the catalytic subunits of its target enzymes and activates them allosterically.

    (c) Cyclic AMP binds the regulatory subunits of its target enzymes and activates them be releasing the catalytic subunits.

    (d) Cyclic AMP must rise to micromolar levels before it activates its target enzymes.

13. During cAMP-mediated hormone activation, what are the three steps at which amplification occurs?

_____

_____

14. What is the molecular basis for the massive diarrhea that occurs in a human infected with *Vibrio cholerae?*

_____

_____

_____

15. A given cell type that uses the G-protein system to transduce hormonal stimuli to adenylate cyclase can be stimulated to produce increased concentrations of cAMP by one hormone or it can be repressed to lower cAMP concentrations by another hormone. How can these diametrically opposed outcomes be effected?

_____

_____

_____

16. Which of the following statements about G proteins are *correct?*

    (a) G proteins carry signals from hormone receptors to adenylate cyclase.

    (b) G proteins comprise a large family of proteins that are involved in regulating enzymes, chemotaxis, visual excitation, and ion channels.

    (c) G proteins cycle between a GTP form and an ATP form by means of a self-catalyzed exchange reaction.

    (d) The hormone-receptor complex leads to the exchange of GTP for GDP on the G protein and to the release of the GTP-G_α subunit.

(e) Many G proteins can participate in a reaction with NAD^+ that leads to their covalent modification.

(f) Mammalian G proteins show sequence similarities to the bacterial translation factor EF-Tu.

The Phosphoinositide Cascade and the Modulation of Ca^{2+} Concentrations

17. Which of the following statements about the phosphoinositide cascade are *correct?*

(a) The phosphoinositide cascade depends upon the hydrolysis of a phospholipid component of the plasma membrane.

(b) A polypeptide hormone interacts with a G_{M1} ganglioside on the cell surface to trigger the phosphoinositide cascade.

(c) The G-protein system probably acts to transduce the stimulus from the receptor to the phosphoinositidase.

(d) Phospholipase C plays a crucial role in the phosphoinositide cascade.

(e) The phosphoinositide cascade produces two different second messengers.

18. Which of the following are the second messengers that are produced by the phosphoinositide cascade?

(a) Phosphatidyl inositol 4,5-bisphosphate (PIP_2)
(b) Inositol 1,4,5-trisphosphate (IP_3)
(c) Inositol 4-phosphate
(d) Inositol 1,3,4,5-tetrakisphosphate
(e) Inositol 1,3,4-trisphosphate
(f) Diacylglycerol (DAG)

19. Which of the following statements about inositol 1,4,5-trisphosphate (IP_3) are *correct?*

(a) IP_3 leads to the uptake of Ca^{2+} by the endoplasmic reticulum and the sarcoplasmic reticulum.

(b) IP_3 is rapidly inactivated by phosphatases or kinases.

(c) IP_3 opens calcium ion channels in the membranes of the endoplasmic reticulum and the sarcoplasmic reticulum.

(d) IP_3 reacts with CTP to form CDP-inositol phosphate, a precursor of PIP_2.

20. Which of the following statements about the actions or targets of the second messengers of the phosphoinositide cascade are *correct?*

(a) Diacylglycerol (DAG) activates protein kinase C.

(b) Most of the effects of IP_3 and DAG are antagonistic.

(c) Protein kinase C shuttles between an inactive soluble form and an active membrane-bound form.

(d) DAG increases the affinity of protein kinase C for Ca^{2+}.

(e) Protein kinase C requires Ca^{2+} for its activity.

(f) The phorbol esters activate protein kinase C because they mimic the action of DAG.

21. Which of the following statements about Ca^{2+} and its roles in the regulation of cellular metabolism are *correct?*

(a) The solubility product of calcium phosphate is small; therefore, low Ca^{2+} levels must be maintained in the cell to avoid precipitation.

(b) Intracellular Ca^{2+} is maintained at concentrations that are several orders of magnitude less than the extracellular concentration by ATP-dependant Ca^{2+} pumps.

(c) The transient opening of ion channels in the plasma membrane or endoplasmic reticulum can rapidly raise cytosolic Ca^{2+} levels.

(d) Ca^{2+} pumps maintain low levels of Ca^{2+} in the sarcoplasmic reticulum relative to the levels in the cytosol.

(e) The binding of Ca^{2+} by a protein can induce a large conformational change because the ion simultaneously coordinates to several groups on the protein.

(f) Ca^{2+} is bound by a family of regulatory proteins that have a characteristic EF-hand, helix-loop-helix structure.

(g) When calmodulin binds Ca^{2+} at its low affinity site, it undergoes a conformational change that allows it to interact with target proteins.

22. Explain how a Ca^{2+} ionophore could mimic the effects of some hormones.

_____

_____

_____

23. If it were incubated with cells in vitro, why would EGTA prevent either a Ca^{2+}-triggering hormone or a Ca^{2+} ionophore from acting?

_____

_____

The Eicosanoid Hormones and the Endorphins

24. Which of the following statements about eicosanoid hormones are *correct?*

(a) The three major classes of eicosanoid hormones are the prostaglandins, the leukotrienes, and the thromboxanes.

(b) The eicosanoid hormones are derived from arachidonic acid.

(c) The eicosanoid hormones are very potent and exert global effects because they are widely distributed by the circulatory system.

(d) The prostaglandins have a variety of diverse physiological effects.

(e) A prostaglandin precursor is derived from phospholipids.

(f) Aspirin inhibits the synthesis of prostaglandin by acetylating prostaglandin synthetase and inhibiting its cyclooxygenase activity.

25. Which of the following statements about endorphins and their actions are *correct?*

(a) The nucleotide sequence that encodes β-endorphin is directly repeated and encodes an identical sequence in the β-lipotropin gene.

(b) Morphine probably is an analgesic because it mimics the actions of an endorphin.

(c) Methionine enkephalin, leucine enkephalin, and β-endorphin share common amino-terminal tetrapeptide sequences.

(d) Endorphins are polypeptides that are derived from longer precursors.

(e) Endorphins share amino acid sequences with some polypeptide hormones that are elaborated by the brain.

(f) Pro-opiocortin is a prohormone that has several pairs of adjacent basic amino acids that serve as sites for proteolysis.

Insulin and the Hormones that Modulate the Tyrosine Kinases

26. Which of the following statements about insulin, its formation, or its hormone activities are *correct*?

(a) Many of the physiological effects of insulin augment those of glucagon and epinephrine.

(b) Two genes encode insulin—one specifies the A chain and the other, the B chain of the protein.

(c) The C-peptide of insulin is sometimes present in a three-polypeptide chain form of the hormone.

(d) Insulin binds with high affinity to receptors in the plasma membrane of its target cells.

(e) By binding, insulin activates its receptor such that it becomes an active tyrosine kinase.

(f) The activated insulin receptor phosphoryates some of its own tyrosine side chains as well as some of those of its target enzymes.

27. Which of the following statements about the tyrosine kinases or hormones that affect them are correct?

(a) Nerve growth factor (NGF) and epidermal growth factor (EGF) stimulate target cells to divide.

(b) NGF and EGF are protein kinases that phosphorylate tyrosine residues.

(c) EGF and insulin share a common mechanism of signal transduction across the plasma membrane.

(d) Receptors for EGF and insulin are integral membrane proteins.

(e) Some oncogenes encode tyrosine kinases.

(f) One oncogene, v-*erb*-B, is a truncated form of an EGF receptor that behaves as though it were an activated EGF receptor–hormone complex.

The Steroid and Thyroid Hormones

28. Which of the following statements about the steroid hormones and their actions are *correct*?

(a) Steroid hormones modulate the activities of their target enzymes by binding to them and acting as allosteric modulators.

(b) Steroid hormones interact with receptors in the plasma membrane of their target cells.

(c) The primary site of action of the steroid hormones is the cell nucleus.

(d) New proteins must be synthesized for steroid hormones to exert their characteristic physiological functions.

(e) Steroid hormone receptors are zinc-finger proteins that bind specifically to DNA when they are complexed with the hormone.

(f) Steroid hormone-receptor complexes are transcriptional enhancers.

29. Match each compound in the left column with its characteristic from the right column.

(a) Thyroxine _____

(b) Glucagon _____

(c) Insulin _____

(d) Cyclic AMP _____

(e) GTP _____

(f) G proteins _____

(g) Adenylate cyclase _____

(h) PIP_2 _____

(i) Ca^{2+} _____

(j) IP_3 _____

(k) DAG _____

(l) A specific phosphodiesterase _____

(m) Glucocorticoid receptor _____

(n) Insulin receptor _____

(o) Glucagon receptor _____

(p) β-Adrenergic receptor _____

(q) Arachidonic acid _____

(1) Cleaved by phospholipase C

(2) Binds the regulatory subunits of specific protein kinases

(3) Has both hormone- and DNA-binding sites

(4) Converts cAMP to AMP

(5) Exchanges with GDP on G_α subunits

(6) Formed from an iodinated polypeptide precursor

(7) An anabolic hormone

(8) A leukotriene precursor

(9) Binds epinephrine

(10) A second messenger that raises intracellular Ca^{2+} levels

(11) Leads to increased cAMP levels when complexed to a polypeptide hormone

(12) A catabolic hormone

(13) Transduces hormone stimulus to adenylate cyclase

(14) Has inducible tyrosine kinase activity

(15) Activated by $G_{s\alpha}$-GTP

(16) Intracellular concentration increased by IP_3

(17) Activates protein kinase C

PROBLEMS

1. Give the major criteria that must be fulfilled by a substance for it to be classified as a hormone.

2. Bee venom is particularly rich in phospholipase A_2, an enzyme that hydrolytically removes the fatty acyl residue at position 2 of phospholipids. The action of phospholipase A_2 on phosphatidyl choline is shown in Figure 38-1. One of the mediators of the inflammatory response following a bee sting (swelling, redness, pain, heat, and loss of function) is lysophosphatidyl choline, the remainder of the phospholipid following the hydrolysis of the fatty acyl residue at position 2. Lysophosphatidyl choline stimulates mast cells to release histamine, which triggers the inflammatory response.

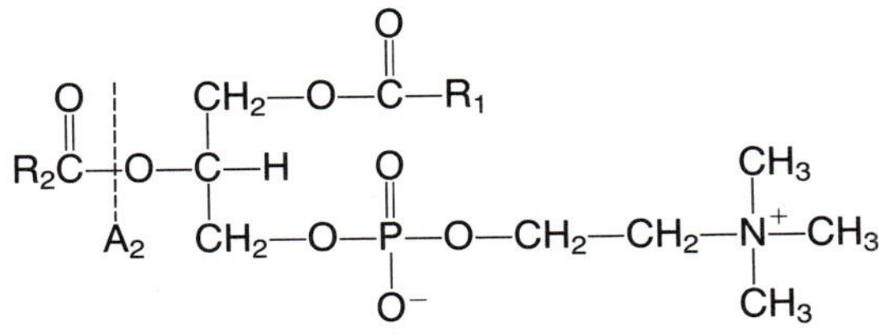

Figure 38-1
Action of phospholipase A_2 on phosphatidyl choline.

(a) Explain the major point of similarity between the system described here and one described by Stryer in Chapter 38.

(b) Suppose that it weren't known which of the hydrolysis products of phosphatidyl choline was important as a mediator of the inflammatory response. Suggest an experiment that might help establish the identity of the active agent.

3. Although Stryer has not yet presented the pathway for the biosynthesis of epinephrine, you should be able to make some educated guesses about the pathway from your knowledge of intermediary metabolism and the structure of epinephrine given in Figure 38-2.

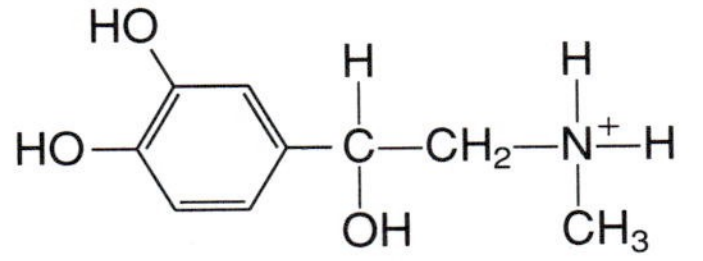

(a) What amino acid most likely serves as a precursor for the biosynthesis of epinephrine?

(b) Outline a plausible metabolic pathway for the synthesis of epinephrine from the precursor you have named in part (a).

Figure 38-2
Structure of epinephrine.

4. Assume that each catalytically active enzyme subunit in the regulatory cascade for glycogen phosphorylase has a turnover number of 1000 s^{-1}. Assume further that 10 G_α-GTP are formed for each molecule of epinephrine bound to receptor. Calculate the theoretical number of glucosyl units of glycogen that would be coverted to glucose 1-phosphate per second as a result of the interaction of one molecule of epinephrine with its receptor on a liver cell membrane. (See p. 458 of Stryer for a discussion of the phosphorylase regulatory cascade and p. 978 for a discussion of G protein.)

5. Aside from the fact that epinephrine has opposite effects on glycogen synthesis and glycogen degradation, what other significant difference exists between the regulatory cascades in the two systems? Would you expect there to be any difference in chemical amplification? Explain. (See Stryer, Chapter 19, p. 449 for a discussion of glycogen metabolism.)

6. The addition of epinephrine or glucagon to liver slices that are incubated in a medium containing $^{32}P_i$ results in the increased incorporation of the label into organic phosphate. Explain these results, and identify the organic phosphate that first becomes labeled.

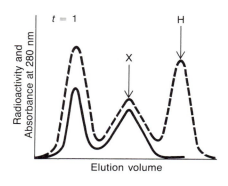

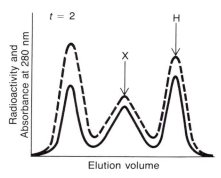

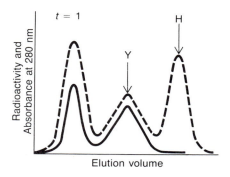

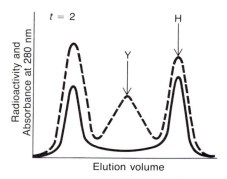

Figure 38-3
Results for two suspected hormone precursors. (The solid line refers to radioactivity; the dashed line to absorbance at 280 nm.)

7. When the experiment described in Problem 6 is repeated on muscle, epinephrine, but not glucagon, is found to stimulate the incorporation of $^{32}P_i$ into organic phosphate. Explain this observation.

8. In the early days of research on insulin action, it was not known whether insulin might enter cells and directly mediate intracellular effects or whether it might act through a second messenger. In a classic experiment, Pedro Cuatrecasas attached insulin covalently to sepharose beads many times the size of fat cells and showed that the addition of the insulin-sepharose complexes to isolated fat cells gave the same stimulation of glucose oxidation as did addition of insulin alone.

    (a) What conclusion might follow from this experiment? Explain.
    (b) What assumption have you made about the effects of adding sepharose alone to fat cells?

9. Suppose that you are trying to isolate a receptor for a polypeptide hormone from liver cells. Suggest an effective means of purification involving specialized-column chromatography and highly purified polypeptide hormone. Also explain how you would get the receptor off of the column.

10. When young rats are placed on a fat-free diet, they fail to thrive and develop dermatitis and a thickening of the skin. This condition can be relieved by the administration of arachidonic acid. Explain these observations.

11. The treatment of porcine proinsulin with trypsin produces a product that has insulinlike activity, but it does not produce insulin itself. Explain the reason for this. (See Figure 38-32 on p. 995 of Stryer.)

12. Speculate how the redundancy of basic residues at the boundaries between prohormones and connecting peptides might provide good evolutionary insurance against potentially damaging mutations.

13. In an attempt to determine if proteins X and Y are possible prohormone precursors of hormone H, pancreatic slices were incubated in a medium containing tritiated leucine for a brief time (the "pulse") and were then transferred to a medium containing unlabeled leucine (the "chase"). At various times after the transfer to the unlabeled medium, the tissues were homogenized and the proteins were separated on a molecular sieving column. The results obtained for proteins X and Y, in two separate experiments, are depicted in Figure 38-3. Which is the more likely precursor of hormone H, X or Y? Explain. Also, explain why the addition of unlabeled leucine is necessary.

14. In early studies on the effect of insulin on tissues, it was difficult to determine whether insulin primarily stimulated glucose transport or whether it had had a stimulating effect on the hexokinase reaction. (After all, a stimulation of the hexokinase reaction could result in the increased uptake of glucose by a tissue.) Finally, a system was developed in which chilled muscle cells could be suspended in a medium containing large concentrations of glucose, and the resulting intracellular concentration of glucose could be measured in the presence and absence of insulin. Predict the effect on the concentration of intracellular glucose when insulin is added to such a preparation if (a) insulin primarily stimulates the transmembrane transport of glucose and (b) if insulin primarily stimulates the hexokinase reaction. (c) Which do you think was actually observed?

15. Some of changes found in diabetes mellitus mimic those found in starvation. For example, elevated plasma levels of ketone bodies may be found in both states. Identify *one* characteristic of diabetes mellitus that is never associated with starvation.

ANSWERS TO SELF-TEST

Introduction

1. b, c, e, f, g

2. a, c, and d. Amino acids are direct precursors of hormones that are small molecules or polypeptides. In the former class, the amino acids are modified. In the latter class, they serve as monomer building blocks in their unmodified forms.

3. (a) 1, 3 (b) 2 (c) 4

Cyclic AMP—Its Formation and Its Role as a Second Messenger

4. a, b, and e. Answer (c) is incorrect because a phosphomonoesterase will not cleave a phosphodiester bond.

5. b and c. Answer (a) is incorrect because the hormone leads to an increase in the activity of adenylate cyclase, not an increase in the amount of the enzyme. Answer (e) is incorrect because the hormone need not enter the cell to carry out its action.

6. The simplest explanation for the tissue specificity of hormones is the presence or absence of receptors for particular hormones on the extracellular surfaces of the tissues. In this case, the targeted thyroid cells have a receptor for TSH whereas muscle cells do not.

7. Yes. Caffeine inhibits cAMP phosphodiesterase and would therefore be expected to increase the steady-state level of cAMP. In the liver, increasing intracellular cAMP levels would stimulate glycogenolysis and thereby increase blood glucose levels.

8. Aside from the likelihood that serum phosphodiesterases might destroy it, cAMP is a polar molecule that does not readily traverse the plasma membrane. Even if a more hydrophobic derivative, such as dibutyryl-cAMP, were used to overcome the permeability problem, there would be no tissue specificity, so all cells would have increased cAMP levels.

Guanyl-Nucleotide-Binding Proteins and Adenylate Cyclase

9. a, d, e, and f. Answer (g) is incorrect because the GTP form of the stimulatory G protein activates adenylate cyclase such that it forms many cAMP molecules; that is, an amplification occurs.

10. c, d, f, and g. Answers (a) and (b) are incorrect because G proteins are peripheral membrane proteins inside cells. They don't bind the hormone but rather the activated hormone receptor-complex, and they carry the stimulus to adenylate cyclase. Answer (e) is incorrect because the hormone receptor must have the hormone bound to it or it must have been activated by hormone binding before the G protein will bind.

11. The phenomenon is called desensitization or adaptation. It allows the system to adapt to a given level of hormone so that it can respond to changes in hormone concentrations rather than to absolute amounts. Desensitization is effected by phosphorylation at multiple sites on the carboxyl-terminal region of the β-adrenergic receptor when it has epinephrine bound to it. These covalent modifications of the hormone-receptor complex inhibit, but do not completely prevent, the GDP-GTP exchange and thereby decrease the activation of adenylate cyclase. The desensitized receptor can still respond to an increase in epinephrine concentrations. A phosphatase reverses the effects of the modification and resensitizes the receptor.

12. a and c. Answer (d) is incorrect because cAMP activates the targeted kinases at concentrations on the order of 0.01 μM.

13. A single hormone molecule combines with a single receptor to form several stimulatory G_α-GTP molecules. Each of these stimulatory molecules activates a adenylate cyclase molecule to form many cAMP molecules. The cAMP molecules activate protein kinases, each of which can phosphorylate many target enzymes.

14. The bacteria elaborate a toxin, a part of which enters the epithelial cells of the intestine. Inside the cell, the toxin acts enzymatically to covalently modify a G_s protein by transferring an ADP-ribose group to it from NAD^+. The covalent modification inhibits the GTPase activity of the $G_{s\alpha}$ subunit such that it remains in its activated form and continues to stimulate adenylate cyclase. The resulting high levels of cAMP stimulate the transport of ions. The resultant extrusion of Na^+ into the gut causes a concomitant efflux of H_2O, which causes diarrhea.

15. The G-protein system comprises either stimulatory G proteins (G_s) or inhibitory G proteins (G_i). Although similar to one another in their subunit structures, the $\beta\gamma$ subunits of G_i proteins, when released from the intact heterotrimer, have high affinity for the α subunit of G_s proteins. By complexing the $G_{s\alpha}$-GTP subunits, they prevent them from activating adenylate cyclase. Thus, a hormone-receptor complex that activates a G_i-protein-linked system antagonizes the effects of hormones that activate a G_s-protein-linked system. Because cells contain many more G_i proteins than G_s proteins, the $G_{s\alpha}$-GTP subunits are effectively sequestered and adenylate cyclase remains inactive, thus preventing the formation of cAMP.

16. a, b, d, e, f

The Phosphoinositide Cascade and the Modulation of Ca²⁺ Concentrations

17. a, c, d, e

18. b and f. Answer (a) is incorrect because PIP_2 is the precursor of the second messengers. The other incorrect choices are all degradation products of IP_3.

19. b and c. Answer (a) is incorrect because IP_3 causes the release, not the uptake, of Ca^{2+}. Answer (b) is correct because not only does a phosphatase act on IP_3 but a specific kinase phosphorylates it to form the inactive tetrakisphosphate derivative. Answer (d) is incorrect because free inositol reacts with CDP-diacylglycerol to form PIP_2 (see p. 549 in Stryer).

20. a, c, d, e, and f. Answer (b) is incorrect because most of the effects of IP_3 and Ca^{2+} are synergistic, not antagonistic.

21. a, b, c, e, f, and g. Answer (d) is incorrect because the pumps work to maintain high levels of Ca^{2+} in the sarcoplasmic reticulum relative to the levels in the cytosol.

22. The ionophore allows Ca^{2+} to enter cells by effectively rendering the membrane permeable to the ion. Since the extracellular Ca^{2+} concentration is higher than the intracellular concentration, the ion enters the cell and the cytosolic level increases. Since some hormones act to raise intracellular Ca^{2+} levels in order to carry out their physiological affects, the ionophore could lead to the same response.

23. EGTA is a specific Ca^{2+} chelator. It would bind tightly to the ion and markedly lower its concentration in the extracellular medium. Consequently, when a hormone or a Ca^{2+} ionophore acted to allow Ca^{2+} influx, none could occur because the concentration gradient of Ca^{2+} would be insufficient.

The Eicosanoid Hormones and the Endorphins

24. a, b, d, e, and f. Answer (c) is incorrect because the eicosanoid hormones have very short half-lives and therefore exert local rather than global effects. Answer (e) is correct because phospholipids supply arachidonate for prostaglandin synthesis.

25. b, c, d, e, and f. Answer (a) is incorrect because one DNA sequence encodes the identical amino acid sequences found in β-endorphin and β-lipotropin. This is possible because β-lipotropin is a precursor of β-endorphin.

Insulin and the Hormones that Modulate the Tyrosine Kinases

26. d, e, and f. Answer (a) is incorrect because the effects of insulin generally antagonize those of glucagon and epinephrine. For example, insulin lowers blood glucose levels whereas epinephrine and glucagon elevate it. Answer (b) is incorrect because a single gene gives rise to a large polypeptide precursor, preproinsulin, that is cut by proteases to form the final two-polypeptide chain product. Answer (c) is incorrect because the C-peptide is removed from the proinsulin by proteolysis. It dissociates because there are no disulfide bonds joining it to the A chain or B chain.

27. a, c, d, e, and f. Answer (b) is incorrect because the hormones themselves do not have tyrosine kinase activities; only their activated receptors are active.

The Steroid and Thyroid Hormones

28. c, d, e, and f. Answer (a) is incorrect because steroid hormones modulate the amount of a target enzyme by stimulating its synthesis rather than by affecting its catalytic activity. Answer (b) is incorrect because the receptors of steroid hormones are in the cytosol and the hormone-receptor complex acts in the nucleus to enhance transcription.

29. (a) 6 (b) 12 (c) 7 (d) 2 (e) 5 (f) 13 (g) 15 (h) 1 (i) 16 (j) 10 (k) 17 (l) 4 (m) 3 (n) 14 (o) 11 (p) 9 (q) 8

1. The major criteria for classifying a substance as a hormone are as follows:

(1) It should be produced by one type of cell and have effects on another type of cell.
(2) Its effects should involve the chemical amplification of the original signal.
(3) It should be produced in response to a stimulus, and its production should cease upon cessation of the stimulus.
(4) The substance should be selectively destroyed following cessation of the stimulus.
(5) The addition of the purified substance to tissues should mimic physiological stimuli.
(6) Specific inhibitors of the physiological response should also abolish the response elicited by the addition of the purified substance to tissues.
(7) Specific receptors for the hormone should exist and should be more abundant in tissues that are more sensitive to the hormone.

2. (a) The bee venom system resembles the phosphoinositide cascade discussed by Stryer on page 985. In that system, a membrane phospholipid is also converted into an active mediator of the response of several hormones.
   (b) One could inject each of the hydrolysis products—lysophosphatidyl choline and the fatty acid—into tissues separately to see which elicits the inflammatory response.

3. (a) Tyrosine serves as a precursor for the biosynthesis of epinephrine. (The biosynthetic pathway for epinephrine is given by Stryer on p. 1025.)
   (b) Hydroxyl groups must be added by hydroxylases, the carboxyl group is lost in a pyridoxal-phosphate-requiring reaction, and the methyl group is donated by S-adenosylmethionine.

4. The theoretical number of glucose 1-phosphates produced per second would be 10^{13}. One molecule of epinephrine would result in the production of 10 G_α-GTP. Each activated α subunit would stimulate adenylate cyclase to produce 1000 cAMP molecules for a total of 10,000 molecules of cAMP. Each of these cAMP molecules would activate one catalytic subunit of protein kinase. (Remember that a molecule of protein kinase exists as an R_2C_2 complex. Two cAMP molecules combine with two R subunits to give two catalytically active C subunits.) Each of the 10,000 active C subunits would result in the production of 10,000 molecules of active phosphorylase kinase, for a total of 10^7 molecules of active phosphorylase kinase. Each molecule of active phosphorylase kinase would in turn activate 1000 molecules of phosphorylase for a total of 10^{10} molecules of active phosphorylase. These, in turn, would lead to the breakdown of 10^{13} glucosyl units per second. (Note: Many simplifying assumptions have been made in this answer, but it illustrates the profound chemical amplification that occurs in systems under hormonal control.)

5. The main difference between the two systems is that there is one less step in the regulatory cascade for glycogen synthase. Protein kinase itself directly phosphorylates glycogen synthase, leading to its inactivation. In the phosphorylase cascade, protein kinase phosphorylates (and activates) glycogen phosphorylase, which in turn phosphorylates (and activates) phosphorylase. Because there is one less step in the regulatory cascade, there is a correspondingly lower chemical amplification. (However, phosphorylase kinase can also phosphorylate glycogen synthase, inactivating it. This has the effect of restoring some of the "missing" amplification. See Stryer, p. 463, Figure 19-14.)

6. Both epinephrine and glucagon act to raise the intracellular cAMP level, leading to the activation of phosphorylase and the increased conversion of the glycosyl units of glycogen to glucose 1-phosphate, which is the first intermediate to be labeled.

7. Both glucagon and epinephrine use cAMP as their intracellular second messenger. Muscle phosphorylase is known to be regulated by a cascade that involves cAMP as a second messenger. The fact that glucagon has no apparent effect on phosphorylase activity in muscle cells can best be explained by postulating that muscle cells have very few glucagon receptors. (Such is, in fact, the case. Glucagon acts primarily on the liver, the first major organ it "sees" after being released by the α cells of the pancreas. Epinephrine has effects on many cells.)

8. (a) The best conclusion is that insulin need not enter the cell to have an effect. The results are consistent with the notion that insulin affects cells by combining with a membrane receptor and releasing some second messenger that mediates its intracellular effects. Note that the experiment doesn't prove that insulin doesn't enter cells.

(b) You very likely assumed that the addition of sepharose alone gave no stimulation of glucose oxidation.

9. The most useful technique would be affinity chromatography. Covalently attach the purified hormone to sepharose beads, and use the complex to fill a column. (See Problem 8.) Homogenize liver cells, and add the homogenate to the column. Receptors will stick on the column because they are complementary in shape to the covalently bound hormone. Elute the receptors from the column as hormone-receptor complexes by adding free hormone to compete with the hormone that is bound to the column.

10. The defects seen in the rats resulted from a deficiency of prostaglandins. The rats on the fat-free diet lacked a supply of polyunsaturated fatty acids, which cannot be synthesized by animals and, hence, must be supplied in the diet. These polyunsaturated fatty acids are required for the synthesis of arachidonic acid. Arachidonic acid, in turn, is required for the synthesis of the prostaglandins.

11. Trypsin cleaves peptides on the carboxyl side of Lys and Arg. Thus, it will cleave proinsulin at the juncture between the connecting peptide and the A chain of insulin (between Arg-63 and Gly-1), but it will cleave one amino acid away from the juncture between the B chain and the connecting peptide (between Arg-31 and Arg-32). The resulting insulinlike compound is one amino acid longer than insulin, having 31 residues in the B chain and a C-terminal Arg rather than a C-terminal Ala. The slightly longer product does have significant insulinlike

activity, however. (There are other possible sites for tryptic cleavage in proinsulin, but the rate of cleavage at many of them is slower because they are less exposed in the three-dimensional structure of proinsulin.)

12. If a trypsinlike enzyme is involved in the conversion of a prohormone to a hormone and if a mutation changes one of the two basic residues, there will still be a basic residue left, so the prohormone can still be cleaved to a product that is very likely to be hormonally active. Another significance of the redundancy in basic residues is that trypsinlike enzymes cleave peptide bonds between two basic residues much more rapidly than they cleave peptide bonds between one basic residue and some other amino acid.

13. The more likely prohormone is Y. Note that the radioactivity disappears from Y as it appears in the hormone, as would be the case for a legitimate precursor in a pulse-chase experiment. Radioactivity does not disappear from X as it appears in H. The results for X could have been obtained if X were a larger protein that was merely synthesized more rapidly than H from the pool of amino acids. The addition of unlabeled leucine is necessary if the precursor is to "disappear" (that is, become unlabeled) as the product "appears" (that is, becomes labeled).

14. (a) If insulin primarily stimulates the transmembrane transport of glucose, the addition of insulin should cause the intracellular glucose concentration to increase.
    (b) If insulin primarily stimulates the hexokinase reaction (the conversion of glucose to glucose 6-phosphate), the addition of insulin should result in a decrease in the intracellular glucose concentration.
    (c) When the experiment was conducted, the result in part (a) was obtained.

15. In diabetes mellitus, blood glucose levels are elevated, due in part to the inability of glucose to enter fat and muscle cells in the absence of insulin and in part to the overproduction (by glycogenolysis and gluconeogenesis) of glucose by the liver in the absence of insulin. In starvation, blood glucose levels are low.

CHAPTER 39

Excitable Membranes and Sensory Systems

In addition to the hormonal signals discussed in Chapter 38, cells respond to a variety of external chemical and physical stimuli through specialized sensory assemblies located in cell membranes. In this chapter, Stryer describes four types of excitable assemblies. First, he discusses chemotaxis, which is the directed movement of bacteria towards chemical substances that are attractants and away from those that are repellents. Next he describes the responses of nerve-axon membranes to electrical stimuli, concentrating on the molecular properties of the Na^+ channel. This section requires a review of the activity and physiological roles of the Na^+-K^+ pump and the concept of electrochemical potential in Chapter 37 because the Na^+ and K^+ gradients across nerve-axon membranes are essential for the transmission of the electrical signals. The third type of excitable assembly transmits nerve impulses from a nerve cell to another nerve cell or to a muscle cell. It is exemplified by the acetylcholine receptor. Finally, Stryer describes the excitable receptors in the retina that are activated by light. In addition to Chapters 37 and 38 of the text, Chapter 12 is a general source of background information about membrane proteins and the methods that are used in their investigation.

LEARNING OBJECTIVES

When you have mastered this chapter, you should be able to complete the following objectives.

Introduction (Stryer pages 1005–1006)

1. List the common features of *excitable assemblies*.

Bacterial Chemotaxis (Stryer, pages 1006–1011)

2. Define *chemotaxis*, and outline the sequence of events in this process.

3. Describe the *flagella* and the *motors* of the chemotactic apparatus of *E. coli*. Relate their actions to the smooth swimming and tumbling motions of a bacterium in response to a *temporal gradient* of an *attractant* or *repellent* substance.

4. Describe the general structure of *methyl-accepting chemotaxis proteins (MCPs)* and their role as *transducers* of chemotactic signals.

5. Discuss the reversible *methylation* of the *MCPs* in the process of *adaptation*.

6. Explain the use of the *che gene mutants* of *E. coli* in the analysis of the *central processing system* of chemotaxis. Outline the proposed switching mechanism and the role of *CheY* in it.

The Na⁺ Channels of Nerve-Axon Membranes (Stryer pages 1011–1016)

7. Describe the transmission of electrical signals by the *action potential*, and explain how this mechanism is based on transient changes in the permeability of the axon membrane to Na^+ and K^+.

8. Explain the molecular basis for the ion selectivity of the *Na⁺ channel* and its inhibition by *tetrodotoxin* and *saxitoxin*.

9. Describe the purification, overall structure, and properties—including *Na⁺ selectivity*, *voltage dependence*, and *spontaneous inactivation*—of Na^+ channels.

The Acetylcholine Receptor (Stryer pages 1017–1027)

10. Discuss the role of *synapses* and *neurotransmitters* in the communication of nerve impulses between excitable cells.

11. List the sequence of events between the synthesis of *acetylcholine* near the *presynaptic* end of axons to its hydrolysis in the *synaptic cleft*.

12. Describe the effects of acetylcholine on the *postsynaptic membrane*. Distinguish between the *nicotinic acetylcholine receptor* and the *muscarinic acetylcholine receptor*.

13. Outline the *patch-clamp technique*, and note its use in the electrical measurement of excitable membranes.

14. Discuss the different conformational states of the acetylcholine receptor and its *desensitization*.

15. Describe the purification of the acetylcholine receptor from *Torpedo*, its specific interaction with α-*bungarotoxin* or *cobratoxin*, its subunit structure, and its low-resolution three-dimensional structure.

16. Discuss the hydrolysis of acetylcholine by *acetylcholinesterase*, and note the high turnover number of this enzyme.

17. Be familiar with the major inhibitors of acetylcholinesterase—*physostigmine, neostigmine, and diisopropyl phosphofluoridate*—and the major inhibitors of the acetylcholine receptor—d-*tubocurarine, decamethonium*, and *succinylcholine*.

18. Summarize the characteristics of a neurotransmitter, and list examples of neurotransmitters.

19. Outline the synthesis of the *catecholamines* from tyrosine, and describe their inactivation.

20. Describe the effects of *glycine* and *γ-aminobutyrate (GABA)* in the central nervous system. Compare the receptors for GABA and glycine with the nicotinic acetylcholine receptor.

21. Give examples and recall the biological roles of signal molecules that are formed in one step from a major metabolic intermediate.

Retinal Photoreceptors (Stryer pages 1027–1038)

22. Distinguish between the *cone* and *rod photoreceptor cells.*

23. Describe the regions of the rod cell, and list their components and functions.

24. Describe the properties of *retinal* and its synthesis from *vitamin A.*

25. Describe the overall structure of *rhodopsin*, its functional regions, and its orientation in the *disc* membrane.

26. List the sequence of events from the absorption of light by 11-cis-*retinal* to the release of *all*-trans-*retinal* from opsin and the regeneration of 11-*cis*-retinal.

27. Discuss the light-induced *hyperpolarization* of the plasma membrane as a result of the closing of *cation-specific channels*. Relate the hyperpolarization effect to the decrease in the concentration of *cyclic GMP* in the *outer segment* of the rod cell.

28. Outline the *enzymatic cascade* leading to the hydrolysis of cyclic GMP. Explain the roles of *photoexcited rhodopsin, transducin*, and *activated phosphodiesterase.*

29. Compare the properties of transducin to those of other members of the *G protein family* (see Stryer, p. 984).

30. Describe the restoration of the dark state by the *GTPase activity* of transducin, *rhodopsin kinase*, the binding of *arrestin* to phosphorylated rhodopsin, and the synthesis of cyclic GMP by *guanylate cyclase.*

31. Describe the functions of the three kinds of cones in color vision.

SELF-TEST

Introduction

1. Which of the following are common features of all excitable membrane assemblies?

   (a) They contain a specific membrane protein that is the receptor for the stimulus.
   (b) The receptor undergoes a conformational change upon stimulation.

(c) The conformational change in the receptor elicits the phosphorylation of a transducing protein.
(d) The conformational change in the receptor and resulting changes in the system are reversible.
(e) The conformational changes in the system lead to the synthesis of cyclic GMP.

Bacterial Chemotaxis

2. Match the major components of the chemotaxis system of *E. coli* in the left column with the appropriate descriptions from the right column.

(a) Chemoreceptor _____

(b) Processing system _____

(c) Motor _____

(d) Flagellum _____

(1) Contains four rings a hook, and a rod
(2) Has binding sites for attractants or repellents
(3) Includes cytosolic peripheral membrane proteins
(4) Contains flagellin

3. For a bacterium moving toward an increasing concentration of an attractant, which of the following statements are *correct*?

(a) Tumbling will be less frequent.
(b) Tumbling will be more frequent.
(c) The counterclockwise rotation of flagella will occur more frequently.
(d) The clockwise rotation of flagella will occur more frequently.

4. The experiment of Koshland and Macnab that demonstrated that bacteria sense temporal gradients of attractants consisted of rapidly mixing a suspension of bacteria devoid of an attractant with a solution containing an attractant. The observed result was that tumbling was suppressed within a second after mixing, indicating that the bacteria detected the change in the concentration of the attractant over time. Explain what the expected result would have been if bacteria detected differences in concentration over space.

_____

_____

5. Which of the following are properties of methyl-accepting chemotaxis proteins?

(a) MCPs may bind attractants or repellents directly.
(b) MCPs may interact with soluble, periplasmic chemosensors.
(c) MCPs contain a motor binding site in the cytosolic region.
(d) MCPs are reversibly methylated in the cytosolic domain.
(e) MCPs include four proteins having highly homologous periplasmic domains.

6. Explain the role of the reversible methylation of MCPs in chemotaxis by *E. coli*.

_____

_____

_____

7. The proposed model for the transduction of chemotactic signals via CheY, the tumble regulator, includes which of the following?

   (a) Activated CheY promotes the counterclockwise rotation of the motor.
   (b) MCPs with bound repellents activate CheY.
   (c) MCPs with bound attractants activate CheY.
   (d) The methylation of MCPs increases the probability of activation of CheY.
   (e) The activation of CheY requires ATP.

The Na⁺ Channels of Nerve-Axon Membranes

8. Which of the following statements about the nerve-axon membrane are *correct?*

   (a) In the resting state, the membrane is more permeable to Na⁺ than to K⁺.
   (b) In the resting state, the membrane potential is approximately $+30$ mV.
   (c) The Na⁺ and K⁺ gradients across the membrane are maintained by the Na⁺-K⁺ pump.
   (d) The equilibrium potential for K⁺ across the membrane is near -75 mV.
   (e) During the action potential, the membrane potential varies between the limits of $+30$ mV and -75 mV.

9. Place the following events of the action potential in their correct sequence.

   (a) The spontaneous closing of Na⁺ channels
   (b) A membrane potential of -75 mV
   (c) The depolarization of the plasma membrane to approximately -40 mV
   (d) The opening of the K⁺ channels
   (e) The opening of Na⁺ channels
   (f) A membrane potential of $+30$ mV
   (g) A membrane potential of -60 mV

10. The K⁺ channel is over 100 times more permeable to K⁺ than to Na⁺. Propose a possible molecular mechanism for this selectivity.

_____

_____

11. Tetradotoxin and saxitoxin are very specific competitive inhibitors of the Na⁺ channel. Explain why.

_____

_____

12. The density of Na^+ channels in the nodes of Ranvier, as determined by the binding of radioactive tetradotoxin, is of the order of 10^4 per μm^2. Calculate the center to center distance between the channels.

13. Which of the following statements about the Na^+ channel, purified and reconstituted in lipid bilayers are *correct?*

(a) It is about 10 times more permeable to Na^+ than to K^+.
(b) It is sensitive to voltage.
(c) It becomes inactivated spontaneously.
(d) It is inhibited by cobratoxin.
(e) It consists of seven hydrophobic transmembrane segments.

The Acetylcholine Receptor

14. Match each of the components of the synapse in the left column with the appropriate description from the right column.

(a) Presynaptic membrane _____
(b) Postsynaptic membrane _____
(c) Synaptic vesicle _____
(d) Synaptic cleft _____

(1) Stores acetylcholine
(2) Contains acetylcholine receptors
(3) Contains acetylcho-linesterase
(4) Fuses with synaptic vesicles in response to Ca^{2+} levels

15. Acetylcholine opens a single kind of cation channel that has a very similar permeability to Na^+ and K^+, yet the influx of Na^+ is much larger than the efflux of K^+. Explain this fact.

_____

_____

16. The equilibrium scheme for the multiple conformational states of the acetylcholine receptor is

$$R \underset{}{\overset{1}{\rightleftharpoons}} AR \underset{}{\overset{2}{\rightleftharpoons}} A_2R \underset{}{\overset{3}{\rightleftharpoons}} A_2R^*$$

$$4 \updownarrow \qquad\qquad \updownarrow 5$$

$$AI \rightleftharpoons A_2I$$

Answer the following questions about these states and their transformations:

(a) Which states are open to cation flow? _____
(b) Which states are closed to cation flow? _____
(c) Which steps are rapid? _____

(d) Which steps are slow? _____

(e) Which states are most stable? _____

17. Explain the use of cobratoxin in the purification of acetylcholine receptor.

_____

_____

18. Which of the following is *not* a characteristic of the structure of the acetylcholine receptor of *Torpedo*.

(a) It has five subunits of four different kinds.

(b) It contains two binding sites for acetylcholine.

(c) It protrudes about 70 Å on the cytoplasmic side.

(d) It can be partly active when the γ or δ chains are missing

(e) It has pentagonal symmetry.

19. Which of the following statements about the catalytic properties of acetylcholinesterase are *correct?*

(a) A glutamate residue is involved in its catalytic mechanism.

(b) It forms an acetyl-enzyme intermediate during catalysis.

(c) It is irreversibly inhibited by diisopropyl phosphofluoridate.

(d) It is competitively inhibited by α-bungarotoxin.

(e) It has a low turnover number.

20. Match each of the three proteins in the left column with the appropriate inhibitors from the right column.

(a) Na^+ channel _____

(b) Acetylcholine receptor _____

(c) Acetylcholinesterase _____

(1) Physostigmine
(2) Decamethonium
(3) Tetrodotoxin
(4) α-Bungarotoxin
(5) Parathion
(6) *d*-Tubocurarine
(7) Saxitoxin
(8) Succinylcholine

21. Which of the following are catecholamine neurotransmitters?

(a) 3,4-Dihydroxyphenylalanine

(b) Acetylcholine

(c) γ-Aminobutyrate

(d) Epinephrine

(e) Glycine

22. The series of reactions in the conversion of tyrosine into epinephrine require which of the following?

(a) O_2

(b) NADPH

(c) Thiamine pyrophosphate

(d) Tetrahydrobiopterin

(e) *S*-Adenosylmethionine

23. Explain the physiological role of γ-aminobutyrate (GABA) in the central nervous system.

_____

_____

Retinal Photoreceptors

24. Match the two main regions of the rod cell in the left column with the appropriate functions or properties from the right column.

    (a) Outer segment _____

    (b) Inner segment _____

    (1) Contains the synaptic terminal
    (2) Contains discs
    (3) Carries out protein synthesis and the generation of ATP
    (4) Contains the photoreceptors that absorb light
    (5) Contains the majority of Na^+-K^+ ATPase pumps
    (6) Contains cation-specific channels that are controlled by cyclic GMP

25. Which of the following statements about retinal, the chromophore of rhodopsin, is *incorrect?*

    (a) It is derived from vitamin A through two reactions, a dehydrogenation and an isomerization.
    (b) In the dark, it is covalently bound to opsin through a Schiff-base linkage.
    (c) In the dark, it is present as the 11-*cis*-retinal isomer.
    (d) When bound to rhodopsin, it absorbs light maximally at 500 nm.
    (e) It becomes the all-*cis*-isomer after absorbing light.

26. With its seven transmembrane helices, rhodopsin has a structure similar to that of which of the following receptors?

    (a) β-Adrenergic receptor
    (b) Nicotinic acetylcholine receptor
    (c) Muscarinic acetylcholine receptor
    (d) LDL receptor
    (e) Chemoreceptor MCPs

27. Place the following events in the excitation of rhodopsin by light in their correct sequence.

    (a) Deprotonation of the Schiff base.
    (b) Conversion of all-*trans*-retinal to 11-*cis*-retinal.
    (c) Hydrolysis of the Schiff base.
    (d) Conversion of 11-*cis*-retinal to all-*trans*-retinal.
    (e) Triggering of the enzyme cascade.
    (f) Series of isomerizations of retinal and conformational changes of the rhodopsin.

28. Explain briefly the major roles of the following participants in the enzymatic cascade that is triggered by the photoexcitation of rhodopsin:

    (a) Transducin _____

    _____

    (b) Cyclic GMP _____

    _____

(c) Activated phosphodiesterase _____

_____

29. How does the photoexcited system return to the dark state?

_____

_____

30. Which of the following statements about color vision are *correct?*

(a) It is mediated by three different chromophores.
(b) It is mediated by three different photoreceptors.
(c) It involves only the cone cells.
(d) It involves only the rod cells.

PROBLEMS

1. *E. coli* cells that are grown in a medium that is deficient in methionine show impaired chemotactic response. Why do you think this is the case?

2. Figure 39-1 depicts the response of bacteria to two chemotactic agents, X and Y. Which agent is likely to be an attractant, and which is likely to be a repellent? Give the reasons for your choices.

3. Figure 39-2 depicts a typical action potential that might be measured in an isolated axon, such as the giant axon of a squid. Give the events that are responsible for (a) the rising phase of the action potential and (b) the falling phase. Specify in each case whether ion flow occurs with or against concentration and/or electrical gradients.

4. When the sciatic nerve is removed from a frog, placed in an isotonic salt solution, and stimulated electrically, it will generate action potentials that can be measured by an electrode placed at some distance from the site of stimulation. When metabolic poisons are added to the preparation, the nerve retains the capability of generating action potentials even though the supply of ATP to drive its Na^+-K^+ pump has been depleted and it is thus incapable of carrying out active transport. Explain how this can be the case.

5. Patients suffering from Parkinson's disease have a deficiency of dopamine in certain areas of the brain (see Stryer, p. 1025). The oral administration of dopamine itself is of little value in managing the symptoms, but the administration of dopa is somewhat effective.

(a) Can you think of a reason why this might be the case?
(b) On page 1025, Stryer points out that patients with Parkinson's disease are given both dopa and inhibitors of dopa decarboxylase. Why are the decarboxylase inhibitors administered?
(c) What characteristic must the decarboxylase inhibitors have in order to be effective in the treatment of Parkinson's disease? Explain.

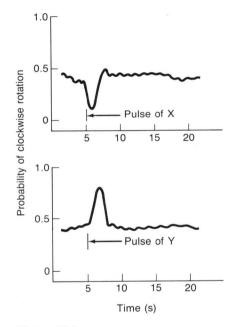

Figure 39-1
Bacterial response to two chemotactic agents.

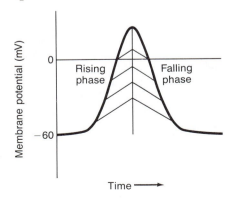

Figure 39-2
An action potential.

6. Write balanced equations for (a) the conversion of tyrosine to dopa and (b) the conversion of norepinephrine to epinephrine. (See Stryer, p. 1025).

7. (a) Suppose that a single, isolated muscle fiber is treated with curare. Will it contract when acetylcholine is added or when it is stimulated electrically?
(b) How would your answer to (a) differ if the muscle fiber had been treated with decamethonium instead? Explain your answers.

8. Suppose that you want to find a drug that would act at the neuromuscular junction to induce muscle relaxation.

(a) Give three possible approaches that you could take, explain the rationale for each, and give a specific example of a known drug that would have the desired effect.
(b) Is there a major potential hazard in the use of any of the three types of agents you have described in part (a)? What is it?

9. Which of the compounds in Figure 39-3 would be expected to have the smallest absorbance per mole in the visible light range? Give the reason for your choice.

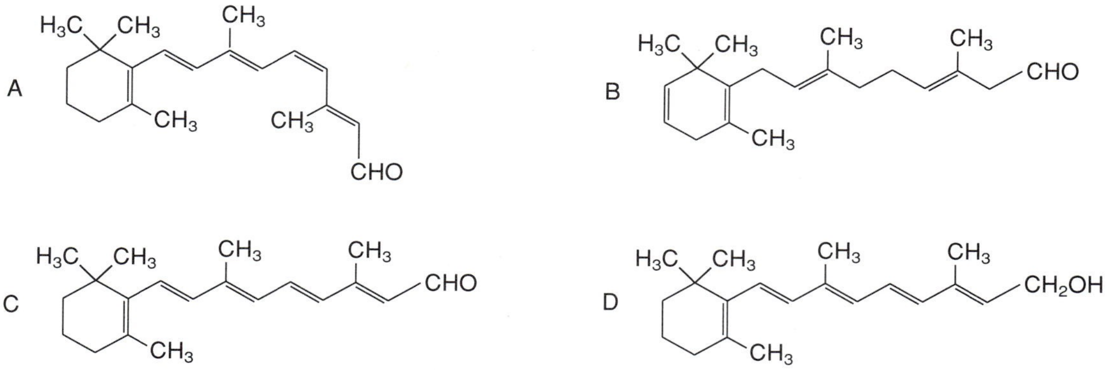

Figure 39-3
Four light-absorbing compounds.

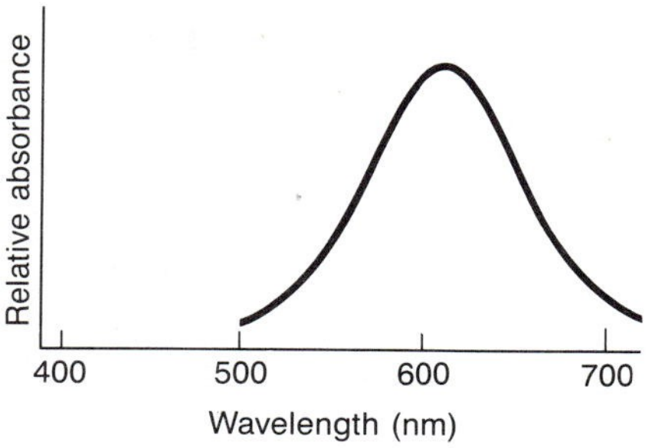

Figure 39-4
Absorption spectrum
of a color-vision pigment.

10. Figure 39-4 shows the absorption spectrum of a light-absorbing pigment that is involved in color vision.

(a) What color light is absorbed by the pigment?
(b) When the pigment is extracted into an organic solvent, what is the color of the resulting solution?

11. 11-*cis*-Retinal is covalently linked to a lysine side chain of opsin by the formation of a Schiff base:

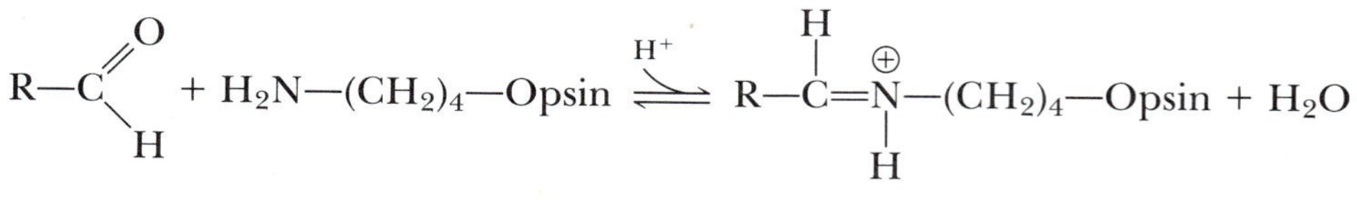

Such linkages can typically be stabilized in the laboratory by reduction with borohydride:

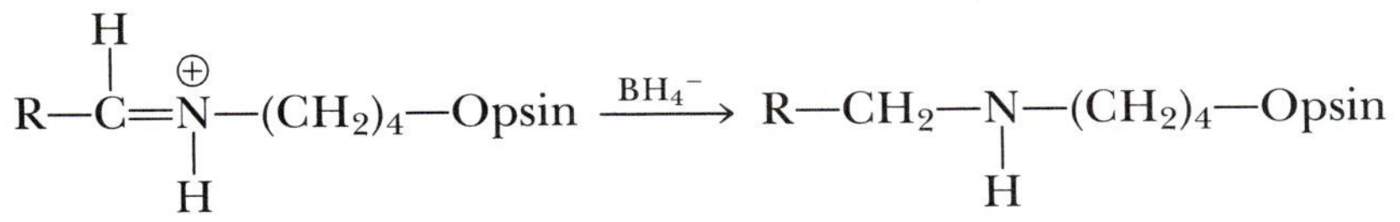

$$R-\overset{\overset{\displaystyle H}{|}}{\underset{\underset{\displaystyle H}{|}}{C}}=\overset{\oplus}{N}-(CH_2)_4-Opsin \xrightarrow{\ BH_4^-\ } R-CH_2-\overset{}{\underset{\underset{\displaystyle H}{|}}{N}}-(CH_2)_4-Opsin$$

Would you expect a rod cell preparation that has been treated with borohydride to be active in the cycle of visual excitation? Why or why not?

12. A color-vision pigment that absorbs red light is chemically cleaved to separate the retinal from the protein. The same is done for a pigment that absorbs blue light. Then a new pigment is constituted using the retinal from the red-absorbing pigment and the opsin from the blue-absorbing pigment. What color of light will be absorbed by the new pigment?

13. Suppose that an experiment on a light-excitable cell involving a patch electrode is carried out as described by Stryer on page 1035 and that the following experimental results are obtained:

[cGMP] (μM)	Y
7.0	0.2
9.0	0.4
10.0	0.5
11.1	0.6
14.1	0.8

Calculate the minimum number of cyclic GMP molecules that are required to open each cation-specific channel of the cell. You should refer to the discussion of Hill plots on page 155 of Stryer.

ANSWERS TO SELF-TEST

Introduction

1. a, b, d

Bacterial Chemotaxis

2. (a) 2 (b) 3 (c) 1 (d) 4

3. a, c

4. After the rapid mixing of the bacteria with a solution containing an attractant, the concentration of the attractant around each bacterium would be uniform, so there would be no spatial gradient. If bacteria sensed spatial rather than temporal gradients, the tumbling rate of the bacteria would not change.

5. a, b, d

6. The reversible methylation of MCPs allows the bacteria to adapt to different basal levels of attractants and repellents and to respond to changes in the relative concentrations of attractants and repellents rather than to their absolute levels.

7. b, d, e

The Na⁺ Channels of Nerve-Axon Membranes

8. c, d, e

9. g, c, e, f, a, d, b, g

10. By analogy with the selective binding of K^+ over Na^+ by valinomycin (see Stryer, p. 967), it is plausible that the discrimination by the K^+ channel is due to the higher stability of hydrated Na^+ relative to hydrated K^+, rather than to different binding affinities of the ions within the channel itself.

11. Both of these toxins contain a guanido group. This positively charged group interacts with a carboxylate group at the entrance of the Na^+ channel on the extracellular side of the membrane to obstruct the conductance pathway.

12. Each Na^+ channel occupies on area equal to $1 \mu m^2/10^4 = 10^{-4} \mu m^2$. Since the length of the side of a square is equal to the square root of the area and the side of the square is equivalent to the distance between the centers of the channels, the distance between Na^+ channels is $10^{-2} \mu m$ or 100 Å.

13. a, b, c

The Acetylcholine Receptor

14. (a) 4 (b) 2 (c) 1 (d) 3

15. The electrochemical gradient for Na^+ influx is steeper than that for K^+ efflux. The concentration gradients across the membrane are similar for both ions, but Na^+ moves from the positive to the negative side of the membrane, whereas K^+ moves from the negative to the positive side; that is, the membrane potential favors Na^+ influx.

16. (a) A_2R^* (b) R, AR, A_2R, AI, A_2I (c) 1, 2, 3 (d) 4, 5 (e) AI, A_2I

17. Cobratoxin binds specifically and with very high affinity to the acetylcholine receptor; therefore, a column with covalently attached cobratoxin can be used in the affinity purification of the receptor from a mixture of macromolecules in the postsynaptic membrane that has been solubilized by adding nonionic detergents.

18. c

19. b, c

20. (a) 3, 7 (b) 2, 4, 6, 8 (c) 1, 5

21. a, d

22. a, b, d, e. Answer (b) is correct because the monooxygenase reactions that introduce hydroxyl groups into the substrate require the NADPH cofactor.

23. γ-Aminobutyrate (GABA) is an inhibitory transmitter in the central nervous system. It causes the hyperpolarization of membranes by increasing their permeability to Cl^-. The hyperpolarization of excitable membranes increases the threshold for the triggering of action potentials.

Retinal Photoreceptors

24. (a) 2, 4, 6 (b) 1, 3, 5

25. e

26. a, c

27. d, f, a, e, c, b

28. (a) Photoexcited rhodopsin binds transducin (T-GDP), catalyzes the exchange of GTP for GDP, and releases T_α-GTP. This form of transducin then activates phosphodiesterase. Hydrolysis of GTP bound to T_α deactivates phosphodiesterase and allows the binding of $T_{\beta\gamma}$, regenerating T-GDP.
    (b) In the dark, cyclic GMP keeps the cation-specific channels open. Activated phosphodiesterase hydrolyzes cyclic GMP to 5'-GMP, leading to the closing of the channels.
    (c) When phosphodiesterase is activated by transducin, it hydrolyzes cyclic GMP, which leads to the closing of the cation-specific channels and to hyperpolarization of the plasma membrane.

29. The return to the dark state requires the deactivation of both phosphodiesterase and rhodopsin and the formation of cyclic GMP. Phosphodiesterase is deactivated by the hydrolysis of GTP bound to T_α. Rhodopsin is deactivated by rhodopsin kinase, which catalyzes the phosphorylation of photoexcited rhodopsin at multiple sites. The phosphorylated rhodopsin binds arrestin, which blocks the binding of transducin. Guanylate cyclase catalyses the synthesis of cyclic GMP.

30. b, c

ANSWERS TO PROBLEMS

1. The chemotactic response depends on the rapid and reversible methylation of several glutamate side chains of the methyl-accepting chemotaxis proteins (MCPs). *S*-Adenosylmethionine is the methyl-group donor in these methylation reactions. A methionine deficiency would result in a decreased pool of *S*-adenosylmethionine and, hence, in an impaired chemotactic response. (See p. 1009 of Stryer.)

2. X is likely to be an attractant and Y is likely to be a repellent. The addition of X causes a decrease in clockwise rotation (an increase in counterclockwise rotation), which leads to smooth swimming. The addition of Y causes an increase in clockwise rotation (a decrease in counterclockwise rotation), which leads to tumbling. (Note that the ordinate for

the plots in Figure 39-1 is the probability of *clockwise* rotation, not the probability of *counterclockwise* rotation as in Figure 39-10 on p. 1010 of Stryer.)

3. Nerve cells, like most animal cells, have a higher concentration of K^+ inside than outside and a higher concentration of Na^+ outside than inside. In addition, there is a membrane potential; that is, the inside of the cell is negative (in this case -60 mV) with respect to the outside.

   (a) The rising phase of the action potential is due to the influx of Na^+ ions down a concentration gradient and an electrical gradient.

   (b) The falling phase is due to the efflux of K^+ ions down a concentration gradient but against an electrical gradient.

4. For each action potential that is generated in an axon, only a very few Na^+ ions enter and a very few K^+ ions depart the cell. Thus, in a poisoned nerve cell, many tens of thousands of impulses may be conducted before ionic equilibrium across the membrane is achieved. The active transport of Na^+ and K^+ across the membrane may be best viewed as necessary in the long run but not in the short run.

5. (a) Many substances are not readily taken up by brain tissues from neighboring capillaries due to the "blood-brain barrier." Dopamine itself can't enter brain cells, but dopa can enter and can be decarboxylated to dopamine within the tissue.

   (b) Dopa decarboxylase is present in many cells of the body. If inhibitors weren't given, peripheral tissues would rapidly decarboxylate dopa to dopamine, which, as we have seen, can't enter brain cells.

   (c) The inhibitors must be able to enter peripheral cells but not brain cells. Otherwise, they would inhibit the dopa decarboxylase in the brain as well, thus causing the administration of dopa to be ineffective.

6. (a)

   Tyrosine + O_2 + tetrahydrobiopterin (BH_4)

   $\qquad$ = dopa + H_2O + dihydrobiopterin (BH_2)

   Note that one atom of molecular oxygen ends up in the hydroxylated aromatic ring and that the other combines with two hydrogens donated by tetrahydrobiopterin to form water.

   (b)

   Norepinephrine + *S*-adenosylmethionine

   $\qquad$ = epinephrine + *S*-adenosylhomocysteine

7. (a) Curare occupies the acetylcholine receptor and inhibits the depolarization of the motor end plate. Therefore, the muscle fiber will be insensitive to the addition of acetylcholine, but it will contract when it is stimulated electrically.

   (b) When decamethonium is added, the acetylcholine receptor is again blocked, but the motor end plate is depolarized. The muscle will therefore be insensitive to both acetylcholine and electrical stimulation.

8. (a) Signal transmission at the neuromuscular junction involves the acetylcholine system. To be effective as a relaxant, a substance must prevent the repetitive transmission of signals by acetylcho-

line. One approach would be to find a drug that binds to the acetylcholine receptor but does not induce the depolarization of the motor end plate (and hence cause muscle contraction). Furthermore, the drug should not be cleaved by acetylcholinesterase so that its effect is (relatively) long-lasting. An example of such an agent is curare (Stryer, p. 1024).

A second approach would be to find an agent, such as succinylcholine or decamethonium (Stryer, p. 1024), that is cleaved very slowly or not at all by acetylcholinesterase and binds to the acetylcholine repector to produce the persistent depolarization of the motor end plate. A muscle treated with such an agent would go through an initial contraction cycle, but it would be refractory to further stimuli because the receptor is blocked and the depolarization of the motor end plate is persistent.

A third approach would be to find an inhibitor of acetylcholinesterase. As a result of adding such an inhibitor, acetylcholine would be maintained at the receptor site and the persistent depolarization of the motor end plate would result. Physostigmine and neostigmine are examples of this third type of agent.

(b) The major hazard in the use of any of the three types of muscle relaxants is that they will also relax the muscle groups used for respiration.

9. Compounds that absorb visible light significantly have long sequences of alternating single and double bonds; that is, they are conjugated. Compound B is unconjugated, so it would have negligible absorbance in the visible range of the spectrum. Compound A is 11-*cis*-retinal, C is all-*trans*-retinal, and D is all-*trans*-retinol.

10. (a) Red light is absorbed by the pigment.
    (b) A solution of the pigment would be blue or blue-green. The pigment absorbs red light, but it transmits light at the blue end of the spectrum.

11. The photoexcitation of rhodopsin leads initially to the isomerization of 11-*cis*-retinal to all-*trans*-retinal and ultimately to the cleavage of all-*trans*-retinal from the protein. In a rod cell preparation that has been treated with borohydride, the Schiff base would be stabilized by reduction, so the removal of the all-*trans*-retinal would be impaired. As a result, the cycle of visual excitation could not occur. (See p. 1032 of Stryer for a discussion of these events.)

12. The new pigment will absorb blue light. The retinal is the same in all color-vision pigments. The protein component, however, varies among the pigments, giving absorption maxima at different wavelengths.

13. We start with the Hill equation from page 155 of Stryer:

$$\log \frac{Y}{1 - Y} = n \log [\text{cGMP}] - n \log [\text{cGMP}]_{50}$$

The second term on the right side of the equation represents the concentration of cGMP that gives 50 percent opening. From inspection of

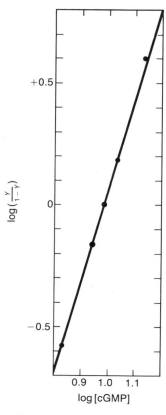

Figure 39-5
Hill plot for Problem 13.

the data obtained, that value is 10 μM. We then get the values in Table 39-1. The resulting Hill plot is shown in Figure 39-5. The Hill coefficient, n, is equal to the slope of the plot and is given by

$$n = \frac{0.8 - (-0.7)}{0.4} = 3.8$$

Since $n = 3.8$, we estimate that four molecules of cGMP must be hydrolyzed to open each channel.

Table 39-1
Values for Hill plot

Y	$1 - Y$	$\dfrac{Y}{(1 - Y)}$	$\log \dfrac{Y}{1 - Y}$	[cGMP]	log [cGMP]
0.2	0.8	0.25	−0.60	7.0	0.85
0.4	0.6	0.67	−0.17	9.0	0.95
0.5	0.5	1.0	0	10.0	1.0
0.6	0.4	1.5	0.18	11.1	1.05
0.8	0.2	4.0	0.60	14.1	1.15

Expanded Solutions to
Text Problems

This section includes additional explanation of the answers in Stryer's text where helpful. Where Stryer's answers needed no expansion, they have been reproduced from his text without change.

Chapter 2

1. Since tropomyosin is double-stranded, each strand will have a mass of 35 kd. If the average residue has a mass of 110 d, there are 318 residues per strand (35,000/110), and the length is 477 Å (1.5 Å/residue × 318).

2. Branching at the β-carbon of the side chain (isoleucine), in contrast to branching at the γ-carbon (leucine), sterically hinders the formation of a helix. This fact can be shown with molecular models.

3. Changing alanine to valine results in a bulkier side chain, which prevents the correct interior packing of the protein. Changing a nearby, bulky, isoleucine side chain to glycine apparently alleviates the space problem and allows the correct conformation to take place.

4. The amino acid sequence of insulin does not determine its three-dimensional structure. By catalyzing a disulfide-sulfhydryl exchange, this enzyme speeds up the activation of scrambled ribonuclease because the native form is the most thermodynamically stable. In contrast, the structure of active insulin is not the most thermodynamically stable form. In fact, the two chains of insulin originate from a single peptide (see Chapter 38) that is folded by disulfide bridges and is later modified by selective proteolysis.

5. Appropriate hydrogen-bonding sites on the protease might form a β pleated sheet with a portion of the target protein. This process would effectively fully extend α helices and other folded portions of the target molecule.

Chapter 3

1. (a) The Edman method is best; hence, phenyl isothiocyanate.
   (b) Since you have a very small amount of sample, sensitivity is important. Hence, dabsyl chloride is the reagent of choice over FDNB (Sanger's reagent).
   (c) Reversible denaturation is usually achieved with 8 M urea. However, if disulfide bonds are present, they must first be reduced with β-mercaptoethanol to obtain a *random-coil* by urea treatment.

The known cleavage specificity of chymotrypsin (d), CNBr (e), and trypsin (f) provides an easy answer.

2. One can rearrange the Henderson-Hasselbach equation to the following:

$$pH - pK = \log \frac{[\text{base}]}{[\text{acid}]}$$

substituting pH = 4 into this equation shows that the log of the [base]/[acid] is −2; therefore, the ratio is 1/100. By similar reasoning, at pH 5, 6, 7, and 8 the respective ratios are 1/10, 1/1, 10/1, and 100/1. Note the logic: the lower the pH, the lower the [base]/[acid]; the higher the pH, the higher the [base]/[acid].

3. Whereas the hydrolysis of peptides yields amino acids, hydrazinolysis yields hydrazides $\left(\begin{array}{c}\text{O}\\\parallel\\-\text{C}-\text{NHNH}_2\end{array}\right)$ of all amino acids *except* the carboxyl-terminal residue. The latter can be separated from the hydrazides by the use of an anion exchange resin. (The hydrazides of aspartic and glutamic acids would also be picked up by the anion exchange resin; thus a further purification step might be necessary.)

4. (a) At pH 7, the only (+) charges will be the amino terminus and the side chains of lysine and arginine; the (−) charges will be the carboxy terminus and the side chains of aspartate and glutamate. By subtracting the total (−) charges from the total (+) charges, one obtains the answer, +1.

   (b) Treatment with CNBr cleaves the hormone at the single methionine, yielding two peptides.

5. Modification of cysteine side chains with ethyleneimine increases their length and adds a plus charge. These modified side chains are now very similar in size and charge to those of lysine and arginine; hence, they are susceptible to trypsin.

6. A 1-mg/ml solution of myoglobin (17.8 kd) is 5.62×10^{-5} M (1/17,800). The absorbance is 0.843 ($15,000 \times 1 \times 5.62 \times 10^{-5}$). Since this is the log of I_0/I, the ratio is 6.96 (the antilog of 0.843). Hence, 14.4% (1/6.96) of the incident light is transmitted.

7. Rod-shaped molecules have larger frictional coefficients than do spherical molecules. Because of this, the rod-shaped tropomyosin has a smaller (slower) sedimentation coefficient than does the spherical hemoglobin even though it has a higher molecular weight.

8. Electrophoretic mobilities are usually proportional to the log of the molecular weight (see Stryer, p. 45). Let x be the difference between the log of 30-kd and the log of the molecular weight of the unknown. Then, the difference between the logs of the molecular weights of the known proteins divided by the difference in their mobilities is equal to x divided by the difference in the mobilities of the unknown and the 30-kd protein. Solving 0.487/0.39 = x/0.18, gives x =

0.224. Then the log of the molecular weight of the unknown is 0.224 + 4.477 = 4.701; and the antilog (molecular weight) equals 50-kd.

9. Although the molecular weight remains essentially the same, reduction of the disulfide bonds apparently makes the molecule less spherical and less compact; hence, it is slower moving.

10. Compare the *diagonal electrophoresis* patterns obtained with the normal and the mutant proteins. If these patterns are essentially identical, the disulfide pairing is the same in both proteins; if they are not, the new cysteine residue is probably involved in the mutant.

11. At pH 7 all lysine side chains carry a plus charge, whereas at pHs above 10 a substantial portion of the side chains are uncharged. Since like charges repel, helix formation is inhibited at pH 7 but not at the higher pHs.

12. The pK of the glutamate side-chain carboxyl is approximately 4.3. Therefore, we predict that at pH 5 or above, where virtually all the side chains are ionized, helix formation will be inhibited due to the repulsion by like charges (COO⁻). Around pH 4, most side chains will be protonated (uncharged) and will not inhibit helix formation.

13. Being the smallest amino acid, glycine can fit into spaces too small to accommodate other amino acids. Thus, if a functionally active conformation requires glycine, no substitute will suffice; hence, glycine is highly conserved.

14. The method of choice is *affinity chromatography* using a column of beads covalently attached to vasopressin. This column should bind the receptor and few other proteins. The receptor can be removed from the washed column by treatment with vasopressin.

15. The slight decrease in the sedimentation coefficient suggests that substrate binding causes a conformational change resulting in a slightly less dense (more buoyant) or less spherical (larger frictional coefficient) molecule or both.

16. E, field strength, corresponds to $\omega^2 r$, centrifugal field; z, charge, corresponds to m′, effective mass. The former, E and $\omega^2 r$, are the forces applied to the molecules; the latter, z and m′, are inherent characteristics of the molecule itself.

17. The shape of a protein will affect its movement in a sedimentation-velocity experiment because shape affects the frictional coefficient. Molecular shape does not affect a sedimentation-equilibrium experiment since frictional coefficients are not involved.

18. *SDS-polyacrylamide gel electrophoresis* separates molecules on the basis of size, with the larger moving more *slowly*. The opposite is true in *gel filtration* because the smaller molecules are retarded as they enter the pores of the gel, while the larger molecules pass through virtually unhindered.

19. Monoclonal antibodies often target specific sites on antigens. These results suggest that the 23-kd and 57-kd proteins have an area of homology that is detected by the first monoclonal antibody, and the

23-kd and 69-kd proteins have a different homology that is detected by the second antibody.

20. The amino-terminal region is not held rigidly in the crystal and is thus free to move somewhat. This causes a fuzziness in the electron-density map, which is a composite of many measurements made over a long time period.

21. One could attach a fluorescent probe to a bacterial degradation product of interest and expose a mixture of cells to this molecule. Cells with a receptor for the degradation product will fluoresce and can be separated by FACS.

Chapter 4

1. By convention, when polynucleotide sequences are written, left to right means $5' \rightarrow 3'$. Since complementary strands are antiparallel, if one wishes to write the complementary sequence without specifically labeling the ends, the order of the bases must be reversed.
   (a) TTGATC
   (b) GTTCGA
   (c) ACGCGT
   (d) ATGGTA

2. (a) Since [A] + [G] account for 0.54 mole-fraction units, [T] + [C] must account for the remaining 0.46 (1 − 0.54).
   (b) Due to base pairing (A:T, G:C) in the complementary strand, [T] = 0.30, [C] = 0.24, and [A] + [G] = 0.46.

3. The length of a DNA segment (in this case 2×10^{-6} m) divided by the distance between the base pairs (3.4×10^{-10} m) gives the answer, 5.88×10^3 base pairs.

4. After 1.0 generation, one-half of the molecules would be ^{15}N-^{15}N, the other half ^{14}N-^{14}N. After 2.0 generations, one-quarter of the molecules would be ^{15}N-^{15}N, the other three-quarters ^{14}N-^{14}N. Hybrid ^{14}N-^{15}N molecules would not be observed in conservative replication.

5. Under intracellular conditions, melting would be more difficult. If it occurred, the mobility of the single strands might be restricted, thus enhancing reannealing to the original molecule.

6. Non-competent strains may not be able to take up DNA. Alternatively, they may have potent deoxyribonucleases, or they may not be able to integrate fragments of DNA into their genome.

7. The Hershey and Chase experiments would have been indecisive if performed with M13 because both ^{35}S (in protein) and ^{32}P (in DNA) would have been found in the bacteria, giving little information as to which component carried genetic information.

8. Thymine is the molecule of choice since it occurs in DNA, is not a component of RNA, and is not readily converted to cytosine or uracil. If they enter the cell, labeled thymidine or dTTP would also be useful molecules.

9. During DNA synthesis, the β- and γ-phosphorous atoms of the nucleoside triphosphates are lost as pyrophosphate. Since the α-phosphorous atom is incorporated into DNA, one should use dATP, dGTP, dTTP, and dCTP labeled with ^{32}P in the α position.

10. Only (c) would lead to DNA synthesis since (a) and (b) have no primer or open end to build on and (d) has no template extending beyond a free 3′—OH.

11. A short polythymidylate chain would serve as a primer since T base pairs with A. Radioactive dTTP labeled in any position except the β- and γ-phosphates would be useful for following chain elongation.

12. After the synthesis of the complementary (−) DNA on the RNA template, the RNA must be disposed of by hydrolysis prior to the completion of the synthesis of the DNA duplex.

13. (a) One should treat the infectious nucleic acid with highly purified ribonuclease or deoxyribonuclease and then determine its infectivity. RNAse will destroy the infectivity if it is RNA; DNAse will destroy it if it is DNA.
    (b) Since the material is infective even after protein removal, it is unlikely that the virus particle carries an enzyme essential for its own replication.

14. Ultimately this mutation results in half the daughter DNA duplexes being normal and half having a TA pair that had been CG. The first two rounds of replication at the mutant site will be as follows:

Chapter 5

1. (a) Deoxyribonucleoside triphosphates versus ribonucleoside triphosphates.
   (b) $5' \rightarrow 3'$ for both.
   (c) Semiconserved for DNA polymerase I, conserved for RNA polymerase.
   (d) DNA polymerase I needs a primer, whereas RNA polymerase does not.

2. Because mRNA is synthesized antiparallel to the DNA template and A pairs with U and T pairs with A, the correct sequence is 5′-UAACGGUACGAU-3′.

3. The 2′-OH group in RNA acts as an intramolecular catalyst. In the alkaline hydrolysis of RNA, it forms a 2′-3′ cyclic intermediate.

4. Cordycepin terminates RNA synthesis. An RNA chain containing cordycepin lacks a 3′-OH group.

5. Since the 5′ end of an mRNA molecule codes for the amino terminus, appropriate use of the genetic code (Stryer, p. 107) leads one to the answer Leu-Pro-Ser-Asp-Trp-Met.

6. Since one has a repeating tetramer (UUAC) and a 3-base code, repetition will be observed at a 12-base interval (3 × UUAC). Comparison of this 12-base sequence with the genetic code leads to the conclusion that a polymer with a repeating tetrapeptide (Leu-Leu-Thr-Tyr) unit will be formed.

7. (a) The probability of getting two mutations in a 3-base sequence is very small.

(b) The more palatable substitutions are those that can result from a *single* base change in the lysine codon AAG. By consulting the genetic code (Stryer, p. 107), one can see that the lysine codon (AAG) can be changed to the codon for Arg, Asn, Gln, Glu, Ile, Met, or Thr by changing a single base.

8. Since three different polypeptides are synthesized, the synthesis must start from three different reading frames. One of these will be in phase with the AAA in the sequence shown in the problem and will therefore have a terminal lysine since UGA is a stop signal. The reading frame in phase with AAU will result in a polypeptide having an Asn-Glu sequence in it, and the reading frame in phase with AUG will have a Met-Arg sequence in it.

9. Highly abundant amino acid residues have the most codons (e.g., Leu and Ser each have six), whereas the least abundant ones have the fewest (Met and Trp each have only one). Degeneracy (a) allows variation in base composition and (b) decreases the likelihood that a substitution of a base will change the encoded amino acid. If the degeneracy were equally distributed, each of the twenty amino acids would have three codons. Benefits (a) and (b) are maximized by assigning more codons to prevalent amino acids than to less frequently used ones.

10. During peptide synthesis, mRNA codons recognize the anticodons of tRNAs, *not* the amino acid the tRNA is carrying. Therefore, the cysteine tRNA (with an anticodon for UGU) carrying alanine will be recognized by UGU on the mRNA during peptide synthesis. The product will therefore be Phe-Cys-His-Val-*Ala*-Ala.

11. Organisms that survive at elevated temperatures are expected to have more GC in their DNA since this raises the melting temperature. The reverse might be expected in colder temperatures. Therefore, one would predict that the hot-spring alga have a higher frequency of the GUC and GUG codons for valine, whereas those from the Antarctic have higher frequencies of GUU and GUA.

12. The degeneracy of the genetic code makes it possible for comparable DNA molecules to have less homology than the proteins they code for since a base change in DNA need not cause a change in the amino acid sequence.

13. (a) In the male, a single defect on the X chromosome could cause color blindness, whereas in the female, two defects would be needed since there are two X chromosomes.

(b) Apparently a portion of the gene was replicated twice at some time in the past. Having two identical visual pigment genes on the X chromosome would protect against color blindness.

Chapter 6

1. The direction of movement on the gel is from top to bottom, with the smallest fragment, in this case G, moving most rapidly. Since the 5′ end carries the ^{32}P label, the 5′→3′ sequence is read from bottom to top, opposite the direction of movement. The sequence is 5′-GGCATAC-3′. It should be noted that the fastest moving spot in the autoradiogram is the radioactive inorganic phosphate resulting from the destruction of the guanine at the 5′ terminus. In the example shown in the text (Figure 6-5, Stryer, p. 120), the P_i spot is not shown.

2.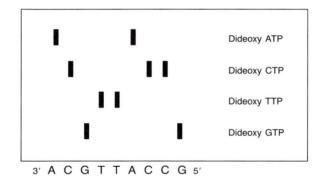

3. Ovalbumin cDNA should be used. *E. coli* lacks the machinery to splice the primary transcript arising from genomic DNA.

4. (a) No, because most human genes are much longer than 4 kb. One would obtain fragments containing only a small part of a complete gene.

(b) No, chromosome walking depends on having *overlapping* fragments. Exhaustive digestion with a restriction enzyme produces nonoverlapping, short fragments.

5. Southern blotting of an MstII digest would distinguish between the normal and mutant genes. The loss of a restriction site would lead to the replacement of two fragments on the Southern blot by a single longer fragment (see Stryer, p. 169). Such a finding would not prove that GTG replaced GAG; other sequence changes at the restriction site could yield the same result.

6. Cech replicated the recombinant DNA plasmid in *E. coli,* and then transcribed the DNA in vitro using bacterial RNA polymerase. He then found that this RNA underwent self-splicing in vitro in the complete absence of any proteins from *Tetrahymena.*

7. (a) Only one of the foreign DNA strands encodes an mRNA specifying a functional protein.

   (b) One could analyze RF molecules from the original single virus infection by restriction enzyme mapping. Asymmetric restriction sites in the foreign DNA fragments and phage chromosome would yield different gel patterns according to the orientation of the foreign DNA with the M13 DNA.

   (c) Single-strand circles from two clones containing the same foreign DNA strand will not hybridize to one another. In contrast, the circles will hybridize to form a duplex if the foreign DNA in them are complements of one another. S1 nuclease digests unpaired DNA strands but it does not digest duplex DNA.

8. Knowledge of the amino acid sequence is essential. It would be helpful to know which bonds are highly susceptible to proteolysis, and which residues are critical for the biological function of the peptide.

Chapter 7

1. (a) $87\ \mu m^3 = 87 \times 10^{-12}\ cm^3 = 8.7 \times 10^{-11}$ ml.
       Since each milliliter contains 0.34 g of Hb, $8.7 \times 10^{-11} \times 0.34 = 2.96 \times 10^{-11}$ g of Hb per cell.

   (b) The molecular weight of Hb is 66,000. Thus,

$$2.96 \times \frac{10^{-11}}{66,000} = 4.49 \times 10^{-14}\ mol\ Hb$$

       Moles of Hb × Avogadro's number = 2.71×10^8 molecules of Hb per cell.

   (c) No. There would be 3.22×10^8 hemoglobin molecules in a red cell if they were packed in a cubic crystalline array. Hence, the actual packing density is about 84% of the maximum possible.

2. $70 \times 70 \times \frac{16}{100} \div 66,000 = 1.19 \times 10^{-2}$ mol Hb.
   Moles of Hb × 4 (4 Fe/Hb) × 55.85 (at. wt. of Fe) = 2.65 g of Fe per adult.

3. (a) $(8/17,800) \times 32$ (mol. wt. of O_2) $= 1.44 \times 10^{-2}$ g O_2 per kg of muscle. Therefore, in whales, 1.44×10^{-1} g O_2 per kg of muscle.

   (b) $\dfrac{80}{17,800} = 4.49 \times 10^{-3}$ mol/kg (or mol/liter) $= 4.49 \times 10^{-3}$ M
       $\dfrac{4.49 \times 10^{-3}\ M}{3.5 \times 10^{-5}\ M} = 128/1$

4. (a) Equilibrium constants express relationships between the concentrations of reactants and products. They also express the ratio of the rate-constant of the forward (off) reaction to the rate constant of the reverse (on) reaction. Therefore,
       $\dfrac{k_{off}}{k_{on}} = \dfrac{k_{off}}{2 \times 10^7\ M^{-1}\ s^{-1}} = 10^{-6}$ M. Solving for k_{off} leads to the answer 20 s^{-1}.

   (b) Mean duration is 0.05 s (the reciprocal of k_{off}).

5. (a) Increasing $[H^+]$ decreases the oxygen affinity (Bohr effect). Since increasing pH decreases $[H^+]$, the oxygen affinity is *increased*.

   (b) Increasing $[CO_2]$ *decreases* oxygen affinity (modified Bohr effect).

   (c) BPG changes the oxygen saturation curves of Hb from hyperbolic to sigmoidal, thus decreasing oxygen affinity. Therefore, increasing BPG decreases oxygen affinity.

   (d) The monomer units of Hb act essentially like Mb; hence, conversion to the monomer *increases* oxygen affinity.

6. Inositol (hexahydroxycyclohexane) hexaphosphate has numerous negative charges on it and phosphate groups whose distance apart approximates that found in BPG. Its charge, size, and shape are sufficiently like those in BPG to mimmick its reaction with hemoglobin.

7. Consider yourself a proton (H^+) trying to escape from a carboxyl (—COOH) or to protonate its anion (—COO$^-$). (Remember that the stronger the acid, the lower the pK and vise versa.) In (a), the positively charged lysine side chain will help repel you (the proton) from the carboxyl, thus making it a stronger acid with a lowered pK. In (b), the presence of another COO$^-$ group will make it more difficult for you to escape from the carboxyl; hence, it will weaken the acid and raise the pK. In (c), the presence of H_2O enhances the ionization of acids by forming H_3O^+. Therefore, if the side chain is placed at a nonpolar site, it is less likely to ionize. The result is a weaker acid with a higher pK.

8. (a) The dissociation constant for the conversion of P to PAB is indeed constant regardless of pathway (P → PB → PAB or P → PA → PAB). Therefore, $K_A \times K_{AB} = K_B \times K_{BA}$. Solving for K_{AB}, one gets 2×10^{-5} M.

   (b) [A] enhances the binding of B because K_B is numerically larger than K_{AB}. It follows that [B] enhances the binding of A because K_A is larger than K_{BA}.

9. Carbon monoxide bound to one heme alters the oxygen affinity of the other hemes in the same hemoglobin molecule. Specifically, CO increases the oxygen affinity of hemoglobin and thereby decreases the amount of O_2 released in actively metabolizing tissues. Carbon monoxide stabilizes the quaternary structure characteristic of oxyhemoglobin. In other words, CO mimics O_2 as an allosteric effector.

10. (a) For maximal transport, $K = 10^{-5}$ M. In general, maximal transport is achieved when $K = ([L_A][L_B])^{0.5}$.

    (b) For maximal transport, $P_{50} = 44.7$ torrs, which is considerably higher than the physiological value of 26 torrs. However, it must be stressed that this calculation ignores cooperative binding and the Bohr effect.

11. (a) Knowledge of the specificity of trypsin leads to the conclusion that, if a hexapeptide is obtained, position 6 must be either Arg or Lys.

(b) The single base change GAG → AAG will change position 6 from Glu to Lys. No Arg codon can be formed from GAG by a single base change.

(c) The mutant hemoglobin moves more slowly toward the anode than does Hb A and Hb S because it is less negatively charged. The isoelectric points of Hb A and Hb S are approximately 6.9 and 7.1, respectively (Stryer, p. 165). Since it has an extra plus charge, this abnormal Hb will have a slightly higher isoelectric point (7.3?). However, at pH 8 it will still have a negative charge and will move toward the anode but at a slower rate than Hb A and Hb S, which have slightly larger negative charges.

12. This does not *prove* the presence of a sickle-cell gene because a mutation of the GAG codon for Glu to GCG (Ala) or GGG (Gly) would give similar experimental results.

13. Mutations in the α gene affect all three hemoglobins because their subunit structures are $\alpha_2\beta_2$, $\alpha_2\delta_2$, and $\alpha_2\gamma_2$. Mutations in the β, δ, or γ genes affect only one of them.

14. Deoxy Hb A contains a complementary site, and so it can add on to a fiber of deoxy Hb S. The fiber cannot then grow further because the terminal deoxy Hb A molecule lacks a sticky patch.

15. The reaction of CO_2 with α-amino groups to form carbamates *lowers* the oxygen affinity of hemoglobin because of the negative charge on the carbamate (Stryer, p. 162). The reaction with cyanate results in an uncharged urealike N-terminus, which *increases* oxygen affinity, thus preventing sickling.

16. (a) The increase in pO_2 will increase the oxygenation of Hb, thus pulling more Fe into the plane of the heme. This causes a conformational change that increases the acidity of certain groups and results in the release of H^+ (Bohr effect).

(b) Cyanate, like CO_2, reacts with the unprotonated form of the α-NH_2 groups (Stryer, p. 162), thus shifting the equilibrium fron $-NH_3^+$ to NH_2, which results in the release of H^+. Since this form of hemoglobin is more readily oxygenated, there is a further release of H^+ (Bohr effect) due to oxygenation.

Chapter 8

1. (a) $(2800 \text{ units} \times 10^{-5} \text{ mol})/(15 \text{ min} \times 60 \text{ s}) =$
$$31.1 \times 10^{-6} \text{ mol/s}$$

(b) $10^{-3} \text{ g}/(20 \times 10^3 \text{ g/mol subunit}) = 5 \times 10^{-8} \text{ mol}$

(c) The turnover number is equal to

$$\frac{31.1 \times 10^{-6} \text{ mol S s}^{-1} \times \text{Avogadro's number}}{5 \times 10^{-8} \text{ mol E} \times \text{Avogadro's number}} =$$
622 molecules S per second per molecule E

The definition of *turnover number* is given by Stryer on page 190.

2. For (a) and (b), proper graphing of the data given will

provide the correct answers: $K_M = 5.2 \times 10^{-6}$ M; $V_{max} = 6.84 \times 10^{-10}$ mol/min.

(c) Turnover = mol S s^{-1}/mol E =
$(6.84 \times 10^{-10})/[(60 \times 10^{-9})/29{,}600] = 337 \text{ s}^{-1}$

3. Penicillinase, like glycopeptide transpeptidase, forms an acyl-enzyme intermediate with its substrate but transfers it to water rather than to the terminal glycine of the pentaglycine bridge.

4. For (a) and (b), proper graphing of the data given will provide the correct answers:

(a) In the absence of inhibitor, V_{max} is 47.6 μmol/min and K_M is 1.1×10^{-5} M. In the presence of inhibitor, V_{max} is the same, and the apparent K_M is 3.1×10^{-5} M.

(b) Competitive.

(c) Since this is competitive inhibition, equation 37 (Stryer, p. 194) applies. The only difference between equation 37 and equation 29 (Stryer, p. 189) is the factor $(1 + [I]/K_i)$. Since in competitive inhibition V_{max} does not change, this factor describes the relationship between K_M and apparent K_M. Hence, the apparent $K_M = K_M (1 + [I]/K_i)$. Therefore, using the data in (a) and (b), 3.1×10^{-5} M $= (1.1 \times 10^{-5}$ M$)(1 + 2 \times 10^{-3}$ M$/K_i)$ and $K_i = 1.1 \times 10^{-3}$ M.

(d) The $[S]/(K_M + [S])$ term in the Michaelis-Menten equation tells us the fraction of enzyme molecules bound to substrate. Thus, $1 \times 10^{-5}/(1 + 3.1)10^{-5} = f_{ES} = 0.243$. Since $K_i = [E][I]/[EI]$, $[EI]/[E] = [I]/K_i = 2 \times 10^{-3}/1.1 \times 10^{-3} = 1.82$. However, the sum of $[EI] + [E]$ is only 0.757 of the total enzyme because the remaining 0.243 is bound to substrate. Therefore, $1.82 = f_{EI}/(0.757 - f_{EI})$. Solving this equation gives $f_{EI} = 0.488$.

(e) Using $[S]/(K_M + [S])$, $3 \times 10^{-5}/(1.1 + 3) \times 10^{-5} = 0.73$; $3 \times 10^{-5}/(3.1 + 3) \times 10^{-5} = 0.49$. This ratio, 0.73/0.49 and 33.8/22.6 (the velocity ratio) are equal.

5. For (a) and (b), proper graphing of the data given will provide the correct answers:

(a) V_{max} is 9.5 μmol/min. K_M is 1.1×10^{-5} M, the same as without inhibitor.

(b) Noncompetitive.

(c) Because this is *noncompetitive* inhibition, use equation 39 (Stryer, p. 195) as follows: 9.5 μmol/min = 47.6 μmol/min/(1 + 10^{-4} M$/K_i$). Solving for K_i one obtains the answer 2.5×10^{-5} M.

(d) Since an inhibitor does not affect K_M, the fraction of enzyme molecules binding substrate = $[S]/(K_M + [S])$, with or without inhibitor. For solution see 4(e).

6. (a) $V = \dfrac{V_{max} [S]}{(K_M + [S])}$

$$\frac{V(K_M + [S])}{[S]} = V_{max}$$

$$V + \frac{VK_M}{[S]} = V_{max}$$

$$V = V_{max} - \frac{VK_M}{[S]}$$

$$V = V_{max} - \frac{K_M V}{[S]}$$

(b) The slope of a straight line is the x-coordinate multiplied by the equation for the straight line. Thus, in the Lineweaver-Burk plot, K_M/V_{max} is the slope (Stryer, p. 189); in the Eadie-Hofstee plot the slope is $-K_M$ because $V/[S]$ is plotted on the x axis; see (a). By inspection, the y-intercept is V_{max}. The x-intercept is V_{max}/K_M because one is extrapolating to $[S] = 0$.

(c) Note that with a competitive inhibitor V_{max} (y-intercept) stays the same but K_M increases (the slope of 2 is greater than the slope of 1). In contrast, with a noncompetitive inhibitor, K_M does not change; 1 and 3 have the same slope (while V_{max} decreases).

7. Potential hydrogen-bond donors at pH 7 are the side chains of the following residues: arginine, asparagine, glutamine, histidine, lysine, serine, threonine, tryptophan, and tyrosine.

8. The rates of utilization of A and B are given by

$$V_A = \left(\frac{k_3}{K_M}\right)_A [E][A]$$

and

$$V_B = \left(\frac{k_3}{K_M}\right)_B [E][B]$$

Hence, the ratio of these rates is

$$V_A/V_B = \left(\frac{k_3}{K_M}\right)_A [A] \Big/ \left(\frac{k_3}{K_M}\right)_B [B]$$

Thus, an enzyme discriminates between competing substrates on the basis of their values of k_3/K_M rather than of K_M alone. Note that the velocity is dependent on the constants (k_3/K_M) *and* the concentrations of enzyme and substrate.

Chapter 9

1. Lysozyme hydrolyzes the glycosidic bond between the C-1 of NAM (M) and the C-4 of NAG (G). Since cleavage occurs between the D and E sugar binding sites (see Stryer, p. 206), G-M-G-M-G-M (b) should be expected to be hydrolyzed rapidly to G-M-G-M and G-M. M-G-M-G-M-G (c) will have more difficulty fitting into the active site, so it will be hydrolyzed more slowly. M-M-M-M-M-M (a), which contains only the bulkier NAM, will have the least access to the active site, so it will have the lowest rate of hydrolysis.

2. (a) G-G will prefer sites B-C because they have binding affinities that have the largest negative $\Delta G^{\circ\prime}$ (see Stryer, p. 209).
   (b) Since NAM does not fit site C and since site D has

a negative binding affinity, G-M will prefer sites A-B or E-F.
   (c) G-G-G-G will prefer sites A-B-C. One sugar residue does not interact with the enzyme, thus avoiding site D, which is energetically unfavorable.

3. During hydrolysis by lysozyme, the oxygen of water appears in the OH group on the anomeric carbon of NAM (see Stryer, p. 206). Therefore, the glycosidic O (the ^{18}O) is in the OH on the C-4 of NAG after hydrolysis.

4. This analog lacks a bulky substituent at C-5 and so it can probably bind to site D without being strained. Consequently, the binding of residue D of this analog is likely to be energetically favorable, whereas the binding of residue D of tetra-NAG costs free energy. See P. van Eikeren and D. M. Chipman, *J. Am. Chem. Soc.* 94(1972):4788.

5. (a) In oxymyoglobin, Fe is bonded to five nitrogens and one oxygen. In carboxypeptidase A, Zn is bonded to two nitrogens and two oxygens.
   (b) In oxymyoglobin, one of the nitrogens bonded to Fe comes from the proximal histidine residue, whereas the other four come from the heme. The oxygen atom linked to Fe is that of O_2. In carboxypeptidase A, the two nitrogen atoms coordinated to Zn come from histidine residues. One of the oxygen ligands is from a glutamate side chain, the other from a water molecule.
   (c) Aspartate, cysteine, and methionine.

6. (a) Yes.
   (b) Histidine 119 in ribonuclease A, histidine 57 in chymotrypsin, glutamic acid 35 in lysozyme, and a zinc-bound water molecule in carboxypeptidase A.

7. (a) Tosyl-L-lysine chloromethyl ketone (TLCK).
   (b) First, determine whether substrates protect trypsin from inactivation by TLCK and, second, ascertain whether the D-isomer of TLCK inactivates trypsin.

8. (a) Serine.
   (b) A hemiacetal between the aldehyde of the inhibitor and the hydroxyl group of the active site serine.

9. The boron atom becomes bonded to the oxygen atom of the active-site serine. This tetrahedral intermediate has a geometry similar to that of the transition state.

10. Precise positioning of catalytic residues and substrates, geometrical strain (distortion of the substrate), electronic strain, and desolvation of the substrate.

11. The zinc ion polarizes the carbonyl group of the scissile bond to make its carbon atom more positively charged. Zn^{2+} also enhances the nucleophilicity of its bound water molecule. Glutamate pulls a proton away from the zinc-bound water molecule, which makes it a stronger nucleophile. See M. A. Holmes and B. W. Matthews, *Biochemistry* 20(1981):6912.

Chapter 10

1. The protonated form of histidine probably stabilizes the negatively charged carbonyl oxygen atom of the scissile bond in the transition state. Deprotonation would lead to a loss of activity. Hence, the rate is expected to be half-maximal at a pH of about 6.5 (the pK of an unperturbed histidine side chain in a protein) and decrease as the pH is raised.

2. (a) Using the same logic as that used in Problem 8 of Chapter 7, one can show that the change in [R]/[T] is the same as the ratio of the substrate affinities of the two forms. For example, the mathematical constant for the conversion of R to T_S is the same whether one proceeds $R \rightarrow T \rightarrow T_S$ or $R \rightarrow R_S \rightarrow T_S$. Let us assume that the constant for the conversion of R to T and R to R_S is 10^3. Since the affinity of R for S is 100 times that of T, it follows that the constant for the conversion of T to T_S is 10. The constant for the conversion of R_S to T_S is therefore equal to $10^3 \times 10/10^3$ or 10. Note that the binding of substrate with a 100-fold tighter binding to R changes the R to T ratio from 1/1000 to 1/10.

   (b) 100. The binding of four substrate molecules changes the [R]/[T] by a factor of $100^4 = 10^8$. The ratio in the absence of substrate is 10^{-6}. Hence, the ratio in the fully liganded molecule is $10^8 \times 10^{-6} = 10^2$.

3. Activation is independent of zymogen concentration because the reaction is intramolecular (Stryer, p. 246).

4. Add blood from the second patient to a sample from the first. If the mixture clots, the second patient has a defect different from that of the first. This type of assay is called a complementation test.

5. Activated factor X remains bound to blood platelet membranes, which accelerates its activation of prothrombin.

6. Antithrombin III is a very slowly hydrolyzed substrate of thrombin. Hence, its interaction with thrombin requires a fully formed active site on the enzyme.

7. Residues a and d are located in the interior of an α-helical coiled coil, near the axis of the superhelix. Hydrophobic interactions between these side chains contribute to the stability of the coiled coil.

8. Replace methionine 358 with leucine, which occupies nearly the same volume and is hydrophobic (Stryer, p. 248).

Chapter 11

1. (a) Every third residue in each strand of a collagen triple helix must be glycine because there is insufficient space for a larger residue.

   (b) Poly(Gly-Pro-Gly) melts at a lower temperature than poly(Gly-Pro-Pro) because the thermal stability of collagen increases with the content of imino acids (proline and hydroxyproline). Apparently the bulkier proline and hydroxyproline side chains are necessary for the cooperative interactions that stabilize the triple-stranded helix. Although glycine is required at every third residue in the triple helix, it results in a less stable helix when it is found at other positions.

   (c) No. Glycine does not occupy every third position.

2. (a and b)

   —Gly—Leu—Pro—Gly—Pro—Pro—Gly—Ala—Pro—Gly

   Susceptible peptide bond

3. (a) Disulfides.

   (b) None.

   (c) Peptide bonds between specific glutamine and lysine side chains.

   (d) Aldol cross-link and hydroxypyridinium cross-link.

   (e) Lysinonorleucine and desmosine.

4. The decarboxylation of α-ketoglutarate is half of the physiological reaction. It seems likely that α-ketoglutarate is first attacked by oxygen to form a peroxy acid, which then reacts with the prolyl substrate. See D. F. Counts, G. J. Cardinale, and S. Udenfriend, *Proc. Nat. Acad. Sci.* 75(1978):2145, for a discussion of this experiment and its mechanistic implications.

5. (a) The aldehyde is more reactive. For example, it can form a hemiacetal with the hydroxyl group of a serine at an active site.

   (b) A tetrahedral intermediate, like the one formed in catalysis by chymotrypsin (Stryer, p. 225), seems likely.

6. The peptide Arg-Gly-Asp-Ser competitively inhibits the binding of one of the modules of fibronectin to integrin, a protein that spans the plasma membrane of fibroblasts.

7. The three-quarter-length and one-quarter-length fragments of collagen formed by the action of tissue collagenases are likely to be more stable than normal. Collagen resorption will probably be impeded.

8. This mutation will probably mimic vitamin C deficiency. Defective hydroxylation of proline will lead to collagen having a lower than normal melting temperature. The abnormal collagen cannot properly form fibers; skin lesions and blood vessel fragility are likely consequences. The clinical picture will resemble scurvy.

Chapter 12

1. $1\ \mu m^2 = (10^{-6}\ m)^2 = 10^{-12}\ m^2.$ $70\ \text{Å}^2 = 70\ (10^{-10}\ m)^2 = 70 \times 10^{-20}\ m^2.$ Since the bilayer has two sides, $(2 \times 10^{-12}/70 \times 10^{-20}) = 2.86 \times 10^6$ molecules.

2. Cyclopropane rings interfere with the orderly packing of hydrocarbon chains, so they increase membrane fluidity.

3. Using the diffusion coefficient equation, $s = (4\ Dt)^{1/2}$, one gets $s = (4 \times 10^{-8} \times 10^{-6})^{1/2}$ or $(4 \times 10^{-8} \times 10^{-3})^{1/2}$ or $(4 \times 10^{-8} \times 1)^{1/2}$. Solving for s gives 2×10^{-7}, 6.32×10^{-6}, and 2×10^{-4} cm, respectively.

4. The gram molecular weight of the protein divided by Avogadro's number $= 10^5\ g/6.02 \times 10^{23} = 1.66 \times 10^{-19}\ g/\text{molecule}.$ $1.66 \times 10^{-19}\ g/1.35$ (density) $= 1.23 \times 10^{-19}\ cm^3/\text{molecule}.$ The volume of a sphere equals $\frac{4}{3}\pi r^3 = 1.23 \times 10^{-19}\ cm^3.$ Solving for r, one gets 3.08×10^{-7} cm. By substituting this value into the equation given,

$$D = 1.38 \times 10^{-16} \times \frac{310}{(6 \times 3.14 \times 1 \times 308 \times 10^{-7})}$$
$$= 7.37 \times 10^{-9}\ cm^2/s$$

Using this value for D and the times given in the problem in the equation shown in the answer to Problem 3, one obtains the distances traversed.

5. The initial decrease in the amplitude of the paramagnetic resonance spectrum results from the reduction of spin-labeled phosphatidyl cholines in the outer leaflet of the bilayer. Ascorbate does not traverse the membrane under these experimental conditions, and so it does not reduce the phospholipids in the inner leaflet. The slow decay of the residual spectrum is due to the reduction of phospholipids that have flipped over to the outer leaflet of the bilayer.

Chapter 13

1. The direction of a reaction when the reactants are initially present in equimolar amounts is dependent on $\Delta G^{\circ\prime}$. Since by convention reactions are written from left to right, if $\Delta G^{\circ\prime}$ is negative, K'_{eq} is positive and the *direction* is to the right because at equilibrium the product concentrations will exceed those of the reactants. If $\Delta G^{\circ\prime}$ is positive, the reverse is true. The $\Delta G^{\circ\prime}$ values for these reactions are (a) $+3$ (left), (b) -5.1 (right), (c) $+7.5$ (left), and (d) -4 kcal/mol (right).

2. Consider a large rock that has been sitting on the side of a mountain for a million years. It has a large amount of potential energy but no kinetic energy—until you push it! Or consider a mixture of H_2 and O_2; it's perfectly stable until you light it! Notice that the thermodynamics of a reaction tell you little, if anything, about its kinetics.

3. (a) Note that phosphoenolpyruvate is *formed* in this reaction; hence, its contribution to $\Delta G^{\circ\prime}$ is *plus* 14.8 kcal/mol. Therefore, $\Delta G^{\circ\prime}$ for the entire reaction is $+14.8 - 7.3 = +7.5$ kcal/mol. $K'_{eq} = 10^{7.5/-1.36} = 3.06 \times 10^{-6}$.

    (b) When ATP/ADP is 10, PEP/Pyr $= 3.06 \times 10^{-6} \times 10 = 3.06 \times 10^{-5}$ and Pyr/PEP $= 1/(3.06 \times 10^{-5}) = 3.28 \times 10^4$.

4. $\Delta G^{\circ\prime} = +5 - 3.3 = 1.7$ kcal/mol. $K'_{eq}\dfrac{[G\text{-}1]}{[G\text{-}6]} = 10^{1.7/-1.36} = 5.62 \times 10^{-2}$. The reciprocal of this is $\dfrac{[G\text{-}6]}{[G\text{-}1]}$ or 17.8.

5. (a) $\Delta G^{\circ\prime} = +7.5 - 7.3 = +0.2$ kcal/mol.

    (b) The hydrolysis of PP_i drives the reaction towards the formation of acetyl CoA by making $\Delta G^{\circ\prime}$ strongly negative ($0.2 - 8.0 = -7.8$ kcal/mol).

6. (a) By definition, $\log K = -pK$. Since $\Delta G^{\circ\prime} = -2.3\ RT \log K$, by substitution one obtains $\Delta G^{\circ\prime} = 2.3\ RT\ pK$.

    (b) $\Delta G^{\circ\prime} = 1.36 \times 4.8 = 6.53$ kcal/mol at 25°C.

7. An ADP unit (or a closely related derivative, in the case of CoA).

8. The activated form of sulfate in most organisms is 3'-phosphoadenosine 5'-phosphosulfate. See P. W. Robbins and F. Lipmann, *J. Biol. Chem.* 229(1957):837.

9. (a) One hertz is a frequency of one cycle per second. 129 MHz $= 129 \times 10^6$ Hz. One *part per million* of this is 129 Hz. Therefore, 2.4 ppm is 2.4×129 Hz $= 310$ Hz.

    (b) Both the protonation and deprotonation rates must be faster than $310\ s^{-1}$.

    (c) The pK' for the equilibrium of $H_2PO_4^{2-}$ and HPO_4^{-} is 7.21 (see Appendix C). Hence, the dissociation constant K is 6.16×10^{-8} M. The rate constant for association k_{on} is equal to k_{off}/K. Because k_{off} is greater than $310\ s^{-1}$, k_{on} must be greater than 5×10^9 M^{-1} s^{-1}.

Chapter 14

1. To answer this problem, one must know the structures of the molecules in question and a couple of definitions. By definition, *epimers* are a pair of molecules that differ from each other only in their configuration at a single asymmetric center. *Anomers* are special epimers that differ only in their configuration at a carbonyl carbon; hence, they are usually acetals or hemiacetals. An aldose-ketose pair is obvious. Inspection of Fischer representations of the molecular pairs leads to the conclusion that (a), (c), and (e) are aldose-ketose pairs; (b) and (f) are epimers; and (e) are anomers.

(Problem 2)

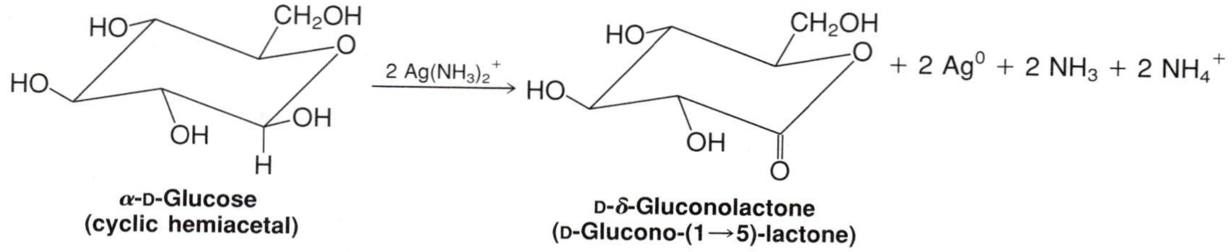

α-D-Glucose
(cyclic hemiacetal)

$2 Ag(NH_3)_2^+$

D-δ-Gluconolactone
(D-Glucono-(1→5)-lactone)

$+ 2 Ag^0 + 2 NH_3 + 2 NH_4^+$

(Problem 3)

Glucose

$H_2N—(Val)Hb$

Schiff base
(aldimine)

Amadori
rearrangement

Amino ketone

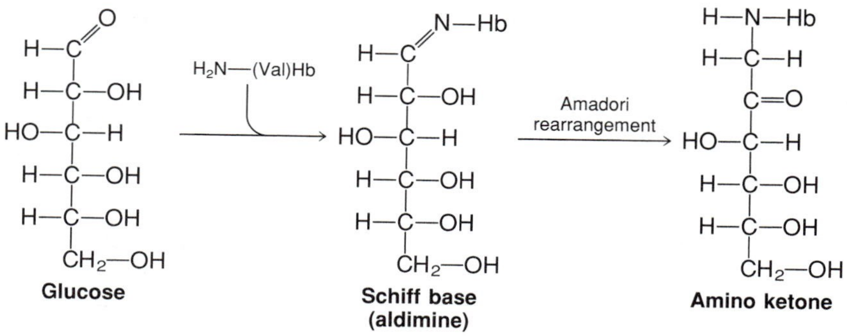

2. A mild oxidant, Tollens' reagent converts aldoses to aldonic acids and free silver as follows:

$$RCHO + 2 Ag(NH_3)_2 + H_2O \longrightarrow$$
$$RCO_2^- + 2 Ag^0 + 3 NH_4^+ + NH_3$$

However, cyclic hemiacetals are oxidized directly to lactones, which are hydrolyzed to the corresponding aldonic acid under alkaline conditions. Thus, in the case of glucose, the major first reaction product is D-δ-gluconolactone. To prepare aldonic acids, Br_2 is usually used as the oxidant because it gives fewer side reactions than does Tollens' reagent.

3. Glucose reacts slowly because the predominant hemiacetal ring form (which is inactive) is in equilibrium with the active straight-chain free aldehyde. The lat-

ter can react with terminal amino groups to form a Schiff base, which can then rearrange to the stable amino ketone, sometimes referred to as Hb A_{1c}, which accounts for approximately 5% to 8% of the hemoglobin in normal adult human red cells. In the diabetic, its concentration may rise to 12% or more due to the elevated concentrations of glucose.

4. Whereas pyranosides have a series of three adjacent hydroxyls, furanosides have only two. Therefore, oxidation of pyranosides uses *two* equivalents of periodate and yields *one* mole of formic acid, whereas oxidation of furanosides uses only *one* equivalent of periodate and yields *no* formic acid.

5. The formation of acetals (such as methylglucoside) is acid catalyzed. In a mechanism similar to that for the

β-D-Methylglucopyranoside

$+ IO_3^- \longrightarrow$

$+ IO_3^-$

2nd equivalent of IO_4^-

(Problem 4)

$IO_3^- + H—C—OH +$

Formic acid

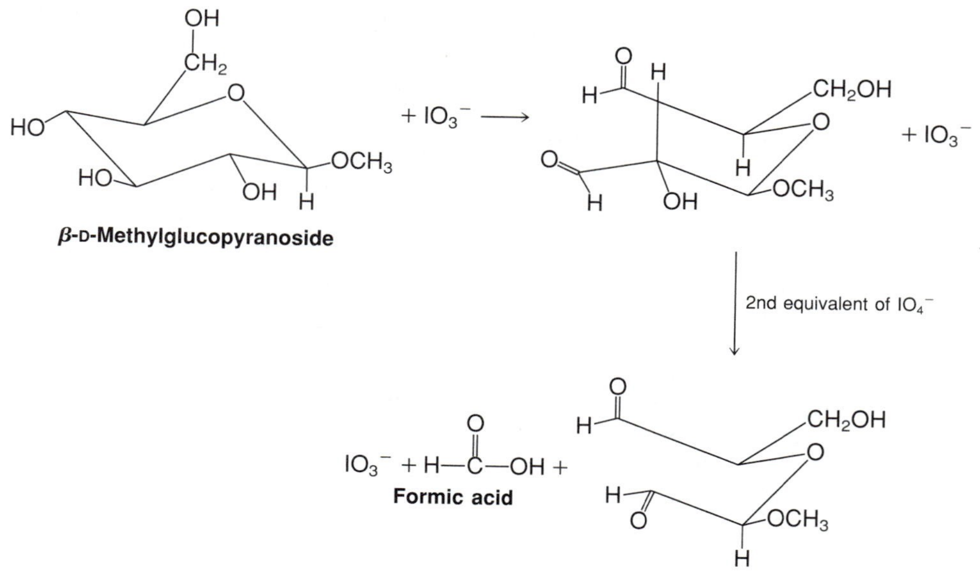

(Problem 4)

$$\text{β-D-Methylfructofuranoside} + IO_4^- \longrightarrow + IO_3^-$$

β-**D**-Methylfructofuranoside

(Problem 5)

D-Glucose
(β-pyranose form)

Electron pair on ring oxygen
can stabilize carbocation at
anomeric position only

esterification of carboxylic acids (shown in most organic texts), the anomeric hydroxyl group is replaced. The resulting carbocation is susceptible to attack by the nucleophilic oxygen of methanol, leading to the incorporation of this oxygen into the methylglucoside molecule.

6. By inspection, A, B, and D are the pyranosyl forms of D-aldohexoses because the CH_2OH is above the plane of the ring. In Haworth projections, OHs above the ring are to the left (Fischer projections) and those below the ring are to the right. Therefore, A is β-D-mannose, B is β-D-galactose, and D is β-D-glucosamine. By similar use of the Haworth projection, C can be identified as β-D-fructose. All these sugars are β because, in Haworth projections, when the CH_2OH attached to the C-5 carbon (the carbon that determines whether the sugar is D or L) is above the ring, the sugar is β when the anomeric hydroxyl is also above the ring.

7. The trisaccharide itself should be a competitive inhibitor of cell adhesion if the trisaccharide unit of the glycoprotein is critical for the interaction.

Chapter 15

1. Glucose is reactive because its open-chain form contains an aldehyde group (see Stryer, p. 642).

2. (a) The key is the aldolase reaction. Note that the carbons attached to the phosphate in glyceraldehyde 3-phosphate and dihydroxyacetone phosphate are interconverted by triose phosphate isomerase and *both* become the terminal carbon of the glyceric acids. Hence, the label is in the methyl carbon of pyruvate.

   (b) 5 mCi/mM. The specific activity is halved because the number of moles of product (pyruvate) is twice that of the labeled substrate (glucose).

3. Glucose + 2 P_i + 2 ADP $\longrightarrow$ 2 lactate + 2 ATP

   (a) To obtain the answer, −29.5 kcal/mol, you add the $\Delta G^{\circ\prime}$ values given in Table 15-2 and the value given for the reduction of pyruvate to lactate. Remember that the values for the three carbon molecules must be doubled since each hexose yields two trioses.

   (b) $\Delta G' = -29.5 + 1.36 \log \dfrac{(5 \times 10^{-5})^2 \times (2 \times 10^{-3})^2}{(5 \times 10^{-3})(10^{-3})^2(2 \times 10^{-4})^2}$

   $= -29.5 + 1.36 \log 50$

   $= -27.2$ kcal/mol

   The concentrations of ATP, ADP, P_i, and lactate are squared because, in reactions such as A → 2B, the $K_{eq} = [B]^2/[A]$.

4. $-7.5 = -1.36 \log \dfrac{[Pyr][ATP]}{[PEP][ADP]}$

   $5.515 = \log \dfrac{[Pyr]}{[PEP]} + \log 10$

   $\dfrac{[Pyr]}{[PEP]} = 10^{4.515}$

   $\dfrac{[PEP]}{[Pyr]} = 10^{-4.515} = 3.06 \times 10^{-5}$

5. Since $\Delta G^{\circ\prime}$ for the aldolase reaction is +5.7 kcal/mol (Stryer, p. 357), $K_{eq} = 10^{5.7/-1.36} = 6.5 \times 10^{-5}$. If we let the concentration of each of the trioses (DHAP and G-3P) formed during the reaction be X, then the concentration of FDP is 10^{-3} M $- X$ at equilibrium since we started with millimolar FDP. Then,

   $$\dfrac{X^2}{10^{-3} - X} = 6.5 \times 10^{-5}$$

   Solving this quadratic equation leads to the answer of 2.24×10^{-4} M for X (DHAP and G-3P). Subtracting this from 10^{-3} gives the FDP concentration of 7.76×10^{-4} M.

6. The 3-phosphoglycerate labeled with ^{14}C accepts the phosphate attached to C-1 of 1,3-BPG (see Stryer, p. 369).

7. Hexokinase has a low ATPase activity in the absence of a sugar because it is in a catalytically inactive conformation (Stryer, p. 271). The addition of xylose closes the cleft between the two lobes of the enzyme.

However, the xylose hydroxymethyl group (at C-5) cannot be phosphorylated. Instead, a water molecule at the site normally occupied by the C-6 hydroxymethyl group of glucose acts as the phosphoryl acceptor from ATP.

8. (a) 2,3-Bisphosphoglycerate (BPG) lowers the oxygen affinity of hemoglobin (Stryer, p. 157). The rate of synthesis of 2,3-BPG is controlled by the level of 1,3-BPG, a glycolytic intermediate.

   (b) The lowered level of glycolytic intermediates leads to less 2,3-BPG, and hence, a higher oxygen affinity.

   (c) Glycolytic intermediates are present at a higher than normal level. The level of 2,3-BPG is increased, which makes the oxygen affinity lower than normal.

9. (a) The fructose 1-phosphate pathway (Stryer, p. 357) forms glyceraldehyde 3-phosphate. Phosphofructokinase, a key control enzyme, is bypassed. Furthermore, fructose 1-phosphate stimulates pyruvate kinase.

   (b) The rapid, unregulated production of lactate can lead to metabolic acidosis.

10. The catalytic site of an activated enzyme molecule contains a phosphorylated serine residue. This phosphoryl group is transferred to either substrate to form a glucose 1,6-bisphosphate intermediate (Stryer, p. 454). The phosphoryl group on the enzyme is slowly lost by hydrolysis; it is regenerated by phosphoryl transfer from 1,6-glucose bisphosphate.

11. The metal ion serves as an electron sink, as does the protonated Schiff base in animal aldolases.

12. EDTA removes the metal ion from the catalytic site of procaryotic aldolases, whereas sodium borohydride reduces the Schiff base intermediate in catalysis by animal aldolases.

Chapter 16

1. To answer this problem one must follow carbon atoms around the citric acid cycle as shown by Stryer on pages 378 and 385. Remember that the randomization of carbon occurs at succinate, a truly symmetrical molecule. Also, this problem (and the answers given) assumes that all pyruvate goes to acetyl CoA. In fact, this is not necessarily true since pyruvate can also enter the cycle at oxaloacetate (Stryer, Chapter 18).

   (a) After one round of the citric acid cycle, the label emerges in C-2 and C-3 of oxaloacetate.

   (b) After one round of the citric acid cycle, the label emerges in C-1 and C-4 of oxaloacetate.

   (c) The label emerges in CO_2 in the formation of acetyl CoA from pyruvate.

   For (d) and (e), the fate is the same as in (a).

2. No, because two carbon atoms are lost in the two decarboxylation steps of the cycle. Hence, there is no *net* synthesis of oxaloacetate.

3. Using the $\Delta G^{\circ\prime}$ values in Table 16-1, one can cal-

culate the equilibrium ratios isocitrate/citrate and *cis*-aconitate/citrate. Thus, we see [iso]/[cit] = $10^{1.5/-1.36}$ = 0.076 and [*cis*-acon]/[cit] = $10^{2/-1.36}$ = 0.034. Therefore, for every 100 citrate molecules, there would be 3.4 aconitate and 7.6 isocitrate molecules or 90% citrate, 3.1% aconitate, and 6.9% isocitrate.

4. Addition of the $\Delta G^{\circ\prime}$ values in Table 16-1 gives the answer −9.8 kcal/mol.

5. The coenzyme stereospecificity of glyceraldehyde 3-phosphate dehydrogenase is the opposite of that of alcohol dehydrogenase (type B versus type A, respectively.)

6. Thiamine thiazolone pyrophosphate is a transition state analog. The sulfur-containing ring of this analog is uncharged, and so it closely resembles the transition state of the normal coenzyme in thiamine-catalyzed reactions (e.g., the uncharged resonance form of hydroxyethyl-TPP, Stryer, p. 380). See J. A. Gutowski and G. E. Lienhard, *J. Biol. Chem.* 251(1976):2863, for a discussion of this analog.

7. $\dfrac{[OAA][NADH]}{[Mal][NAD^+]} = 10^{7/-1.36} = 7.08 \times 10^{-6}$
   Since [NADH]/[NAD$^+$] = 1/8, [OAA]/[Mal] = $7.08 \times 10^{-6} \times 8 = 5.67 \times 10^{-5}$. The reciprocal of this is 1.75×10^4, the smallest [Mal]/[OAA] ratio permitting net OAA formation.

8. Methane is first oxidized by a monooxygenase to methanol; NADH is the reductant. Methanol is then oxidized to formaldehyde; PQQ, a novel quinone, is the electron acceptor in this step. Formaldehyde is oxidized to formic acid, which is in turn oxidized to CO_2. NADH is formed in each of these steps. About 5 ATP are formed (3 ATP from NADH and 2 ATP from $PQQH_2$). See G. Gottschalk, *Bacterial Metabolism* (2nd ed.) (Springer-Verlag, 1986), p. 163.

9. The enolate anion of acetyl CoA attacks the carbonyl carbon atom of glyoxylate to form a C–C bond. This reaction is like the condensation of oxaloacetate with the enolate anion of acetyl CoA (Stryer, p. 383). Glyoxylate contains a hydrogen atom in place of the —CH_2COO^- of oxaloacetate; the reactions are otherwise nearly identical.

Chapter 17

1. (a) 15 (b) 2 (c) 38 (d) 16 (e) 36 (f) 19. These answers are readily obtained if one remembers (1) the ATP yields from the various parts of glycolysis, (2) cytoplasmic NADH yields only 2 ATPs, (3) pyruvate $\longrightarrow$ acetyl CoA + 1 NADH (3 ATPs), and (4) each acetyl CoA traversing the citric acid cycle yields 12 ATPs (3 NADH, 1 FADH$_2$, 1 GTP).

2. Remember that the values in Table 17-1 are *reduction* potentials.

   (a) The reaction is $2\,G—SH + \frac{1}{2}O_2 \rightleftharpoons G—S—S—G + H_2O$. $\Delta E_0'$ for the *oxidation* of glutathione is 0.82 + 0.23 (the oxidation potential) = 1.05 V. Then, $\Delta G^{\circ\prime} = -1.05 \times 2 \times 23 = -48.4$ kcal/mol.

(b) Since in the reduction of glutathione NADH is *oxidized*, $\Delta E_0' = -0.23 + 0.32 = +0.09$ V and $\Delta G^{\circ\prime} = -0.09 \times 2 \times 23 = -4.15$ kcal/mol.

3. (a) Blocks electron transport and proton pumping at site 3.

   (b) Blocks electron transport and ATP synthesis by inhibiting the exchange of ATP and ADP across the inner mitochondrial membrane.

   (c) Blocks electron transport and proton pumping at site 1.

   (d) Blocks ATP synthesis without inhibiting electron transport by dissipating the proton gradient.

   (e) Blocks electron transport and proton pumping at site 3.

   (f) Blocks electron transport and proton pumping at site 2.

4. Oligomycin inhibits ATP formation by interfering with the utilization of the proton gradient. It does not block electron transport.

5. For oxidation by NAD^+, $\Delta E_0' = -0.32 - 0.03 = -0.35$ V and $\Delta G^{\circ\prime} = -(-0.35) \times 2 \times 23 = +16.1$ kcal/mol. For oxidation by FAD, $\Delta E_0' = 0 - 0.03 = -0.03$ V and $\Delta G^{\circ\prime} = -(-0.03) \times 2 \times 23 = +1.38$ kcal/mol.

6. Cyanide can be lethal because it binds to the ferric form of cytochrome $(a + a_3)$ and thereby inhibits oxidative phosphorylation. Nitrite converts ferrohemoglobin to ferrihemoglobin, which also binds cyanide. Thus, ferrihemoglobin competes with cytochrome $(a + a_3)$ for cyanide. This competition is therapeutically effective because the amount of ferrihemoglobin that can be formed without impairing oxygen transport is much greater than the amount of cytochrome $(a + a_3)$.

7. $\Delta G^{\circ\prime} = -0.2 \times 2$ (or 3 or 4) $\times 23 = -9.23, -13.8,$ or -18.5 kcal/mol.

   These become positive values (energy input) for the $\Delta G'$ of ATP synthesis. Therefore,

   $$\Delta G' = \Delta G^{\circ\prime} + 1.36 \log \frac{[\text{ATP}]}{[\text{ADP}][\text{P}_i]}$$

   for 2 protons,

   $$9.23 = 7.3 + 1.36 \log \frac{[\text{ATP}]}{[\text{ADP}][\text{P}_i]}$$

   $$\frac{[\text{ATP}]}{[\text{ADP}][\text{P}_i]} = 10^{1.93/1.36} = 26.2$$

   Similar calculations with 3 and 4 protons gives ratios of 6.04×10^4 and 1.72×10^8, respectively.

8. Biochemists use E_0', the value at pH 7, whereas chemists use E_0, the value in 1 N H^+. The prime denotes that pH 7 is the standard state.

9. Such a defect (called Luft's syndrome) was found in a thirty-eight-year-old woman who was incapable of performing prolonged physical work. Her basal metabolic rate was more than twice normal, but her thyroid function was normal. A muscle biopsy showed that her mitochondria were highly variable and atypical in structure. Biochemical studies then revealed that oxidation and phosphorylation were not tightly coupled in these mitochondria. In this patient, much of the energy of fuel molecules was converted into heat rather than ATP. See R. Luft, D. Ikkos, G. Palmieri, L. Ernster, and B. Afzelius. *J. Clin. Invest.* 41(1962):1776.

10. The absolute configuration of thiophosphate is opposite to that of ATP in the reaction catalyzed by ATP synthase. This result is consistent with an in-line phosphoryl transfer reaction occurring in a single step. The retention of configuration in the Ca^{2+}-ATPase reaction points to two phosphoryl transfer reactions—inversion by the first, and a return to the starting configuration by the second. The Ca^{2+}-ATPase reaction proceeds by a phosphorylated enzyme intermediate. See M. R. Webb, C. Grubmeyer, H. S. Penefsky, and D. R. Trentham, *J. Biol. Chem.* 255(1980):255.

11. Dicylohexylcarbodiimide reacts readily with carboxyl groups, as was discussed earlier in regard to its use in peptide synthesis (Stryer, p. 65). Hence, the most likely targets are aspartate and glutamate side chains. In fact, aspartate 61 of subunit c of *E. coli* F_0 is specifically modified by this reagent. Conversion of this aspartate to an asparagine by site-specific mutagenesis also eliminates proton conduction. See A. E. Senior, *Biochim. Biophys Acta* 726(1983):81.

Chapter 18

1. (a) To make six pentoses, four glucose 6-phosphates must be converted to fructose 6-phosphate (no ATP required) and one glucose 6-phosphate must be converted to two molecules of glyceraldehyde 3-phosphate (this requires one ATP). These are converted to pentoses by the following reactions (Stryer, p. 431):

   2 fructose 6-phosphate + 2 glyceraldehyde 3-phosphate $\longrightarrow$ 2 erythrose 4-phosphate + 2 xylulose 5-phosphate

   2 fructose 6-phosphate + 2 erythrose 4-phosphate $\longrightarrow$ 2 glyceraldehyde 3-phosphate + 2 sedoheptulose 7-phosphate

   2 glyceraldehyde 3-phosphate + 2 sedoheptulose 7-phosphate $\longrightarrow$ *2 xylulose 5-phosphate + 2 ribose 5-phosphate*

   (b) What really happens is that six molecules of glucose 6-phosphate are converted to six CO_2 + six ribulose 5-phosphates. The ribulose phosphates are then converted back to five molecules of glucose 6-phosphate.

2. Since the C-1 of glucose is lost during the conversion to pentose, carbon atoms 2 through 6 of glucose become carbon atoms 1 through 5 of the pentose. That is, each pentose carbon is numerically 1 less than its counterpart in glucose.

3. Oxidative decarboxylation of isocitrate to α-keto-glutarate. A β-keto acid intermediate is formed in both reactions.

4. Ribose 5-P is first converted to xylulose 5-P (labeled in C-1) via ribulose 5-P. Transketolase can then catalyze the conversion of ribose 5-P + xylulose 5-P to sedoheptulose 7-P (labeled in C-1 and C-3) and glyceraldehyde 3-P. Transaldolase then transfers carbons 1-3 of sedoheptulose 5-P to glyceraldehyde 3-P, forming erythrose 4-P, which is unlabeled (from carbons 4-7 of sedoheptulose) and fructose 6-P, which is labeled in C-1 and C-3 (from C-1 and C-3 of sedoheptulose).

5. Pyruvate carboxylase has a covalently bound biotin cofactor and is therefore inhibited by avidin. Since oxceloacetate is an intermediate in the conversion of pyruvate to glucose, reactions (b) and (e) are inhibited.

6. Form a Schiff base between a ketose substrate and transaldolase, reduce it with titrated $NaBH_4$, and fingerprint the labeled enzyme.

7. From Stryer, page 1056:

$$K_{eq} = \frac{[\text{GSH}]^2[\text{NADP}^+]}{[\text{GSSG}][\text{NADPH}]} = 1126$$

After substituting for GSH and GSSG,

$$\frac{[\text{NADP}^+]}{[\text{NADPH}]} = 1.126 \times 10^4$$

$$\frac{[\text{NADPH}]}{[\text{NADP}^+]} = \frac{1}{1.126 \times 10^4}$$

$$= 8.9 \times 10^{-5}$$

8. The lactate level in the maternal circulation, and hence in the fetal circulation, increases during pregnancy because the mother is carrying a growing fetus with its own metabolic demands. The shift to H_4 in the fetal heart enables the fetus to use lactate as a fuel. Consequently, there is less need for gluconeogenesis by the liver and kidneys of the mother.

9. Fructose 2,6-bisphosphate, present at a high concentration when glucose is abundant, normally inhibits gluconeogenesis by blocking fructose 1,6-bisphosphatase. In this genetic disorder, the phosphatase is active irrespective of the glucose level. Hence, substrate cycling is increased, which generates heat. The level of fructose 1,6-bisphosphate is consequently lower than normal. Less pyruvate is formed, resulting in less acetyl CoA. In turn, less ATP is formed by the citric acid cycle and oxidative phosphorylation.

Chapter 19

1. Galactose + ATP + UTP + H_2O + glycogen$_n$ ⟶
   glycogen$_{n+1}$ + ADP + UDP + 2 P_i + H^+
2. Fructose + 2 ATP + 2 H_2O ⟶
   glucose + 2 ADP + 2 P_i
3. There is a deficiency of the branching enzyme.

4. The concentration of glucose 6-phosphate is elevated in von Gierke's disease. Consequently, the phosphorylated D form of glycogen synthetase is active.

5. Glucose is an allosteric inhibitor of phosphorylase *a*. Hence, crystals grown in its presence are in the T state. The addition of glucose 1-phosphate, a substrate, shifts the R ⇌ T equilibrium toward the R state. The conformational differences between these states are sufficiently large that the crystal shatters unless it is stabilized by chemical cross-links. The shattering of a crystal caused by an allosteric transition was first observed by Haurowitz in the oxygenation of crystals of deoxyhemoglobin.

6. H. G. Hers [*Ann. Rev. Biochem.* 45(1976):167] suggested that these kinetics would ensure a lag in the dephosphorylation of subunit B, which would allow glycogen to be degraded before phosphorylase kinase is inactivated by its phosphatase.

7. (a) The control of glycogen phosphorylase and synthase will be impaired. Specifically, epinephrine will not trigger the breakdown of glycogen and the cessation of glycogen synthesis.

   (b) Calmodulin mediates the activation of phosphorylase kinase by elevated Ca^{2+} during muscle contraction. Hence, glycogen will not be degraded in concert with contraction.

   (c) Protein phosphatase 1 will be continually active. Hence, the level of phosphorylase *b* will be higher than normal, and glycogen will be less readily degraded.

Chapter 20

1. (a) Glycerol + 2 NAD^+ + P_i + ADP ⟶
      pyruvate + ATP + H_2O + 2 NADH + H^+
   (b) Glycerol kinase and glycerol phosphate dehydrogenase.

2. Stearate + ATP + $13\frac{1}{2}$ H_2O + 8 FAD +
      8 NAD^+ ⟶ $4\frac{1}{2}$ acetoacetate + $12\frac{1}{2}$ H^+
      + 8 $FADH_2$ + 8 NADH + AMP + 2 P_i

3. (a) Oxidation in mitochondria, synthesis in the cytosol.
   (b) Acetyl CoA in oxidation, acyl carrier protein for synthesis.
   (c) FAD and NAD^+ in oxidation, NADPH for synthesis.
   (d) L-isomer of 3-hydroxyacyl CoA in oxidation, D-isomer in synthesis.
   (e) Carboxyl to methyl in oxidation, methyl to carboxyl in synthesis.
   (f) The enzymes of fatty acid synthesis, but not those of oxidation, are organized in a multienzyme complex.

4. Because mammals lack the enzymes to introduce double bonds at carbon atoms beyond C-9 but can increase the length of the fatty acid chain at the carboxyl end, the easiest way to determine which unsaturated fatty acid is the precursor is to note the number of carbons from the ω end (CH_3 end) to the nearest double bond. Thus, in (a) this number is 7 carbons;

hence, palmitoleate is the precursor. In (b) it is 6 carbon atoms; hence, linoleate. In (e) it is 9 carbon atoms; hence, oleate; etc. (See Stryer, p. 490.)

5. During fatty acid biosynthesis, the carbon chain grows two carbons at a time by the condensation of an acyl-ACP with malonyl-ACP, with the malonyl-ACP becoming, in every case, the carboxyl end of the new acyl-ACP. Thus, the chain grows from methyl to carboxyl. Since ^{14}C labeled malonyl CoA was added a short time before synthesis was stopped, the fatty acids whose synthesis was completed during this short period will be heavily labeled towards the carboxyl end (the last portion synthesized) and less heavily labeled, if at all, on the methyl end.

6. The enolate anion of one thioester attacks the carbonyl carbon atom of the other thioester to form a C—C bond.

7. Adipose cell lipase is activated by phosphorylation. Hence, overproduction of the cAMP-activated kinase will lead to accelerated breakdown of triacylglycerols and depletion of fat stores.

8. When the blood glucose level is low, acetyl CoA carboxylase is switched off by phosphorylation. Impaired phosphorylation will lead to persistent activation of the carboxylase. Malonyl CoA will be synthesized even when glucose is scarce.

Chapter 21

1. (a) Pyruvate, (b) oxaloacetate, (c) α-ketoglutarate, (d) α-ketoisocaproate, (e) phenylpyruvate, and (f) hydroxyphenylpyruvate.

2. Aspartate + α-ketoglutarate + GTP + ATP + 2 H_2O + NADH + H^+ $\longrightarrow$ $\frac{1}{2}$ glucose + glutamate + CO_2 + ATP + GDP + NAD^+ + 2 P_i

3. Aspartate + CO_2 + NH_4^+ + 3 ATP + NAD^+ + 4 H_2O $\longrightarrow$ oxaloacetate + urea + 2 ADP + 4 P_i + AMP + NADH + H^+

4. (a) Label the methyl carbon atom of L-methylmalonyl CoA with ^{14}C. Determine the location of ^{14}C in succinyl CoA. The group transferred is the one bonded to the labeled carbon atom.

   It should be noted that one must distinguish between the free and thioester carboxyls in succinyl CoA. In free succinic acid, the carboxyls are indistinguishable. A possible strategy for distinguishing them is to convert succinyl CoA to the half amide of succinate (thioester → amide) and then run a Hoffmann degradation.

   (b) The proton that is abstracted from the methyl group of L-methylmalonyl CoA is directly transferred to the adjacent carbon atom.

5. Thiamine pyrophosphate.

6. It acts as an electron sink. See C. Walsh, *Enzymatic Reaction Mechanisms*, (W. H. Freeman, 1979), p. 178.

7. Deuterium is abstracted by the radical form of the coenzyme. The methyl group rotates before hydrogen is returned to the product radical. (See Stryer, p. 508.)

8. A carbanion or a carbonium ion. (See Stryer, p. 508.)

Chapter 22

1. $\Delta E_0' = -0.32 - (-0.43) = +0.11\ V.$ $\Delta G^{0'}$ (to reduce 1 mol of $NADP^+$) $= -2 \times 0.11 \times 23.06 = -5.08$ kcal/mol.

2. Aldolase participates in the Calvin cycle, whereas transaldolase participates in the pentose phosphate pathway.

3. The conversion of ribulose 1,5-bisphosphate to 3-phosphoglycerate does not require ATP, so it will continue until the ribulose 1,5-bisphosphate is largely depleted.

4. When the concentration CO_2 is drastically decreased, the rate of conversion of ribulose 1,5-bisphophate to 3-phosphoglycerate will greatly decrease whereas the rate of utilization of 3-phosphoglycerate will not be diminished.

5. Phycoerythrin, the most peripheral protein in the phycobilisome.

6. (a) It expresses a key aspect of photosynthesis—namely, that water is split by light. The evolved oxygen in photosynthesis comes from water.

   (b) The 12 H_2O come from the oxidation of 10 mol of NADH (2 from glycolysis, 2 from pyruvate → acetyl CoA, and 6 from the TCA cycle) and 2 moles of $FADH_2$ (TCA cycle). The H_2O used are 2 at glyceraldehyde 3-phosphate dehydrogenase, 2 to hydrolyze acetyl CoA during citrate synthesis, and 2 by fumarase.

7. The addition of pyridine increases the proton storage capacity of the thylakoid space. More pumped protons can then flow through the ATP-synthesizing complex in the dark. For a discussion of this experiment, see M. Avron, *Ann. Rev. Biochem.* 46(1977):145.

8. DCMU inhibits electron transfer between Q and plastoquinone in the link between photosystems II and I. O_2 evolution can occur in the presence of DCMU if an artificial electron acceptor such as ferricyanide can accept electrons from Q.

9. CABP resembles the addition compound formed in the reaction of CO_2 and ribulose 1,5-bisphosphate (p. 534). As predicted, CABP is a potent inhibitor of the enzyme.

**2-Carboxyarabinitol
1,5-bisphosphate
(CABP)**

10. Aspartate + glyoxylate $\longrightarrow$ oxaloacetate + glycine
11. (a) The energy of photons is inversely proportional to the wavelength. Since 600-nm photons have an energy content of 47.6 kcal/einstein, 1000-nm light will have an energy content of 600/1000 × 47.6 = 28.7 kcal/einstein.
    (b) -28.7 kcal/mol ($\Delta G^{\circ\prime}$) $= -1 \times V \times 23.06$. Therefore, $V = -28.7/-23.06 = 1.24$ volts.
    (c) One 1000-nm photon has the free energy content of 2.39 ATP. A minimum of 0.42 photon is needed to drive the synthesis of an ATP.

Chapter 23

1. Glycerol + 4 ATP + 3 fatty acids + 4 H_2O $\longrightarrow$
   triacylglycerol + ADP + 3 AMP + 7 P_i + 4 H^+
2. Glycerol + 3 ATP + 2 fatty acids + 2 H_2O +
   CTP + serine $\longrightarrow$ phosphatidyl serine +
   CMP + ADP + 2 AMP + 6 P_i + 3 H^+
3. (a) CDP-diacylglycerol, (b) CDP-ethanolamine, (c) acyl CoA, (d) CDP-choline, (e) UDP-glucose or UDP-galactose, (f) UDP-galactose, and (g) geranyl pyrophosphate.
4. (a and b) None, because the label is lost as CO_2.

Chapter 24

1. Glucose + 2 ADP + 2 P_i + 2 NAD^+ +
   2 glutamate $\longrightarrow$ 2 alanine + 2 α-ketoglutarate +
   2 ATP + 2 NADH + H^+
2. $N_2 \longrightarrow NH_4^+ \longrightarrow$ glutamate $\longrightarrow$ serine $\longrightarrow$
   glycine $\longrightarrow$ δ-aminolevulinate $\longrightarrow$
   porphobilinogen $\longrightarrow$ heme
3. (a) Tetrahydrofolate, (b) tetrahydrofolate, and (c) N^5-methyltetrahydrofolate.
4. γ-Glutamyl phosphate may be a reaction intermediate.
5. The administration of glycine led to the formation of isovalerylglycine. This water-soluble conjugate, in contrast with isovaleric acid, is excreted very rapidly by the kidneys. See R. M. Cohn, M. Yudkoff, R. Rothman, and S. Segal, *New Engl. J. Med.* 299(1978):996.
6. H-D exchange points to the existence of a diimide intermediate. See W. A. Bulen, *Proc. Int. Symp. N_2 Fixation* (Washington State University Press, 1976).

7. They carry out nitrogen fixation. The absence of photosystem II provides an environment in which O_2 is not produced. Recall that the nitrogenase is very rapidly inactivated by O_2. See R. Y. Stanier, J. L. Ingraham, M. L. Wheelis, and P. R. Painter, *The Microbial World,* 5th ed. (Prentice-Hall, 1986), pp. 356–359, for a discussion of heterocysts.
8. The cytosol is a reducing environment, whereas the extracellular milieu is an oxidizing environment. Glutathione is the major sulfhydryl buffer in the cytosol.

9. The sulfoximine moiety, erroneously recognized as the γ-carboxylate of glutamate, is enzymatically phosphorylated to form methionine sulfoximine phosphate. This phosphorylated product binds very tightly to glutamine synthetase. See R. Rando, *Accts. Chem. Res.* 8(1975):281.
10. Glutathione normally feedback-inhibits γ-glutamylcysteine synthetase, the enzyme catalyzing the first step in its biosynthesis. The absence of glutathione leads to high levels of γ-glutamylcysteine, which is converted into 5-oxoproline by γ-glutamyl cyclotransferase.

Chapter 25

1. Glucose + 2 ATP + 2 $NADP^+$ + H_2O $\longrightarrow$
   PRPP + CO_2 + ADP + AMP + 2 NADPH + H^+
2. Glutamine + aspartate + CO_2 + 2 ATP +
   NAD^+ $\longrightarrow$ orotate + 2 ADP + 2 P_i +
   glutamate + NADH + H^+
3. (a, c, d, and e) PRPP, (b) carbamoyl phosphate.
4. PRPP and formylglycinamide ribonucleotide.
5. dUMP + serine + NADPH + H^+ $\longrightarrow$
   dTMP + $NADP^+$ + glycine
6. There is a deficiency of N^{10}-formyltetrahydrofolate. Sulfanilamide inhibits the synthesis of folate by acting as an analog of *p*-aminobenzoate, one of the precursors of folate.
7. PRPP is the activated intermediate in the synthesis of (a) phosphoribosylamine in the de novo pathway of purine formation, (b) purine nucleotides from free bases by the salvage pathway, (c) orotidylate in the formation of pyrimidines, (d) nicotinate ribonucleotide, (e) phosphoribosyl-ATP in the pathway leading to histidine, and (f) phosphoribosyl-anthranilate in the pathway leading to tryptophan.
8. It seems likely that glutamine yields ammonia as a result of the catalytic action of the small subunit. The nascent ammonia would then react with an activated form of CO_2 that is formed by the large subunit. The bicarbonate-dependent ATPase activity implies that this activated species is carbonyl phosphate. Reaction of this carbonic-phosphoric mixed anhydride with NH_3 yields carbamate, which would then react with ATP to give carbamoyl phosphate. For a discussion of this enzymatic mechanism, see C. Walsh, *Enzymatic Reaction Mechanisms* (W. H. Freeman, 1979), pp. 150–154.
9. Analogous reactions occur in the urea cycle (Stryer, p. 501)—from citrulline to arginosuccinate, and then to arginine.
10. Tyrosine 122 was converted into phenylalanine 122 by site-specific mutagenesis. This mutant enzyme has the same size, iron content, and iron-sensitive absorption spectrum as the wild type but is totally devoid of enzymatic activity. This experiment by A. Larsson and B. M. Sjoberg (*EMBO J.* 5(1986):2037) provides strong evidence that tyrosine 122 is the site of the free radical in ribonucleotide reductase.

11. (a) Though often termed *binding constants,* the values given are really *dissociation constants.* Therefore, to calculate the free energy of binding, one uses the reciprocals of the values given. Thus, for the wild type,

$$\Delta G^{\circ\prime} = -1.36 \log \frac{1}{7 \times 10^{-11}} = -1.36 \times 10.15$$
$$= -13.8 \text{ kcal/mol}$$

(b) The mutants bind the unprotonated form of methotrexate. The side chain amide of asparagine 27 donates a proton to the N-1 nitrogen of methotrexate to form a hydrogen bond. The other side chain hydrogen atom is hydrogen-bonded to the oxygen atom of a water molecule. In the serine 27 mutant, the oxygen atom of a water molecule accepts a proton from the serine OH group, and a hydrogen atom of water is donated to the N-1 nitrogen of methotrexate.

Mode of binding of methotrexate to the serine 27 mutant of dihydrofolate reductase. [After E. H. Howell, J. E. Villafranca, M. S. Warren, S. J. Oatley, and J. Kraut. *Science* 231(1986):1125.]

12. (a) *S*-Adenosylhomocysteine hydrolase activity is markedly diminished in ADA-deficient patients because of reversible inhibition by adenosine and suicide inactivation by 2′-deoxyadenosine.

(b) The activated methyl cycle (Stryer, p. 583) is blocked in these patients. *S*-Adenosylhomocysteine is a potent inhibitor of methyl transfer reactions involving *S*-adenosylmethionine. For a discussion of adenosine deaminase deficiency, see N. M. Kredich and M. S. Hershfield in Stanbury, J. B., Wyngaarden, J. B., Fredrickson, D. S., Goldstein, J. L., and Brown, M. S., (eds.), *The Metabolic Basis of Inherited Disease* (5th ed., McGraw-Hill, 1983), pp. 1157–1183.

13. (a) Cell A cannot grow in a *HAT* medium because it cannot synthesize dTMP either from thymidine or dUMP. Cell B cannot grow in this medium because it cannot synthesize purines either by the de novo pathway or the salvage pathway. Cell C can grow in a *HAT* medium because it contains active thymidine kinase (enabling it to phosphorylate thymidine to dTMP) and hypoxanthine-guanine phosphoribosyl transferase (enabling it to synthesize purines from hypoxanthine by the salvage pathway).

(b) Transform cell A with a plasmid containing foreign genes of interest and a functional thymidine kinase gene. The only cells that will grow in a *HAT* medium are those that have acquired a thymidylate kinase gene; nearly all of these transformed cells will also contain the other genes on the plasmid.

14. These patients have a high level of urate because of the breakdown of nucleic acids. Allopurinol prevents the formation of kidney stones and blocks other deleterious consequences of hyperuricemia by inhibiting the formation of urate (Stryer, p. 621).

Chapter 26

1. Liver contains glucose 6-phosphatase, whereas muscle and brain do not. Hence, muscle and brain, in contrast with liver, do not release glucose. Another key enzymatic difference is that liver has little of the transferase needed to activate acetoacetate to acetoacetyl CoA. Consequently, acetoacetate and 3-hydroxybutyrate are exported by the liver for use by heart muscle, skeletal muscle, and brain.

2. (a) Adipose cells normally convert glucose to glycerol 3-phosphate for the formation of triacylglycerols. A deficiency of hexokinase would interfere with the synthesis of triacylglycerols.

(b) A deficiency of glucose 6-phosphatase would block the export of glucose from liver following glycogenolysis. This disorder (called von Gierke's disease) is characterized by an abnormally high content of glycogen in the liver and a low blood glucose level.

(c) A deficiency of carnitine acyltransferase I impairs the oxidation of long-chain fatty acids. Fasting and exercise precipitate muscle cramps in these individuals.

(d) Glucokinase enables the liver to phosphorylate glucose even in the presence of a high level of glucose 6-phosphate. A deficiency of glucokinase would interfere with the synthesis of glycogen.

(e) Thiolase catalyzes the formation of two molecules of acetyl CoA from acetoacetyl CoA and CoA. A deficiency of thiolase would interfere with the utilization of acetoacetate as a fuel when the blood sugar level is low.

(f) Phosphofructokinase will be less active than normal because of the lowered level of F-2,6-BP. Hence, glycolysis will be much slower than normal.

3. (a) A high proportion of fatty acids in the blood are

bound to albumin. Cerebrospinal fluid has a low content of fatty acids because it has little albumin.

(b) Glucose is highly hydrophilic and soluble in aqueous media, in contrast with fatty acids, which must be carried by transport proteins such as albumin. Micelles of fatty acids would disrupt membrane structure.

(c) Fatty acids, not glucose, are the major fuel of resting muscle.

4. (a) A watt is equal to 1 joule per second (0.239 calorie per second). Hence, 70 watts is equivalent to 0.07 kJ/s or 0.017 kcal/s.

(b) A watt is a current of one ampere across a potential of one volt. For simplicity, let us assume that all of the electron flow is from NADH to O_2 (a potential drop of 1.14 V). Hence, the current is 61.4 amperes, which corresponds to 3.86×10^{20} electrons per second (1 ampere = 1 coulomb/s = 6.28×10^{18} charges/s).

(c) Three ATP are formed per NADH oxidized (two electrons). Hence, one ATP is formed per 0.67 electrons transferred. A flow of 3.86×10^{20} electrons per second therefore leads to the generation of 5.8×10^{20} ATP per second or 0.96 mmole per second.

(d) The molecular weight of ATP is 507. The total body content of ATP of 50 grams is equal to 0.099 mole. Hence, ATP turns over about once per hundred seconds when the body is at rest.

5. The store of ATP at rest is used in a half second. Creatine phosphate is the major source of ~P during the first four seconds of the sprint. Glycolysis provides most of the additional ATP that is consumed.

6. The 1748 kcal available in glucose or glycogen would be consumed in 8740 seconds, or 146 minutes, if the rate of energy expenditure during the marathon were 0.2 kcal/s (12 times the basal level of 0.017 kcal/s given in problem 4).

7. A high blood glucose level would trigger the secretion of insulin, which would stimulate the synthesis of glycogen and triacylglycerols. A high insulin level would impede the mobilization of fuel reserves during the marathon.

Chapter 27

1. DNA polymerase I uses deoxyribonucleoside triphosphates; pyrophosphate is the leaving group. DNA ligase uses a DNA-adenylate (AMP joined to the 5'-phosphate) as a reaction partner; AMP is the leaving group. Topoisomerase I uses a DNA-tyrosyl intermediate (5'-phosphate linked to the phenolic OH); the tyrosine residue of the enzyme is the leaving group.

2. At pH 6, the histidine residue at the active site is likely to be protonated and hence unable to accept a proton to activate the bound water molecule.

3. DNase I does not cleave Z-DNA because it lacks the minor groove of B-DNA. The arginine and lysine side chains at the active site are not sterically complementary to phosphate groups in Z-DNA.

4. FAD, CoA, and NADP$^+$ are plausible alternatives.

5. DNA ligase relaxes supercoiled DNA by catalyzing the cleavage of a phosphodiester bond in a DNA strand. The attacking group is AMP, which becomes attached to the 5'-phosphoryl group at the site of scission. AMP is required because this reaction is the reverse of the final step in the joining of pieces of DNA (see Stryer, Figure 27-20 on p. 659).

6. Positive supercoiling resists the unwinding of DNA. The melting temperature of DNA increases in going from negatively supercoiled to relaxed to positively supercoiled DNA. Positive supercoiling is probably an adaptation to high temperature.

7. (a) The twisting number changes by -2 when a turn of right-handed B-DNA ($T = +1$) switches into a turn of left-handed Z-DNA ($T = -1$). Because L stays the same, W changes by $+2$. Hence, $L = 100$, $T = 102$, and $W = -2$ after the transition.

(b) The DNA becomes less supercoiled and hence less compact. It will move more slowly.

(c) The torsional energy stored in a supercoiled DNA molecule is proportional to the square of the superhelix density. Free energy is released when the degree of negative supercoiling decreases. The endergonic B-to-Z transition is driven by the accompanying exergonic decrease in negative supercoiling.

(d) This protein stabilizes B-DNA. Hence, the midpoint of the B-to-Z transition will occur at higher degree of negative supercoiling.

8. (a) Long stretches of each occur because the transition is highly cooperative.

(b) B-Z junctions are energetically highly unfavorable.

(c) A—B transitions are less cooperative than B—Z transitions because the helix stays right-handed at an A-B junction but not at a B-Z junction.

9. ATP hydrolysis is required to release DNA gyrase after it has acted on its DNA substrate. Negative supercoiling requires only the binding of ATP, not its hydrolysis.

10. (a) Pro (CCC), Ser (UCC), Leu (CUC) and Phe (UUC). Alternatively, the last base of each of these codons could be U.

(b) These C → U mutations were produced by nitrous acid.

11. (a) No, it was produced by the deletion of the first base in the sequence shown below and the insertion of another base at the end of this sequence.

(b) -AGUCCAUCACUUAAU-

Chapter 28

1. The dnaB protein and rep protein participate in DNA replication. The recB component of the recBCD complex generates single-stranded DNA in

general recombination. The recA protein binds to single-stranded DNA and catalyzes its invasion of duplex DNA and a switch in base pairing.

2. (a) The recA-DNA complex contains 18.6 base pairs per turn, compared with 10.4 for B-DNA. Hence, the linking number is 56, compared with 100 for the relaxed circle formed in the absence of recA protein.

   (b) The DNA molecule after removal of recA protein is highly supercoiled and hence much more compact than its relaxed counterpart.

3. Proteolysis of the lexA protein at the onset of the SOS response relieves the repression of synthesis of lexA mRNA. Increased formation of lexA protein will terminate the SOS response.

4. (a) In the presence of ATPγS, recA protein binds to DNA but its dissociation is blocked because this analog is not readily hydrolyzed. Strand exchange does not take place.

   (b) RecB protein binds to DNA in the presence of this analog but its release is blocked. Consequently, single-stranded DNA is not formed.

Chapter 29

1. Heparin, a glycosaminoglycan (Stryer, p. 276), is highly anionic. Its negative charges, like the phosphodiester bridges of DNA templates, bind to lysine and arginine residues of β′.

2. This mutant sigma would competitively inhibit the binding of holoenzyme and prevent the specific initiation of RNA chains at promoter sites.

3. The core enzyme without sigma binds more tightly to the DNA template than does the holoenzyme. The retention of sigma after chain initiation would make the mutant RNA polymerase less processive. Hence, RNA synthesis would be much slower than normal.

4. A 100-kd protein contains about 910 residues, which are encoded by 2730 nucleotides. At a maximal transcription rate of 50 nucleotides per second, the protein would be synthesized in 54.6 seconds.

5. Initiation at strong promoters occurs every two seconds. In this interval, 100 nucleotides are transcribed. Hence, centers of transcription bubbles are 340 Å apart.

6. Note that the DNA strands shown in Figure 29-30 are, by convention, complementary to the DNA strand coding for mRNA. If every T is changed to U, the polarity and base sequence of the DNA strands shown are identical to the mRNA synthesized from the complementary strand. Thus, TCT becomes UCU (Ser), and so forth.

7. A change from U to C in the recognition sequence AAUAAA for the endonuclease caused this defect in a thalasemic patient. Cleavage occurred at the AAUAAA 900 nucleotides downstream from this mutant AACAAA site. See S. H. Orkin, T-C. Cheng,

S. E. Antonarakis, and H. H. Kazazian, Jr., *EMBO J.* 4(1985):453.

Chapter 30

1. The enzyme-bound Ile-AMP intermediate is necessary for the $^{32}PP_i$ exchange into ATP. Since isoleucine is a requirement, labeled ATP will be formed only in (c).

2. Four bands: light, heavy, a hybrid of light 30S and heavy 50S, and a hybrid of heavy 30S and light 50S.

3. About 799 high-energy phosphate bonds are consumed—400 to activate the 200 amino acids, 1 for initiation, and 398 to form 199 peptide bonds.

4. (b, c, and f) Type 1; (a, d, and e) type 2.

5. The simplest hypothesis is that the CCA anticodon of a tryptophan tRNA has mutated to UCA, which is complementary to UGA. However, analysis of this altered tRNA produces a surprise. Its anticodon is unaltered. Rather, there is a substitution of A for G at position 24. Thus, a residue far from the anticodon in the linear base sequence can influence the fidelity of codon recognition.

6. One approach is to synthesize a tRNA charged with a reactive amino acid analog. For example, bromoacetyl-phenylalanyl-tRNA is an affinity-labeling reagent for the P site of *E. coli* ribosomes. See H. Oen, M. Pellegrini, D. Eilat, and C. R. Cantor, *Proc. Nat. Acad. Sci.* 70(1973):2799.

7. The sequence GAGGU is complementary to a sequence of five bases at the 3′ end of 16S rRNA and is located several bases on the 5′ side of an AUG codon. Hence this region is a start signal for protein synthesis. The replacement of G by A would be expected to weaken the interaction of this mRNA with the 16S rRNA and thereby diminish its effectiveness as an initiation signal. In fact, this mutation results in a tenfold decrease in the rate of synthesis of the protein specified by this mRNA. See J. J. Dunn, E. Buzash-Pollert, and F. W. Studier, *Proc. Nat. Acad. Sci.* 75(1978):2741, for a discussion of this informative mutant.

8. The nitrogen atom of the deprotonated α-amino group of aminoacyl-tRNA is the nucleophile in peptide-bond formation.

9. Leu-Gly-Trp-tRNA occupies the P site because peptide bond formation and translocation have occurred. The entry of the next aminoacyl-tRNA into the A site is blocked because fusidic acid prevents EF-G-GDP from dissociating.

10. Proteins are synthesized from the amino to the carboxyl end on ribosomes, and in the reverse direction in the solid-phase method. The activated intermediate in ribosomal synthesis is an aminoacyl-tRNA; in the solid-phase method, the adduct of the amino acid and dicyclohexylcarbodiimide.

11. The error rates of DNA, RNA, and protein synthesis are of the order of 10^{-10}, 10^{-5}, and 10^{-4} per nucleo-

tide (or amino acid) incorporated. The fidelity of all three processes depends on the precision of base pairing to the DNA or mRNA template. No error correction occurs in RNA synthesis. In contrast, the fidelity of DNA synthesis is markedly increased by the $3' \rightarrow 5'$ proofreading nuclease activity and by post-replicative repair. In protein synthesis, the mischarging of some tRNAs is corrected by the hydrolytic action of the aminoacyl-tRNA synthetase. Proofreading also takes place when aminoacyl-tRNA occupies the A site on the ribosome; the GTPase activity of EF-Tu sets the pace of this final stage of editing.

Chapter 31

1. (a) Cleavable amino-terminal signal sequences are usually 13 to 36 residues long and contain a highly hydrophobic central region 10 to 15 residues long. The amino-terminal part has at least one basic residue. The cleavage site is preceded (at -1 and -3) by small neutral residues.
   (b) Procaryotic signal sequences are similar to eucaryotic ones. In addition, a stop-transfer sequence is needed to keep the protein in the plasma membrane.
   (c) Mannose 6-phosphate residues direct proteins to lysosomes.
   (d) Integral membrane proteins initially in the endoplasmic reticulum go to the plasma membrane unless they carry instructions to the contrary. No specific signal is needed.
   (e) An amino-terminal sequence containing positively charged residues, serine, and threonine, in addition to hydrophobic residues.
   (f) An amino-terminal arginine, aspartate, leucine, lysine, or phenylalanine.

2. The cytosolic portion of the receptor, which enables it to interact with coated pits, is likely to be altered.

3. (a) Endocytosis mediated by a receptor specific for mannose 6-phosphate residues. Endocytic vesicles containing the added lysosomal enzymes then fuse with lysosomes.
   (b) The addition of mannose 6-phosphate to the extracellular medium should prevent lysosomal enzymes from reaching their destination if the normal pathway involves secretion outside the cell and import back into the cell. However, mannose 6-phosphate does not inhibit normal lysosomal targeting, showing that lysosomal enzymes reach their destination without leaving the cell.

4. A specific targeting sequence is not required. Rather, amino-terminal sequences are scanned for overall features such as the presence of hydrophobic, basic, and hydroxyl amino acids. Sequence (d) is ineffective because it contains too many polar residues (glutamines) in place of hydrophobic residues (such as leucine and phenylalanine).

5. (a) The chimeric protein will probably be found in the cytosol. The transmembrane sequence of a membrane-bound immunoglobulin functions as a stop-transfer sequence and not as a signal sequence.
   (b) The chimeric protein will probably be found in the plasma membrane because chymotrypsinogen is synthesized with a signal sequence.

6. Chloroplasts contain three kinds of membranes (outer, inner, and thylakoid), whereas mitochondria contain only two (outer and inner). The inner membrane of chloroplasts, in contrast with that of mitochondria, is not energized. Hence, chloroplasts use ATP instead of a proton-motive force as the energy source for importing proteins.

7. The addition of a permanent base such as chloroquine will prevent acidification of endosomes, which should block the entry of virus into the cytosol.

8. Secretory proteins that are erroneously targeted to the cytosol will be rapidly degraded because their amino termini mark them for rapid destruction (Stryer, Table 31-2, on p. 794). Rapid elimination of mistargeted secretory proteins is important because some of them (e.g., trypsinogen) could wreak havoc in the cytosol.

Chapter 32

1. (a) The *lac* repressor is missing in an i^- mutant. Hence, this mutant is constitutive for the proteins of the *lac* operon.
   (b) This mutant is constitutive for the proteins of the *trp* operon because the *trp* repressor is missing.
   (c) The arabinose operon is not expressed in this mutant because the P2 form of the *araC* protein is needed to activate transcription.
   (d) This mutant is lytic but not lysogenic because it cannot synthesize the λ repressor.
   (e) This mutant is lysogenic but not lytic because it cannot synthesize the N protein, a positive control factor in transcription.

2. One possibility is that an i^s mutant produces an altered *lac* repressor that has almost no affinity for inducer but normal affinity for the operator. Such a *lac* repressor would bind to the operator and block transcription even in the presence of inducer.

3. This mutant has an altered *lac* operator that fails to bind the repressor. Such a mutator is called O^c (operator constitutive).

4. The cyclic AMP binding protein (CAP) is probably defective or absent in this mutant.

5. An *E. coli* cell bearing a λ prophage contains λ repressor molecules, which also block the transcription of the immediate-early genes of other λ viruses.

6. (a) Translation of the p_{RF} transcript is from five to ten times as rapid as that of the p_{RM} transcript because it contains the full protein synthesis initiation signal (as discussed in Stryer, p. 753).
   (b) The more effective translation of the p_{RF} transcript provides a burst of λ repressor molecules needed to establish the lysogenic state. See

M. Ptashe, K. Backman, Z. Humagun, A. Jeffrey, R. Maurer, B. Meyer, and R. T. Sauer, *Science* 194(1976):156, for a discussion.

7. The rate constant for association of wild-type *lac* repressor with the operator site is near the diffusion-controlled limit. Hence, a 100-fold increase in binding affinity implies that the mutant dissociates from the operator site about 100-fold more slowly than does the wild-type. The lag between addition of inducer and the initiation of transcription of the *lac* operon will be much longer in the mutant than in the wild type.

8. A repressor finds its target site by first binding to a nonspecific site anywhere in the DNA molecule and then diffusing along the DNA to reach the specific site. The repressor binds less tightly to nonspecific DNA at low ionic strength than at high ionic strength because of increased electrostatic repulsion. In contrast, binding to the operator is nearly independent of ionic strength because the interaction is mediated by hydrogen-bond and van der Waals interactions with the bases of the operator site.

9. The mutant protein acts as a repressor even when tryptophan is not bound. Expression of the *trp* operon is permanently repressed in this mutant bacterium.

10. When RF2 is present, UGA is read as a stop codon, leading to the formation of a truncated 25-residue polypeptide.

- Gly - Tyr - Leu - Stop

GGG UAU CUU UGA ↙Read by RF2

When the level of RF2 is very low, termination at UGA does not occur. Instead, a shift of the reading frame occurs, resulting in continued protein synthesis with Asp 26 as the next residue. See W. Craigen, R. Cook, W. Tate, and C. Caskey. *Proc. Nat. Acad. Sci.* 82(1985):3619.

- Gly - Tyr - Leu - Asp - Tyr - Asp -

GGG UAU CUU UGAC UAC GAC
↑
Frameshift in absence of RF2. U is skipped. GAC is first codon of new reading frame.

Chapter 33

1. (a) In procaryotes, translation begins while transcription is still in progress. In eucaryotes, these processes are separate in space and time.

(b) Eucaryotic transcripts are monogenic, whereas procaryotic transcripts are typically polygenic.

(c) In procaryotes, several proteins can be specified by a single primary transcript, one for each gene encoding the mRNA. In eucaryotes, a primary transcript encoded by a single gene can give rise to multiple proteins through alternative splicing. Cleavage of a polyprotein can also yield multiple proteins in eucaryotes.

(d) Most of procaryotic DNA encodes proteins and functional RNAs. In contrast, most of eucaryotic DNA does not encode functional macromolecules.

(e) Most procaryotic genes are clustered in operons. Eucaryotes do not have operons.

2. Centromeres, telomeres, and *ars* (autonomous replicating sequences serving as origins of replication) are needed to fashion a synthetic yeast chromosome. See A. Murray and J. W. Szostak, *Nature* 305(1983):189.

3. The mitochondrial genome is much smaller than the nuclear genome. Furthermore, a cell contains a large number of mitochondria. A substantial proportion of them, say 5%, could be nonfunctional without injuring the cell.

4. Mitochondrial DNA is evolving at a very rapid rate (see R. L. Cann, M. Stoneking, and A. C. Wilson, *Nature* 325(1986):31.). It is possible but not likely that a human mitochondrial gene will be transferred to the nuclear genome in the next ten million years. The existence of different genetic codes in the mitochondrion and the cytosol is a formidable barrier to gene transfer. In particular, UGA specifies tryptophan in mitochondrial proteins but is a stop signal in cytosolic protein synthesis.

5. One experimental approach is to add histones to linear duplex DNA and then form closed circles using DNA ligase. The histones are then removed from the supercoiled circles. Topoisomerase I is added to an aliquot to form the relaxed counterpart. The melting temperature of the supercoiled DNA is compared with that of the relaxed circle. A left-handed superhelix has a lower melting temperature than the relaxed circle, whereas a right-handed superhelix has a higher melting temperature. Recall that left-handed supercoiled DNA (negatively supercoiled; W is negative) is poised to be unwound (Stryer, p. 660).

6. The broken end invades internal C_{1-3} sequences that are complementary. The intact strand serves as a template in repair synthesis.

7. 5S RNA has the same base sequence (except for U in place of T) as the coding strand of its gene, the one that strongly binds TFIIIA. Hence, 5S RNA competes with its gene for the binding of TFIIIA. When 5S RNA is abundant, little TFIIIA is bound to the internal control region, and so transcription is slowed. The binding of TFIIIA to 5S RNA as well as the internal control region of the gene is the basis of a feedback loop that regulates the amount of 5S RNA.

8. (a) DNA inversions lead to alternating expression of a pair of flagellar genes (H1 and H2) in *Salmonella*. A flip-flop circuit changes an exposed bacterial protein at a frequency determined by the activity of a recombinase (Stryer, p. 818).

(b) One possible mechanism for changing *MAT* is to transfer *HML* or *HMR* into the *MAT* site. Alternatively, *HML* or *HMR* can serve as a template for the synthesis of a new *MAT* gene. In fact, *HML* and *HMR* are preserved at their original sites; they serve as templates without being physically removed from their original loci. See J. Rine, R. Jensen, D. Hagen, L. Blair, and I. Herskowitz. *Cold Spring Harbor Symp. Quant. Biol.* 45(1981):951.

9. One possibility is that the transcript of the 35-nucleotide sequence primes the transcription of the rest of the other segment of the gene. Alternatively, the two RNAs could be joined by an unusual *trans* (intermolecular) splicing reaction.

Chapter 34

1. The nucleotide sequence is searched for an open reading frame—a sequence that has no internal stop codons between the codon for the initiating methionine and the stop codon after the carboxyl-terminal residue. Peptides about ten residues long corresponding to portions of the deduced amino acid sequence are synthesized by the solid-phase method and used as immunogens. Antibodies against these peptides then serve as specific reagents for identifying the encoded protein. A Western blot would be a good way to start. The antibodies could also be used to immunoprecipitate the protein and to isolate it by affinity chromatography. See R. F. Doolittle, *Of Urfs and Orfs: A Primer on How to Analyze Derived Amino Acid Sequences* (University Books, 1986) for a concise and interesting discussion.

2. Long concatamers are formed by the association of complementary single-stranded tails of nascent duplexes. Complementarity arises from the redundancy in base sequence of the ends. The 3′ end of one duplex in the concatamer then serves as the primer to fill the gap at the 5′ end of the adjoining duplex in the repeating chain of new DNA molecules.

```
      3′ 5′
    a|   |A
    b|   |B
    c|   |C
    x|   |X
    y|   |Y
     |   |Z
    a|   |A
    b|   |B
    c|
    d|   |D
    e|   |E
    f|   |F
    a|   |A
    b|   |B
      5′ 3′
```

3. (a) 5′-UGGACUUUGUGGGAUACCCUCGCUUU-3′
   (b) The second leucine in the sequence becomes phenylalanine.
   (c) Each sequence is highly constrained by the other.

4. The infected cell would be lysed before progeny virions are fully synthesized and assembled.

5. Polyamines neutralize the negatively charged RNA inside the virion. See S. S. Cohen and F. P. McCormick, *Adv. Virus Res.* 24(1979):331–387.

6. Influenza hemagluttinin (HA) binds sialic acid (*N*-acetylneuraminate) residues on the surface of cells in the respiratory tract. This binding enables influenza virus to be endocytosed into susceptible cells. Erythrocytes, like cells of the respiratory tract, contain sialic acid residues on their cell-surface glycoproteins. The binding of influenza virus to erythrocytes in vitro can be inhibited by adding glycopeptides containing sialic acid.

7. (a) DNA viruses use the replication machinery of their eucaryotic hosts. Hence, their genomes must pass through the nucleus of infected cells. DNA viruses also exploit the transcriptional and splicing apparatus of their hosts. Some RNA viruses can replicate and form mRNA in the cytosol because they encode their own transcriptase and carry it in the virion. The genomes of other RNA viruses (e.g., tetroviruses) must pass through the nucleus.
   (b) Influenza virus exploits the splicing apparatus of the host cell to form ten mRNAs from its eight genomic segments.

8. (a) The CD4 cell-surface receptor is present on brain cells as well as T4 lymphocytes.
   (b) Prepare an antibody to the CD4 protein of T4 lymphocytes, and use immunofluorescence microscopy to determine whether this receptor is present on other kinds of cells.

9. RNA-assisted cleavage assures that virus particles are not formed of a protein coat containing no RNA. A closed icosahedral shell is formed only after RNA is packaged inside.